D1698424

Analytical Methods in Supramolecular Chemistry

Edited by
Christoph Schalley

1807–2007 Knowledge for Generations

Each generation has its unique needs and aspirations. When Charles Wiley first opened his small printing shop in lower Manhattan in 1807, it was a generation of boundless potential searching for an identity. And we were there, helping to define a new American literary tradition. Over half a century later, in the midst of the Second Industrial Revolution, it was a generation focused on building the future. Once again, we were there, supplying the critical scientific, technical, and engineering knowledge that helped frame the world. Throughout the 20th Century, and into the new millennium, nations began to reach out beyond their own borders and a new international community was born. Wiley was there, expanding its operations around the world to enable a global exchange of ideas, opinions, and know-how.

For 200 years, Wiley has been an integral part of each generation's journey, enabling the flow of information and understanding necessary to meet their needs and fulfill their aspirations. Today, bold new technologies are changing the way we live and learn. Wiley will be there, providing you the must-have knowledge you need to imagine new worlds, new possibilities, and new opportunities.

Generations come and go, but you can always count on Wiley to provide you the knowledge you need, when and where you need it!

William J. Pesce
President and Chief Executive Officer

Peter Booth Wiley
Chairman of the Board

Analytical Methods in Supramolecular Chemistry

Edited by
Christoph Schalley

WILEY-VCH Verlag GmbH & Co. KGaA

The Editor

Prof. Dr. Christoph A. Schalley
Freie Universität Berlin
Inst. f. Chemie u. Biochemie
Takustr. 3
14195 Berlin
Germany

1st Edition 2006
1st Reprint 2008

Library of Congress Card No.: applied for
British Library Cataloguing-in-Publication Data
A catalogue record for this book is available from the British Library.

Bibliographic information published by the Deutsche Nationalbibliothek
Die Deutsche Nationalbibliothek lists this publication in the Deutsche Nationalbibliografie; detailed bibliographic data are available in the Internet at ⟨http://dnb.d-nb.de⟩.

Printed in the Federal Republic of Germany.
Printed on acid-free paper.

Cover Design Adam Design, Weinheim
Typesetting Asco Typesetters, Hong Kong
Printing Strauss GmbH, Mörlenbach
Binding Litges & Dopf GmbH, Heppenheim

ISBN: 978-3-527-31505-5

Contents

Analytical Methods in Supramolecular Chemistry. Edited by Christoph Schalley

ISBN: 978-3-527-31505-5

Preface

Supramolecular Chemistry, conceptually founded as a research field in its own in the 1960s, is a rapidly growing field at the borderline of several disciplines such as bio(organic) chemistry, material sciences, and certainly the classical chemistry topics, i.e. (in)organic and physical chemistry. Historically, the development of supramolecular chemistry certainly depended on the development of analytical methods which could solve the questions associated to the complex architectures held together by noncovalent bonds and those arising from weak intermolecular bonding and the highly dynamic features of many supramolecular species. However, not only supramolecular chemistry benefited from the methodological development. Vice versa, the problems faced by supramolecular chemists led to specific methodological solutions and thus mediated their development to a great extent.

Several excellent textbooks on Supramolecular Chemistry exist, starting with Fritz Vögtle's seminal best-seller "Supramolekulare Chemie" [1], which was translated into several languages and thus found a broad international readership not only among experts in the field, but also among chemistry students. Other authors: Jean-Marie Lehn [2], Jerry Atwood and Johnathan Steed [3], and most recently Katsuhiko Ariga and Toyoki Kunitake [4], have provided expertly written textbooks. These textbooks focus on and are organized along the chemistry involved, but do not focus much on the methods utilized to study this chemistry – with one notable exception: Hans-Jörg Schneider's and Anatoly Yatsimirski's fine introduction into the "*Principles and Methods in Supramolecular Chemistry*" [5]. The present book aims at a more in-depth description of different methods utilized in this branch of chemical research.

Clearly, a choice had to be made as to which of the many methods available today should be included. This choice is likely biased to some extent by the editor's own preferences and a reader might arrive at the conclusion that another choice would have been better. Some chapters deal with methods of fundamental importance. For example, Chapter 2 provides a practical guide to the determination of binding constants by NMR and UV methods and thus covers an aspect imminently important to the field, which deals with noncovalent binding and weak interactions. Similar arguments hold for the next two chapters on isothermal titration calorimetry and extraction methods. The following chapters on mass spectrometry, diffusion-ordered NMR spectroscopy, photochemistry, and circular dichroism do

Analytical Methods in Supramolecular Chemistry. Edited by Christoph Schalley

ISBN: 978-3-527-31505-5

not primarily provide an in-depth introduction into these so-to-say classical methods, but focus on exciting new achievements in the context of Supramolecular Chemistry. Chapters 9 and 10 discuss methods used to address quite complex architectures, for example generated by crystal engineering and through surface attachment and self-assembly processes at surfaces. The functional aspect of many supramolecular systems appears most pronounced in Chapter 11 which introduces methods for the characterization of membrane channels. The book terminates with a discussion of the contributions that theory can make to Supramolecular Chemistry, a field which adds valuable insight, although it is often believed to be very limited due to the sheer size of the complexes and architectures under study.

Each of the chapters introduces the reader to a particular method. However, the reader will probably need to have at least some basic knowledge of supramolecular chemistry itself. Although the book begins with a short introductory chapter to provide some necessary background, it is impossible to give a concise and comprehensive overview after more than four decades of quick growth in the field. In that sense, it aims at an already somewhat experienced readership.

I am grateful to all authors of the individual chapters for their excellent contributions to the book. Particularly, I would like to thank Dr. Steffen Pauly from Wiley-VCH for his great help in preparing the final manuscript and his guidance through the production process. It was great joy to assemble this book and I sincerely hope that it is fun to read.

Berlin, October 2006 *Christoph A. Schalley*

References

1 First German text: F. Vögtle, *Supramolekulare Chemie*, Teubner, Stuttgart 1989. First English text: F. Vögtle, *Supramolecular Chemistry: An Introduction*, Wiley, Chichester 1991.

2 J.-M. Lehn, *Supramolecular Chemistry – Concepts and Perspectives*, Verlag Chemie, Weinheim 1995.

3 J. W. Steed, J. L. Atwood, *Supramolecular Chemistry*, Wiley, New York 2000.

4 K. Ariga, T. Kunitake, *Supramolecuar Chemistry – Fundamentals and Applications*, Springer, Berlin 2003.

5 H.-J. Schneider, A. Yatsimirsky, *Principles and Methods in Supramolecular Chemistry*, Wiley, Chichester 2000.

List of Contributors

Dipl.-Ing. (FH) Bianca Antonioli
Technische Universität Dresden
Fachrichtung Chemie
Bergstr. 66
01069 Dresden
Germany

Dr. Liat Avram
Tel Aviv University
School of Chemistry
The Sackler Faculty of Exact Sciences
Ramat Aviv 69978
Tel Aviv
Israel

Prof. Mário Nuno Berberan-Santos
Instituto Superior Técnico
Centro de Química-Física Molecular
1049-001 Lisboa
Portugal

Prof. Yoram Cohen
Tel Aviv University
School of Chemistry
The Sackler Faculty of Exact Sciences
Ramat Aviv 69978
Tel Aviv
Israel

Dr. Tamar Evan-Salem
Tel Aviv University
School of Chemistry
The Sackler Faculty of Exact Sciences
Ramat Aviv 69978
Tel Aviv
Israel

Dr. Limor Frish
Tel Aviv University
School of Chemistry
The Sackler Faculty of Exact Sciences
Ramat Aviv 69978
Tel Aviv
Israel

Dr. Kerstin Gloe
Technische Universität Dresden
Fachrichtung Chemie und
Lebensmittelchemie
Bergstr. 66
01069 Dresden
Germany

Prof. Dr. Karsten Gloe
Technische Universität Dresden
Fachrichtung Chemie und
Lebensmittelchemie
Bergstr. 66
01069 Dresden
Germany

Prof. Dr. B. A. Hermann
LMU Munich, Walther-Meissner-Institute of
the Bavarian Academy of Science
Center for Nano Science (CeNS)
Walther-Meissner-Str. 8
85748 Garching
Germany

Prof. Keiji Hirose
Osaka University
Division of Frontier Materials Science
Department of Materials Engineering Science
Graduate School of Engineering Science
1-3 Machikaneyama Toyonaka
Osaka 560-8531
Japan

Dipl.-Ing. Stefanie Juran
Forschungszentrum Rossendorf e.V.
Institut für Radiopharmazie
Bautzner Landstraße 128
01328 Dresden
Germany

Priv.-Doz. Dr. Barbara Kirchner
Universität Bonn
Lehrstuhl für Theoretische Chemie
Wegelerstr. 12
53115 Bonn
Germany

Dipl.-Chem. Michael Kogej
Universität Bonn
Kekulé-Institut für Organische Chemie und Biochemie
Gerhard-Domagk-Str. 1
53121 Bonn
Germany

Dr. Petr Maloň
Academy of Sciences of the Czech Republic
Institute of Organic Chemistry and Biochemistry
Flemingovo n. 2
166 10 Praha 6
Czech Republic

Dr. Monique M. Martin
Ecole Normale Supérieure
UMR CNRS-ENS 8640, Département de Chimie
24 rue Lhomond
75231 Paris Cedex 05
France

Prof. Dr. Stefan Matile
University of Geneva
Department of Organic Chemistry
Quai Ernest-Ansermet 30
1211 Geneva 4
Switzerland

Prof. Dr. Markus Reiher
ETH Zurich, Honggerberg Campus HCI
Laboratorium für Physikalische Chemie
Wolfgang-Pauli-Str. 10
8093 Zurich
Switzerland

Prof. Dr. Kari Rissanen
University of Jyväskylä
NanoScience Center, Department of Chemistry
Survontie 9
40014 Jyväskylä
Finland

Dr. Naomi Sakai
University of Geneva
Department of Organic Chemistry
30, quai Ernest Ansermet
1211 Geneva
Switzerland

Prof. Dr. Christoph A. Schalley
Freie Universität Berlin
Inst. für Chemie und Biochemie
Takustr. 3
14195 Berlin
Germany

Prof. Dr. Franz P. Schmidtchen
Technical University of Munich
Department of Chemistry
Lichtenbergstr. 4
85747 Garching
Germany

Dr. Holger Stephan
Forschungszentrum Rossendorf e.V.
Institut für Radiopharmazie
Bautzner Landstrasse 128
01328 Dresden
Germany

Prof. Marie Urbanová
Institute of Chemical Technology
Department of Physics and Measurement
Technická 5
166 28 Praha 6
Czech Republic

Prof. Bernard Valeur
Conservatoire National des Arts et Métiers
Laboratoire de Chimie Générale
292 rue Saint-Martin
75141 Paris Cedex
France
and
ENS-Cachan
UMR CNRS 8531
Laboratoire de Photophysique et Photochimie Supramoléculaires et Macromoléculaires
Département de Chimie
61 Avenue du Président Wilson
94235 Cachan Cedex
France

1
Introduction

Christoph A. Schalley

1.1
Some Historical Remarks on Supramolecular Chemistry

The fundaments of Supramolecular Chemistry date back to the late 19th century, when some of the most basic concepts for this research area were developed. In particular, the idea of coordination chemistry was formulated by Alfred Werner (1893) [1], the lock-and-key concept was introduced by Emil Fischer (1894) [2], and Villiers and Hebd discovered cyclodextrins, the first host molecules (1891) [3]. A few years later, Paul Ehrlich devised the concept of receptors in his *Studies on Immunity* (1906) [4] by stating that any molecule can only have an effect on the human body, if it is bound ("Corpora non agunt nisi fixata"). Several of these concepts were refined and modified later. Just to provide one example, Daniel Koshland formulated the induced fit concept (1958) for binding events to biomolecules which undergo conformational changes in the binding event [5]. The induced fit model provides a more dynamic view of the binding event, compared with the rather static key-lock principle and is thus more easily able to explain phenomena such as cooperativity. Even the German word for "Supramolecule" appeared in the literature as early as 1937, when Wolf and his coworkers introduced the term "Übermolekül" to describe the intermolecular interaction of coordinatively saturated species such as the dimers of carboxylic acids [6].

The question immediately arising from this brief overview on the beginnings of supramolecular chemistry is: Why hasn't it been recognized earlier as a research area in its own right? Why did it take more than 40 years from the introduction of the term "Übermolekül" to Lehn's definition of supramolecular chemistry [7] as the "chemistry of molecular assemblies and of the intermolecular bond" [8]?

There are at least two answers. The first relates to the perception of the scientists involved in this area. As long as chemistry accepts the paradigm that properties of molecules are properties of the molecules themselves, while the interactions with the environment are small and – to a first approximation – negligible, there is no room for supramolecular chemistry as an independent field of research. Although solvent effects were already known quite early, this paradigm formed the basis of the thinking of chemists for a long time. However, with an increasing number of

Analytical Methods in Supramolecular Chemistry. Edited by Christoph Schalley

ISBN: 978-3-527-31505-5

examples of the importance of the environment for the properties of a molecule, a paradigm shift occurred in the late 1960s. Chemists started to appreciate that their experiments almost always provided data about molecules in a particular environment. It became clear that the surroundings almost always have a non-negligible effect. Consequently, the intermolecular interactions became the focus of research and a new area was born. With this in mind, chemists were suddenly able to think about noncovalent forces, molecular recognition, templation, self-assembly and many other aspects into which supramolecular chemistry meanwhile diversified.

The second answer is not less important, although somewhat more technical in nature. Supramolecules are often weakly bound and highly dynamic. Based on intermolecular interactions, complex architectures can be generated, often with long-range order. All these features need specialized experimental methods, many of which still had to be developed in the early days of supramolecular chemistry. As observed quite often, the progress in a certain research area – here supramolecular chemistry – depends on the development of suitable methods. An emerging new method on the other hand leads to further progress in this research field, since it opens new possibilities for the experimenters. It is this second answer which prompted us to assemble the present book in order to provide information on the current status of the methods used in supramolecular chemistry. It also shows how diverse the methodological basis is, on which supramolecular chemists rely.

1.2 The Noncovalent Bond: A Brief Overview

Before going into detail with respect to the analytical methods that are applied in contemporary supramolecular chemistry, this brief introduction to some basic concepts and research topics within supramolecular chemistry is intended to provide the reader with some background. Of course, it is not possible to give a comprehensive overview. It is not even achievable to review the last 40 or so years of supramolecular research in a concise manner. For a more in-depth discussion, the reader is thus referred to some excellent text books on supramolecular chemistry [7].

Noncovalent bonds range from coordinative bonds with a strength of several hundreds of kJ mol^{-1} to weak van der Waals interactions worth only a few kJ mol^{-1}. They can be divided in to several different classes. Attractive or repulsive interactions are found, when two (partial) charges interact either with opposite polarity (attraction) or the same polarity (repulsion). Ion–ion interactions are strongest with bond energies in the range of ca. 100 to 350 kJ mol^{-1}. The distance between the charges and the extent of delocalization over a part of a molecule or even the whole molecule have an effect on the strength of the interaction. Consequently, the minimization of the distance between two oppositely charged ions will be a geometric factor, when it comes to the structure of the supramolecular aggregate – even though there is no particular directionality in the ion–ion interaction. Interactions between ions and dipoles are somewhat weaker (ca. 50–200 kJ mol^{-1}). Here, the orientation of the dipole with respect to the charge is important. A typical example for such an ion–dipole complex is the interaction of alkali metal ions with crown ethers. Other coordination complexes with transition metal ions as the cores are often used in supramolecular assembly. Here, the dative bond has a greater covalent contribution, which makes it difficult to clearly draw the line between supramolecular and molecular chemistry. Even weaker than ion–dipole forces (5–50 kJ mol^{-1}) are the interactions between two dipoles. Again, the relative orientation of the two interacting dipoles plays an important role.

Hydrogen bonding [9] is pivotal in biochemistry (e.g. in the formation of double stranded DNA and protein folding) and was also greatly employed in artificial supramolecules. One reason is that many host–guest complexes have been studied in noncompetitive solvents where the hydrogen bonds can become quite strong. Another, maybe equally important reason is the directionality of the hydrogen bond which allows the chemist to control the geometry of the complexes and to design precisely complementary hosts for a given guest (see below). One should distinguish between strong hydrogen bonds with binding energies in the range of 60–120 kJ mol^{-1} and heteroatom–heteroatom distances between 2.2 and 2.5 Å, moderate hydrogen bonds (15–60 kJ mol^{-1}; 2.5–3.2 Å), and weak hydrogen bonds with binding energies below ca. 15 kJ mol^{-1} and long donor–acceptor distances of up to 4 Å. This classification is also expressed in the fact that strong hydrogen bonds have a major covalent contribution, while moderate and weak ones are mainly electrostatic in nature. Also, the range of possible hydrogen bond angles is narrow in strong H bonds (175°–180°) so that there is excellent spatial control here, while moderate (130°–180°) and weak (90°–150°) hydrogen bonds are more flexible. Furthermore, one should always make a difference between hydrogen bonding between neutral molecules and charged hydrogen bonds. The latter ones are usually significantly stronger. For example, the $F{-}H\cdots F^-$ hydrogen bond has a bond energy of ca. 160 kJ mol^{-1} and thus is the strongest hydrogen bond known.

Noncovalent forces also involve π-systems, which can noncovalently bind to cations or other π-systems. The cation-π interaction [10] amounts to ca. 5–80 kJ mol^{-1} and plays an important role in biomolecules. Aromatic rings such as benzene bear a quadrupole moment with a partially positive σ-scaffold and a partially negative π-cloud above and below the ring plane. Consequently, alkali metal and other cations can form an attractive interaction when located above the center of the aromatic ring. The gas-phase binding energy of a K^+ cation to benzene (80 kJ mol^{-1}) is higher than that of a single water molecule to the same cation (75 kJ mol^{-1}). Consequently, one may ask why potassium salts don't dissolve in benzene. One answer is that the cation is stabilized by more than one or two water molecules in water and the sum of the binding energies is thus higher than that of a K^+ solvated by two or three benzenes. Another oft forgotten, but important point is the solvation of the corresponding anion. Water is able to solvate anions by forming hydrogen bonds. In benzene such an interaction is not feasible. Again, we touch the topic discussed in the beginning: the effects of the environment.

π-systems can also interact favorably with other π-systems. The interactions usually summarized with the term π-stacking are, however, quite complex. Two similarly electron-rich or electron-poor π-systems (e.g. benzene as a prototype) tend not to interact in a perfect face-to-face manner [11], because the two partially negative π-clouds would repulse each other. Two options exist to avoid this repulsion: in the crystal, benzene forms a herringbone-packing. Each benzene molecule is thus positioned with respect to its next neighbors in an edge-to-face orientation. This causes an attractive interaction between the negative π-cloud of one benzene with the positive σ-scaffold of the other. Larger aromatic molecules, for example porphyrins, may well crystallize in a face-to-face orientation. However, they reduce the repulsive forces by shifting sideways. The picture changes significantly, when two aromatics interact one of which is electron-rich (prototypically a hydroquinone), one electron-deficient (prototypically a quinone). These two molecules can then undergo charge transfer interactions which can be quite strong and usually can be identified by a charge-transfer band in the UV/vis spectrum.

On the weak end of noncovalent interactions, we find van der Waals forces (<5 kJ mol^{-1}) which arise from the interaction of an electron cloud polarized by adjacent nuclei. Van der Waals forces are a superposition of attractive dispersion interactions, which decrease with the distance r in a r^{-6} dependence, and exchange repulsion decreasing with r^{-12}.

A particular case, finally, which perfectly demonstrates the influence of the environment, is the hydrophobic effect which relies on the minimization of the energetically unfavorable surface between polar/protic and unpolar/aprotic molecules. Hydrophobic effects play an important role in guest binding by cyclodextrins, for example. Water molecules residing inside the unpolar cavity cannot interact with the cavity wall strongly. If they are replaced by an unpolar guest, their interaction with other water molecules outside the cavity is much stronger, resulting in a gain in enthalpy for the whole system. In addition to these enthalpic contributions, entropy changes contribute, when several water molecules are replaced by one guest molecule, because the total number of translationally free molecules increases.

There are more noncovalent interactions which cannot all be introduced here. Forces between multipoles have been expertly reviewed recently [12]. Also, weak interactions exist between nitrogen and halogen atoms [13], and dihydrogen bridges [14] can be formed between metal hydrides and hydrogen bond donors. Finally, close packing in crystals is an important force in crystallization and crystal engineering. The present introductory chapter will not discuss these, but rather focus on the most important ones mentioned above.

1.3
Basic Concepts in Supramolecular Chemistry

The following sections discuss some fundamental concepts in supramolecular chemistry. The list is certainly not comprehensive and the reader is referred to textbooks for a broader scope of examples. However, the selection reveals that supramolecular research developed from its heart, i.e. the examination and understanding of the noncovalent bond, to more advanced topics which make use of that knowledge to build large, complex architecture, to understand the action of biomolecules, to implement function into molecular devices such as sensors, to control mechanical movement, to passively and actively transport molecules, and to use supramolecules as catalysts.

Clearly, molecular recognition processes are the prototypical supramolecular reactions on which the other aspects are based. Without molecular recognition, there are no template effects, no self-assembly, and certainly no self-replication. In contrast to opinions sometimes encountered among chemists from other areas, supramolecular chemistry did not come to a halt with the examination of hosts and guests and their interactions. Sophisticated molecular devices are available which not only are based on, but go far beyond mere molecular recognition.

1.3.1 Molecular Recognition: Molecular Complementarity

After these remarks, the first question is: What is a good receptor for a given substrate? How can we design a suitable host which binds a guest with specificity? According to Fischer's lock-and-key model, complementarity is the most important factor. Most often, it is not one noncovalent interaction alone which provides host–guest binding within a more or less competing environment, but the additive or even cooperative action of multiple interactions. The more complementary the binding sites of the host to those of the guest, the higher the binding energy. This refers not only to individual noncovalent bonds, but to the whole shape and the whole electrostatic surface of both molecules involved in the binding event. Selective binding is thus a combination of excellent steric fit with a good match of the charge distributions of guest surface and the hosts cavity and a suitable spatial arrangement of, for example, hydrogen bond donors and acceptors, thus maximizing the attractive and minimizing the repulsive forces between host and guest.

Cation recognition developed quickly early on, due to the combination of the often rather well-defined coordination geometry of most cationic species and the usually higher achievable binding energies coming from ion–dipole interactions. Actually, many of the basic concepts in supramolecular chemistry have been derived from studies in cation recognition. The design of neutral hosts for neutral guests and in particular anion recognition [15] are still a challenge nowadays.

1.3.2 Chelate Effects and Preorganization: Entropy Factors

A binding event in which one complex forms from two molecules is entropically disfavored. The entropic costs need to be paid from the reaction enthalpy released upon host–guest binding. However, strategies exist which can reduce these costs to a minimum.

One approach is to incorporate more than one binding site in one host molecule. When the first bond is formed, the entropic costs of combining two molecules are taken care of. The second and all following binding events between the same two partners will not suffer from this effect again and thus contribute more to the free enthalpy of binding. This effect is called the chelate effect and has long been known from coordination chemistry, where ethylene diamine or 2,2′-bipyridine ligands easily replace ammonia or pyridine in a transition metal complex. Bidentate binding generates rings and the chelate effect depends on their sizes. Optimal are five membered rings as formed by the ethylene diamine or bipyridine ligands discussed above. Smaller rings suffer from ring strain, larger rings need a higher degree of conformational fixation compared with their open-chain forms and are thus entropically disfavored. The latter argument can be refined. If the same number of binding sites are incorporated in a macrocycle or even macrobicycle, guest binding will again become more favorable, because each cyclization reduces the conformational flexibility for the free host and thus the entropic costs stemming from conformational fixation during guest binding. These effects have entered the literature as the macrocyclic and macrobicyclic effect. Donald Cram developed these ideas into the preorganization principle [16]. A host which is designed to display the binding sites in a conformationally fixed way, perfectly complementary to the guest's needs, will bind significantly more strongly than a floppy host which needs to be rigidified in the binding event. This becomes strikingly clear, if one compares conformationally flexible 18-crown-6 with the spherand shown in Fig. 1.1 which displays the six oxygen donor atoms in a preorganized manner. The alkali binding constants of the two host molecules differ by factors up to 10^{10}!

While discussing entropic effects, it should not be forgotten that examples exist for enthalpically disfavored, entropy-driven host–guest binding. This is possible, if the free host contains more than one solvent molecule as the guests, which upon guest binding are replaced by one large guest as discussed for cyclodextrins above. In this case, a host–solvent complex releases more molecules than it binds and the overall reaction benefits entropically from the increase in particle number.

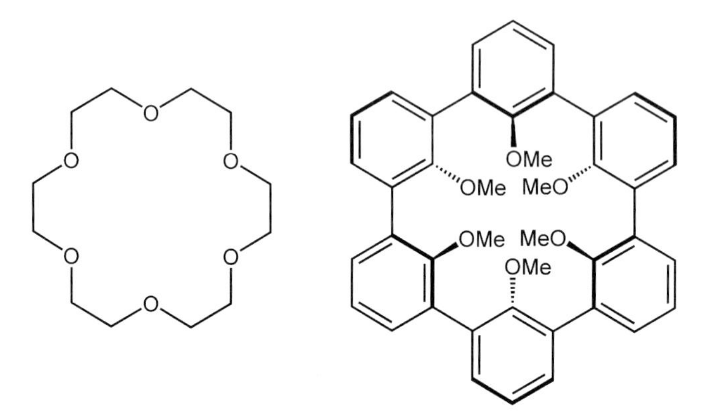

Fig. 1.1. Preorganization does matter. A comparison of 18-crown-6 and the spherand on the right with respect to alkali metal ion binding reveals that the spherand has an up to 10 orders of magnitude higher binding constant.

1.3.3
Cooperativity and Multivalency

Cooperativity and multivalency are phenomena arising in molecular recognition at hosts with more than one binding site. In order to avoid misunderstandings, one should clearly distinguish the two terms. Cooperativity describes the influence of binding a guest at the host's binding site A on the second binding step occurring at site B of the same host. Cooperativity can be positive, which means that binding strength of the second guest is increased by the first one and the sum of both binding energies is more than twice the binding energy of the first guest. Cooperativity can also be negative, if the first binding event decreases the binding of the second guest. Many examples for cooperativity are known from biochemistry, the most prominent one certainly oxygen binding at hemoglobin [17]. This protein is a $\alpha_2\beta_2$ tetramer with four oxygen binding hemes as the prosthetic groups, one in each subunit. Upon binding the first oxygen molecule to one of the heme groups, conformational changes are induced in the protein tertiary structure which also affect the other subunits and prepare them for binding oxygen more readily. From this example, it becomes clear that cooperativity does not necessarily rely on interactions between a multivalent host and a multivalent guest, but that there may well be mechanisms to transmit the information of the first binding event to the second one, even if both are monovalent interactions. The concept of cooperativity has been applied to supramolecular chemistry and was recently discussed in the context of self-assembly [18] (see below).

Conceptually related to the chelate effect, multivalency [19] describes the unique thermodynamic features arising from binding a host and a guest *each* equipped with more than one binding site. Although sometimes not used in a stringent way in the chemical literature, one should use the term "multivalency" only for those host–guest complexes, in which the dissociation into free host and guest requires at least the cleavage of two recognition sites. The concept of multivalency has been introduced to adequately describe the properties of biomolecules [20]. For example, selectivity and high binding strengths in recognition processes at cell surfaces usually require the interaction of multivalent receptors and substrates. Due to the complexity of many biological systems, limitations exist for a detailed analysis of the thermochemistry and kinetics of multivalent interactions between biomolecules. For example, the monovalent interaction is usually unknown and thus, a direct comparison between the mono- and multivalent interaction is often not feasible. The sometimes surprisingly strong increase of binding energy through multivalency is thus not fully understood in terms of enthalpy and entropy.

Recently, this concept was applied convincingly to artificial supramolecules. The examination of artificial, designable, and less complex multivalent systems provides an approach which easily permits analysis of the thermodynamic and kinetic effects in great detail. As an example, the binding of a divalent calixarene ligand bearing two adamantane endgroups on each arm binds more strongly to a cyclodextrin by a factor of 260 compared with the monovalent interaction – a much

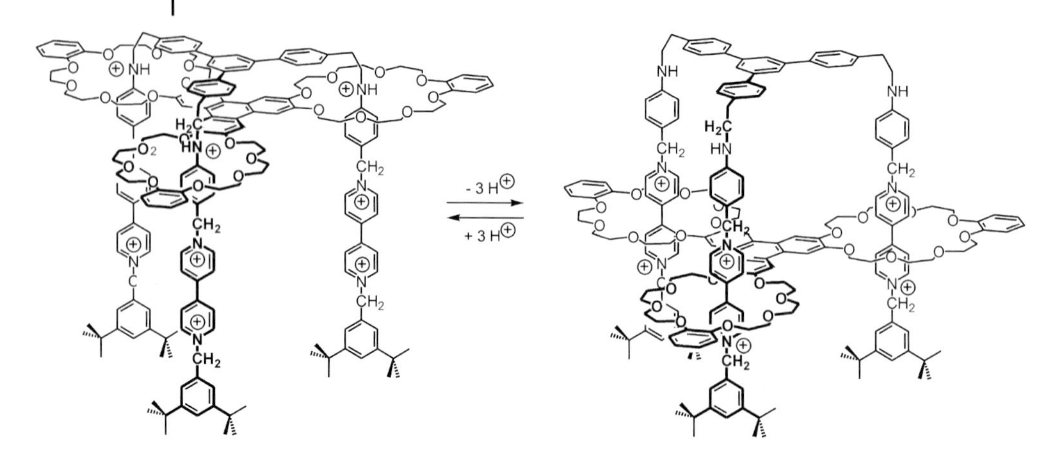

Fig. 1.2. Molecular elevator synthesized by utilizing multivalency. The position of the wheel component can be controlled by protonation/deprotonation.

higher increase than expected for merely additive interactions. If offered many cyclodextrin hosts on a surface, the binding constant again increases by 3 orders of magnitude [21]. Another example is shown in Fig. 1.2 [22]. A three-armed guest is capable of forming a triply threaded pseudorotaxane with the tris-crown derivative. Attachment of stoppers at the ends of each arm prevents deslippage of the axle components. The trivalent interaction increases the yield of the synthesis through favorable entropic contributions. At the same time, the function of a "molecular elevator" is implemented: depending on protonation and deprotonation of the dialkyl amines, the crown ethers move back and forth between two different stations along the axle.

1.3.4
Self-assembly and Self-organization

Self-assembly [23] is a strategy used by supramolecular chemists to reduce the efforts required for the generation of complex structures and architectures. Instead of tedious multistep covalent syntheses, simple building blocks are programmed with the suitably positioned binding sites and upon mixing the right subunits, they spontaneously assemble without any additional contribution from the chemist. Several requirements must be met: (i) the building blocks must be mobile, but this requirement is almost always fulfilled with molecules in solution due to Brownian motion; (ii) the individual components must bear the appropriate information written into their geometrical and electronic structure during synthesis to provide the correct binding sites at the right places. Since their mutual recognition requires specificity, self-assembly is a matter of well pre-organized building blocks (see above); (iii) the bonds between different components must be reversibly formed. This means that the final aggregate is generated thermodynamically controlled under equilibrium conditions. This aspect is important, because kinetically controlled processes do not have the potential for error correction and thus usually lead to mixtures. The reversibility of self-assembly processes also results in quite dynamic aggregates prone to exchange reactions of their building blocks.

Self-assembly is ubiquitous in nature [24] and often occurs on several hierarchy levels simultaneously in order to generate functional systems. For example, the shell-forming protein building blocks of the tobacco mosaic virus [25] need to fold into the correct tertiary protein structure before they can be organized around a templating RNA strand. All these processes are mediated by noncovalent forces which guide the formation of secondary structure elements on the lowest hierarchy level. These form the tertiary structure on the next level which displays the necessary binding sites for the assembly of the virus from a total of 2131 building blocks to occur as programmed on the highest level. Other examples for hierarchical self-assembly are multienzyme complexes, the formation of cell membranes with all the receptors, ion channels, or other functional entities embedded into them, or molecular motors such as ATP synthase. Self-assembly is thus an efficient strategy to create complexity and – together with it – function in nature.

Self-assembly has also been applied to numerous different classes of complexes in supramolecular chemistry [26]. Since we cannot discuss them all here, Fig. 1.3 shows only one example of a capsule reversibly formed from two identical self-complementary monomers which are bound to each other by hydrogen bonding

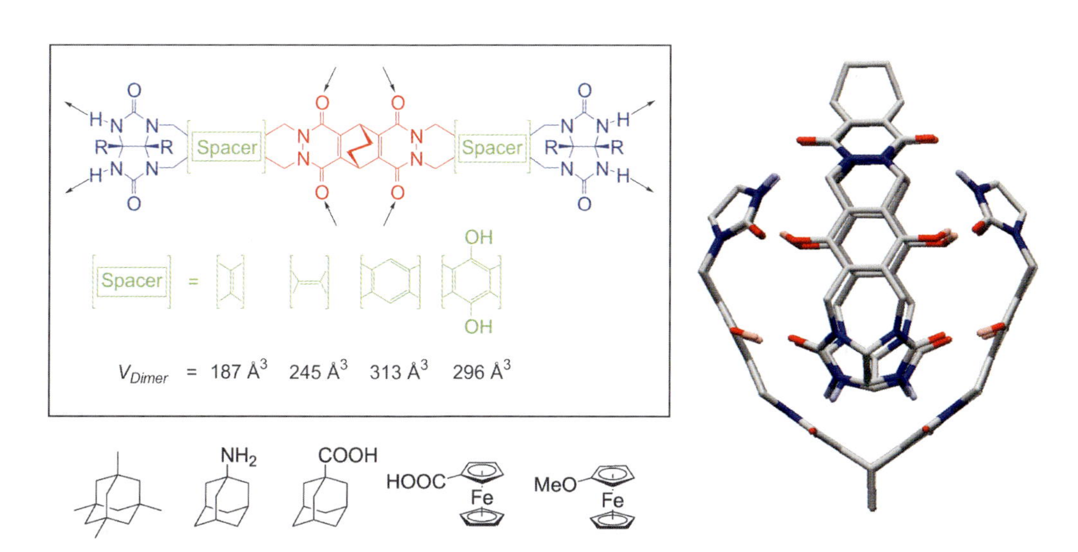

Fig. 1.3. Self-assembling "softball". Right: Computer model of the softball bearing the hydroquinone spacer (side chains are omitted). Box (left): Different monomers which form dimers with cavities of volumes between 187 and 313 Å^3 depending on the spacer length. Left: A selection of good guest molecules which can occupy the cavity inside the capsule.

[27]. The two monomers can encapsulate guests in the interior cavity of the capsule. Even more than one guest can be encapsulated, and reactions can be catalyzed inside.

Another term which is often used in the literature as synonymous with self-assembly is self-organization. However, again, we should be precise with respect to the meaning of the terms we use. One suggestion for definitions would be to distinguish processes which lead to the thermodynamic minimum and thus lead to chemical equilibria. These processes should be called self-assembly processes. On the other hand is the broad variety of spontaneous organization which occurs far away from the thermodynamic equilibrium. Many processes in living organisms are examples for self-organization in this sense. The major difference between self-assembly and self-organization is that self-assembly occurs even in a closed system while self-organization can be characterized as a steady state in which a system remains without falling to the thermodynamic minimum, because energy is constantly flowing through it. This definition has the advantage that it makes a clear difference between the two terms. This advantage however comes at the price that it is experimentally difficult to determine which is which by simple criteria.

1.3.5
Template Effects

One way to control the outcome of a reaction is templating. Like in the macroscopic world, a chemical template organizes reaction partners and thus allows the chemist to control their reactivity to achieve the formation of a desired product. However, it is almost impossible to give a concise definition of the term "template" [28]. Templates span the whole range from biochemistry with its complex apparatus for DNA replication [29] to the formation of structured inorganic materials [30] to the templated synthesis of macrocycles [31] to the preparation of supramolecular catalysts [32] – just to name a few examples. Nevertheless, all these have in common that a template must serve different purposes: (i) it organizes reaction partners for the formation of a desired product whose synthesis cannot be achieved in the absence of the template. Thus, a template controls reactivity and produces form; (ii) the template needs to bind to the reaction partners. Molecular recognition is thus a necessary prerequisite for template syntheses and the binding sites of the components must be complementary to each other. Usually, binding is due to noncovalent bonds, although examples for covalent templates exist; (iii) the control of reactivity and the recognition of the reaction partners imply information to be programmed into the template which is transferred to the product of the reaction.

There are different ways to categorize templates. One could for example try to distinguish template effects according to the (non-)covalent interactions involved. This classification remains ambiguous for templates operating through different forces at the same time. A maybe better way to classify templates relates to their topography. The early templated crown ether syntheses utilized alkali metal ions around which macrocycles form with size selectivity [33]. Such templates are convex, because of their convex surface mediating the template effect. In contrast, a receptor binding two molecules which react inside a cavity is concave. This is true for many templates leading to mechanically interlocked species. One of the most prominent natural templates, i.e. single-stranded DNA, could be called a linear template according to this classification. Finally, a surface on which molecules self-assemble into an ordered array [34] may be considered as a planar template. [35].

Although there certainly is some overlap, one should distinguish between a reactant, a template, and a catalyst [36]. A strict definition would stress that the template must be removable after a successful reaction, while a reactant at least in part remains in the product. However, these definitions become blurred. For example, the synthesis of rotaxanes, catenanes, and knots [37] often relies on macrocycles which in their cavity bind an axle component in a pseudorotaxane fashion. Thus, the macrocycle acts as the template which organizes the axle in a threaded geometry. Ring closure of the axle or the attachment of stoppers lead to catenanes or rotaxanes, respectively. The macrocyclic template finally becomes part of the product according to the strict definition would be considered as a reactant rather than a template. Nevertheless, this view on the synthesis of interlocked molecules – one out of many examples is shown in Fig. 1.4 – neglects the organization of the two pseudorotaxane components which is essential for the formation of the mechanical bond. Thus, these syntheses are widely accepted as template-mediated in the chemical literature, although the use of removable transition metal ions for the synthesis of mechanically interlocked molecules [38] is probably the only true template synthesis for interlocked molecules in the strict sense. It is similarly difficult to separate templates from catalysts: on one hand, many templates do not promote catalytic reactions, because the template does not generate turnover. They need to be used in stoichiometric amounts and have to be separated from the product. On the other hand, some catalysts do not organize the reactants in space but rather change their intrinsic reactivity as for example encountered in general acid or base catalysis. Thus, they cannot be regarded as templates. These are the clear-cut cases. However, mixed forms exist, where a template is bound reversibly to the product or where a catalyst organizes the reactants with respect to their geometry. We therefore put forward a more abstract view of what a template is and consider a template as the sum of all connections between the species reacting with each other which are involved in geometrically controlling the reactivity in the desired way. It is the array of interactions and their spatial arrangement that count.

1.3.6 Self-replication and Supramolecular Catalysis

While multivalency, self-assembly, and template effects provide strategies aiming at generating more and more complex architectures, supramolecular chemistry can also be utilized for controlling reactivity and even catalyzing reactions. Closely related to organocatalysis, supramolecular catalysts [39] accelerate reactions by

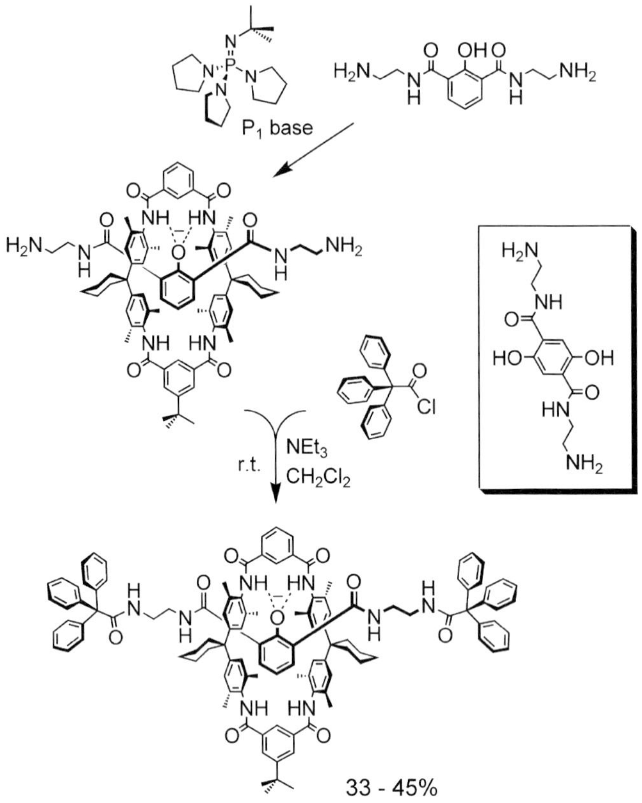

Fig. 1.4. Anion-templated rotaxane synthesis. The axle center piece is threaded through the macrocycle's cavity by hydrogen bonding. Stopper attachment to both axle ends traps the wheel on the axle. Inset: hydroquinone-based center piece which can also be used, but with lower efficiency.

lowering the barriers. The principles by which they fulfil the task are very different. Increasing the local concentration of the reactands by encapsulation is one example (see Fig. 1.3 above), increasing the intrinsic reactivity of carbonyl compounds through hydrogen bonding [40] is another and many more exist.

Originating from the question how the living organisms came into existence, self-replication is a special, but certainly intriguing case of supramolecular catalysis. If one thinks about the complex ribosome, which nowadays transscribes genetic information stored in nucleic acids into proteins, which then become involved in the duplication of DNA, it is immediately clear that this apparatus is much too complex to self-organize accidentially at the beginning of life. Instead, much simpler mechanisms must have existed in the early world. In order to find an answer, several research groups provided evidence that short DNA oligomers are indeed able to self-replicate in the presence of the appropriate template [41].

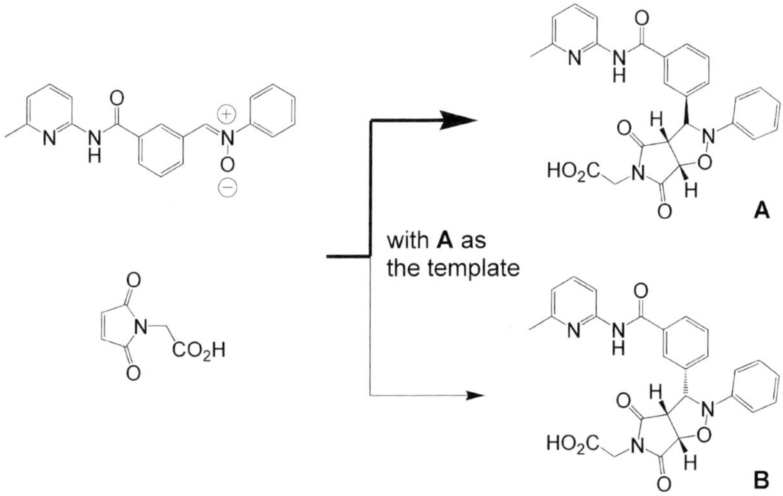

Fig. 1.5. A minimal self-replicating system. In the presence of template **A**, the two reactands on the left are organised in a way suitable for a 1,3-dipolar cycloaddition reaction. The pyridineamide part of the template recognizes the acid substituent in the reactand, while the second reactand is recognized by the carboxylic acid incorporated in the template. Particularly interesting is the fact that template **A** favors its own formation, while the other stereoisomer **B** is formed only in low amounts.

Later, suitable α-helical peptides have been shown to self-replicate as well [42]. In the context of supramolecular chemistry most interesting are however organic minimal-replicators [43] which are not based on biomolecules. Figure 1.5 shows an example for a minimal self-replicating system, which operates even in a chiro-selective way. One given enantiomer of the template catalyzes its own formation, while the other enantiomer is by and large suppressed.

1.3.7 Molecular Devices and Machines: Implementing Function

Early supramolecular chemistry certainly focussed on the noncovalent bond and the beauty of structures which can be generated employing it. This is certainly the case for topologically interesting molecules such as rotaxanes, catenanes, knotanes and Borromean rings [37]. It also holds for the generation of self-assembling capsules, helicates [44], or metallo-supramolecular tetrahedra, octahedra and the like [45]. However, the focus has shifted in contemporary supramolecular chemistry towards the implementation of function into noncovalent architectures. The scope of function is broad and ranges from light-induced energy and electron transfer processes [46] and molecular wires [47] to switches [48], molecular "motors" [46], and devices for the active pH-driven transport of molecules through membranes. This area is too broad to give a satisfying introduction here and thus, the reader is referred to the literature cited.

1.4 Conclusions: Diverse Methods for a Diverse Research Area

The admittedly short and simplified considerations above make clear that one aim of supramolecular chemistry is to mimic natural processes. The above sections deliberately chose examples from biochemistry as well as the multitude of artificial supramolecules in order to point to the relations which exist between the two fields. Understanding the details of noncovalent binding is much more difficult in a complex biomolecule, and thus simple model systems provide the basis for a more profound analysis. However, supramolecular chemistry goes beyond merely creating model systems for naturally occurring species. In contrast to biomolecules, supramolecular chemistry can utilize the whole range of conditions achievable, for example with respect to the use of organic solvents, in which many biomolecules would lose their integrity, because they are designed for an aqueous surrounding. Higher or lower temperatures or different pressures can also be applied. Supramolecules may even find their applications under conditions where biomolecules would not have the necessary long-term stability. The implementation of function also aims at new functions which are not realized in nature. In particular, the latter two aspects lead us to the second research area to which supramolecular chemistry contributes significantly: material sciences. Self-assembly, for example, is a strategy to create long-range order and has even been applied to particles on a micro- to millimeter scale [49].

If one thinks about function, in particular switches, logic gates, and molecular wires, it becomes clear that supramolecular chemistry is also about information processing. However, it is not only its potentially upcoming use in microelectronics: information processing begins at a much more fundamental level. Templates transfer spatial information between molecules; in order to achieve correctly self-assembling species, the building blocks of the assembly need to be programmed with the appropriate binding sites. Information transfer and information processing already starts at the molecular level.

A view back on the last few decades makes perfectly clear that supramolecular chemistry has become a highly diverse field which requires the interdisciplinary use of a huge variety of methods to answer the scientific questions addressed. Diversity however is not the only challenge for the methods that are needed. The complexity of the architectures meanwhile realized requires sophisticated structure analysis tools. The highly dynamic features of supramolecules need kinetic methods able to address many different time scales. Gathering evidence for the functions implemented is impossible without a sound methodological basis. Finally, the wish to image and influence single molecules led to the application of scanning probe microscopy to supramolecular systems. The present book intends to take this into account and to provide an overview on methods used in supramolecular chemistry – even though it is probably not possible to be comprehensive.

References and Notes

1 A. Werner, *Zeitschr. Anorg. Chem.* **1893**, 3, 267.

2 E. Fischer, *Ber. Deutsch. Chem. Ges.* **1894**, 27, 2985. Also, see: J.-P. Behr (ed.), *The Lock and Key Principle. The State of the Art – 100 Years On*, Wiley, Chichester 1994.

3 a) A. Villiers, C. R. Hebd, *Seances Acad. Sci.* **1891**, 112, 435; b) A. Villiers, C. R. Hebd, *Seances Acad. Sci.* **1891**, 112, 536.

4 P. Ehrlich, *Studies on Immunity*, Wiley, New York 1906.

5 D. E. Koshland, Jr., *Proc. Natl. Acad. Sci. USA* **1958**, 44, 98.

6 K. L. Wolf, H. Frahm, H. Harms, *Z. Phys. Chem. (B)* **1937**, 36, 237.

7 For textbooks, see: a) F. Vögtle, *Supramolekulare Chemie*, Teubner, Stuttgart **1992**; b) J.-M. Lehn, *Supramolecular Chemistry*, Verlag Chemie, Weinheim **1995**; c) H.-J. Schneider, A. Yatsimirsky, *Principles and Methods in Supramolecular Chemistry*, Wiley, New York **2000**; d) J. W. Steed, J. L. Atwood, *Supramolecular Chemistry*, Wiley, New York **2000**.

8 J.-M. Lehn, *Pure Appl. Chem.* **1979**, 50, 871.

9 G. A. Jeffrey, *An Introduction to Hydrogen Bonding*, Oxford University Press, Oxford 1997.

10 J. C. Ma, D. A. Dougherty, *Chem. Rev.* **1997**, 97, 1303.

11 C. A. Hunter, J. K. M. Sanders, *J. Am. Chem. Soc.* **1990**, 112, 5525.

12 R. Paulini, K. Müller, F. Diederich, *Angew. Chem.* **2005**, 117, 1820; *Angew. Chem. Int. Ed.* **2005**, 44, 1788.

13 P. Auffinger, F. A. Hays, E. Westhof, P. S. Ho, *Proc. Natl. Acad. Sci. USA* **2004**, 101, 16789.

14 R. H. Crabtree, P. E. M. Siegbahn, O. Eisenstein, A. L. Rheingold, T. F. Koetzle, *Acc. Chem. Res.* **1996**, 29, 348.

15 a) C. Seel, A. Galán, J. de Mendoza, *Top. Curr. Chem.* **1995**, 175, 101; b) F. P. Schmidtchen, M. Berger, *Chem. Rev.* **1997**, 97, 1609; c) P. D. Beer, P. A. Gale, *Angew. Chem.* **2001**, 113, 502–532; *Angew. Chem. Int. Ed.* **2001**, 40, 487; d) J. J. Lavigne, E. V. Anslyn, *Angew. Chem.* **2001**, 113, 3212–3225; *Angew. Chem. Int. Ed.* 2001, 40, 3119; e) J. L. Sessler, J. M. Davis, *Acc. Chem. Res.* **2001**, 34, 989; f) K. Bowman-James, *Acc. Chem. Res.* **2005**, 38, 671.

16 D. J. Cram, *Angew. Chem.* 1986, 98, 1041; *Angew. Chem. Int. Ed.* **1986**, 25, 1039.

17 W. A. Eaton, E. R. Henry, J. Hofrichter, A. Mozzarelli, *Nat. Struct. Biol.* **1999**, 6, 351.

18 G. Ercolani, *J. Am. Chem. Soc.* **2003**, 125, 16097.

19 For reviews, see: a) N. Röckendorf, T. K. Lindhorst, *Top. Cur. Chem.* **2002**, 217, 201; b) S.-K. Choi, *Synthetic Multivalent Molecules*, Wiley-Interscience, Hoboken, USA, 2004; c) A. Mulder, J. Huskens, D. N. Reinhoudt, *Org. Biomol. Chem.* **2004**, 2, 3409; d) J. D. Badjic, A. Nelson, S. J. Cantrill, W. B. Turnbull, J. F. Stoddart, *Acc. Chem. Res.* **2005**, 38, 723.

20 M. Mammen, S.-K. Choi, G. M. Whitesides, *Angew. Chem.* **1998**, 110, 2908; *Angew. Chem. Int. Ed.* **1998**, 37, 2754.

21 a) J. Huskens, M. A. Deij, D. N. Reinhoudt, *Angew. Chem.* **2002**, 114, 4647; *Angew. Chem. Int. Ed.* **2002**, 41, 4467; b) A. Mulder, T. Auletta, A. Sartori, A. Casnati, R. Ungaro, J. Huskens, D. N. Reinhoudt, *J. Am. Chem. Soc.* **2004**, 126, 6627; c) T. Auletta, M. R. de Jong, A. Mulder, F. C. J. M. van Veggel, J. Huskens, D. N. Reinhoudt, S. Zou, S. Zapotoczny, H. Schönherr, G. J. Vancso, L. Kuipers, *J. Am. Chem. Soc.* **2004**, 126, 1577.

22 J. D. Badjic, V. Balzani, A. Credi, S. Silvi, J. F. Stoddart, *Science* **2004**, 303, 1845.

23 Reviews: a) J. S. Lindsey, *New J. Chem.* **1991**, 15, 153; b) G. M. Whitesides, J. P. Mathias, C. T. Seto, *Science* **1991**, 254, 1312; c) D. Philp, J. F. Stoddart, *Angew. Chem.* 1996, 108, 1243; *Angew. Chem. Int. Ed.* **1996**, 35, 1154. (d) C. A. Schalley, A. Lützen,

M. Albrecht, *Chem. Eur. J.* **2004**, 10, 1072.

24 T. D. Hamilton, L. R. MacGillivray, *Self-Assembly in Biochemistry in: Encyclopedia of Supramolecular Chemistry*, J. L. Atwood, J. W. Steed (Eds.), Dekker, New York, 2004, 1257.

25 A. Klug, *Angew. Chem.* **1983**, 95, 579; *Angew. Chem., Int. Ed. Engl.* **1983**, 22, 565.

26 L. F. Lindoy, I. M. Atkinson, *Self-Assembly in Supramolecular Chemistry*, Royal Society of Chemistry, Cambridge 2000.

27 For a review, see: F. Hof, S. L. Craig, C. Nuckolls, J. Rebek, Jr., *Angew. Chem.* **2002**, 114, 1556; *Angew. Chem. Int. Ed.* **2002**, 41, 1488.

28 a) D. H. Busch, N. A. Stephensen, *Coord. Chem. Rev.* **1990**, 100, 119; b) R. Cacciapaglia, L. Mandolini, *Chem. Soc. Rev.* **1993**, 22, 221; c) N. V. Gerbeleu, V. B. Arion, J. Burgess, *Template Synthesis of Macrocyclic Compounds*, Wiley-VCH, Weinheim 1999; d) T. J. Hubin, A. G. Kolchinski, A. L. Vance, D. H. Busch, *Adv. Supramol. Chem.* **1999**, 5, 237; e) F. Diederich, P. J. Stang (eds.) *Templated Organic Synthesis* Wiley-VCH, Weinheim **2000**; f) T. J. Hubin, D. H. Busch, *Coord. Chem. Rev.* **2000**, 200–202, 5.

29 See, for example: D. Voet, J. G. Voet, *Biochemistry*, Wiley, Chichester 1990.

30 K. J. C. van Bommel, A. Friggeri, S. Shinkai, *Angew. Chem.* **2003**, 115, 1010; *Angew. Chem. Int. Ed.* **2003**, 42, 980.

31 B. C. Gibb, *Chem. Eur. J.* **2003**, 9, 5181.

32 J. K. M. Sanders, *Pure Appl. Chem.* **2000**, 72, 2265.

33 C. J. Pedersen, *J. Am. Chem. Soc.* **1967**, 89, 7019.

34 C. Safarowsky, L. Merz, A. Rang, P. Broekmann, B. A. Hermann, C. A. Schalley, *Angew. Chem.* **2004**, 116, 1311; *Angew. Chem. Int. Ed.* **2004**, 43, 1291.

35 J. F. Hulvat, S. I. Stupp, *Angew. Chem.* **2003**, 115, 802; *Angew. Chem. Int. Ed.* **2003**, 42, 778.

36 S. Anderson, H. L. Anderson, *Templates in Organic Synthesis: Definitions and Roles* in ref. [28e], p. 1.

37 J.-P. Sauvage, C. Dietrich-Buchecker (eds.), *Molecular Catenanes, Rotaxanes, and Knots*, Wiley-VCH, Weinheim 1999.

38 a) J.-P. Sauvage, *Acc. Chem. Res.* **1990**, 23, 319; b) D. A. Leigh, P. J. Lusby, S. J. Teat, A. J. Wilson, J. K. Y. Wong, *Angew. Chem.* **2001**, 113, 1586; *Angew. Chem. Int. Ed.* **2001**, 40, 1538; c) P. Mobian, J.-M. Kern, J.-P. Sauvage, *J. Am. Chem. Soc.* **2003**, 125, 2016.

39 a) F. Diederich, *J. Chem. Educ.* **1990**, 67, 813; b) P. Scrimin, P. Tecilla, U. Tonellato, *J. Phys. Org. Chem.* **1992**, 5, 619; c) J.-M. Lehn, *Appl. Catal. A*, **1994**, 113, 105; d) M. C. Feiters, in: *Comprehensive Supramolecular Chemistry*, Eds: J. L. Atwood, J. E. D. Davies, D. D. MacNicol, F. Vögtle, Pergamon Press, Oxford 1996, Vol. 11, p. 267; e) J. K. M. Sanders, *Chem. Eur. J.* **1998**, 4, 1378.

40 P. R. Schreiner, *Chem. Soc. Rev.* **2003**, 32, 289.

41 a) L. E. Orgel, *Nature* **1992**, 358, 203; b) T. Li, K. C. Nicolaou, *Nature* **1994**, 369, 218; c) D. Sievers, G. von Kiedrowski, *Nature* **1994**, 369, 221.

42 K. Severin, D. H. Lee, J. A. Martinez, M. R. Ghadiri, *Chem. Eur. J.* **1997**, 3, 1017.

43 a) E. A. Wintner, M. M. Conn, J. Rebek, Jr., *Acc. Chem. Res.* **1994**, 27, 198; b) A. Robertson, A. J. Sinclair, D. Philp, *Chem. Soc. Rev.* **2000**, 29, 141.

44 M. Albrecht, *Chem. Rev.* **2001**, 101, 3457.

45 J.-P. Sauvage (ed.), *Transition Metals in Supramolecular Chemistry*, Wiley, Chichester 1999.

46 V. Balzani, M. Venturi, A. Credi, *Molecular Devices and Machines – A Journey into the Nanoworld*, Wiley-VCH, Weinheim 2003.

47 L. De Cola (Ed.), *Molecular Wires* in: *Topics in Current Chemistry*, vol. 257, Springer, Heidelberg 2005.

48 B. Feringa (ed.), *Molecular Switches*, Wiley-VCH, Weinheim **2001**.

49 N. B. Bowden, M. Weck, I. S. Choi, G. M. Whitesides, *Acc. Chem. Res.* **2001**, 34, 231.

2
Determination of Binding Constants

Keiji Hirose

2.1
Theoretical Principles

2.1.1
The Binding Constants and Binding Energies

Generally, the formation of a complex between a host and a guest is a basic and important process in supramolecular chemistry. Selectivity in the complexation is an important property in determining the molecular recognition ability of the host molecule that discriminates among different guest species. The ratio of the binding constants of the corresponding complexations is usually treated as a measure of selectivity. Because binding constants have been used as a basic criterion for the evaluation of the host–guest complexation process, a binding constant (K) has to be determined for the quantitative analysis of a complex formation [1, 2, 3]. In addition to the ratio of binding constants, the temperature dependence of selectivity is also important to gain insight into the origin of supramolecular functions. Indeed, some supramolecular systems have been reported to show a large influence of temperature on selectivity [4] including temperature dependent inversion of enantioselectivity [5]. Therefore, thermodynamic parameters [enthalpy (ΔH), entropy (ΔS) and Gibbs free energy (ΔG)] are more suitable criteria in order to express the molecular recognition abilities.

Through Eqs. (2.1) and (2.2), the thermodynamic parameters, binding constant, and temperature (T) are related to each other as described in Fig. 2.1 and Eq. (2.3), the van't Hoff equation. Theoretically, the determination of binding constants at different temperature offers these thermodynamic parameters from the slope and the intercept of line in Fig. 2.1. The binding energy (ΔG) at a distinct temperature is calculated using the obtained ΔH, ΔS and T according to Eq. (2.2), the Gibbs–Helmholtz equation. *Therefore, the important point in the quantitative analysis of host–guest complexation is how to determine the binding constant with high reliability.*

$$\Delta G = -RT \ln K \tag{2.1}$$

Analytical Methods in Supramolecular Chemistry. Edited by Christoph Schalley

ISBN: 978-3-527-31505-5

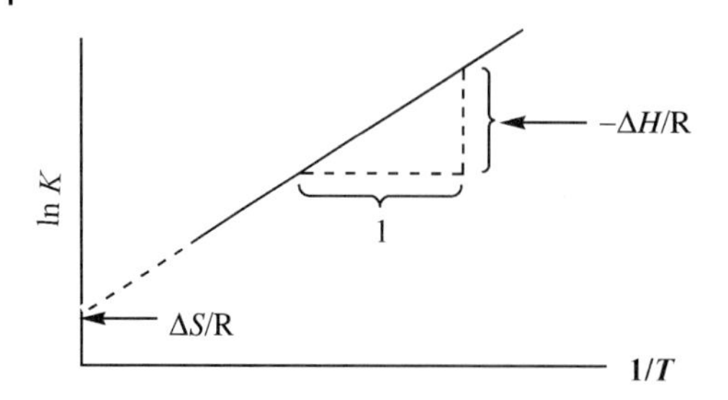

Fig. 2.1. Van't Hoff plot, i.e. the correlation of ΔH, ΔS, K and temperature according to the van't Hoff equation.

$$\Delta G = \Delta H - T\Delta S \tag{2.2}$$

$$\Rightarrow \ln K = -\frac{\Delta H}{R} \cdot \frac{1}{T} + \frac{\Delta S}{R} \tag{2.3}$$

2.1.2
A General View on the Determination of Binding Constants

A general way to determine the binding constant is based on a simple binding equilibrium model, i.e. Eq. (2.4). The terms binding constant, equilibrium constant, and stability constant are synonymous with each other. The activity coefficients are generally unknown and the stability constant K, based on the concentrations, is usually employed. Judging from this situation, the question of the activity coefficients of the solutes is disregarded here in order to simplify the discussion. Nevertheless, it should be remembered that this point is not always insignificant.

The basic equations for the host–guest complexation are the following four Eqs. (2.4)–(2.7).

$$a \cdot \mathrm{H} + b \cdot \mathrm{G} \rightleftarrows \mathrm{C} \tag{2.4}$$

$$K = \frac{[\mathrm{C}]}{[\mathrm{H}]^a \cdot [\mathrm{G}]^b} \tag{2.5}$$

$$[\mathrm{H}]_0 = [\mathrm{H}] + a \cdot [\mathrm{C}] \tag{2.6}$$

$$[\mathrm{G}]_0 = [\mathrm{G}] + b \cdot [\mathrm{C}] \tag{2.7}$$

with H: host; G: guest; C: complex ($\mathrm{H}_a \cdot \mathrm{G}_b$)
a, b: stoichiometry as shown in Eq. (2.4)
$[\mathrm{H}]_0$: initial (total) concentration of host molecule
$[\mathrm{G}]_0$: initial (total) concentration of guest molecule
[H], [G], [C]: equilibrium concentrations of host, guest, and complex, respectively

Equation (2.8) is derived from Eqs. (2.5)–(2.7).

$$K = \frac{[C]}{([H]_0 - a \cdot [C])^a \cdot ([G]_0 - b \cdot [C])^b} \tag{2.8}$$

The parameters can be classified as follows:

K, a, b: Constants (a and b are integers larger than or equal to 1.)
$[H]_0$, $[G]_0$: Variables which can be set up as experimental conditions
$[H]$, $[G]$, $[C]$: Variables dependent on the equilibrium

2.1.3 Guideline for Experiments

From Eq. (2.8) and the classification of its parameters, we can deduce the following guideline for experiments to determine the binding constant. When $[C]$ is obtained at the equilibrium in which a and b are known, K can be derived directly according to the Eq. (2.8) from the experimental conditions, $[H]_0$ and $[G]_0$. Consequently, in order to determine the binding constants, the following four steps have to be carried out:

- determination of stoichiometry, namely, a and b
- evaluation of complex concentration, $[C]$
- setting up the concentration conditions, $[H]_0$ and $[G]_0$
- data treatment.

The following sections deal with the principle and also the practical issues necessary for an understanding and performance of the above four points in this order.

2.2 A Practical Course of Binding Constant Determination by UV/vis Spectroscopy

2.2.1 Determination of Stoichiometry

There are different methods to determine the stoichiometry, e.g. Continuous Variation Methods [6], the Slope Ratio Method [7], the Mole Ratio Method [8], and others. Because the Continuous Variation Method is the most popular among these, this method is adopted here to determine the stoichiometry.

In order to determine the stoichiometry by the Continuous Variation Method, the following four points have to be considered and carried out:

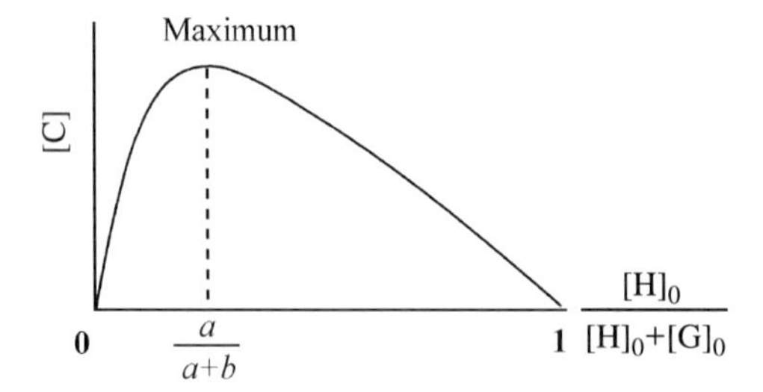

Fig. 2.2. Determination of stoichiometry (a, b) from the x-coordinate at the maximum of the curve in Job's Plot.

- keeping the sum of $[H]_0$ and $[G]_0$ constant (α)
- changing $[H]_0$ from 0 to α
- measuring [C]
- data treatment (Job's plot).

The stoichiometry $\left(\frac{a}{a+b}\right)$ is obtained from the x-coordinate at the maximum in Job's curve (Fig. 2.2), where the y-axis is [C] and x-axis is $\frac{[H]_0}{([H]_0 + [G]_0)}$.

For the comprehension of the theoretical background of the Continuous Variation Method, the required equations are (2.4)–(2.7) and (2.9)–(2.11).

$$\alpha = [H]_0 + [G]_0 \tag{2.9}$$

$$x = \frac{[H]_0}{([H]_0 + [G]_0)} \tag{2.10}$$

$$y = [C] \tag{2.11}$$

$[H]_0$ and $[G]_0$ will be substituted by the function of x and α from Eqs. (2.9) and (2.10).

$$[H]_0 = \alpha \cdot x \tag{2.12}$$

$$[G]_0 = \alpha - \alpha \cdot x \tag{2.13}$$

From Eqs. (2.4)–(2.7) and (2.11)–(2.13):

$$K = \frac{y}{\{(\alpha \cdot x - a \cdot y)^a \cdot (\alpha - b \cdot y - \alpha \cdot x)^b\}}$$

$$\Rightarrow K \cdot (\alpha - b \cdot y - \alpha \cdot x)^b \cdot (\alpha \cdot x - a \cdot y)^a = y \tag{2.14}$$

The Eq. (2.14) is then differentiated, and the $\frac{dy}{dx}$ is substituted by zero. Then the x-coordinate at the maximum in the curve is obtained.

$$K \cdot [(\alpha - b \cdot y - \alpha \cdot x)^b \cdot \{(\alpha \cdot x - a \cdot y)^a\}' + \{(\alpha - b \cdot y - \alpha \cdot x)^b\}' \cdot (\alpha \cdot x - a \cdot y)^a] = \frac{dy}{dx}$$

$$K \cdot \left[(\alpha - b \cdot y - \alpha \cdot x)^b \cdot a \cdot (\alpha \cdot x - a \cdot y)^{a-1} \cdot \left(\alpha - a \cdot \frac{dy}{dx}\right) + b \cdot (\alpha - b \cdot y - \alpha \cdot x)^{b-1} \cdot \left(-b \cdot \frac{dy}{dx} - \alpha\right) \cdot (\alpha \cdot x - a \cdot y)^a\right] = \frac{dy}{dx}$$

Substitution of $\frac{dy}{dx}$ by zero yields:

$$K \cdot [(\alpha - b \cdot y - \alpha \cdot x)^b \cdot a \cdot (\alpha \cdot x - a \cdot y)^{a-1} \cdot \alpha + b \cdot (\alpha - b \cdot y - \alpha \cdot x)^{b-1} \cdot (-\alpha) \cdot (\alpha \cdot x - a \cdot y)^a] = 0$$

Division by $K \cdot (\alpha - b \cdot y - \alpha \cdot x)^{b-1} \cdot (\alpha \cdot x - a \cdot y)^{a-1} \cdot \alpha$ produces

$$a \cdot (\alpha - b \cdot y - \alpha \cdot x) - b \cdot (\alpha \cdot x - a \cdot y) = 0$$

$$a \cdot \alpha - a \cdot b \cdot y - a \cdot \alpha \cdot x - b \cdot \alpha \cdot x + b \cdot a \cdot y = 0$$

$$a \cdot \alpha - a \cdot \alpha \cdot x - b \cdot \alpha \cdot x = 0$$

Division by α.

$$a - ax - bx = 0$$

$$\Rightarrow x = \frac{a}{a + b} \tag{2.15}$$

Equation (2.15) means that $\frac{a}{a+b}$ is the x-coordinate at the maximum $\left(\frac{dy}{dx} = 0\right)$ in the curve of Eq. (2.14). It thus provides the correlation between the stoichiometry and the x-coordinate at the maximum in Job's plot. For example, when 1:1 complexation is predominant at the equilibrium, the maximum appears $x = 0.5$ $(a = b = 1)$. In the case of 1:2 complexation, the maximum is at $x = 0.333$.

The practically important point here is the following: Even if the concentration of complex ([C]) could not be measured directly, the [C] (y-axis) can be replaced with a parameter proportional to [C]. Then the same x-coordinate and with it the same stoichiometry is obtained at the maximum as that in Job's plot. This means the stoichiometry can even then be determined even when [C] cannot be obtained

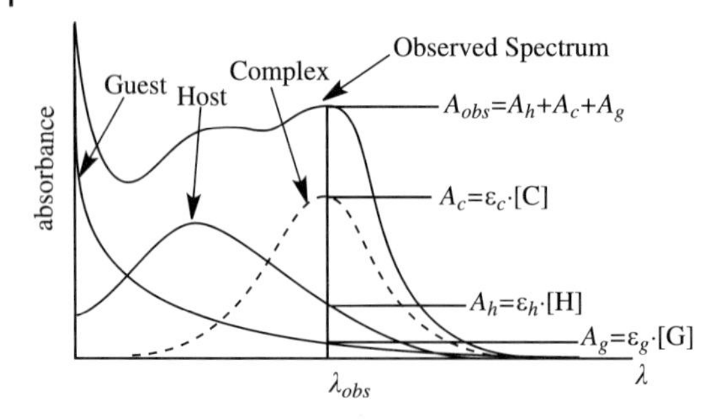

Fig. 2.3. Representative UV/vis spectra to show correlation of observed spectra and each component.

directly. The practical question for each individual experiment is then how to modify the y-coordinate.

Depending on the selected experiment, a suitable observable property should be selected and the complex concentration [C] should be replaced by it in Job's plot. In the following, UV/vis spectroscopy is discussed as a representative example.

The concentrations and absorbances of each species are related as the following Eqs. (2.16)–(2.18) in the case of investigation by means of UV/vis spectroscopy. The observed absorbance is expressed as Eq. (2.19) and Fig. 2.3. The length of the optical cell is fixed here to 1 cm as a premise. The definitions of the abbreviations are given below. The definitions of other abbreviations (a, b, $[H]_0$, $[G]_0$, [H], [G], [C]) are the same as described before.

$$A_h = \varepsilon_h \cdot [H] = \varepsilon_h \cdot ([H]_0 - a \cdot [C]) \tag{2.16}$$

$$A_g = \varepsilon_g \cdot [G] = \varepsilon_g \cdot ([G]_0 - b \cdot [C]) \tag{2.17}$$

$$A_c = \varepsilon_c \cdot [C] \tag{2.18}$$

$$A_{obs} = A_h + A_g + A_c \tag{2.19}$$

A_{obs}: observed absorbance
A_h, A_g, A_c: absorbances of host, guest, and complex respectively
ε_h, ε_g, ε_c: molar absorptivities of host, guest, and complex respectively

Equation (2.19) is transformed to Eq. (2.20) by using Eqs. (2.16)–(2.18).

$$A_{obs} = \varepsilon_h \cdot ([H]_0 - a \cdot [C]) + \varepsilon_g \cdot ([G]_0 - b \cdot [C]) + \varepsilon_c \cdot [C]$$

$$\Rightarrow A_{obs} - \varepsilon_h \cdot [H]_0 - \varepsilon_g \cdot [G]_0 = (\varepsilon_c - a \cdot \varepsilon_h - b \cdot \varepsilon_g) \cdot [C] \tag{2.20}$$

Equation (2.20) shows that $A_{obs} - \varepsilon_h \cdot [H]_0 - \varepsilon_g \cdot [G]_0$ is proportional to [C] because $(\varepsilon_c - a \cdot \varepsilon_h - b \cdot \varepsilon_g)$ is constant. The molar absorptivities ε_h, ε_g can be determined from independent measurements using the pure host and the pure guest, respectively. The concentrations $[H]_0$, $[G]_0$, are known because they are the experi-

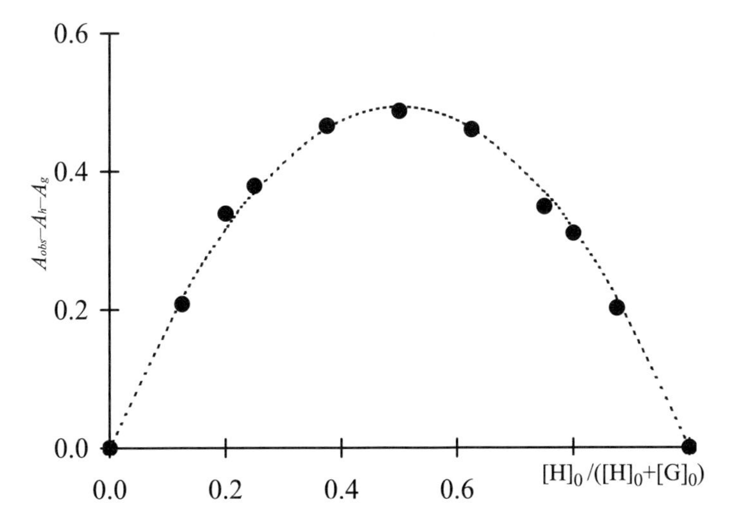

Fig. 2.4. Modified Job's Plot for complexation of host and guest by UV/vis spectroscopy: • observed; -------: calculated.

mental conditions set up by the experimenter. Consequently, $(A_{obs} - \varepsilon_h \cdot [\mathrm{H}]_0 - \varepsilon_g \cdot [\mathrm{G}]_0)$ is determined from the experiments by means of UV/vis spectroscopy. The stoichiometry is determined from the *x*-coordinate at the maximum in the curve which might be called a modified Job's plot where $(A_{obs} - \varepsilon_h \cdot [\mathrm{H}]_0 - \varepsilon_g \cdot [\mathrm{G}]_0)$ is plotted as the *y*-coordinate instead of [C]. An actual example of this modified Job's plot is shown in Fig. 2.4. The *x*-coordinate at the maximum in the curve is 0.5. This supports the 1:1 host–guest complexation. To get a better feeling of the practical experiment, a spreadsheet for the Continuous Variation Method is attached as Appendix 2.1 [9].

2.2.2
Evaluation of Complex Concentration

When the observed property is the complex concentration ([C]) at equilibrium itself, there is no difficulty. But the actual complex concentration cannot be observed directly in most cases. Thus, the question how to evaluate [C] is an important practical issue. The practical way depends on the property that can be observed in each experiment. In this section, two typical cases for the evaluation of the complex concentration at the equilibrium by UV/vis spectroscopic methods are discussed.

Case 1: the absorption bands of host, guest and complex overlap From Eq. (2.20), the following Eq. (2.21) is derived.

$$[\mathrm{C}] = \frac{A_{obs} - \varepsilon_h \cdot [\mathrm{H}]_0 - \varepsilon_g \cdot [\mathrm{G}]_0}{\varepsilon_c - a \cdot \varepsilon_h - b \cdot \varepsilon_g} \tag{2.21}$$

If all constants (a, b, ε_h, ε_g and ε_c) could be obtained, [C] could be determined using A_{obs} and the experimental conditions, i.e. $[H]_0$ and $[G]_0$. Since the molar absorptivity of the complex (ε_c) is not measurable directly, a titration experiment and regression are necessary for the evaluation of complex concentration.

This is the most complicated case of host–guest complexation detected by means of UV/vis spectroscopy because the absorption bands of all components, host, guest and complex, overlap. However, quite often, the situation can be simplified by choosing a detection wavelength, at which one component (e.g. the guest) has a $\varepsilon = 0$. This scenario is discussed next (Case 2).

Case 2: the absorption bands of only two components overlap As a typical example, the complexation of a chromophoric chiral crown ether and an amino alcohol in chloroform is shown in Fig. 2.5. In the visible region, 2-amino-1-propanol (**2**) has no absorption. However, both host **1** and complex **3** show clear absorption bands which overlap. In Fig. 2.5, UV/vis spectra of a chloroform solution of pure **1** and

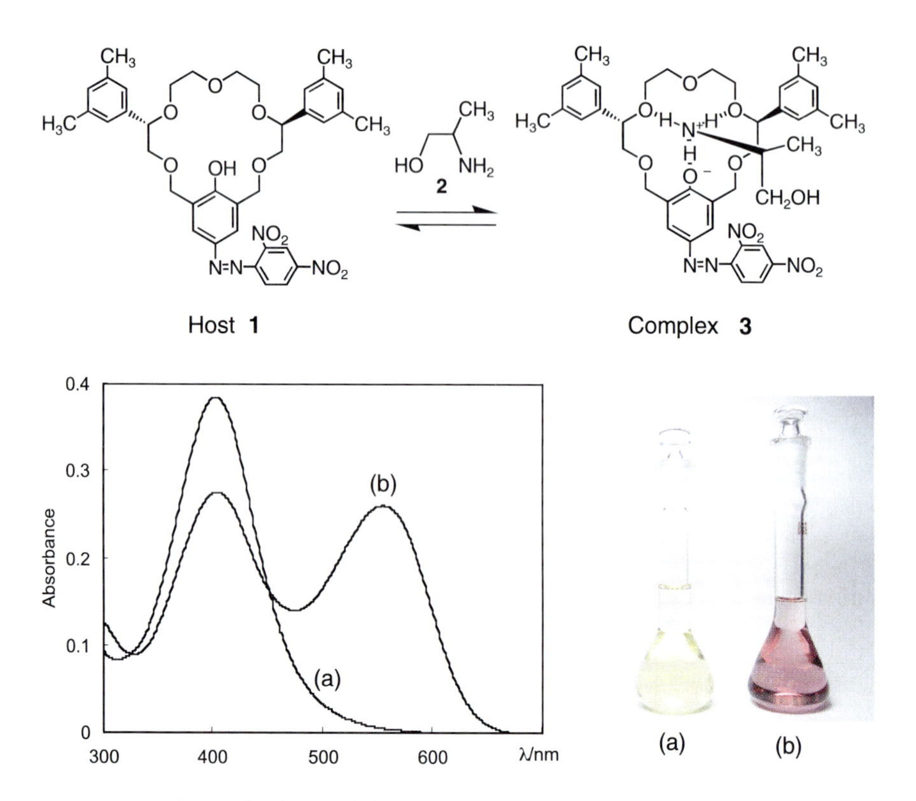

Fig. 2.5. Typical example of spectral changes caused by host–guest complexation: UV/vis spectra and corresponding pictures: (a) A chloroform solution of host 1 (1.85×10^{-5} M) at 25 °C (solid line); (b) The same solution as (a) containing 2 equivalents of (R)-2 (dotted line).

its mixture with **2** are shown. In order to avoid the complexity arising from overlapping absorption bands and to maximize the spectral change, a detection wavelength of 555 nm was chosen for the measurements. In this example, ε_g is zero, ε_h is small but not negligible, and ε_c is large at λ_{max} of complex **3** (555 nm).

For this case, Eq. (2.21) can be simplified and yields Eq. (2.22) just by substitution of ε_g by zero.

$$[C] = \frac{A_{obs} - \varepsilon_h \cdot [H]_0}{\varepsilon_c - a \cdot \varepsilon_h} \tag{2.22}$$

Compared with Eq. (2.21), Eq. (2.22) is significantly simplified. Because three parameters (b, ε_g, and $[G]_0$) disappear from Eq. (2.21), data treatment is also much simpler. If all constants (a, ε_h and ε_c) were obtained, [C] would be determined using A_{obs} and the experimental condition, $[H]_0$. *Nevertheless, Since the molar absorptivity of the complex (ε_c) is not measurable directly, a titration experiment and regression are necessary for the evaluation of complex concentration in this case.*

2.2.3
Precautions to be Taken when Setting Up Concentration Conditions of the Titration Experiment

Each method for binding analysis has limitations. There are dangerous sources of systematic errors that are often encountered in host–guest complexation, that is, the danger of carrying out titrations at concentrations unsuitable for the equilibrium being measured. The origin of this error will be discussed and methods for avoiding these problems will be presented in the following three sections.

2.2.3.1 Correlation between $[H]_0$, $[G]_0$, x and K

The experimental conditions that can be set up by the experimenter are $[H]_0$ and $[G]_0$ (see Eq. (2.8) and classification of variables). How should the experimental conditions, $[H]_0$ and $[G]_0$, be changed for the titration? There are so many possibilities described graphically in Fig. 2.6 containing a dilution experiment, a continuous variation experiment, a constant $[H]_0$ with different $[G]_0$ experiment and two more complicated examples. The criteria to decide the way of change might be as follows;

- reliability
- easy ways for experiment and calculation
- applicability
- acceptability.

Let us consider how to choose the experimental conditions, $[H]_0$ and $[G]_0$ using a 1:1 host–guest complexation stoichiometry for simplicity.

$$H + G \rightleftarrows C \tag{2.23}$$

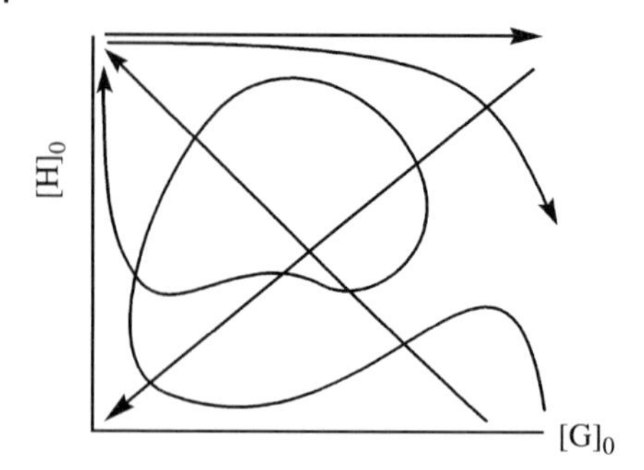

Fig. 2.6. Graphical representation showing possible ways to change $[H]_0$ and $[G]_0$ for a titration experiment.

From Eq. (2.8)

$$K = \frac{[C]}{([H]_0 - [C]) \cdot ([G]_0 - [C])} \tag{2.24}$$

Let us figure the correlation between K and complexation ratio (x).

$$y = K, \quad x = \frac{[C]}{[H]_0} \tag{2.25}$$

then Eq. (2.24) is transformed to

$$y = \frac{x}{(1 - x) \cdot ([G]_0 - [H]_0 \cdot x)} \tag{2.26}$$

Figure 2.7 is the graph of Eq. (2.26) where x (x-coordinate) is 0 to 1, K (y-coordinate) is 10 to 100 000 000, and $[H]_0$ is 0.01 to 0.000 001, $[H]_0 = [G]_0$ as a premise. In general, caution is expressed as: "measurements below 20% and above 80% complexation ratio (x) yield uncertain values." This caution is interpreted with Fig. 2.7 as follows.

The steep rises of K at the complexation ratio ranges less than 20% and more than 80% cause the transfer of magnified errors from the complexation ratio into K. When K is determined based on the measurement of a property directly connected to or proportional to the complexation ratio, the obtained errors in K are magnified compared with those for the observed property. In the case where $[H]_0 = 0.0001$ M as an example, the complexation ratio is between 0.2 and 0.8 when K is between 3000 and 200 000 M^{-1}, so an accurate experiment is carried out. This brief discussion shows, how the accuracy of K is governed by the choice of concentrations, $[H]_0$, $[G]_0$ and also K itself.

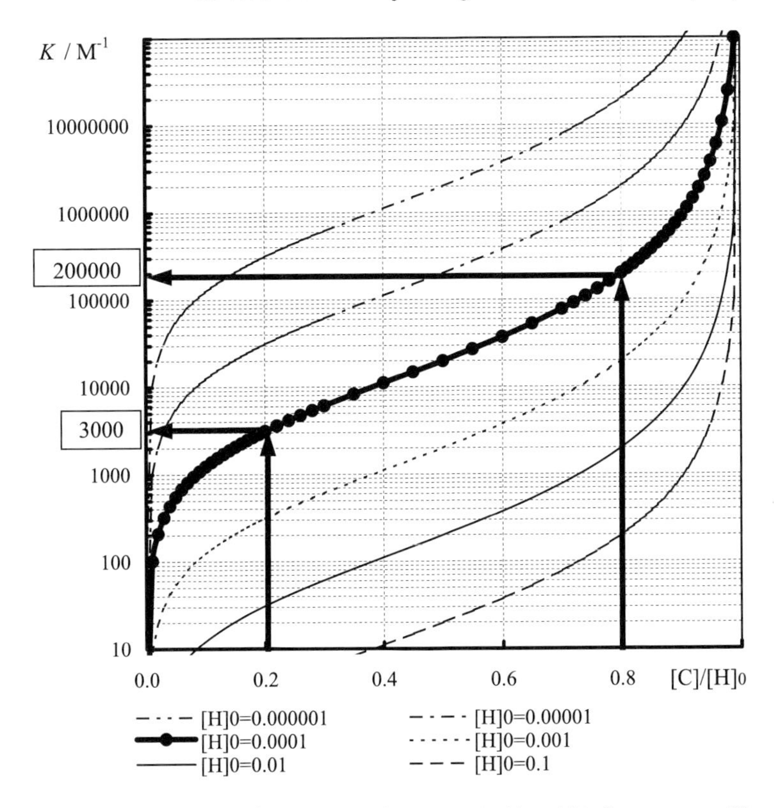

Fig. 2.7. The correlation between complexation ratio (x) and binding constant (K).

2.2.3.2 How to Set Up $[H]_0$

Setting up the concentration of host $[H]_0$ is limited by the measured properties, the apparatus, and other features of the experiment. For example, $[H]_0$ for NMR spectroscopy is roughly in the range of 0.01 M with one or two orders of magnitude variation. $[H]_0$ for UV/vis spectroscopy, which depends severely on the molar absorptivity, is roughly in the range of 0.0001 M. Therefore, $[G]_0$ is often the only variable which can be set up in a wide range, because $[H]_0$ is usually governed by the experimental method.

2.2.3.3 How to Set Up $[G]_0$

In order to consider how to set up the concentration of $[G]_0$, Fig. 2.8 is drawn based on Eq. (2.26) where $[H]_0 = 0.0001$ and $\frac{[G]_0}{[H]_0}$ is changed from 0.1 to 1000. The correlation between complexation ratio x and accurately obtainable K range by changing $[G]_0$ with constant $[H]_0$ ($= 0.0001$ M) is clear when based on Fig. 2.8. Considering the suitable x range ($0.2 < x < 0.8$) for reliable measurement in Fig. 2.8, the combination of $[H]_0$, $[G]_0$ and K is determined. For example, when

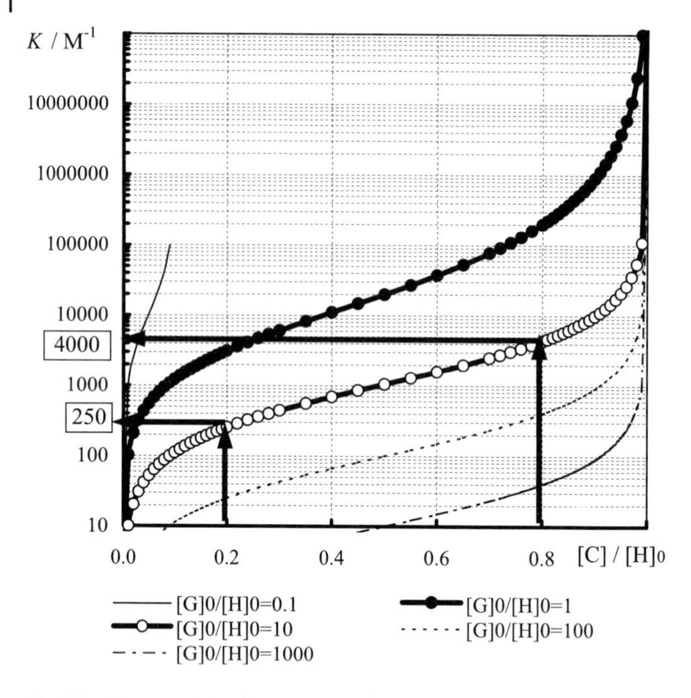

Fig. 2.8. The correlation between complexation ratio (x) and binding constant (K).

$[G]_0 = 0.001$ M, and $[H]_0 = 0.0001$ M, then $[G]_0/[H]_0 = 10$, consequently, a reliable range of K of 250 to 4000 M^{-1} is obtained by following the arrows in Fig. 2.8.

By repeating these procedures for several combinations of $[H]_0$, $[G]_0$, the obtained K ranges are summarized in Fig. 2.9. This figure is useful for a preliminary check of the experimental concentration conditions and for choosing a suitable experimental method.

In most cases K is determined with a titration experiment followed by regression of the obtained data based on the theoretical equations above. In the commonly used titration experiment, $[G]_0$ is changed while the range of $[H]_0$ change is limited. In such a case, the important point is how to set up the range of $[G]_0$. A representative answer to this question is obtained by considering the correlation between $[G]_0/[H]_0$ and the complexation ratio based on Fig. 2.10 where the x-coordinate is the concentration ratio of guest over host mixed in the cell and the y-coordinate is complexation ratio. The graph in Fig. 2.10 is based on Eq. (2.29) which is derived from Eq. (2.24) by multiplying both sides of the equation by $[H]_0$, dividing the denominator and numerator by $[H]_0{}^2$, then substituting with y and x according to Eq. (2.27).

$$y = \frac{[C]}{[H]_0}, \quad x = \frac{[G]_0}{[H]_0} \tag{2.27}$$

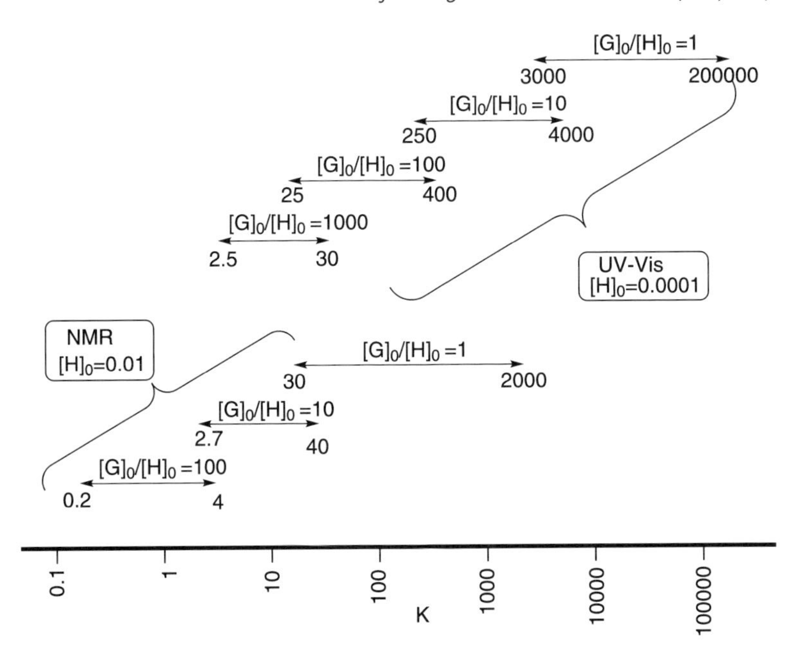

Fig. 2.9. Reliable regions of $[H]_0$ and $[G]_0$ for K determination shown for representative concentrations of UV/vis and NMR spectroscopic experiments.

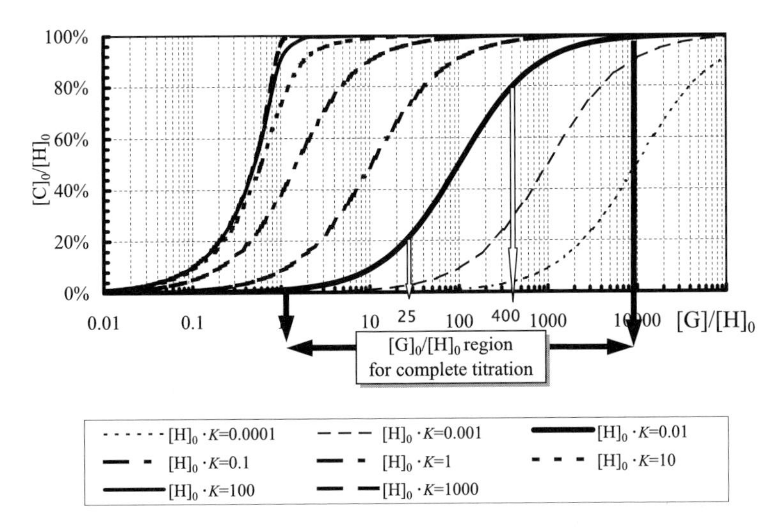

Fig. 2.10. Calculated $[C]/[H]_0$ over $[G]_0/[H]_0$ plots: Useful graphs for $[G]_0$ range determination of titration experiments.

$$[\mathrm{H}]_0 \cdot K = \frac{y}{(1-y)\cdot(x-y)} \tag{2.28}$$

Displacement using the equation $\alpha = [\mathrm{H}]_0 \cdot K$ and transformation produces

$$\alpha \cdot y^2 - (\alpha + \alpha \cdot x + 1) \cdot y + \alpha \cdot x = 0 \tag{2.29}$$

Figure 2.10 is obtained by changing α from 0.0001 to 1000, which corresponds to the change of K from $K = \frac{0.0001}{[\mathrm{H}]_0}$ to $K = \frac{1000}{[\mathrm{H}]_0}$. Though tracing from the bottom to the top of the S-curve in Fig. 2.10 is necessary for complete identification of each equilibrium, it is possible to determine the binding constant by plotting the data $\frac{[\mathrm{C}]}{[\mathrm{H}]_0}, \frac{[\mathrm{G}]_0}{[\mathrm{H}]_0}$ as Fig. 2.10 which are obtained from experiment, followed by curve-fitting using Eq. (2.29). When $[\mathrm{H}]_0\ K = 0.01$ is picked as an example, the range of $[\mathrm{G}]_0$ for a complete titration is 1 $[\mathrm{H}]_0$ to 10 000 $[\mathrm{H}]_0$ M as indicated in Fig. 2.10. In order to reduce error, the $[\mathrm{G}]_0$ area should be avoided where lines are close together. When the experimental conditions are in an area of this figure crowded with graphs, a small error in $[\mathrm{G}]_0$ causes a plot on a different S-curve whose K is much different. Then this unsuitable concentration setting results in the low reliability of the calculated K value. From this consideration, the range of the complexation ratio between 0.2 to 0.8 is suitable here again for a reliable measurement. The suitable range of $[\mathrm{G}]_0$ could be obtained from Fig. 2.10. For this example, the suitable range of $[\mathrm{G}]_0$ is that from 25 $[\mathrm{H}]_0$ to 400 $[\mathrm{H}]_0$. This is the range expressed relative to $[\mathrm{H}]_0$. In order to obtain this suitable range of $[\mathrm{G}]_0$, Fig. 2.11 is drawn as follows.

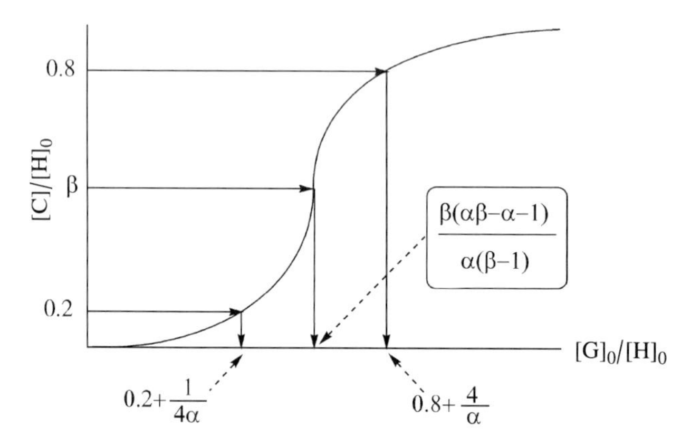

Fig. 2.11. Useful graph for $[\mathrm{G}]_0$ range determination of titration experiments ($\alpha = [\mathrm{H}]_0 \cdot K$, $\beta = [\mathrm{C}]/[\mathrm{H}]_0$).

From Eq. (2.29) it is possible to express the x-coordinate using α and β as follows,

$$x = \frac{\beta \cdot (\alpha \cdot \beta - \alpha - 1)}{\alpha \cdot (\beta - 1)} \tag{2.30}$$

$$x = \frac{[G]_0}{[H]_0}, \quad \alpha = [H]_0 \cdot K, \quad \beta = \frac{[C]}{[H]_0}$$

The complexation ratio here is β. With Eq. (2.30), the $[G]_0$ range for the titration experiment where the complexation ratio is 0.2 to 0.8 is obtained as functions of α just by entering $\beta = 0.2$ or $\beta = 0.8$ into Eq. (2.30). The result is summarized in Fig. 2.11. On entering $\alpha = 0.01$, the suitable x range is obtained easily from Fig. 2.11 as

$$25.2 = 0.2 + \frac{1}{4 \cdot 0.01} \leq \frac{[G]_0}{[H]_0} \leq 0.8 + \frac{4}{0.01} = 400.8$$

One more consideration for the $[G]_0$ should be mentioned here by going back to Fig. 2.10. When $[H]_0$ K is larger than one, the curves are close together even if the complexation ratio is between 0.2 and 0.8. Consequently, as a premise for a reliable experiment, $[H]_0$ K should be *smaller than one*. When $[H]_0$ K is larger than one, $[H]_0$ should be reduced. When $[H]_0$ cannot be reduced, another observable should be chosen together with an appropriate spectroscopic method.

From Fig. 2.11, the $[G]_0$ range is limited as described below.

$$0.2 + \frac{1}{4 \cdot \alpha} \leq \frac{[G]_0}{[H]_0} \leq 0.8 + \frac{4}{\alpha} \quad \text{where } \alpha = [H]_0 \cdot K \tag{2.31}$$

Multiplying by $[H]_0$, followed by transformation results in

$$(0.2 \cdot [H]_0 \cdot K + 0.25) \cdot \frac{1}{K} \leq [G]_0 \leq (0.8 \cdot [H]_0 \cdot K + 4) \cdot \frac{1}{K} \tag{2.32}$$

When $[H]_0$ K is set up smaller than one, the range of $[H]_0$ K is between 0 to 1. Then,

$$0.25 \leq (0.2 \cdot [H]_0 \cdot K + 0.25) \leq 0.45$$

$$4 \leq (0.8 \cdot [H]_0 \cdot K + 4) \leq 4.8$$

$$\Rightarrow 0.25 \cdot K_{diss} \leq [G]_0 \leq 4.8 \cdot K_{diss} \tag{2.33}$$

Equation (2.33) clearly shows that a suitable $[G]_0$ range is connected to the extent of dissociation constant $K_{diss}(= K^{-1})$. Wilcox [10] has also shown clearly the importance of the p-value, originally introduced by Weber [11], which is defined as

$$p = \frac{[\mathrm{C}]}{[\mathrm{G}]_0} \quad \text{when } [\mathrm{H}]_0 \geq [\mathrm{G}]_0 \tag{2.34}$$

$$p = \frac{[\mathrm{C}]}{[\mathrm{H}]_0} \quad \text{when } [\mathrm{G}]_0 \geq [\mathrm{H}]_0 \tag{2.35}$$

The criterion for the best condition can be written as

$$0.2 \leq p \leq 0.8 \tag{2.36}$$

Based on this criterion (Eq. (2.36)), the suitable range of concentration for titration experiment is shown as below.

$$\frac{1}{10} \cdot \frac{1}{K} \leq [\mathrm{G}]_0 \leq 10 \cdot \frac{1}{K} \tag{2.37}$$

This range (Eq. (2.37)) covers the range defined by Eq. (2.33). The concentration range of the titration experiment must be carefully chosen based on a preliminary estimation of K.

Summing up, we can conclude with respect to the concentration ranges of host and guest:

For a reliable experiment, the magnitude of K should be predicted with an educated guess, then the method decided, e.g. NMR spectroscopy or UV/vis spectroscopy or fluorometry, etc., which decides roughly the range of $[\mathrm{H}]_0$, and finally decides the range of $[\mathrm{G}]_0$ using Fig. 2.11 and/or Eq. (2.32).

2.2.4 Data Treatment

2.2.4.1 General View

After this discussion of how to perform the titration experiments in order to collect data for reliable K values, the next step is how to treat the data collected, to obtain the K values.

Some data treatment methods are general, some rely on approximations and thus are subject to some premises, and some are just regression methods. Typical examples of the approximate methods are the Benesi–Hildebrand [12], Ketelaar [13], Nagakura–Baba [14], Scott [15], Scatchard, and Hammond [16] methods which approximate $[\mathrm{G}]$ by $[\mathrm{G}]_0$.

From Eqs. (2.5) and (2.7) and $a = b = 1$, we derive

$$\begin{aligned} &[\mathrm{G}]_0 = [\mathrm{G}] + K \cdot [\mathrm{H}] \cdot [\mathrm{G}] \\ &\Rightarrow [\mathrm{G}]_0 = [\mathrm{G}](1 + K \cdot [\mathrm{H}]) \end{aligned} \tag{2.38}$$

If $K \cdot [\mathrm{H}] \ll 1$, it can safely be assumed that $[\mathrm{G}]_0 = [\mathrm{G}]$. This condition is frequently encountered for weak complexation, where K is small. Furthermore,

$[G]_0 \gg [H]_0$ is usually employed for the practical titration which is thought to be essential. However, not all systems can be investigated under this condition $[G]_0 \gg [H]_0 (K \cdot [H] \ll 1)$.

When the assumption $[G]_0 = [G]$ can not be applied, other approximation or regression methods have to be employed. Here, the regression method is shown. Typical examples of the regression methods are the Rose–Drago [17], Nakano [18], and Creswell–Allred [19] methods. Because of its wide applicability, a practical guide based on the Rose–Drago method is presented here for an example from UV/vis spectroscopy.

Originally, the Rose–Drago method was used for UV/vis spectroscopy for evaluating acid–base equilibria and complexation of iodine. The only assumption for the original method is that there are at most two species observed which obey Beer's law in the concentration range employed. There is no other assumption so that the Rose–Drago method is widely applicable. The results are presented graphically in this method and by closer inspection one can quantitatively determine the precision. An example case is described in the following section.

2.2.4.2 **Rose–Drago Method for UV/vis Spectroscopy**

The Rose–Drago method is described here using a 1:1 host–guest complexation stoichiometry detected by UV/vis spectroscopy. The observed property is the absorbance which is followed in a titration experiment to collect the necessary absorbance data. For the data treatment with the Rose–Drago method, a spreadsheet program is attached as Appendix 2.2 [20].

Into Eq. (2.8), $a = b = 1$ is substituted. Then the reciprocal is

$$\frac{1}{K} = [C] - ([H]_0 + [G]_0) + \frac{[H]_0 \cdot [G]_0}{[C]} \tag{2.39}$$

Combining Eq. (2.21) with Eq. (2.39) gives

$$\frac{1}{K} = \frac{A_{obs} - \varepsilon_h \cdot [H]_0 - \varepsilon_g \cdot [G]_0}{\varepsilon_c - \varepsilon_h - \varepsilon_g} - ([H]_0 + [G]_0) + \frac{\varepsilon_c - \varepsilon_h - \varepsilon_g}{A_{obs} - \varepsilon_h \cdot [H]_0 - \varepsilon_g \cdot [G]_0} \cdot [H]_0 \cdot [G]_0 \tag{2.40}$$

This is the most complicated host–guest complexation detected by means of UV/vis spectroscopy because the absorption bands of all components, host, guest, and complex overlap.

First of all, the constants ε_h, and ε_g in Eq. (2.40) have to be obtained from independent measurements, because they are molar absorptivities of the pure host and the pure guest, respectively. Then, A_{obs} needs to be measured at different combinations of $[H]_0$ and $[G]_0$ followed by regression of the obtained data using Eq. (2.40). Theoretically, A_{obs} values at more than two different combinations of $[H]_0$ and $[G]_0$ give two unknowns, K and ε_c.

Measurement of absorbance at different combinations of $[H]_0$ and $[G]_0$ supplies a matrix $\{A_{obsn}, [H]_{0n}, [G]_{0n}\}$ consisting of 3 elements.

A_{obsn} Observed absorbance of n-th measurement
$[H]_{0n}$ Concentration of host molecule at initial stage for n-th measurement
$[G]_{0n}$ Concentration of guest molecule at initial stage for n-th measurement

Combining Eq. (2.40) and definition (2.41)–(2.45) leads to Eq. (2.46).

$$Y = \frac{1}{K} \tag{2.41}$$

$$X = \varepsilon_c - \varepsilon_h - \varepsilon_g \tag{2.42}$$

$$a_n = A_{obsn} - \varepsilon_h \cdot [H]_{0n} - \varepsilon_g \cdot [G]_{0n} \tag{2.43}$$

$$b_n = [H]_{0n} + [G]_{0n} \tag{2.44}$$

$$c_n = \frac{[H]_{0n} \cdot [G]_{0n}}{A_{obsn} - \varepsilon_h \cdot [H]_{0n} - \varepsilon_g \cdot [G]_{0n}} \tag{2.45}$$

Then

$$Y = \frac{a_n}{X} - b_n + c_n \cdot X \tag{2.46}$$

According to Eq. (2.46), one combination of data (e.g. $\{A_{obs1}, [H]_{01}, [G]_{01}\}$ and $\{A_{obs2}, [H]_{02}, [G]_{02}\}$) supplies a matrix of answers $\{X, Y\}$. A representative solution is as follows.

As an example, one combination of data where $n = 1$ and $n = 2$ (e.g. $\{A_{obs1}, [H]_{01}, [G]_{01}\}$ and $\{A_{obs2}, [H]_{02}, [G]_{02}\}$) is used here.

$$Y = \frac{a_1}{X} - b_1 + c_1 \cdot X \tag{2.47}$$

$$Y = \frac{a_2}{X} - b_2 + c_2 \cdot X \tag{2.48}$$

Subtraction of both sides, followed by multiplication by X results in

$$(c_1 - c_2) \cdot X^2 + (b_1 - b_2) \cdot X + (a_1 - a_2) = 0 \tag{2.49}$$

$$\Rightarrow X = \frac{-(b_1 - b_2) \pm \sqrt{(b_1 - b_2)^2 - 4 \cdot (c_1 - c_2) \cdot (a_1 - a_2)}}{2 \cdot (c_1 - c_2)} \tag{2.50}$$

Substituting Eq. (2.47) with Eq. (2.50) yields an expression for *Y*.

The obtained $\{X, Y\}$ is merely an answer which satisfies both Eqs. (2.47) and (2.48), but it is not necessarily the chemically correct answer. For example, a chemically reasonable *Y* should have a positive sign. Based on such chemical limitation, correct sets of answers should be collected.

The maximum number of obtainable answer pairs $\{X, Y\}$ is ${}_nC_2$ pairs for *n* combinations of concentration conditions. For example, 5 pairs of $\{[H]_{0n}, [G]_{0n}\}$ give 10 $(= {}_5C_2)$ pairs of $\{X, Y\}$. *These* $\{X, Y\}$ *are obtained under the premise where 1 to 1 complexation. No approximation is introduced into this solution.* The reciprocal of the obtained Y's is the binding constant *K*. The number of obtained *K* at this stage might be ${}_nC_2$.

2.2.4.3 Estimation of Error

Statistics teaches that the deviation of data based on less than 30 measurements is not a normal distribution but Student's t-distribution. So it is suitable to express the binding constant *K* with 95% confidence interval calculated by applying by Student's t-distribution. Student's t-distribution includes the normal distribution. When the number of measurements is more than 30, Student's t-distribution and the normal distribution are practically the same. The actual function of Student's t-distribution is very complicated so that it is rarely used directly. A conventional way to apply Student's t-distribution is to pick up data from the critical value table of Student's t-distribution under consideration of 'degree of freedom', 'level of significance' and 'measured data.' It is troublesome to repeat this conventional way many times. Most spreadsheet software even for personal computers has the function of Student's t-distribution. Without any tedious work, namely, picking up data from the table, statistical treatment can be applied to experimental results based on Student's t-distribution with the aid of a computer. In Fig. 2.12, an example is shown. When the measurement data are input into the gray cells, answers can be obtained in the cell D18 and D19 instantaneously.

When the confidence interval obtained after statistical treatment is very wide, there is high probability that a precise experiment has not been carried out. The experimental conditions and also each procedure should be checked then.

2.2.5 Conclusion for UV/vis Spectroscopic Method

The method described here includes no approximation at the data treatment step, so it can be used generally. In addition, the required level of mathematical knowledge is not high, only a formula for polynomials of degree 2, therefore the logical basis can be easily understood. Moreover, statistical treatment of the obtained data is understandable with primary statistics. These are the merits to use this method at first in order to understand the way of determination of binding constants.

When stoichiometry of the complex is not 1 to 1 or when other premises are not satisfied, the way of data treatment should be changed or modified. Nonlinear least

A1	B	C	D	E	F	G
2					**How to use this sheet**	
3		Data	34. 72		1.Input data at D3 to D9	
4			36. 84		2.input level of significance (α=5%)	
5			33. 25		3.find answer in D18 and D19	
6			37. 78			
7			39. 16			
8			35. 08			
9			34. 54			
10		Average	35.91		=AVERAGE(D3:D9)	
11		s	1. 93		=STDEVP(D3:D9)	
12						
13		α	0. 05		0. 05	
14		degree of freedom	6		=COUNTA(D3:D9)-1	
15		$t_{\alpha/2}$	2.447		=TINV(D13,D14)	
16						
17		95%	confidence interval		confidence interval	
18		*K*=	35.91		=AVERAGE(D3:D9)	
19		±	1.78		=D15*D11/COUNTA(D3:D9)^0.5	
20						
21						

Fig. 2.12. Spreadsheet for statistical data treatment based on Student's t-distribution.

square data treatment is one of the best approximations. Using eqautions (2.8) and (2.21) for UV/vis spectroscopy, other complexations may be applied even if a, b are not one.

2.3 Practical Course of Action for NMR Spectroscopic Binding Constant Determination

With regard to the NMR spectroscopic binding constant determination, we should distinguish two cases differing by the guest exchange rate at equilibrium. In the case where the host–guest complexation equilibrium has a similar exchange rate compared with the NMR time scale, the NMR peaks broaden and/or disappear, so it is impossible to perform any meaningful experiment here. The following two cases are suitable for measurement by NMR spectroscopy [21].

Case 1: The host–guest complexation equilibrium, which has a very slow exchange rate compared with the NMR time scale.

Case 2: The host–guest complexation equilibrium, which has a very fast exchange rate compared with the NMR time scale.

It should be briefly mentioned here that temperature and solvent changes may help to circumvent problems arising if the process under study occurs with a rate comparable to the time scale of the NMR instrument. In such a case, cooling leads to slow exchange (case 1), while heating increases the exchange reaction rate (case 2). Similarly, a less competitive solvent usually slows down the process (case 1), while a more competing one accelerates it (case 2).

2.3.1 Determination of Stoichiometry

In case 1 (slow exchange), the peaks which are assigned to the host parts in the complex and those to the free host are observed separately in the same NMR spectrum. Those peaks appear at individual chemical shifts. In Fig. 2.13, there is a representative NMR spectrum where the peaks, which are assigned to a free or complexed host, are observed individually with the integration ratio m to n. The composition of the complex is H_aG_b. Then, the integration of host parts in the complex over total integration of the host parts is as follows.

$$\frac{a \cdot [C]}{[H]_0} = \frac{n}{m+n}$$

$$\Rightarrow \frac{n}{m+n}[H]_0 = a \cdot [C] \tag{2.51}$$

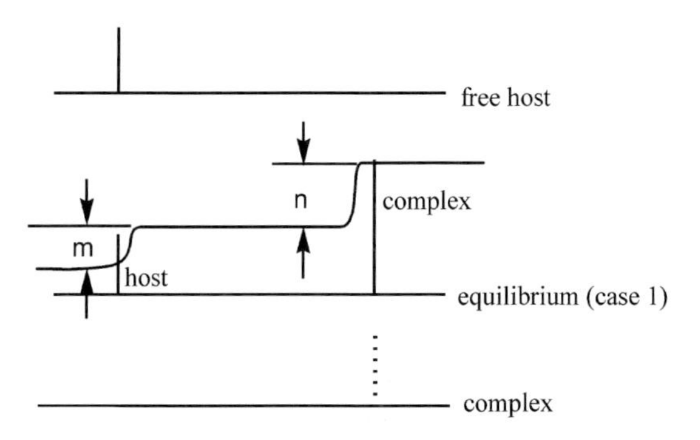

Fig. 2.13. Representative NMR spectra for slow guest exchange indicating the correlation of peak integrations and binding constant K.

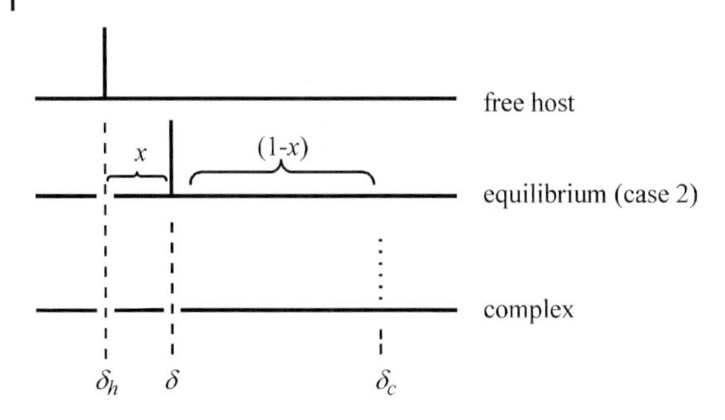

Fig. 2.14. Representative NMR spectra for fast guest exchange indicating the correlation of complexation ratio *x* and chemical shift.

The stoichiometry is determined from the *x*-coordinate at the maximum in the curve which might be called modified Job's plot where $\frac{n}{m+n}[\mathrm{H}]_0$ is plotted as the *y*-coordinate instead of [C], for the following reason.

Equation (2.51) means that $\frac{n}{m+n}[\mathrm{H}]_0$ is proportional to [C] since a is constant.

As experimental condition $[\mathrm{H}]_0$ can be set up.

The ratio of $\frac{n}{m+n}$ is obtained from the NMR spectral data.

In case 2 (fast exchange), the peaks which are assigned to the host parts in the complex and those to the free host are averaged.

Figure 2.14 shows a representative NMR spectrum where the peaks, which are assigned to the free and complexed host part, appear at the weighted average chemical shift of the free host and complexed host. In this case

δ: observed chemical shift
δ_h, δ_c: chemical shifts of the free and complexed host, respectively
x: ratio of complexed host at equilibrium over total host

$$\delta = \delta_h \cdot (1-x) + \delta_c \cdot x \quad \text{and} \quad x = \frac{a \cdot [\mathrm{C}]}{[\mathrm{H}]_0}$$

$$\Rightarrow [\mathrm{H}]_0 \cdot (\delta - \delta_h) = a \cdot [\mathrm{C}] \cdot (\delta_c - \delta_h) \tag{2.52}$$

The stoichiometry is determined from the *x*-coordinate at the maximum in the modified Job's plot where $[\mathrm{H}]_0 \cdot (\delta - \delta_h)$ is plotted as *y*-coordinate instead of [C] for the following reason.

Equation (2.52) means that $[\mathrm{H}]_0 \cdot (\delta - \delta_h)$ is proportional to [C], since $a \cdot (\delta_c - \delta_h)$ is constant.

As the experimental condition, $[\mathrm{H}]_0$ can be set up.

The $(\delta - \delta_h)$ are obtained from the NMR spectral data. This case is often observed for the complexation with crown ether and amine.

2.3.2
Evaluation of Complex Concentration

For case 1 (slow exchange), Eq. (2.53) is derived from Eq. (2.52) just by a simple transformation. With the stoichiometry (a) obtained as described above, [C] is determined using the experimental conditions ($[\mathrm{H}]_0$) according to Eq. (2.53), since $\frac{n}{m+n}$ is obtained from the NMR measurement.

$$[\mathrm{C}] = \frac{1}{a} \cdot \frac{n}{m+n} \cdot [\mathrm{H}]_0 \tag{2.53}$$

For case 2 (fast exchange), Eq. (2.54) is derived from Eq. (2.52) just by a simple transformation.

$$[\mathrm{C}] = \frac{1}{a} \cdot \frac{\delta - \delta_h}{\delta_c - \delta_h} \cdot [\mathrm{H}]_0 \tag{2.54}$$

If all constants (a, δ_h, δ_g and δ_c) were obtainable, [C] would be determined from δ and the experimental conditions ($[\mathrm{H}]_0$). Since δ_c cannot be obtained directly, a titration experiment and regression are necessary for the evaluation of complex concentration.

2.3.3
Data Treatment for NMR Method

2.3.3.1 Rose–Drago Method for NMR Spectroscopy

In this section, the way to apply the original Rose–Drago method for UV/vis spectroscopy to NMR titration data is described, especially for host–guest systems on fast exchange.

Using the equilibrium of 1 to 1 host–guest complexation as example, the way to apply the original Rose–Drago method to NMR spectroscopy is given here. As mentioned above, host–guest complexations are classified into two for the determination of binding constants by means of NMR spectroscopy.

When the host–guest complexation equilibrium has a very slow exchange rate compared with the NMR time scale, the concentration of the complex is expressed as follows ($a = 1$ in Eq. (2.53)).

$$[\mathrm{C}] = \frac{n}{m+n} \cdot [\mathrm{H}]_0 \tag{2.55}$$

When the host–guest complexation equilibrium has a very fast exchange rate compared with the NMR time scale, the concentration of the complex is expressed as follows ($a = 1$ in Eq. (2.54)).

$$[\mathrm{C}] = \frac{\delta - \delta_h}{\delta_c - \delta_h} \cdot [\mathrm{H}]_0 \tag{2.56}$$

From Eq. (2.39) and Eq. (2.56), the following equation is derived.

$$\frac{1}{K} = \frac{(\delta - \delta_h) \cdot [\mathrm{H}]_0}{(\delta_c - \delta_h)} - ([\mathrm{H}]_0 + [\mathrm{G}]_0) + \frac{(\delta_c - \delta_h)}{(\delta - \delta_h)} \cdot [\mathrm{G}]_0 \tag{2.57}$$

Here we carried out the substitution using following definitions (2.58)–(2.62).

$$Y = \frac{1}{K} \tag{2.58}$$

$$X = \delta_c - \delta_h \tag{2.59}$$

$$a_n = (\delta_n - \delta_h) \cdot [\mathrm{H}]_{0n} \tag{2.60}$$

$$b_n = [\mathrm{H}]_{0n} + [\mathrm{G}]_{0n} \tag{2.61}$$

$$c_n = \frac{[\mathrm{G}]_{tn}}{(\delta_n - \delta_h)} \tag{2.62}$$

Then the Eq. (2.57) is expressed as follows.

$$Y = \frac{a_n}{X} - b_n + c_n \cdot X \tag{2.63}$$

This Eq. (2.63) is the same as Eq. (2.46). So from this stage, the same procedures for UV/vis spectroscopy can be applied for NMR spectroscopy.

2.3.3.2 Estimation of Error for NMR Method

Estimation of the error of the Rose–Drago method for NMR spectroscopic data is essentially same as that for UV/vis spectroscopy, which is described in corresponding sections.

2.3.3.3 Nonlinear Least Square Data Treatment of NMR Titration Method

This method is widely applicable and generally acceptable but includes an approximation. In this section, the equilibrium of 1:1 host–guest complexation detected by NMR spectroscopy is treated. The observed property is the chemical shift (δ). The chemical shift data of the titration experiment were collected.

Equation (2.8), where $a = b = 1$, translates into Eq. (2.64).

$$[\mathrm{C}] = \frac{\left([\mathrm{H}]_0 + [\mathrm{G}]_0 + \frac{1}{K}\right) \pm \sqrt{\left([\mathrm{H}]_0 + [\mathrm{G}]_0 + \frac{1}{K}\right)^2 - 4 \cdot [\mathrm{H}]_0 \cdot [\mathrm{G}]_0}}{2} \tag{2.64}$$

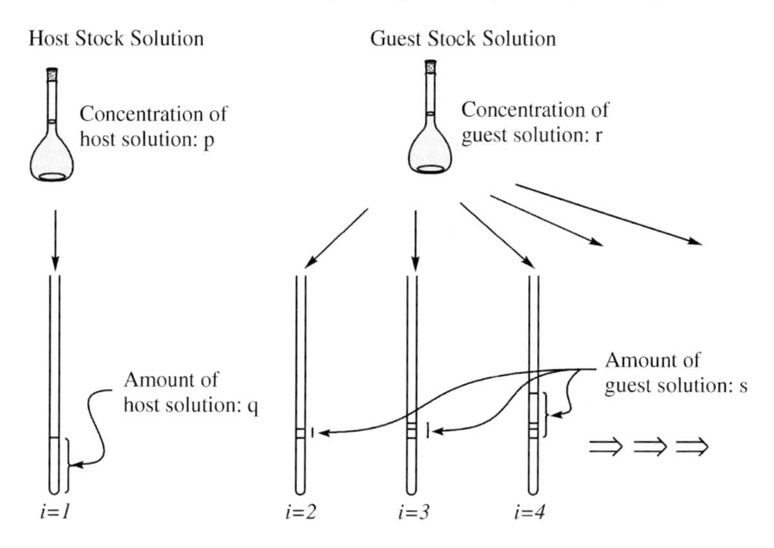

Fig. 2.15. Illustration of a typical titration experiment to show procedures and constants.

This Eq. (2.64) is modified with Eq. (2.56) (rapid exchange NMR and 1:1 stoichiometry) and the following three definitions.

$$y = \delta - \delta_h \tag{2.65}$$

$$a = \delta_c - \delta_h \tag{2.66}$$

$$b = K \tag{2.67}$$

Then,

$$\begin{aligned} y &= a \cdot \frac{[\mathrm{C}]}{[\mathrm{H}]_0} \\ &= a \cdot \frac{\left([\mathrm{H}]_0 + [\mathrm{G}]_0 + \frac{1}{b}\right) \pm \sqrt{\left([\mathrm{H}]_0 + [\mathrm{G}]_0 + \frac{1}{b}\right)^2 - 4 \cdot [\mathrm{H}]_0 \cdot [\mathrm{G}]_0}}{2 \cdot [\mathrm{H}]_0} \\ &= \frac{a}{2} \cdot \left\{ \left(\frac{[\mathrm{G}]_0}{[\mathrm{H}]_0} + 1 + \frac{1}{b \cdot [\mathrm{H}]_0}\right) \pm \sqrt{\left(\frac{[\mathrm{G}]_0}{[\mathrm{H}]_0} + 1 + \frac{1}{b \cdot [\mathrm{H}]_0}\right)^2 - 4 \cdot \frac{[\mathrm{G}]_0}{[\mathrm{H}]_0}} \right\} \end{aligned} \tag{2.68}$$

A typical titration experiment is graphically expressed in Fig. 2.15. The host solution in NMR tube is titrated by the addition of guest stock solution. Equation (2.68) is modified with the following experimental constants and parameters according to the typical experimental method for NMR titration.

p: concentration of host solution
q: amount of host solution
r: concentration of guest solution
s: amount of guest solution

Then,

$$[\mathrm{H}]_0 = \frac{p \cdot q}{s+q} \tag{2.69}$$

$$[\mathrm{G}]_0 = \frac{r \cdot s}{s+q} = x \tag{2.70}$$

$$\Rightarrow s = \frac{x \cdot q}{(r-x)} \tag{2.71}$$

Equations (2.72) and (2.73) are derived from Eqs. (2.69), (2.70) and (2.71).

$$[\mathrm{H}]_0 = \frac{p \cdot (r-x)}{r} \tag{2.72}$$

$$\frac{[\mathrm{G}]_0}{[\mathrm{H}]_0} = \frac{r \cdot x}{p \cdot (r-x)} \tag{2.73}$$

Then basic Eq. (2.68) is expressed with experimental constants and variables as follows.

$$y = \frac{a}{2} \cdot \left\{ \left(\frac{r \cdot x}{p \cdot (r-x)} + 1 + \frac{r}{b \cdot p \cdot (r-x)} \right) \pm \sqrt{\left(\frac{r \cdot x}{p \cdot (r-x)} + 1 + \frac{r}{b \cdot p \cdot (r-x)} \right)^2 - 4 \cdot \frac{r \cdot x}{p \cdot (r-x)}} \right\} \tag{2.74}$$

$$\Rightarrow \frac{\partial y}{\partial a} = \frac{1}{2} \cdot \left\{ \left(\frac{r \cdot x}{p \cdot (r-x)} + 1 + \frac{r}{b \cdot p \cdot (r-x)} \right) \pm \sqrt{\left(\frac{r \cdot x}{p \cdot (r-x)} + 1 + \frac{r}{b \cdot p \cdot (r-x)} \right)^2 - 4 \cdot \frac{r \cdot x}{p \cdot (r-x)}} \right\} \tag{2.75}$$

$$\Rightarrow \frac{\partial y}{\partial b} = \frac{a}{2} \cdot \left(-\frac{r}{b^2 \cdot p \cdot (r-x)} \right) \cdot \left\{ 1 \pm \sqrt{\left(\frac{r \cdot x}{p \cdot (r-x)} + 1 + \frac{r}{b \cdot p \cdot (r-x)} \right)^2 - 4 \cdot \frac{r \cdot x}{p \cdot (r-x)}} \cdot \left(\frac{r \cdot x}{p \cdot (r-x)} + 1 + \frac{r}{b \cdot p \cdot (r-x)} \right) \right\} \tag{2.76}$$

The approximation procedure of this nonlinear method is described below.

It is assumed a_0 and b_0 where α and β (correction of a and b) are small enough so that y is approximately equal to Eq. (2.80) where higher-order parts could be omitted from the Taylor series expansion (2.78) of y at $(a, b) = (a_0, b_0)$. Then α and β are calculated which minimize the sum of square deviation. Following is the step-by-step procedure.

First of all, proper a_0 and b_0 are assumed.

$$\begin{cases} a = a_0 + \alpha \\ b = b_0 + \beta \end{cases} \tag{2.77}$$

(a, b: desired parameters; a_0, b_0: assumed parameters; α, β: correction of a, b)

Practically the meaning of word 'proper' here might be considered to be expressed as follows.

α and β are small enough ($\frac{\alpha}{a_0} \leq 10^{-1}$ and $\frac{\beta}{b_0} \leq 10^{-1}$).

Secondly, the equation of y is transferred to a linear expression by using this approximation.

Taylor series expansion of y at $(a, b) = (a_0, b_0)$ is the following Eq. (2.78).

$$y = y_0 + \left(\frac{\partial y}{\partial a}\right)_0 \cdot \alpha + \left(\frac{\partial y}{\partial b}\right)_0 \cdot \beta + \left(\frac{\partial^2 y}{\partial a^2}\right)_0 \cdot \alpha^2 + \left(\frac{\partial^2 y}{\partial b^2}\right)_0 \cdot \beta^2 + \cdots \tag{2.78}$$

The values, y_0, $\left(\frac{\partial y}{\partial a}\right)_0$, and $\left(\frac{\partial y}{\partial b}\right)_0$ are obtained based on the above-mentioned Eqs. (2.74), (2.75), (2.76) where $a = a_0$, $b = b_0$. As an approximation, the higher order parts can be replaced with zero, because α and β are small enough by the assumption.

$$\left(\frac{\partial^2 y}{\partial a^2}\right)_0 \cdot \alpha^2 + \left(\frac{\partial^2 y}{\partial b^2}\right)_0 \cdot \beta^2 + \cdots \approx 0 \tag{2.79}$$

Then the following Eq. (2.80) after approximation is derived from Eqs. (2.78) and (2.79).

$$y \approx y_0 + \left(\frac{\partial y}{\partial a}\right)_0 \cdot \alpha + \left(\frac{\partial y}{\partial b}\right)_0 \cdot \beta \tag{2.80}$$

The next step is to minimize the sum of square deviations for titration data by determining proper α and β.

Using this Eq. (2.80), deviation d_i is defined as follows for each titration step $(i = 1, 2, 3, \ldots)$, where i is the running index.

$$d_i = y_i - y_{0i} - \left(\frac{\partial y}{\partial a}\right)_{0i} \cdot \alpha - \left(\frac{\partial y}{\partial b}\right)_{0i} \cdot \beta \tag{2.81}$$

The value y_i is obtained from Eq. (2.65) for each titration step ($i = 1, 2, 3, \ldots$). The values, y_{0i}, $\left(\frac{\partial y}{\partial a}\right)_{0i}$, and $\left(\frac{\partial y}{\partial b}\right)_{0i}$ are obtained based on the above-mentioned Eqs. (2.74), (2.75), (2.76), where $a = a_0$, $b = b_0$, and constants (p, q, r, s) for each titration step ($i = 1, 2, 3, \ldots$).

In order to obtain the value α and β where the sum of the square deviation ($\sum d_i^2$) is minimized, the following equations are the necessary conditions.

$$\frac{\partial}{\partial \alpha}\sum d_i^2 = 0 \tag{2.82}$$

$$\frac{\partial}{\partial \beta}\sum d_i^2 = 0 \tag{2.83}$$

The d_i is the function of y_i, y_0, $\left(\frac{\partial y}{\partial a}\right)_0$, $\left(\frac{\partial y}{\partial b}\right)_0$, α and β. The values y_i, y_0, $\left(\frac{\partial y}{\partial a}\right)_0$, and $\left(\frac{\partial y}{\partial b}\right)_0$ are obtained from Eqs. (2.65), (2.74)–(2.76) so that Eqs. (2.82) and (2.83) are expressed as functions of two parameters (α, β). The two equations with two parameters (α, β) are easily solved and give only one pair of answers (α, β). When the values of obtained α, β are small enough ($\frac{\alpha}{a_0} \leq 10^{-3}$ and $\frac{\beta}{b_0} \leq 10^{-3}$), the assumed a_0 and b_0 are considered to have been proper. Consequently, $a = a_0 + \alpha$ and $b = b_0 + \beta$ are thought to be what was desired to be determined. When α and β are not small enough, the assumption of proper a_0 and b_0 must be repeated until small enough α and β are obtained.

The practically important point here is how to assume proper a_0 and b_0. Theoretically, there is no general rule on how to assume proper a_0 and b_0. A generally recommended way is as follows. First of all, plausible a_0 and b_0 may be used based on the information of similar experimental results; then the sum of obtained α and a_0 may be used as a_0, sum of β and b_0 as b_0 for the second trial. This trial is repeated until the small enough α and β are obtained. One other way is shown in Appendix 2.3 [22], which requires only few repetitions. Another easy way is to use the software function, e.g. SOLVER [23], to minimize $\sum d_i^2$ by changing a_0 and b_0. In any event, the finally obtained α and β give δ_c and K using Eqs. (2.66) and (2.67).

The above-mentioned nonlinear least square method for the case with two parameters (α, β) is the basic one and easily extended to the cases with more parameters. Considering the possibility of obtaining the reliable δ_h of this NMR titration experiment, data treatment should be carried out with three parameters for the better regression. The programs of spreadsheet software for this three-parameters-method are developed and shown in Appendix 2.4 [24].

2.3.3.4 Estimation of Error for Nonlinear Least Square Method of NMR Spectroscopy

The above-mentioned nonlinear least square method carried out with three parameters may provide more reliable constants, δ_h, δ_c, and K. The constants are obtained very quickly and efficiently by starting from an assumption of proper a_0, b_0, and c_0 by hand or by SOLVER. However, standard deviation of each parameter, which is indispensable in order to estimate error for each constant, is not provided. The standard deviations of the regression parameters are described in the literature mathematically [25]. The elucidation of the standard deviation is a purely mathematical issue. One of the convenient ways to obtain the standard deviation is to apply the "SOLVSTAT" macro [26] to the above mentioned spreadsheet, which is easy to use and provides reliable results compared with the commercial program. It is also possible to use commercial programs, such as Igor®. The way to elucidate the standard deviations is not important but the estimation of error for titration experiment is indispensable in order to appreciate the reliability of the experiments.

2.4
Conclusion

This chapter describes in detail the theoretical principles underlying binding constant determinations using only a basic level of mathematics, statistics, and programs of spreadsheet software. As a concrete example, a practical measurement and a practical data treatment of an UV/vis and NMR titration experiment are discussed. The programs attached as Appendices would function with commonly available spreadsheet software on personal computers and provide another way to understand the contents described in this chapter. The appendices are also useful for the reader when running an actual experiment.

	A	B	C	D	E	F	G	H	I	J	K	L
1	**Appendix 2.1. Spreadsheet for continuous variation method**											
2												
3	1. Preparation of host stock solution and measure absorbance											
4		Host	Wcr									
5		Weight(mg)	7.85 mg		Host Stock Soln.				λobs=	576 nm		
6		Molecular Weight	710.74		4.42E-04	mol/l		Absorbance at 576nm is		0.051		
7		Volume(ml)	25.0 ml									
8												
9	2. Preparation of guest stock solution				picked up		diluted to					
10		Guest	2APO		0.98 ml		25.0 ml					
11		Weight(mg)	6.88 mg									
12		Purity(%)	100%					Absorbance at 576nm is		0.029		
13		Molecular Weight	61.08		Guest Stock Soln.							
14		Volume(ml)	10.0 ml		4.42E-04	mol/l						
15												
16												

Appendix 2.1. (cont.)

	A	B	C	D	E	F	G	H	I	J	K	L
17	3. Input the amount of host and guest stock solution mixed in UV cell and observed absorbance *Aobs*											
18						ε_h		ε_g	K	ε_c		average
19		Added amount of stock soln.				115.0		65.678	266	38000		4.42E-04
20	Run	Host(ml)	Guest(ml)	*Aobs*	[H]	*Ah* parts	[G]	*Ag* parts	forecasted[C]	forecasted*Ac*	forecasted*Aobs*	[H]+[G]
21	1	3.5	0.5	0.2500	3.87E-04	0.0445	5.52E-05	3.63E-03	5.68E-06	0.2157	0.2637	4.42E-04
22	2	3.2	0.8	0.3568	3.53E-04	0.0406	8.83E-05	5.80E-03	8.30E-06	0.3155	0.3619	4.42E-04
23	3	3.0	1.0	0.3944	3.31E-04	0.0381	1.10E-04	7.25E-03	9.73E-06	0.3697	0.4151	4.42E-04
24	4	2.5	1.5	0.5028	2.76E-04	0.0318	1.66E-04	1.09E-02	1.22E-05	0.4621	0.5048	4.42E-04
25	5	2.0	2.0	0.5267	2.21E-04	0.0254	2.21E-04	1.45E-02	1.30E-05	0.4929	0.5328	4.42E-04
26	6	1.5	2.5	0.5027	1.66E-04	0.0191	2.76E-04	1.81E-02	1.22E-05	0.4621	0.4993	4.42E-04
27	7	1.0	3.0	0.4136	1.10E-04	0.0127	3.31E-04	2.18E-02	9.73E-06	0.3697	0.4042	4.42E-04
28	8	0.8	3.2	0.3723	8.84E-05	0.0102	3.53E-04	2.32E-02	8.30E-06	0.3155	0.3488	4.42E-04
29	9	0.5	3.5	0.2400	5.52E-05	0.0064	3.86E-04	2.54E-02	5.68E-06	0.2157	0.2474	4.42E-04
30												
31	4. For the calcd line in Figure, do Tool-Solver: minimize E34 by changing I19&J19											
32		Graph Data										
33	Run	[H]o/[H]o+[G]o	forecasted. *Ac*	*Aobs-Ah-Ag*	Σ d^2							
34		1.00	0.0000	0.0000	1.39E-03							
35	1	0.88	0.2157	0.2019	1.89E-04							
36	2	0.80	0.3155	0.3104	2.63E-05							
37	3	0.75	0.3697	0.3490	4.27E-04							
38	4	0.63	0.4621	0.4602	3.87E-06							
39	5	0.50	0.4929	0.4868	3.78E-05							
40	6	0.38	0.4621	0.4655	1.15E-05							
41	7	0.25	0.3697	0.3791	8.91E-05							
42	8	0.20	0.3155	0.3389	5.50E-04							
43	9	0.13	0.2157	0.2083	5.46E-05							
44		0.00	0.0000	0.0000								

Figure 4. Modified Job's Plot for complexation of host and guest by UV visible spectroscopy ● : observed; ------- : calculated

	A	B	C	D	E	F	G	H	I	J	K
1	**Appendix 2.2. Spreadsheet for Rose-Drago**					1. Basic Equations					
2	**Method by UV-visible Spectroscopy**					$1/K=(An-Ahn-Agn)/(\varepsilon_c-\varepsilon_h-\varepsilon_g)-([H]_{0n}+[G]_{0n})$					
3						$+[H]_{0n}[G]_{0n}(\varepsilon_c-\varepsilon_h-\varepsilon_g)/(An-Ahn-Agn)$					
4	Instruction					*an*	$=An-Ahn-Agn$		$=An-\varepsilon_h\cdot[H]_{0n}-\varepsilon_g\cdot[G]_{0n}$		
5	1. Input data into gray cells at 2!					*bn*	$=-([H]_{0n}+[G]_{0n})$				
6	2. Remove improper data at 8!					*cn*	$=[H]_{0n}[G]_{0n}/(An-Ahn-Agn)$				
7	3. Check reliability in the graph at 9!					$Y=1/K$					
8	4. That's all!					$X=\varepsilon_c-\varepsilon_h-\varepsilon_g$					
9						$\therefore Y=an/X+bn+cn*X$					
10	2. Data of Titration Experiment										
11	Host Abb.	SR{24D}		$[H]_{0n}$	$[G]_{0n}$	*An*	*Ahn*	*Agn*	*An-Ahn-Agn*		
12	Guest Abb.	1APO	n=1	1.88E-05	2.15E-02	0.6477	1.88E-05	2.15E-02	0.626		
13	Temp/℃	30	n=2	1.88E-05	1.62E-02	0.6026	1.88E-05	1.62E-02	0.586		
14	λobs/nm	580	n=3	1.88E-05	1.08E-02	0.5376	1.88E-05	1.08E-02	0.527		
15	ε_h	1	n=4	1.88E-05	7.54E-03	0.4700	1.88E-05	7.54E-03	0.462		
16	ε_g	1	n=5	1.88E-05	4.31E-03	0.3554	1.88E-05	4.31E-03	0.351		
17											
18	3. Calculation of *an*, *bn*, and *cn*.				4. Calcd. X1, X2				5. Ordering of X1, X2(X1>X2)		
19	n	*an*	*bn*	*cn*	combination of n		X1	X2	X1-X2	X1	X2
20	n=1	0.6261	-2.16E-02	6.48E-07	1	2	7.37E+00	4.17E+04	-4.17E+04	4.17E+04	7.37E+00
21	n=2	0.5864	-1.62E-02	5.19E-07	2	3	1.11E+01	4.02E+04	-4.02E+04	4.02E+04	1.11E+01
22	n=3	0.5268	-1.08E-02	3.85E-07	3	4	1.99E+01	4.14E+04	-4.14E+04	4.14E+04	1.99E+01
23	n=4	0.4624	-7.56E-03	3.07E-07	4	5	3.45E+01	4.25E+04	-4.25E+04	4.25E+04	3.45E+01
24	n=5	0.3511	-4.33E-03	2.31E-07	5	1	1.60E+01	4.13E+04	-4.13E+04	4.13E+04	1.60E+01
25	n=1	0.6261	-2.16E-02	6.48E-07	1	3	9.22E+00	4.10E+04	-4.10E+04	4.10E+04	9.22E+00
26	n=2	0.5864	-1.62E-02	5.19E-07	2	4	1.44E+01	4.07E+04	-4.07E+04	4.07E+04	1.44E+01
27	n=3	0.5268	-1.08E-02	3.85E-07	3	5	2.72E+01	4.20E+04	-4.19E+04	4.20E+04	2.72E+01
28	n=4	0.4624	-7.56E-03	3.07E-07	4	1	1.17E+01	4.11E+04	-4.11E+04	4.11E+04	1.17E+01
29	n=5	0.3511	-4.33E-03	2.31E-07	5	2	1.99E+01	4.12E+04	-4.11E+04	4.12E+04	1.99E+01
30										4.13E+04	1.71E+01

Appendix 2.2. (cont.)

	A	B	C	D	E	F	G	H	I	J	K
31	6. Data for Graph										
32	$Y=1/K$	$X=\varepsilon c-\varepsilon h-\varepsilon g$									
33	$Y=an/X+bn+cnX$	50	1.2E+02	2.9E+02	7.1E+02	1.7E+03	4.2E+03	1.0E+04	2.5E+04	4.2E+04	6.0E+04
34	n=1	-9.0E-03	-1.6E-02	-1.9E-02	-2.0E-02	-2.0E-02	-1.9E-02	-1.5E-02	-5.5E-03	5.9E-03	1.7E-02
35	n=2	-4.4E-03	-1.1E-02	-1.4E-02	-1.5E-02	-1.5E-02	-1.4E-02	-1.1E-02	-3.3E-03	5.8E-03	1.5E-02
36	n=3	-2.4E-04	-6.4E-03	-8.9E-03	-9.8E-03	-9.8E-03	-9.0E-03	-6.8E-03	-1.2E-03	5.5E-03	1.2E-02
37	n=4	1.7E-03	-3.7E-03	-5.9E-03	-6.7E-03	-6.8E-03	-6.2E-03	-4.4E-03	5.2E-05	5.5E-03	1.1E-02
38	n=5	2.7E-03	-1.4E-03	-3.1E-03	-3.7E-03	-3.7E-03	-3.3E-03	-1.9E-03	1.4E-03	5.5E-03	9.5E-03
39											
40	7. Y1, Y2 and *K1*, *K2* from X1, X2					8. Statistical treatment of X1(εc−εh−εg) and *K*					
41	Y1	Y2	*K1=1/Y1*	*K2=1/Y2*		combination of n		X1 Data	Check Data Here	K1=1/Y1 Data	Check Data Here
42	5.49E-03	#######	1.82E+02	1.58E+01		1	2	4.17E+04	4.17E+04	1.82E+02	1.82E+02
43	4.72E-03	#######	2.12E+02	2.72E+01		2	3	4.02E+04	4.02E+04	2.12E+02	2.12E+02
44	5.17E-03	#######	1.93E+02	6.39E+01		3	4	4.14E+04	4.14E+04	1.93E+02	1.93E+02
45	5.51E-03	#######	1.82E+02	1.71E+02		4	5	4.25E+04	4.25E+04	1.82E+02	1.82E+02
46	5.23E-03	#######	1.91E+02	5.66E+01		5	1	4.13E+04	4.13E+04	1.91E+02	1.91E+02
47	5.00E-03	#######	2.00E+02	2.16E+01		1	3	4.10E+04	4.10E+04	2.00E+02	2.00E+02
48	4.94E-03	#######	2.02E+02	4.07E+01		2	4	4.07E+04	4.07E+04	2.02E+02	2.02E+02
49	5.38E-03	#######	1.86E+02	1.17E+02		3	5	4.20E+04	4.20E+04	1.86E+02	1.86E+02
50	5.06E-03	#######	1.98E+02	3.13E+01		4	1	4.11E+04	4.11E+04	1.98E+02	1.98E+02
51	5.19E-03	#######	1.93E+02	7.49E+01		5	2	4.12E+04	4.12E+04	1.93E+02	1.93E+02
52			1.94E+02	6.19E+01			Average	41312	41312	194	194
53	9. Graphical Expression						standard deviation	619	619	9	9
54											
55							level of significance α	0.05	0.05	0.05	0.05
56							degree of freedom	9	9	9	9
57							$t_{a/2}$	2.262	2.262	2.262	2.262
58											
59							confidence interval	95%		95%	95%
60							x	41312	41312	194	K=193.9
61							±	443	443	6	±6.4
62							Upper limit	41755	41755	200	200
63							Lower limit	40870	40870	187	187
64											
65								εc=	41314		
66									±443		
67											
68											

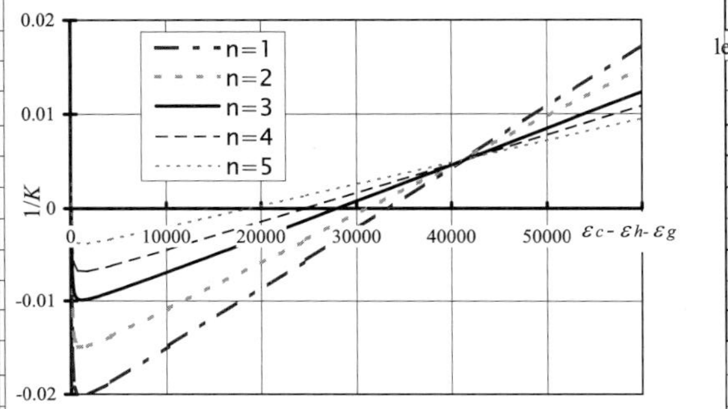

Figure Graphical expression to appreciate the determined *K* according to Rose-Drago method

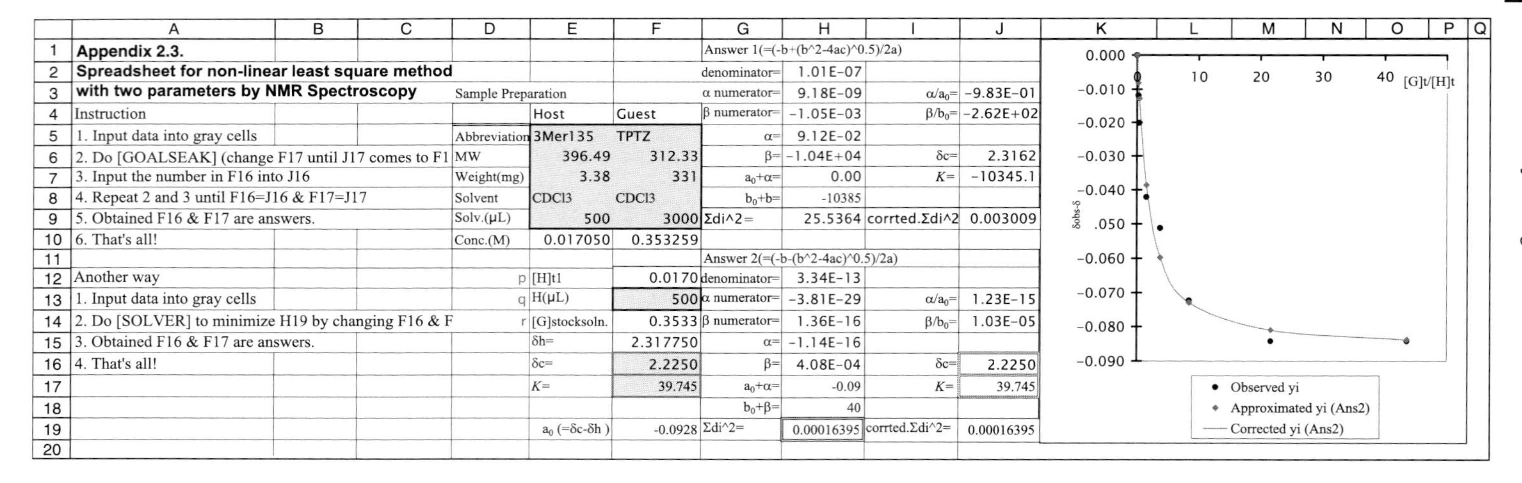

	A	B	C	D	E	F	G	H	I	J
1	**Appendix 2.3.**						Answer 1(=(-b+(b^2-4ac)^0.5)/2a)			
2	**Spreadsheet for non-linear least square method**						denominator=	1.01E-07		
3	**with two parameters by NMR Spectroscopy**			Sample Preparation			α numerator=	9.18E-09	α/a_0=	-9.83E-01
4	Instruction				Host	Guest	β numerator=	-1.05E-03	β/b_0=	-2.62E+02
5	1. Input data into gray cells			Abbreviation	3Mer135	TPTZ		α=	9.12E-02	
6	2. Do [GOALSEAK] (change F17 until J17 comes to F1			MW	396.49	312.33	β=	-1.04E+04	δc=	2.3162
7	3. Input the number in F16 into J16			Weight(mg)	3.38	331	$a_0+\alpha$=	0.00	*K*=	-10345.1
8	4. Repeat 2 and 3 until F16=J16 & F17=J17			Solvent	CDCl3	CDCl3	b_0+b=	-10385		
9	5. Obtained F16 & F17 are answers.			Solv.(μL)	500	3000	Σdi^2=	25.5364	corrted.Σdi^2	0.003009
10	6. That's all!			Conc.(M)	0.017050	0.353259				
11							Answer 2(=(-b-(b^2-4ac)^0.5)/2a)			
12	Another way			p	[H]t1	0.0170	denominator=	3.34E-13		
13	1. Input data into gray cells			q	H(μL)	500	α numerator=	-3.81E-29	α/a_0=	1.23E-15
14	2. Do [SOLVER] to minimize H19 by changing F16 & F			r	[G]stocksoln.	0.3533	β numerator=	1.36E-16	β/b_0=	1.03E-05
15	3. Obtained F16 & F17 are answers.				δh=	2.317750	α=	-1.14E-16		
16	4. That's all!				δc=	2.2250	β=	4.08E-04	δc=	2.2250
17					*K*=	39.745	$a_0+\alpha$=	-0.09	*K*=	39.745
18							$b_0+\beta$=	40		
19					a_0 (=δc-δh)	-0.0928	Σdi^2=	0.00016395	corrted.Σdi^2=	0.00016395
20										

Appendix 2.3. (cont.)

	A	B	C	D	E	F	G	H	I	J	K	L	M	N	O	P	Q
21	Running Index (i)	1	2	3	4	5	6	7	8	memorandum							
22	Added amount of G soln. (μ L)	0.0	5.6	3.6	28.0	54.8	111.5	318.1	529.9								
23	Integrated amount of G soln. (μ L)	0	5.6	9.2	37.2	91.9	203.4	521.5	1051.4	s (=[G]t)							
24	δobs./ppm	2.317750	2.305850	2.297620	2.275650	2.266490	2.245440	2.233540	2.233540								
25	[G]ti	0.00000	0.00388	0.00636	0.02444	0.05487	0.10216	0.18035	0.23941		c						
26	[H]ti	0.01705	0.01686	0.01674	0.01587	0.01440	0.01212	0.00835	0.00549								
27	**[G]t/[H]t**	**0.0000**	**0.2300**	**0.3800**	**1.5400**	**3.8100**	**8.4300**	**21.6100**	**43.5700**								
28	**Observed yi**	**0.000**	**-0.012**	**-0.020**	**-0.042**	**-0.051**	**-0.072**	**-0.084**	**-0.084**	δobs.-δh	(b^2-4c)^0.5						
29	Approximated yi (Ans1)	-0.230	-0.244	-0.255	-0.344	-0.549	-0.995	-2.297	-4.476	y_0i	(-b+(b^2-4c)^0.5)/2			Answer 1			
30	**Approximated yi (Ans2)**	**0.000**	**-0.008**	**-0.013**	**-0.039**	**-0.060**	**-0.073**	**-0.081**	**-0.084**		(-b-(b^2-4c)^0.5)/2			Answer 2			
31	Corrected yi (Ans1)	-0.029	-0.030	-0.031	-0.034	-0.038	-0.044	-0.064	-0.100	y_0i+(∂ /∂ a)$_0$	α+(∂ /∂ b)$_0$·β			Answer 1			
32	**Corrected yi (Ans2)**	**0.000**	**-0.008**	**-0.013**	**-0.039**	**-0.060**	**-0.073**	**-0.081**	**-0.084**					Answer 2			
33	deviation (di^2)	0.000851	0.000338	0.000117	0.000061	0.000188	0.000811	0.000396	0.000247	(yi-corrected y0i)^2				Answer 1			
34		0.000000	0.000014	0.000053	0.000013	0.000073	0.000000	0.000010	0.000000					Answer 2			
35	**calculation area**	2.476	2.722	2.883	4.125	6.557	11.506	25.625	49.149		-b			Answer 1			
36		6.129	6.490	6.791	10.859	27.756	98.672	570.195	2241.342		(b^2-4c)			Answer 2			
37	yi-y0i	0.230	0.233	0.234	0.302	0.497	0.922	2.212	4.392					Answer 1			
38		0.0000	-0.0038	-0.0073	-0.0036	0.0085	0.0007	-0.0032	-0.0004					Answer 2			
39	(yi-y0i)^2	0.0528	0.0541	0.0550	0.0913	0.2473	0.8506	4.8943	19.2909					Answer 1			
40		0.0000	0.0000	0.0001	0.0000	0.0001	0.0000	0.0000	0.0000					Answer 2			
41	(δ y/δ a)0i	2.476	2.635	2.744	3.710	5.913	10.720	24.752	48.246					Answer 1			
42		0.000	0.087	0.138	0.415	0.644	0.786	0.873	0.903					Answer 2			
43	(δ y/δ a)0i^2	6.129	6.942	7.531	13.767	34.960	114.913	612.653	2327.667	3124.563	1		2	Answer 1			
44		0.000	0.008	0.019	0.172	0.415	0.618	0.762	0.816	2.811				Answer 2			
45	(δ y/δ b)0i	2.43E-06	2.52E-06	2.56E-06	2.74E-06	2.73E-06	2.63E-06	2.52E-06	2.48E-06	2.06E-05				Answer 1			
46		0.00E+00	-8.34E-08	-1.29E-07	-3.07E-07	-2.98E-07	-1.93E-07	-8.90E-08	-4.65E-08	-1.15E-06				Answer 2			
47	(δ y/δ b)0i^2	5.92E-12	6.34E-12	6.57E-12	7.51E-12	7.46E-12	6.90E-12	6.37E-12	6.16E-12	5.32E-11	2	2		Answer 1			
48		0.00E+00	6.96E-15	1.67E-14	9.40E-14	8.86E-14	3.71E-14	7.93E-15	2.16E-15	2.54E-13				Answer 2			
49	(yi-y0i)(δ y/δ a)0i	5.69E-01	6.13E-01	6.44E-01	1.12E+00	2.94E+00	9.89E+00	5.48E+01	2.12E+02	2.82E+02		1	4	Answer 1			
50		0.00E+00	-3.32E-04	-1.01E-03	-1.49E-03	5.49E-03	5.14E-04	-2.80E-03	-3.79E-04	-2.51E-10				Answer 2			
51	(yi-y0i)(δ y/δ b)0i	5.59E-07	5.85E-07	6.01E-07	8.28E-07	1.36E-06	2.42E-06	5.58E-06	1.09E-05	2.28E-05		4	1	Answer 1			
52		0.00E+00	3.17E-10	9.42E-10	1.10E-09	-2.54E-09	-1.26E-10	2.85E-10	1.95E-11	1.03E-16				Answer 2			
53	(δ y/δ a)0i*(δ y/δ b)0i	6.03E-06	6.63E-06	7.03E-06	1.02E-05	1.62E-05	2.82E-05	6.25E-05	1.20E-04	2.56E-04	3	3	3	Answer 1			
54		0.00E+00	-7.28E-09	-1.79E-08	-1.27E-07	-1.92E-07	-1.52E-07	-7.77E-08	-4.19E-08	-6.16E-07	denominator	α numerator	β numerator	Answer 2			

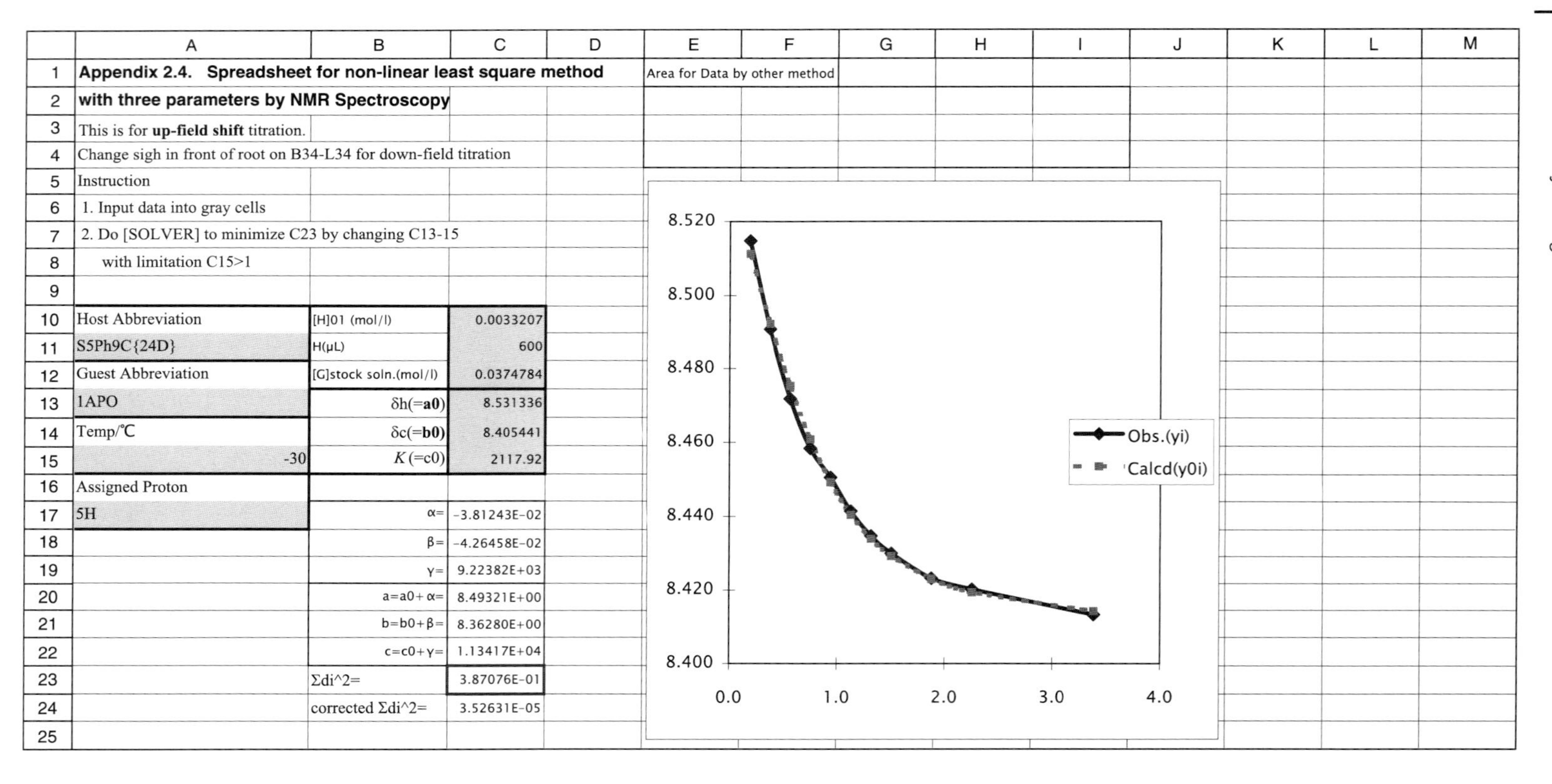

	A	B	C	D	E
1	**Appendix 2.4. Spreadsheet for non-linear least square method**				Area for Data by other method
2	**with three parameters by NMR Spectroscopy**				
3	This is for **up-field shift** titration.				
4	Change sigh in front of root on B34-L34 for down-field titration				
5	Instruction				
6	1. Input data into gray cells				
7	2. Do [SOLVER] to minimize C23 by changing C13-15				
8	with limitation C15>1				
9					
10	Host Abbreviation	[H]01 (mol/l)	0.0033207		
11	S5Ph9C{24D}	H(μL)	600		
12	Guest Abbreviation	[G]stock soln.(mol/l)	0.0374784		
13	1APO	δh(=**a0**)	8.531336		
14	Temp/℃	δc(=**b0**)	8.405441		
15	-30	*K* (=c0)	2117.92		
16	Assigned Proton				
17	5H	α=	-3.81243E-02		
18		β=	-4.26458E-02		
19		γ=	9.22382E+03		
20		a=a0+α=	8.49321E+00		
21		b=b0+β=	8.36280E+00		
22		c=c0+γ=	1.13417E+04		
23		Σdi^2=	3.87076E-01		
24		corrected Σdi^2=	3.52631E-05		
25					

Appendix 2.4. (cont.)

	A	B	C	D	E	F	G	H	I	J	K	L	M
26	Running Index (i)	1	2	3	4	5	6	7	8	9	10	11	
27	Integrated amount of G soln. (μL)	10	20	30	40	50	60	70	80	100	120	180	
28	δobs./ppm(=yi)	8.514643	8.490785	8.471848	8.458405	8.450565	8.441468	8.434718	8.429970	8.423220	8.420243	8.413263	
29		7.77360E+02	7.78360E+02	7.79360E+02	7.80360E+02	7.81360E+02	7.82360E+02	7.84360E+02	7.85360E+02	7.86360E+02	7.85360E+02	7.85360E+02	Sum
30	[G]oi	0.000614	0.001209	0.001785	0.002342	0.002883	0.003407	0.003916	0.004409	0.005354	0.006246	0.008649	0.040815
31	[H]oi	0.003266	0.003214	0.003163	0.003113	0.003065	0.003019	0.002974	0.002930	0.002846	0.002767	0.002554	0.032911
32	**l=[G]oi/[H]oi**	0.2	0.4	0.6	0.8	0.9	1.1	1.3	1.5	1.9	2.3	3.4	14.296178
33	**Obs.(yi)**	8.515	8.491	8.472	8.458	8.451	8.441	8.435	8.430	8.423	8.420	8.413	92.949125
34	**Calcd(y0i)**	8.511	8.492	8.475	8.461	8.449	8.440	8.434	8.429	8.423	8.419	8.414	92.949125
35	Complexation Ratio	0.133	0.322	0.473	0.579	0.642	0.714	0.767	0.805	0.859	0.882	0.938	7.113664

References and Notes

1 R. M. Izatt, K. Pawlak, J. S. Bradshaw, *Chem. Rev.* **1991**, 91, 1721.

2 R. M. Izatt, J. S. Bradshaw, K. Pawlak, R. L. Bruening, B. J. Tarbet, *Chem. Rev.* **1992**, 92, 1261.

3 X. X. Zhang, J. S. Bradshaw, R. M. Izatt, *Chem. Rev.* **1997**, 97, 3313.

4 As examples for temperature-dependent enantiomer-selective binding of chiral guests to chiral hosts evaluated by determining thermodynamic parameters (enthalpy, entropy): a) K. Ogasahara, K. Hirose, Y. Tobe, K. Naemura, *J. Chem. Soc., Perkin Trans. 1* **1997**, 3227; b) K. Naemura, K. Nishioka, K. Ogasahara, Y. Nishikawa, K. Hirose, Y. Tobe, *Tetrahedron: Asymmetry* **1998**, 9, 563; c) K. Naemura, K. Matsunaga, J. Fuji, K. Ogasahara, Y. Nishikawa, K. Hirose, Y. Tobe, *Anal. Sci.* **1998**, 14, 175; d) K. Hirose, P. Aksharanandana, M. Suzuki, K. Wada, K. Naemura, Y. Tobe, *Heterocycles* **2005**, 66, 405.

5 Typical examples: a) K. Naemura, J. Fuji, K. Ogasahara, K. Hirose, Y. Tobe, *Chem. Commun.* **1996**, 2749; b) K. Naemura, T. Wakebe, K. Hirose, Y. Tobe, *Tetrahedron: Asymmetry* **1997**, 8, 2585; c) K. Hirose, J. Fuji, K. Kamada, Y. Tobe, K. Naemura, *J. Chem. Soc., Perkin Trans. 2* **1997**, 1649.

6 a) Y. Shibata, B. Inoue, Y. Nakatuka, *Nihon Kagakukaishi* **1921**, 42, 983; b) R. Tsuchida, *Bull. Chem. Soc. Jpn.* **1935**, 10, 27; c) P. Job, *Compt. Rend.* **1925**, 180, 928; d) P. Job, *Ann. Chim. Phys.* **1928**, 9, 113.

7 a) A. E. Harvey, D. L. Manning, *J. Am. Chem. Soc.* **1950**, 72, 4488; b) H. E. Bent, C. L. French, *J. Am. Chem. Soc.* **1941**, 63, 568.

8 J. H. Yoe, A. L. Jones, *Ind. Eng. Chem. Anal. Ed.* **1944**, 16, 111.

9 K. Hirose: spreadsheet software. Simple instruction of this software is shown on the sheet.

10 C. S. Wilcox, Design, Synthesis, and Evaluation of an Efficacious Functional Group Dyad. Methods and Limitations in the Use of NMR for Measuring Host-Guest Interactions in: Frontiers in *Supramolecular Organic Chemistry and Photochemistry*, Eds. H.-J. Schneider, H. Dürr, VCH-Verlag, Weinheim 1991, pp. 123–143.

11 G. Weber in *Molecular Biophysics*, Eds. B. Pullman, M. Weissbluth, Academic Press, New York 1965, pp 369–97.

12 H. A. Benesi, J. H. Hildebrand, *J. Am. Chem. Soc.* **1949**, 71, 2703.

13 J. A. A. Ketelaar, C. van de Stolpe, A. Goudsmit, W. Dzcubas, *Rec. Trav. Chim. Pays-Bas* **1952**, 71, 1104.

14 a) S. Nagakura, *J. Am. Chem. Soc.* **1954**, 76, 3070; b) S. Nagakura: *J. Am. Chem. Soc.* 1958, 80, 520; c) H. Baba, *Bull. Chem. Soc. Jpn.* **1958**, 31, 169.

15 R. L. Scott, *Rec. Trav. Chim. Pays-Bas* **1956**, 75, 787.

16 P. R. Hammond, *J. Chem. Soc.* **1964**, 479.

17 a) N. J. Rose, R. S. Drago, *J. Am. Chem. Soc.* **1959**, 81, 6138; b) S. Nagakura, *J. Am. Chem. Soc.* **1958**, 80, 520.

18 M. Nakano, N. I. Nakano, T. Higuchi, *J. Phys. Chem.* **1967**, 71, 3954.

19 a) C. J. Creswell, A. L. Allred, *J. Phys. Chem.* **1962**, 66, 1469; b) H. Stamm, W. Lamberty, J. Stafe, *Tetrahedron* **1976**, 32, 2045.

20 K. Hirose: spreadsheet software. Simple instruction of this software is shown on the spreadsheet.

21 Diffusion–ordered NMR spectroscopy is another NMR method which is used to analyze binding properties. This method is discussed later in this book.

22 K. Hirose: spreadsheet software. Simple instruction of this software is shown on the spreadsheet.

23 Add-in function of spreadsheet software on a personal computer (Microsoft Excel 2004).

24 K. Hirose, Y. Nishikawa, K. Matsunaga: unpublished software. Simple instruction of this software is shown on the spreadsheet.

25 K. J. Johnson, *Numerical Methods in Chemistry*, Marcel Dekker, New York 1980, pp 278ff.

26 E. J. Billo, *Excel for Chemists*, 2nd Edn, Wiley, New York 2001, pp. 233ff.

3
Isothermal Titration Calorimetry in Supramolecular Chemistry

Franz P. Schmidtchen

3.1
Introduction

Every compound in solution interacts with all others – in principle. In reality, the extent of this interaction at any moment in time can vary within a wide range depending on the exact chemical nature of all individual partners participating and their concentration. In this scenario, supramolecular chemistry specifically addresses the high-end, strongly interacting fraction of compounds, evaluating the origin of their stickyness for each other with the perspective to exploit this comprehension for some application. Despite their strength in interaction, the fundamental and common feature in this ensemble is the ready break-up of mutual adhesions due to thermal collisions, leading to quite rapid interchange of all possible pairwise relationships. As a consequence of this reversibility in bonding, a time invariant equilibrium is eventually reached, characterized by uneven amounts of all molecular species present, yet subject to their intrinsic preferences to generate a uniform chemical potential. The approach to this equilibrium and any later perturbation by chemical (addition of compounds) or physical (temperature, pressure) stimuli is accompanied by an energetic response that can be sensed and quantified in relation to the initial perturbation. It is this universality and independence on material peculiarities (presence/absence of an indicator probe, transparency, homogeneity, etc.) that renders the measurement of heat energy (calorimetry) traded in a process in solution, an indispensable and powerful tool to learn about and characterize supramolecular interactions [1–11].

We must realize at this point that the calorimetric technique entirely builds upon the establishment of equilibrium conditions enabling the use and help from equilibrium thermodynamics. Many facets of supramolecular chemistry, however, do not comply with this prerequisite. Molecular recognition, for instance, in particular as a crucial property of all living matter, which exists because of nonequilibrium conditions, must be considered a process that relies on kinetic selectivity and thus *per se* is not open to an all-encompassing description of the phenomenon using this technique [12]. Similar arguments limit the utility of calorimetry in other vectorial processes like membrane transport, signaling, catalysis or locomotion. Never-

Analytical Methods in Supramolecular Chemistry. Edited by Christoph Schalley

ISBN: 978-3-527-31505-5

theless, though full characterization of flow processes is not possible by calorimetry alone, some partial steps may well be, and thus can contribute to the illumination of the entire sequence of events. A prominent example illustrating the virtue of calorimetric measurements in the biological context is drug design. Though the molecular basis of enzyme catalytic function is unlikely to be clarified by calorimetry, the destruction of function by creating high affinity inhibitors is greatly aided, because this method exclusively addresses ground state energies and is unable to address transient equilibria with transition states. The energetic signature of enzyme-inhibitor association determined by calorimetry may even serve as a guideline to combat the development of drug resistance caused by high mutation rates in the active centers of the target enzymes [13]. Fortunately, in many flow processes conditions can be found that slow down the overall conversion, allowing the energetic determination of transient pre-equilibria states. Compared with most spectroscopic methods, calorimetry exhibits a rather slow response to a triggered change. Yet, many kinetically controlled phenomena can be adjusted to allow equilibrium measurements in the seconds-to-minutes time domain and, hence, are amenable to calorimetric investigation. These cases add to the vast number of applications that are based on true equilibrium supramolecular association as in ion sequestration or bulk phase extraction [14].

Two decades after the advent of highly sensitive and user-friendly commercial microcalorimeters [1–3] the measurement of heat occupies a well established place in the investigation of molecular association processes in solution. Understandably, biological studies claim the largest share and a number of excellent review articles document our comprehension of the subtleties, benefits and pitfalls of the method in these fields [4–11]. A number of experts commendably also evaluate the progress in applications and elaboration of the technique at regular (yearly) intervals [15]. With a phase-lag of about a decade other chemical sciences which are traditionally devoted to the construction of irreversibly connected covalent frameworks discover the virtues and profits of calorimetry as their focus shifts towards more fragile and even switchable (e.g. bistable) structural entities [16]. Recruiting microcalorimetry as a standard technique in nonaqueous applications also appears warranted.

3.2
The Thermodynamic Platform

Supramolecular interactions by definition involve the reversible formation and cleavage of bonds not only between the molecular species in focus, let it be for example an artificial host–guest pair, the association of an enzyme substrate/inhibitor complex, or the strand pairing of nucleic acids, but include also the rearrangement of the solvation shell of all participating partners encompassing the secondary interactions like ion-pairing of charged participants or the solvophobic aggregation. The heat evolved or consumed in the shuttling of bonds, thus, is the integral result of manifold simultaneous events that add up to a truly global response. Owing to the huge number of individual species present even in the most diluted sample solution, the energetic output is constant for defined compositions and framing conditions (p, T) and potential fluctuations are many orders of magnitude smaller than observable by our most sensitive detectors. In essence, such a system behaves much like a black box where input and output can be controlled and determined, yet the distribution of influx inside or the origin of eflux eludes our immediate inspection. However, correlating the observable energetic output with changes happening at the molecular level represents the ultimate goal and objective of supramolecular science. The role of calorimetry in this theatre is to provide a reliable basis of experimental observables leaving the connection to molecular behavior and structural detail open to the creativity in experimental design and to the interpretation of the questioner.

Unlike other useful methods employed in supramolecular chemistry (e.g. mass spectrometry, chemical force microscopy), calorimetry reports on ensembles averaged over time and individual energies of their members. This sets the stage to employ the framework of thermodynamics for a full energetic characterization of the system under study. Heat as the primary observable in calorimetry is commonly measured at constant (atmospheric) pressure and thus represents an enthalpy change ΔH. If enthalpy is measured in response of a change in total composition of the system, the output depends on the extent of interaction between the components permitting access to molecular affinity. Provided just one 1:1 stoichiometric binding process dominates the molecular bond rearrangement the corresponding constant of mass action K_{assoc} can be determined. The affinity constant K_{assoc} relates to the Gibbs enthalpy of association ΔG^o according to Eq. (3.1):

$$\Delta G^o = -RT \ln K_{\text{assoc}} \tag{3.1}$$

$$\Delta S^o = \frac{\Delta H^o}{T} - \frac{\Delta G^o}{T} \tag{3.2}$$

With enthalpy ΔH^o and Gibbs enthalpy ΔG^o at hand the change in standard entropy ΔS^o is easily calculated from the Gibbs–Helmholtz equation (Eq. (3.2)). From a single calorimetric experiment at constant temperature the main state functions ΔH, ΔG and ΔS of the binding process are accessible, if presumptions with respect to the singularity and stoichiometry of the process apply. Conducting such measurements at a range of temperatures yields the heat capacity ΔC_p (Eq. (3.3)).

$$\Delta C_p{}^o = \frac{d\Delta H^o}{dT} = \frac{T\,d\Delta S^o}{dT} \tag{3.3}$$

In principle, ΔC_p itself can be a function of temperature, but in the narrow span of temperatures of interest in supramolecular binding there is little risk to approximate ΔC_p by Eq. (3.4) where T_1 and T_2 denote two different absolute temperatures furnishing the respective standard enthalpies $\Delta H^o{}_1$ and $\Delta H^o{}_2$.

$$\Delta C_p{}^{o} = \frac{\Delta H^o{}_2 - \Delta H^o{}_1}{T_2 - T_1} = \frac{\Delta S^o{}_2 - \Delta S^o{}_1}{\ln T_2 - \ln T_1} \tag{3.4}$$

The heat capacity ΔC_p occupies a pivotal position in understanding supramolecular interactions because it represents the temperature gradient of the energetic components composing the Gibbs enthalpy ΔG and thus allows the affinity to be calculated at various temperatures [17]. Quite unlike most covalent bond formations in preparative chemistry which are largely enthalpy-dominated and frequently possess vanishing heat capacities (cf. "click-chemistry" [18]) all supramolecular processes come along with a substantial variation of the standard enthalpy with temperature. Associations in general feature negative ΔC_p values which for aqueous media in many cases can be correlated with the change in surface area buried from solvent on complexation. A typical temperature plot of the state functions is depicted in Fig. 3.1.

The two panels refer to supramolecular complexations having endothermic (A, left) or exothermic (B, right) enthalpies at room temperature. Either case is characterized by the same constant and negative heat capacity ($\Delta C_p = -200$ J mol^{-1} K^{-1}) causing the enthalpy ΔH to change sign in the temperature span investigated. At $\Delta H = 0$ the complexation is solely entropy driven and the association constant hits a maximum at this point. From the diagram it is obvious that ΔH and the entropic component $T\Delta S$ compensate each other leading to a very flat temperature depen-

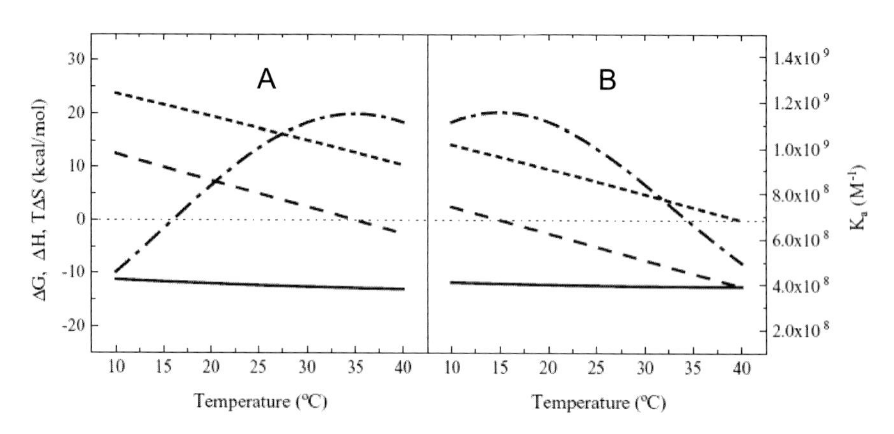

Fig. 3.1. Temperature profiles of the association constant K_{assoc} (dash-dotted line), the enthalpy ΔH (dashed line), the entropy as $T\Delta S$ (dotted line) and the resultant Gibbs enthalpy G (solid line) for a host–guest complexation characterized by: (panel A) K_{assoc} (298 K) $= 10^9$ M^{-1}, ΔH (298 K) $= +20.9$ kJ mol^{-1}, $\Delta C_p = -210$ J K^{-1} mol^{-1}; (panel B) K_{assoc} (298 K) $= 10^9$ M^{-1}, ΔH (298 K) $= -20.9$ kJ mol^{-1}, $\Delta C_p = -210$ J K^{-1} mol^{-1}; both enthalpy and entropy possess substantial temperature dependencies, yet, ΔG does not change by more than 8 kJ over the span from 10 to 40 °C rendering the estimation of binding enthalpies or heat capacity changes by noncalorimetric methods quite problematic. Reproduced with permission from A. Velasquez-Campoy et al., *Biophys. Chem.* **2005**, 115, 115.

dence of the Gibbs enthalpy ΔG. Such behavior is a form of enthalpy–entropy compensation that appears as an intrinsic property of weak (i.e. supramolecular) interactions [19, 20]. The similarity in slopes of the enthalpy and entropy components implies that the change in heat capacity ΔC_p which defines the temperature gradients according to Eqs. (3.3) and (3.5) must be much greater than the entropy ΔS.

$$\frac{d\Delta H}{dT} = \Delta C_p \tag{3.3}$$

$$\frac{dT\Delta S}{dT} = \Delta C_p + \Delta S \tag{3.5}$$

$$\frac{d\Delta H}{dT} \approx \frac{dT\Delta S}{dT} \, only, \, if \, \Delta C_p \gg \Delta S$$

$$\ln \frac{K_{\text{assoc}}}{K_{\text{assoc}}(T_r)} = \frac{\Delta H(T_r) - T_r \Delta C_p}{R} \left(\frac{1}{T_r} - \frac{1}{T} \right) + \frac{\Delta C_p}{R} \ln \frac{T}{T_r} \tag{3.6}$$

Consequently, there is no justification for disregard of this quantity as it is frequently executed in the spectroscopic determination of the binding enthalpy using rudimentary van't Hoff relationships (e.g. $\ln K_{\text{assoc}} = -\frac{\Delta H}{R}\frac{1}{T} + \frac{\Delta S}{R}$). Instead a formula (Eq. (3.6)) must be used for calculation that honors a nonzero heat capacity ΔC_p and allows the derivation of $\Delta H(T_r)$ from the affinity constant $K_{\text{assoc}}(T_r)$ at the reference temperature T_r and a series of binding constants K_{assoc} obtained at various temperatures T by a nonlinear fitting process. The substantial discrepancies found between association enthalpies ΔH obtained by direct calorimetry or via the temperature dependence of the association constant using a van't Hoff treatment most likely arise from lack of precision in the original measurements eventually combined with inadequate data evaluation [21, 22]. Because calorimetry is the only method yielding enthalpies as direct experimental observables, the values derived on this basis appear more credible than from any other alternative method.

Another result apparent from Fig. 3.1 is the lack of correlation between the change in enthalpy ΔH and Gibbs enthalpy ΔG. Again this disparity emerges as a consequence of enthalpy–entropy compensation, yet it seems fair to state that appreciation of this fact in supramolecular chemistry is severely neglected. Common customs try to correlate measured affinities e.g. in abiotic host–guest systems with a goodness of fit (i.e. with binding strength) that is often derived and supported by energy-minimized structures delivered by molecular mechanics. Irrespective of the consideration of solvent influence which adds another level of complexity such comparisons are bound to lead to erroneous conclusions because the contribution of the entropy component is excluded *a priori*. In some instances this may be justifiable, however, as a rule the omission of the entropic influence means an amputation of an essential property that distinguishes supramolecular from covalently connected systems. Because of weaker structural definition of the former, the energetic signature represents a welcome marker for additional characterization. Two host–guest systems having identical affinity (ΔG_{ass}) may greatly differ in composition of the enthalpic and entropic contribution and thus possess sharply distinguished suitability for certain supramolecular functions. Our striving for comprehension of structure and function of supramolecular complexation mandates the appreciation of the entire span of thermodynamic state functions. Isothermal titration calorimetry is currently the most accurate, sensitive, fast and convenient technique that can obtain these quantities.

3.3
Acquiring Calorimetric Data

The measurement of heat conveniently takes temperature as an indicator. In modern calorimeters, the instrumental design employs two principal approaches that are easily distinguished with respect to the effect on the temperature output: In the adiabatic mode, heat evolution or consumption by the chemical process under investigation leads to a permanent increase or decrease of temperature. The extent of the change depends on the heat capacity of the instrument, which must be calibrated in separate experiments. Alternatively, the heat effect may be dissipated to a heat sink, so that the measurement following an initial perturbation falls back on a constant temperature baseline (isothermal mode). The heat flow may be observed directly using a pile of thermocouples (heat conduction device) or may be actively regulated to maintain a fixed level (power compensation). Both instrumental designs reach sensitivities in the nanowatt regime by means of a differential measurement relative to an internal reference. The schematic blue print of a power compensation titration calorimeter is shown in Fig. 3.2. Two coin-shaped identical cells (each holding about 1.7 ml) are permanently seated in an insulated compartment typically regulated 5–10 degrees above the environmental temperature to allow a cooling heat flow. Both cells are completely filled, the reference cell containing pure solvent and the measurement cell is filled with the solution of one partner of the binding reaction to be studied. The other reactant, usually prepared in 10–20-fold higher concentration is delivered from a syringe that is coaxially inserted through the long access tube. The tip of the syringe is deformed to a paddle to allow rapid mixing of the cell contents when the syringe device is rotated at a stirring rate of 200–400 min^{-1}. The reference cell is continuously heated to set a temperature difference of about 0.01 degrees over the nominal temperature of the insulating jacket. A similar electrical power heater is attached to the sample cell and is automatically regulated by a feedback mechanism to minimize the temperature difference ΔT between the cells. On injecting aliquots of several microliters from the syringe, the association of the binding partners produces a heat effect that raises or lowers the temperature in the sample cell. The deflection of temperature triggers the feedback regulator to adjust the electrical power needed to maintain identical temperatures in both cells. The change in the respective feedback current is the primary signal observed and corresponds to a heat pulse (heat production

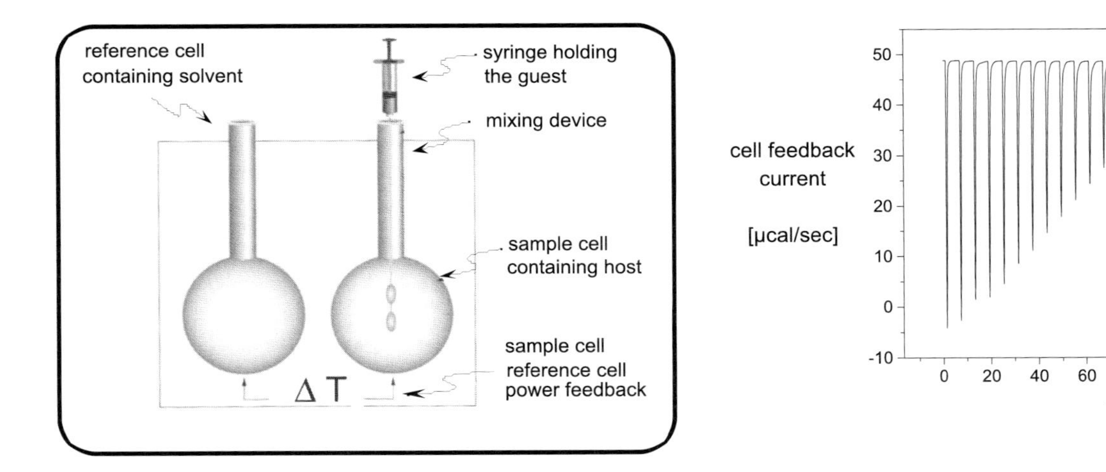

Fig. 3.2. The left panel schematically depicts the instrumental setup of a power compensation calorimeter. Both cells are completely filled, the reference cell with pure solvent, the sample cell with a solution of one of the host–guest partners (e.g. the host). On addition of microliter aliquots of the guest solution delivered from the computer-driven syringe a heat effect occurs that is counter-regulated by the cell feedback current to maintain ΔT at zero. The right panel illustrates the data output consisting in a number of heat pulses that decrease in magnitude following the progressive saturation of the host binding site by the incremental addition of the guest species.

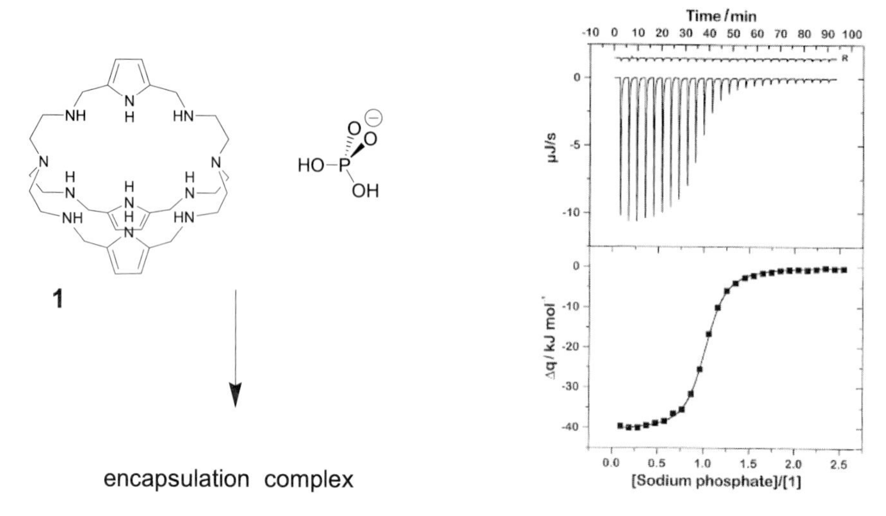

Fig. 3.3. ITC titration of 0.047 mM cryptand **1** in 50 mM MOPSO-buffer, pH 7.0 with 26 × 3 μl 2.0 mM sodium phosphate in the same buffer. The blank titration of the phosphate into plain buffer is shown at the top deliberately shifted by 1.5 μJ s^{-1} for better viewing. Reproduced with permission from E. Grell et al., *J. Therm. Anal. Cal.* **2004**, 77, 483.

over time). Integration with respect to time gives the energy that was traded on injecting the known amount of the reaction partner to the sample cell. If a series of injections is made, the compound in the cell is progressively converted to the molecular complex leading to diminishing heat effects as the association approaches completion. A typical output picture showing the exothermic encapsulation of phosphate by cryptand **1** in aqueous buffer at pH 7.0 is depicted in Fig. 3.3.

The upper panel shows downward directed pulses indicating the diminution of the feedback current necessary to keep a zero temperature difference to the reference cell as the heat from the exothermic association reaction makes up for the rest. The integration of the heat pulses when plotted versus the nominal molar ratio of the injected compound over the one contained in the cell yields a titration curve that exhibits a characteristic shape. In the case shown in Fig. 3.3 the sigmoidal appearance reflects the adequate choice of absolute concentration relations to allow the determination of the molar enthalpy ΔH^o from the extrapolated step height of the curve, the stoichiometry n of the binding process from the position of the inflection point along the molar ratio axis (sodium phosphate/**1**) and the affinity constant K_{assoc} from the slope in the inflection point. Modern calorimeters offer the advantage of having these quantities determined by software routines that use nonlinear curve fitting to find the most probable parameters describing the supramolecular association with regard to the specifics of the instrument (cell volume, volume displacement etc).

Of course, a decisive prerequisite in any meaningful evaluation of calorimetric data is the judicious choice of experimental conditions, the appropriate correction of the data with respect to ubiquitous nonspecific contributions like the heat of dilution or mixing and above all the adequate choice of a model representing the relevant processes in solution. Similar to many other data evaluations where several individual contributions convene to generate a singular output (as e.g. in kinetics) adherence of the experimental data to a certain model does not ultimately prove the model, but surely disproves all non-fitting alternatives. The various options are discussed further below. At this point we shall focus the attention on the production of good quality data that in the end will be all decisive on the success of interpretative attempts.

Paramount to the experimental setup is the purpose of the calorimetric titration as it dictates the selection of the dimensionless c-value (Eq. (3.7)) [1].

$$c = n \times [\mathrm{A}] K_{\mathrm{assoc}} \tag{3.7}$$

$[\mathrm{A}]$ = concentration of the titrate compound in the cell [M].

K_{assoc} = affinity constant $[\mathrm{M}^{-1}]$; n = stoichiometric factor.

In the case depicted in Fig. 3.3. The c-value amounts to 80 and falls well within the range of 5–500 that displays clear sigmoidal curvature and is best suited for the calculation of ΔH, K_{assoc} and n in a single experiment.

Balancing the c-value frequently is a game with bold restrictions emerging from instrumental sensitivity or unspecific interferences alike. If affinity is high (in artificial host–guest systems $K_{\mathrm{assoc}} > 10^6\ \mathrm{M}^{-1}$), the concentration required to place the c-value in the prospected range may be so small as to cause detection problems and insufficient signal-to-noise ratios, especially when the molar enthalpies are not great either ($|\Delta H| \sim 0$–$10\ \mathrm{kJ\ mol}^{-1}$). On the contrary, if affinities are quite limited, high concentrations of the interacting compounds are needed which may saturate the responsiveness of the instrument and bear the risk of covering the effect of interest by an overwhelming unspecific background response. In many cases, the problems at either borderline can be relieved by raising or lowering the temperature as modern calorimeters can be used between zero and 80 °C without special equipment. The substantial change in heat capacity ΔC_p endemic in supramolecular interactions may easily shift the enthalpy and Gibbs enthalpy into the desirable range.

In some instances, the precise estimation of the interaction enthalpy ΔH or the stoichiometry n rather than the affinity constant K_{assoc} is desired. Then, raising the c-value well over 1000 by an increase of the initial concentration is a beneficial option. The titration curve will then appear as a step (jump) function as in Fig. 3.4 because the titrant added in aliquots from the syringe will be totally converted to the complex in each addition until the reaction partner in the cell is consumed completely. The subsequent injections will only show the spurious heats of dilution and mixing and thus will end in parallel to the molar ratio axis. The jump

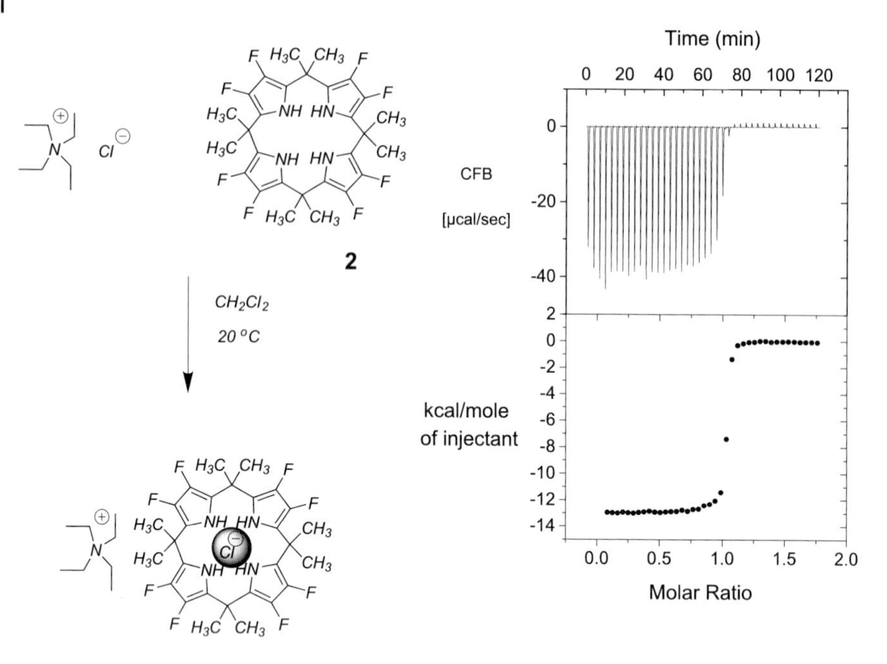

Fig. 3.4. Titration of fluoro-calixpyrrole **2** in dichloromethane (40 × 4 μl) into a 1.2 mM solution of tetraethylammonium chloride in the same solvent. The abrupt step appearance indicates a *c*-value of ca. 4000.

event marks the molar ratio of the components in the complex and the step height gives the molar enthalpy referring to the compound delivered from the syringe. This procedure is more accurate than the previous case because no extrapolation (fitting) is required.

Of course, these statements apply for the ideal case when the complex formation adheres to a fixed stoichiometry and no other noticeable processes occur simultaneously in solution. Regrettably, this situation is much less frequent than desired. Especially in artificial host–guest binding systems, which have not undergone evolutionary optimization, the observation of nonintegral stoichiometries is the rule rather than the exception. If impurities or compound degradation can be excluded as a cause (see below), nonintegral stoichiometry ratios always indicate a higher complexity in the system that can emerge from the participation of more complex species (higher-order complexation) or competing equilibria (e.g. ion pairing). The options for remediation from unresolvable interference then are quite limited. In addition to changing the temperature which in supramolecular systems is a powerful strategy to arrive at altered and potentially more fortuitous conditions, massive dilution to suppress effects from unspecific low-affinity binding has been successful in some cases. However, the heat pulses in very diluted solutions will lead to sensitivity problems as they approach the experimental noise level. Some relief can then be expected when covering the entire titration curve with only 7–10 injec-

Tab. 3.1. Optimal experimental setting of the molar ratio according to Eq. (3.8).

c [Eq. (3.7)]	R_m
1	20.0
5	7.3
10	5.3
50	3.2
100	2.7
500	1.9

tions. An illuminating discussion of the tuning of experimental parameters can be found in the literature [23].

$$R_m = 6.4/c^{0.2} + 13/c \tag{3.8}$$

Provided the stoichiometry n is secured by supplementary noncalorimetric evidence and, furthermore, is unperturbed by high concentrations of the interaction partners, an analysis of the statistical error allows some recommendations to be made for parameter adjustment to optimize the precision [24]. Hence, under these prerequisites (which admittedly seldom apply in artificial host–guest systems) the enthalpy ΔH and association constant K_{assoc} can be determined to less than 1% relative error in a broad range ($10 < K_{assoc} < 10^5$; valid for ΔH, too, when $K_{assoc} > 30$ M^{-1}) just following a simple recipe [23]: (i) Use no more than 10 injections of the titrant solution; (ii) set the final molar ratio R_m in accord with the empirically determined Eq. (3.8), but not smaller than 1.1 where c is as defined in Eq. (3.7); this requires a crude estimate of the binding constant K_{assoc}. Table 3.1 lists the digest of Eq. (3.8) revealing the range of the molar ratio R_m that needs to be covered. Obviously, the excess of one host–guest partner over the other is considerably greater than customary, in particular if low c-values cannot be avoided; (iii) The initial concentration of the titrate partner in the cell should be as large as possible, however, preventing c (Eq. (3.7)) from exceeding 1000. If the regime of low c-values ($c < 1$) cannot be avoided because of experimental restrictions, a fair accuracy in K_{assoc} and ΔH can still be achieved provided the interaction stoichiometry is known from another source [28]. In this case, the titration curve appears featureless in the limit of c almost as a straight line with little inclination and no inflection point (see Fig. 3.5 for an example). Finding the error minimum then can present a problem as the error hypersurface becomes rather flat and parameter fits easily diverge. A redefinition of the fit parameters can help in the latter instance, yet this requires a modification of the standard evaluation software and thus is unappealing to many experimentalists.

We need to emphasize at this point that the meticulous analysis of methodological errors is certainly mandatory to evaluate the potential of the instruments. Most

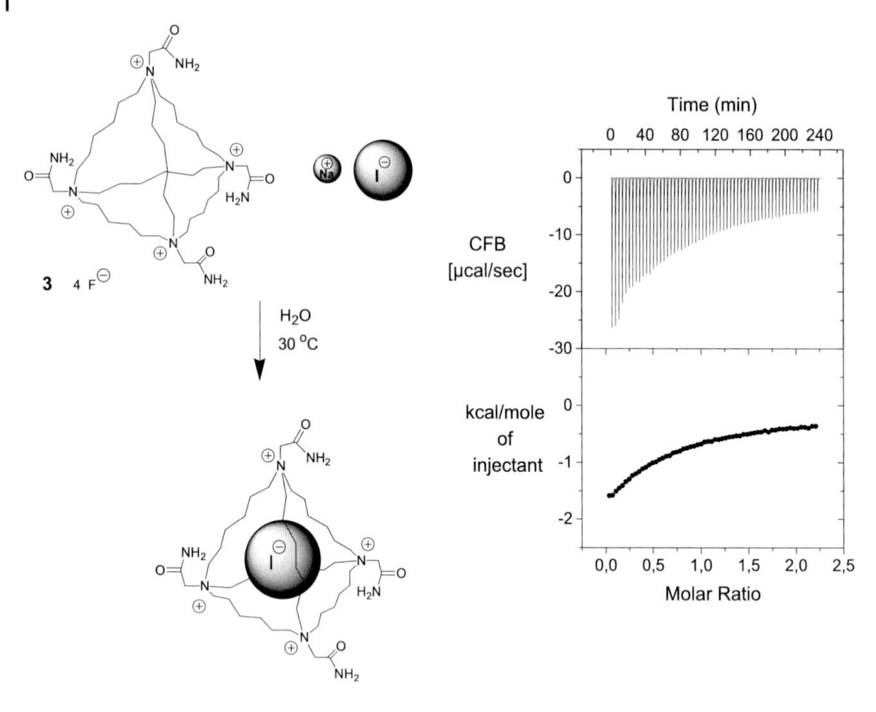

Fig. 3.5. ITC-titration of the macrotricyclic cavity host **3** (8.3 mM) with sodium iodide (0.1 M) in water at 303 K. The *c*-value (Eq. 3.7) in this case is around 1.

applications of the calorimetric method, however, do not require accuracy and precision to be pushed to the extreme limit, as other sources of error outmatch by far the methodological uncertainties. First in line of the factors that interfere with the reproducibility of calorimetric results is the purity of the compounds used. Owing to the ubiquity of heat effects even small deviations from a nominal composition of a compound may result in dramatic differences in the calorimetric output. A frequent problem of this kind in abiotic host–guest binding is the presence of solvent of crystallization. Ordinarily, this is not considered an impurity as it is often present in a fixed stoichiometric ratio and can be accounted for in elemental analyses and in the spectroscopic evaluation. In calorimetry, however, solvent of crystallization adds a heat contribution of unknown size which does not emerge from the interaction under study and thus tends to falsify the results. The problem is amplified by the polarity difference between the solvent of recrystallization and the one used in the supramolecular investigation. Thus, the worst case is met when compounds in a hydrated form are employed in nonpolar or aprotic solvents like chloroform, dichloromethane or acetone. Experience tells us that even in polar solvents like acetonitrile or dimethylsulfoxide (DMSO) the heat evolution on introduction of protic solvents like water is strongly nonlinear with concentration. Moreover, many host–guest interactions in organic solvents involve hydrogen bonding and eventually respond in a nonmonotonous heat output on a gradual increase of the water content at low concentrations ($<1\%$). Such effects are hard to correct urging utmost care in the handling and storage of the solvent to ensure exclusion of traces of humidity. For instance, the highest quality of commercial acetonitrile dried to <10 ppm of water content still contains ca. 1 mM of hydrogen bond donors that may interfere with host–guest binding studied, generally at very similar concentrations.

Of course, the same arguments named for the interference by solvent also hold for more adventitious impurities. Reproducible results in calorimetry in general mandate a higher purity of the compounds than is usual with methods based on some structural signal. As oily, honey-like compounds which are frequently the primary products of chromatographic purification very tenaciously hold back solvents or remnants from the stationary phase causing puzzling perturbations in the concrete measurements, recrystallized compounds are much preferred in calorimetric investigations. Even if all precautions are met and the host–guest systems behave like an ideal 1:1 stoichiometric interaction (the simplest case), the accuracy of calorimetric determinations very rarely reaches the high level claimed in many articles in the literature. Considering all the errors involved with respect to the purity of chemicals, but in particular in the handling of milligram/microliter quantities of the compounds to prepare the dilute solutions in a minimum volume of the costly high quality solvents, a realistic estimate of the reproducibility of the affinity constant in independent runs must be set at a factor of 3 corresponding to a $\Delta\Delta G$ of $\sim$3 kJ mol^{-1}. Differences larger than this threshold are deemed significant and deserve an interpretation. The uncertainty of an enthalpy as the directly measured quantity ordinarily is somewhat smaller at $\Delta\Delta H \sim$ 1–2 kJ mol^{-1} (depending on the c-value of the determination, see above) whilst the entropic component represented by $T\Delta\Delta S$ is more blurred at $\sim$6 kJ mol^{-1} as a result from its origin by subtraction according to the Gibbs–Helmholtz equation.

This setting of calorimetric significance is a conservative estimate that may give a guideline to assess the merit in achievement in the design of molecular hosts. On comparing two hosts, any discussion on the origin of differences in state function of associations are not likely to be physically meaningful, if they fall within these uncertainty limits.

In addition to the uncertainty introduced by experimental indetermination, the data evaluation itself has bearing on the correctness of the thermodynamic parameters. One can find an abundance of cases in the recent literature where all too naïve use of the fitting software provided with the calorimeters did indeed furnish numerical values for the state functions, yet inspection of the corresponding graph disclosed a rather poor fit to the pertinent data. The fitting algorithms have been optimized to rapidly converge on an error minimum in parameter space. However, there is no guarantee for reaching the global (lowest) one. It is a quite tedious, yet mandatory task to challenge the validity of the minimum parameters by the input of different initial values. Only numerous and very liberal variation of the starting parameters in a minimization calculation can generate confidence in a particular result.

Most commercial nonlinear regression software allows the simulation of the fitting graph from deliberately chosen numerical values of the target parameters. It is helpful and economical to find a set of starting parameters by trial and error or by systematic variation that generates a similar shape of the graph to fit the experimental data. The ultimate goal is to obtain their trustworthy representation. Subsequent minimization then most likely shrinks the statistical error without affecting the overall shape of the fit function.

In modern calorimeters, the extraction of the target molar enthalpy ΔH^o and Gibbs enthalpy ΔG^o from the experimental raw data after correction for ubiquitous heat contributions (dilution, mixing) employs theoretical binding models. The fitting process *a priori* supposes that no other factors significantly influence the heat output and all systematic errors have been accommodated. In practice, this premise is sometimes violated owing to e.g. baseline drift, temperature change, systematic imbalance of titrate and titrant solutions due to degradation or evaporation etc, hence aggravating conclusions on the validity of a certain binding model. In order to arrive at an exact and physically meaningful representation of the data, a straightforward approximation to account for all systematic deviations from ideality that are not already covered by the blank titration is to add or subtract a constant amount of heat. Consequently, the entire data set is offset with respect to the ΔH axis (y-translation) potentially with a dramatic improvement with respect to statistical error and convergence. Such a transformation by a constant shift is legitimate as the desired parameters depend on the shape of the titration curve (K_{assoc} on the slope in the inflection point and ΔH^o on the step height) rather than on its absolute position [25]. A perfect fit following such reasonable manipulation is a strong argument in favor of the validity of a particular binding model.

Occasionally, the participation of more than just one supramolecular process is obvious from the inspection of the heat profile. A fortuitous case is depicted in Fig. 3.6 describing the heat response observed on titrating α-cyclodextrin to (*R*)-camphor in water. Clearly, the pronounced minimum in the titration curve results from two binding processes of opposite enthalpy representing the first and second step in the assembly of a 2:1 stoichiometric host–guest complex. The switch from an endothermic to an exothermic heat output in the successive steps indicates the mutual interdependence of these sequential processes suggesting a 2-site sequential binding model. Such a theoretical model of higher order binding is implemented in the commercial evaluation software and features only four parameters (the molar enthalpy and microscopic association constant in each of the two steps) that need to be derived from the fitting procedure. An alternative model featuring two independent binding sites would additionally require two parameter values characterizing the stoichiometry. In general, one strives to fit the data with the least number of adjustable parameters which is equivalent to the selection of the simplest model that can represent the data set. There is no rational justification to this premise as it relates to the famous philosophical problem of "Occam's Razor". Nevertheless, it is a sensible means to restrict the possible model options.

In artificial host–guest systems, the range in Gibbs enthalpy is rather limited (K_{assoc} varies between 10 and 10^8 M^{-1} corresponding to ΔG^o of -6 to -40

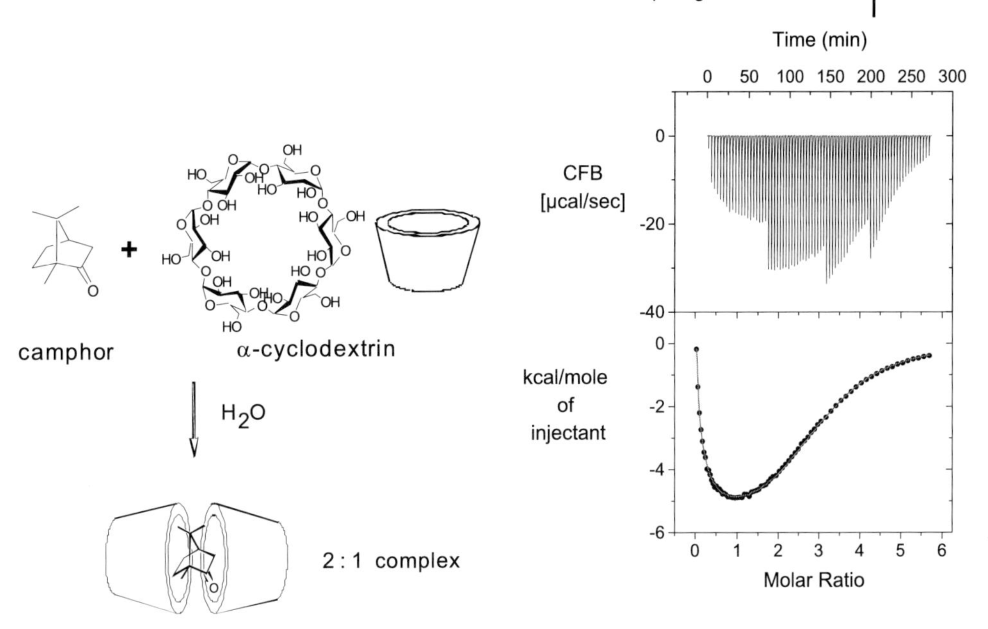

Fig. 3.6. ITC-titration of 2.1 mM camphor with 50 mM α-cyclodextrin in water at 303 K. The rugged profile of the heat pulses (top panel on the right) derives from the portionwise increase of the injection volume to cover a large span in molar ratio [26].

kJ mol^{-1}). In combination with the rather low specificity (discrimination against competing guests) frequently observed multiple simultaneous random sequence and sequential binding steps must be anticipated and, in fact, are quite common. The fundamental notion must be appreciated that mutual interactions of all components do indeed occur and the exclusive dominance of the interaction process under study is not automatically warranted. Hence, it is of prime importance to choose a model that reflects the available experimental data and allows their physically meaningful and verifiable extrapolation.

In some host–guest systems, none of the available theoretical binding models can account for the entire data range. The complexity in the number and interdependence of binding events may prevent the deduction of an all-encompassing coherent scheme. A typical example shown by a putatively primitive abiotic receptor is depicted in Fig. 3.7. The appearance of the heat output in this titration is well reproducible, but eludes immediate comprehension. In similar cases, the truncation of the range of system composition (a restriction on the molar ratio axis) may furnish a section that can be adapted to a more fundamental binding model. As the overlap of this part with the nonaccountable fraction of the entire set is unknown, the quantitative reliability of the parameters derived from partial fitting is at high risk.

Obtaining useful experimental data requires not only an optimal setting of variables, but also the recognition of systematic instrumental misbehavior. Some

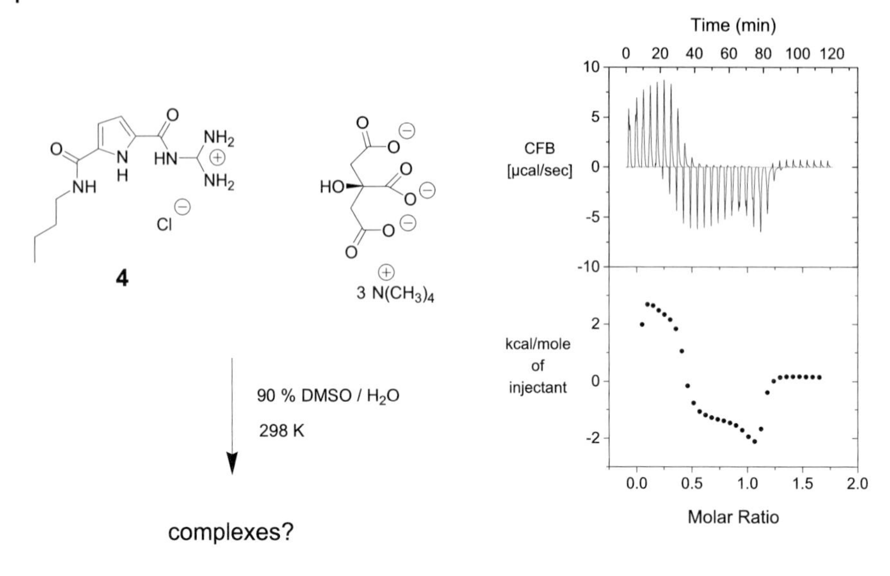

Fig. 3.7. ITC-titration of the guanidinium salt **4** (2.8 mM) with TMA citrate (20.5 mM) in 90% vol DMSO/water at 298 K.

typical examples of odd data output that could be traced to a distinct cause are collected in Fig. 3.8. As remediation from malfunction may be both frustrating and time consuming, the proper assignment of a strange data appearance to a probable cause seems all-important especially if the measurement cell is inaccessible due to its permanent fixation in the instrument. The consultation of the website of the manufacturer of the instrument may also prove helpful in the diagnosis of an unfamiliar response. The automated operation of modern calorimeters is highly beneficial in pinpointing the origin of experimental problems because mistakes in handling can largely be excluded. The "black box" design thus reduces errors in operational manipulation and fosters the reliability and comparability of calorimetric results through standard protocols.

3.4 Extending the Applicability

The necessity to adhere to a *c*-value ranging from 1 to 1000 (Eq. (3.7)) in order to enable the determination of the affinity constant, the molar enthalpy and the stoichiometry from a single experiment effectively limits K_{assoc} to the span between 100 and 10^8 M^{-1}. Though this range certainly suffices to cover the vast majority of artificial host–guest binding, special circumstances in abiotic receptor chemistry or the adoption of biological systems may require the expansion at both the low end and the high end of the affinity range. The limitations primarily arise from instrumental restraints: At the minute concentrations of host and guest partners necessary to meet the *c*-value criterion at very high affinities the heat output in each

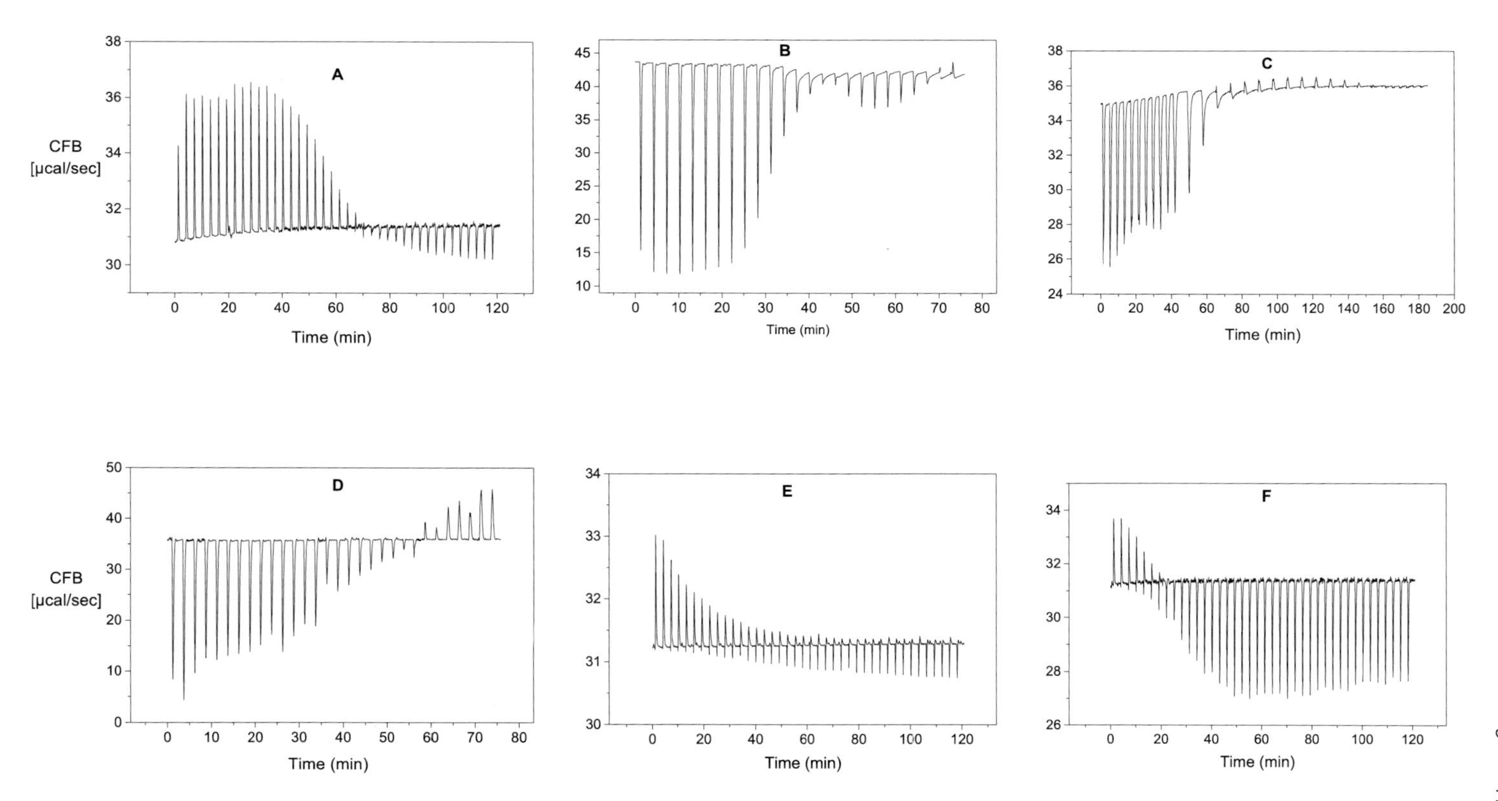

Fig. 3.8. Malfunctional ITC-output: A: high noise level due to insufficient degassing; B: additional heat output emerging from a subsequent chemical reaction (degradation); C: considerably delayed equilibration owing to aggregation phenomena; D: variable pulse height caused by bubbles in the syringe; E: extra heat pulses caused by worn-out syringe plunger grip; F: high noise level due to incorrect syringe height setting.

titration step will produce an insufficient signal-to-noise ratio. In contrast, low affinities will require high substrate concentrations and consequently will give massive heat effects that can saturate the responsiveness of the calorimeter. At much lower substrate concentrations that would remedy the heat overflow or which are dictated by the solubility limits of the substrate the heat pulse will not change much with the molar ratio of the compounds thwarting the elucidation of the thermodynamic parameters. The easy way out from this dilemma is to tune the composition of the system. First choice is the arrangement of the solvation conditions to shift the affinity into the measurement window. In biological supramolecular systems polar solvents like DMSO, dimethylformamide, methanol or acetone are frequently used to enhance the solubility of organic substrates or dim down the hydrophobicity and micellation tendency in aqueous buffers. Such additions very effectively diminish the affinity of organic substrates in water, however, at the expense of an increased complexity arising from a much less transparent interplay of effects on the molecular level. Since solvation energies at least in more polar solvents by far outmatch most bilateral host–guest interactions a change in solvation conditions will tend to cover the more subtle host–guest interactions of interest. For the comprehension of the molecular scene in general, the gain in the reliability of the measurement derived from solvent manipulation is often annihilated by less lucid and harder to interpret calorimetric results rendering this approach less recommendable.

Another alternative makes use of the temperature dependence of supramolecular interactions. Most (but by no means all) host–guest associations feature negative enthalpies opening the option to transpose, for instance, a too high affinity into the measurement window by mere temperature elevation. Assuming typical values ($\Delta H = -40$ kJ mol^{-1}, $\Delta C_p = -200$ J K^{-1} mol^{-1}) for simple host–guest systems one can diminish the affinity constant by a factor of 13 on raising the temperature by 50 degrees from ambient temperatures. For the more stable artificial systems such a temperature enhancement may be tolerable, yet it is clear that no dramatic extension in the accessible range of affinities can be expected from temperature variation.

In many supramolecular associations, protons are traded simultaneously with some external buffer species as acidic or basic sites of the host or guest become intimately involved in mutual binding. The heat of deprotonation/reprotonation depends on the extent and the number of protons exchanged with the buffer and of the molar enthalpies of all sites participating in the process. Thus, the observable heat ΔH_{obs} is composed of the supramolecular fraction ΔH_{supr} supplemented by the portions attributable to the intrinsic change in the protonation state of the binding partners (ΔH_{prot}) and the corresponding release or uptake of the number n of protons by the buffer ($n \times \Delta H_{buff}$)

$$\Delta H_{obs} = \Delta H_{supr} + \Delta H_{prot} + n \times \Delta H_{buff}$$

It is obvious from this relation that the heat output ΔH_{obs} can be tuned by the prudent choice of the buffer compound which is not immediately involved in the binding reaction. The molar enthalpies of protonation of about 100 buffer species have been compiled [27] and for a certain pH value (e.g. pH 7) can vary from moderately negative (e.g. $\Delta H^\circ(HSO_3^-$, p$K = 7.2) = -3.65$ kJ mol^{-1}) or weakly positive ($\Delta H^\circ(H_2PO_4^-$, p$K = 7.2) = +3.6$ kJ mol^{-1}) to strongly positive (e.g. ΔH°(HEPES, p$K = 7.56) = +20.4$ kJ mol^{-1}; ΔH°(Tris, p$K = 8.0) = +47.5$ kJ mol^{-1}). Hence, a weakly exothermic association involving a proton loss to the buffer that is hardly recognizable in an aqueous Tris buffer will produce a substantial heat signal at the same concentrations, however, measured in a phosphate buffer. Eventually this effect may allow the concentrations to be reduced in order to meet the *c*-value requirement necessary for affinity determination. From the systematic variation of buffers with differing molar enthalpies of ionization the number of protons transferred on binding can be deduced giving access to the structural interpretation of the binding process.

The back-and-forth shuttling of protons between the host–guest system and the buffer may also prove advantageous from a standpoint of experimental convenience. Polyions like proteins frequently possess their isoelectric points (pI values) in the pH region of interest in the binding experiments and hence show very limited solubility hampering the due experiments. A change of pH then can greatly improve solubility and ease the experimental handling. Yet, the data obtained at the altered pH value have to be translated back to the original pH conditions of interest. A thermodynamic cycle as depicted in Fig. 3.9 provides the computational basis for this transformation. In addition to the measurement of host–guest binding at the altered pH, the changes in the state functions of the individual mo-

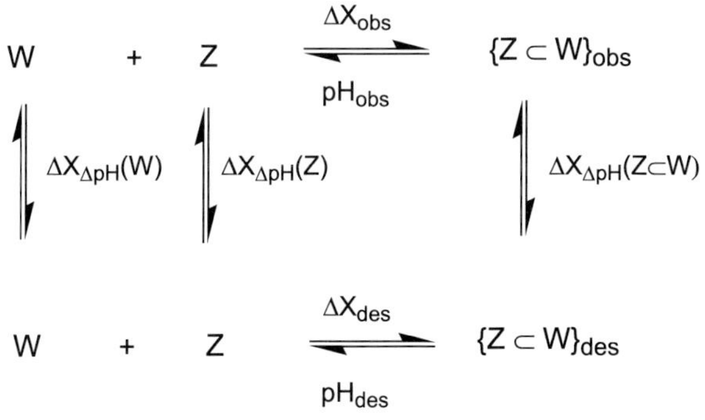

$$W + Z \underset{pH_{obs}}{\overset{\Delta X_{obs}}{\rightleftharpoons}} \{Z \subset W\}_{obs}$$

$\Delta X_{\Delta pH}(W)$ $\Delta X_{\Delta pH}(Z)$ $\Delta X_{\Delta pH}(Z \subset W)$

$$W + Z \underset{pH_{des}}{\overset{\Delta X_{des}}{\rightleftharpoons}} \{Z \subset W\}_{des}$$

$$\Delta X_{des} = \Delta X_{obs} + \Delta X\Delta_{pH}(Z \subset W) - (\Delta X_{\Delta pH}(Z) + \Delta X_{\Delta pH}(W))$$

Fig. 3.9. Thermodynamic cycle of the interaction of host W with guest Z at two different pH values, the experimentally accessible one pH_{obs} and the desired one pH_{des}. The cycle connects the various desired state functions (ΔX_{des}) to the directly observed ones in the supramolecular association (ΔX_{obs}) and the changes occuring with the individual species ($\Delta X_{\Delta pH}$). These changes have to be determined separately.

lecular species happening with the shift in pH have to be determined. The change in the desired state functions ΔX_{des} under the experimentally inaccessible condition then results from an additive correction of the observable parameter ΔX_{obs} by the difference of these state functions on the right and left hand sides in Fig. 3.9.

The linkage of chemical processes in thermodynamic cycles arising from the definition of the state functions (the state of a system is independent on the way on which it was reached) is a great virtue that can immediately be exploited by calorimetry. In this sense, making use of the protonation properties is just a special case in the much broader field of competition analysis [25]. In fact, any molecular species that is expelled or taken up by the host as a result of binding, the guest of interest may serve as a probe in calorimetric determinations. Prominent examples include the binding or release of metal cations from proteins (e.g. Ca^{2+}, Mg^{2+}) which can be linked to their complexation by complexones (ethylenediaminetetraacetic acid (EDTA) or similar).

A special case, which allows the accessible affinity range to be considerably extended, builds upon the direct expulsion of a guest from its binding site within a host compound by another guest. Such mandatory displacement presumes total absence of mixed ternary complexes comprising, for example, both guests simultaneously bound to the host. Absence of mixed complexes is readily conceived for evolutionary optimized biological hosts, but meets considerable risk with the less specific abiotic analogs. In principle, a competition assay to determine high affinity binding of a guest Z to a host W that is beyond direct titration involves at least two individual measurements which should be supplemented by an exploratory verification titration (see Fig. 3.11 below for an example). (i) From the direct titration of Z into a solution of W only the molar enthalpy ΔH^o can be determined, because at all concentrations of W necessary to read a signal the c-value (Eq. (3.7)) exceeds 1000 and the calorimetric titration curve appears as an abrupt step; (ii) A second guest Y exhibiting moderate affinity is titrated into the host furnishing the association constant $K_{assoc}(Y)$ and the respective molar enthalpy $\Delta H^o(Y)$ from this run; (iii) In a third titration the guest Z is titrated into a solution of the host containing a fixed concentration of guest Y (this requires for instance, to load the same concentration of Y into the solution of Z; most evaluation software, however, can also account for the changing concentration of Y if a solution of the pure guest Z is titrated in). The apparent association constant and molar enthalpy can be digested by a nonlinear regression procedure using the parameters obtained in the previous titration with plain Y to yield the desired state functions of the high affinity process. An implementation of this competition scheme into the commercial regression software Origin is available from its author.

An illustrative example of this technique is shown in Fig. 3.10. A similar scheme for the determination of low-affinity association where neither the binding constant nor the molar enthalpy can be deduced directly (Fig. 3.11). Again, a competitor of intermediate affinity is employed to serve as a relay system. Determining the state functions in the absence and presence of a known concentration of the weak affinity guest gives two data sets. When they are fed to the regression software a

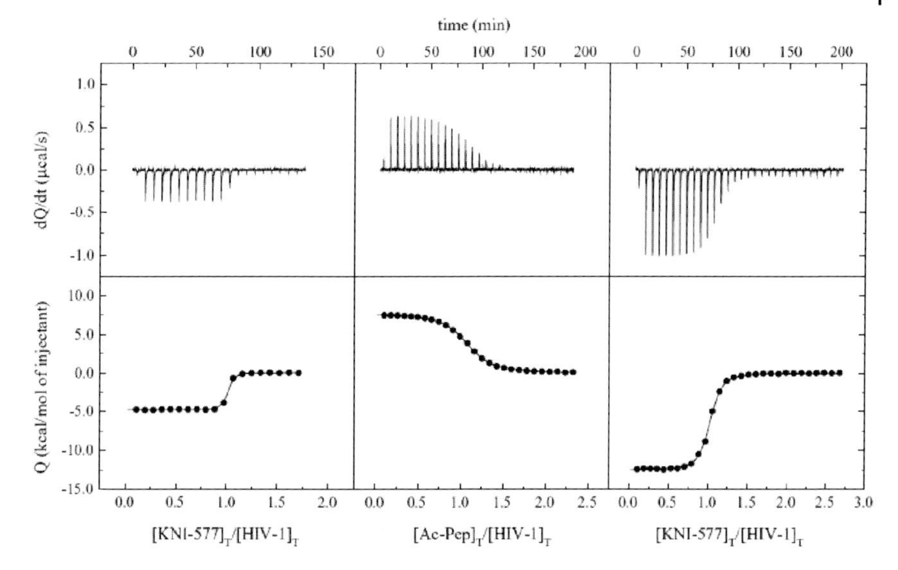

Fig. 3.10. Determining very high binding affinity. Experimental scheme aimed at determining the thermodynamic parameters for the binding of KNI-577 to HIV-1-protease. Three titrations are shown: (left) KNI-577 binding to HIV-1, (center) acetyl-pepstatin binding to HIV-1 and (right) KNI-577 binding to HIV-1 prebound to acetyl-pepstatin. In the first titration, the only binding parameter estimated is: $\Delta H = -19.7 \pm 0.8$ kJ mol^{-1}, because affinity cannot be reliably determined. In the second titration, the estimated parameters are: $K_{assoc} = 2.3 \pm 0.8 \times 10^6$ M^{-1}, $\Delta H = +24.5 \pm 0.8$ kJ mol^{-1}. Applying the displacement analysis to the third titration yields the following parameters for the binding of KNI-577 to HIV-1 protease: $K_{assoc} = 4.9 \pm 1.2 \times 10^9$ M^{-1}, $\Delta H = -19.7 \pm 0.8$ kJ mol^{-1}. Reproduced with permission from A. Velázquez-Campoy et al. Biophys. Chem. **2005**, 115, 115–124.

best estimate of the energetics of the weakly binding guest is furnished. By virtue of adjuvant displacement binding the usable range of ITC is considerably expanded and now allows the elucidation of the binding energetics in supramolecular associations over the entire span of affinities of potential utility in abiotic applications at a tolerable sacrifice in accuracy. Finding suitable competitors to promote this technique in a concrete case may nevertheless prove to present a respectable hurdle.

3.5 Perspectives

Isothermal titration calorimetry has already reached a state of the art that makes it the method of choice for most solution phase binding studies. The development of the technique does not yet seem to be exhausted, as the sensitivity of the instru-

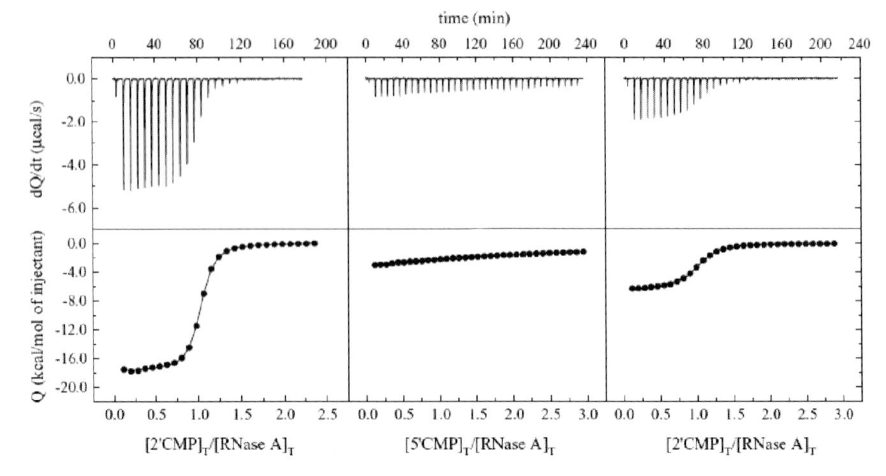

Fig. 3.11. Determining very low binding affinity. Experimental scheme aimed at determining the thermodynamic parameters for the binding of 5′CMP to RNase A. In the case of the CMP inhibitors of RNase A, a change in the position of the hydroxyl group from 2′ to 5′ position causes a dramatic decrease in the strength of the interaction, thus, illustrating the effect of minor modifications in the ligand on the thermodynamic parameters of binding. Three titrations are shown: (left) 2′CMP binding to RNase A, (center) 5′CMP binding to RNase A and (right) 2′CMP binding to RNase A pre-bound to 5′CMP. In the first titration, the binding parameters are: $K_{assoc} = 8.3 \pm 0.8 \times 10^6$ M^{-1} and $\Delta H = -74.5 \pm 0.4$ kJ mol^{-1}. It is not possible to obtain a reliable value for either the affinity or the enthalpy of binding from the second titration. Applying the displacement analysis to the third titration yields the following thermodynamic parameters for the binding of 5VCMP to RNase A: $K_{assoc} = 4.2 \times 10^3$ M^{-1}, $\Delta H = -68.2$ kJ mol^{-1}. Reproduced with permission from A. Velásquez-Campoy et al. Biophys.Chem. **2005**, 115, 115–124.

ments to investigate even smaller quantities of material and the robotic automation to handle large sample numbers increase with each calorimeter generation. Improvements on the methodological side run in parallel and feature great time savings in operation in addition to enhanced power for discrimination between alternative binding models [29] or obtaining core thermodynamic parameters like the heat capacity [30]. New applications that also encompass kinetic parameters as in the characterization of biocatalysts [31, 32] or the analysis of low molecular weight host–guest dynamics [33] open wide areas that will surely profit from a relatively fast, nondestructive, noninvasive, accurate and experimentally convenient to use instrumental method like ITC.

Acknowledgement

The continuous and generous support from Hans-Fischer-Gesellschaft, Munich, is gratefully acknowledged.

References

1 Wiseman, T. S., Williston, S., Brandts, J. F., Lin, L. N. Rapid measurement of binding constants and heats of binding using a new titration calorimeter; *Anal. Biochem.* **1989**, 179, 131–137.

2 Hallén, D., Wadsö, I. Solution microcalorimetry; *Pure Appl. Chem.* **1989**, 61, 123–132.

3 Wadsö, I. Trends in isothermal microcalorimetry; *Chem. Soc. Rev.* **1997**, 79–86.

4 Velázquez Campoy, A., Freire, E., ITC in the post-genomic era ...? Priceless; *Biophys. Chem.* **2005**, 115, 115–125.

5 Lewis, E. A., Murphy, K. P., Isothermal titration calorimetry; *Meth. Molec. Biol.* **2005**, 305, 1–15.

6 Weber, P. C., Salemme, F. R. Applications of calorimetric methods to drug discovery and the study of protein interactions; *Curr. Opin. Struct. Biol.* **2003**, 13, 115–121.

7 Ladbury, J. Isothermal titration calorimetry: application to structure-based drug design; *Thermochim. Acta* **2001**, 380, 209–215.

8 Holdgate, G. A. Making cool drugs hot; The use of isothermal titration calorimetry as a tool to study binding energetics; *BioTechniques* **2001**, 31, 164–184.

9 Jelesarov, I., Bosshard, H. R. Isothermal titration calorimetry and differential scanning calorimetry as complementary tools to investigate the energetics of biomolecular recognition; *J. Mol. Recognit.* **1999**, 12, 3–18.

10 Cooper, A. Thermodynamic analysis of biomolecular interactions; *Curr. Opin. Chem. Biol.* **1999**, 3, 557–563.

11 Schmidtchen, F. P. Calorimetry: An indispensible tool in the design of molecular hosts; in: *Macrocyclic Chemistry, Current Trends and Future Perspectives*, (K. Gloe ed.), Springer Dordrecht, **2005**, pp. 291–302.

12 van Regenmörtel, M. H. V. Molecular recognition in the post-reductionist era; *J. Mol. Recognit.* **1999**, 12, 1–2.

13 Velázquez Campoy, A., Freire, E. Incorporating target heterogeneity in drug design; *J. Cell. Biochem.* **2001**, supp. 37, 82–88.

14 Moyer, B. A., Bonnesen, P. V., Custelcean, R., Delmau, L. H., Hay, B. P. Strategies for using host–guest chemistry in the extractive separations of ionic guests; *Kemija u Industriji* **2005**, 54, 65–87.

15 a) Ababou, A., Ladbury, J. E. Survey of the year 2004: literature on applications of isothermal titration calorimetry; *J. Mol. Recognit.* **2006**, 19, 79–89; b) Cliff, M. J., Gutierrez, A., Ladbury, J. E., A survey of the year 2003: literature on applications of isothermal titration calorimetry; *J. Mol. Recognit.* **2004**, 17, 513–523; c) Cliff, M., Ladbury, J. E., A survey of the year 2002 literature on applications of isothermal titration calorimetry; *J. Mol. Recognit.* **2003** 16, 383–391.

16 cf. Corbellini, F., van Leeuwen, F. W. B., Bejleveld, H., Hooiman, H., Spek, A. L., Verboom, W., Crego-Calama, M., Reinhoudt, D. N. Multiple ionic interactions for noncovalent synthesis of molecular capsules in polar solvents; *New J. Chem.* **2005**, 29, 243–248.

17 Cooper, A., Johnson, C. M., Lakey, J. H., Nöllmann, M., Heat does not come in different colours: entropy–enthalpy compensation, free energy windows, quantum confinement, pressure perturbationcalorimetry, solvation and the multiple causes of heat capacity effects in biomolecular interactions; *Biophys. Chem.* **2001**, 93, 215–230.

18 Kolb, H. C., Finn, M. G., Sharpless, K. B., Click Chemistry: diverse chemical function from a few good reactions; *Angew. Chem. Int. Ed.* **2001**, 40, 2004–2021.

19 Dunitz, J. D., Win some, lose some: Enthalpy–entropy compensation in

weak intermolecular interactions; *Chem & Biol.* **1995**, 2, 709–712.

20 Exner, O., Entropy–enthalpy compensation and anticompensation: solvation and ligand binding; *Chem. Commun.* **2000**, 1655–1656.

21 Tellinghuisen, J., Van't Hoff analysis of Ko(T): How good ... or bad?; *Biophys. Chem.* **2006**, 120, 114–120.

22 Horn, J. R., Russel, D., Lewis, E. A., Murphy, K. P. van't Hoff and calorimetric enthalpies from isothermal titration calorimetry: are there significant discrepancies?; *Biochemistry* **2001**, 40, 1774–1778.

23 Tellinghuisen, J., Optimizing experimental parameters in isothermal titration calorimetry; *J. Phys. Chem. B* **2005**, 109, 20027–20035.

24 a) Tellinghuisen, J., A study of statistical error in isothermal titration calorimetry; *Anal. Biochem.* **2003**, 321, 79–88; b) Tellinghuisen, J., Statistical error in isothermal titration calorimetry: Variance function estimation from generalized least squares; *Anal. Biochem.* **2005**, 343, 106–115.

25 Sigurskjold, B. W. Exact analysis of competition ligand binding by displacement isothermal titration calorimetry; *Anal. Biochem.* **2000**, 277, 260–266.

26 Schmidtchen, F. P. The anatomy of the energetics of molecular recognition by calorimetry: chiral discrimination of camphor by α-cyclodextrin; *Chem. Eur. J.* **2002**, 8, 3522–3529.

27 Goldberg, R. N., Kishore, N., Lennen, R. N., Thermodynamic quantities for the ionization reactions of buffers; *J. Phys. Chem. Ref. Data* **2002**, 31, 231–370.

28 Turnbull, W. B., Daranas, A. H., On the value of c: Can low affinity systems be studied by isothermal titration calorimetry? *J. Am. Chem. Soc.* **2003**, 125, 14859–14866.

29 Markova, N., Hallén, D., The development of a continuous isothermal titration calorimetric method for equilibrium studies; *Anal. Biochem.* **2004**, 331, 77–88.

30 Chavelas, E. A., Zubillaga, R. A., Pulido, N. O., García-Hernández, E., Multithermal titration calorimetry: a rapid method to determine binding heat capacities; *Biophys. Chem.* **2006**, 120, 10–14.

31 Bianconi, M. L., Titration calorimetry as a tool to determine thermodynamic and kinetic parameters of enzymes; in: *Biocalorimetry* (eds. Ladbury, J. E., Doyle, M. L.) Wiley, Chichester, **2004**, p. 175–185.

32 O'Brien, R., Haq, I., Applications of biocalorimetry: binding, stability and enzyme kinetics; in: *Biocalorimetry* (eds. Ladbury, J. E., Doyle, M. L.) Wiley, Chichester, **2004**, p. 3–34.

33 Cooper, A., Nutley, M., MacLean, E. J., Cameron, K., Fielding, L., Mestres, J., Palin, R. Mutual induced fit in cyclodextrin–rocuronium complexes; *Org. Biomolec. Chem.* **2005**, 3, 1863–1871.

4
Extraction Methods

Holger Stephan, Stefanie Juran, Bianca Antonioli, Kerstin Gloe and Karsten Gloe

4.1
Introduction

Molecular recognition, binding and transport of different chemical species represents an aspect of supramolecular chemistry that has relevance to a number of areas that include biochemical processes, analytical techniques, recycling and environmental processes as well as aspects of catalysis and medicine. Over the years, a large number of both efficient and selective receptors for cations, anions, salts and zwitterions based on different architectures and binding modes have been developed and studied [1–4].

Among the manifold experimental techniques employed for the application of such receptors has been the investigation of the distribution of species between two immiscible solutions, normally an aqueous and an organic phase, under the influence of the receptor in the organic phase. Such a procedure has often allowed characterization of the receptor's complexation behavior towards individual species as well as enabling an evaluation of its suitability for species monitoring, separation and/or concentration; especially with respect to possible analytical applications as well as for use in extraction and membrane transport processes. For example, in early studies Pedersen adapted the extraction technique for the determination of the relative cation binding strengths of the first reported synthetic crown compounds [5], while Cram et al. exploited the widely used picrate extraction method enabling a quantitative treatment of cation extraction [6].

The state of the art with respect to supramolecular extraction has been summarized in several recent reviews from different perspectives, including in terms of the underlying fundamental chemical processes involved as well as with respect to possible applications in research and industry [7–15].

It is the intention of this chapter to present an overview of the potential of the extraction technique for characterizing the binding strengths, selectivity, speciation and phase transfer behavior of supramolecular receptors towards selected cations, anions, salts and zwitterions in aqueous–organic, two-phase systems. The role of

Analytical Methods in Supramolecular Chemistry. Edited by Christoph Schalley

ISBN: 978-3-527-31505-5

various factors that influence the behavior of such (often) complicated multicomponent systems will be discussed from a supramolecular chemistry point of view. Finally, a general discussion of the experimental technique employed along with promising potential applications in different fields will be presented.

4.2 The Extraction Technique

The general principle of solvent or (liquid–liquid) extraction is illustrated in Fig. 4.1. The first step, the extraction (a), starts with two solutions, aqueous and organic, which are immiscible and where the aqueous phase contains the solute and the organic phase the complexing agent – the so-called extractant. The phase transfer of the solute is driven by the complex formation between the solute and the extractant and is normally accomplished by intensive mixing of the two phases. After phase separation, the winning of the separated solute and the regeneration of the extractant occurs from the organic phase in a second step by means of a back extraction (b) [16].

Different analytical methods such as atomic absorption, UV/vis or inductively coupled plasma spectroscopy are routinely used for the determination of solute concentrations mainly in the aqueous phase. In contrast, the radiotracer technique allows a rapid and straight forward measurement of these concentrations in the two phases [17]. A further advantage of this last method is its high detection sensitivity which allows the use of very small amounts of ligand for the individual experiments. This can be especially helpful when only a few milligrams of the complexation reagent are available as is often the case when novel supramolecular systems are involved. A multitude of useful radionuclides are available, allowing extraction studies involving ionic or molecular species of relevance to different fields that range from the biochemical and medical areas, through analytical monitoring, material winning and recycling as well as environmental protection processes.

In the author's laboratories the radio-tracer, micro-extraction technique shown in Fig. 4.2 is employed for extraction studies. In brief, the procedure employed usually involves 500 μl of each phase (organic by transferpettor, aqueous by transferpette) being transferred into 2-cm^3 polypropylene microcentrifuge tubes. Normally, a small quantity of the radionuclide corresponding to the species of interest (cationic, anionic or molecular) is added to the aqueous phase. Intensive mixing of the two phases is achieved by mechanical shaking (overhead, thermomixer). All samples are centrifuged after extraction is complete and the concentration of the species of interest is determined by means of radioactivity measurements using a NaI(Tl)-γ-scintillation counter or a α/β-liquid-scintillation counter, depending on the radiation of the radionuclide employed. Processing of the experimental data provides the distribution ratio D_{solute}, which is the key parameter for describing the distribution behavior of the solute between the two phases under the experi-

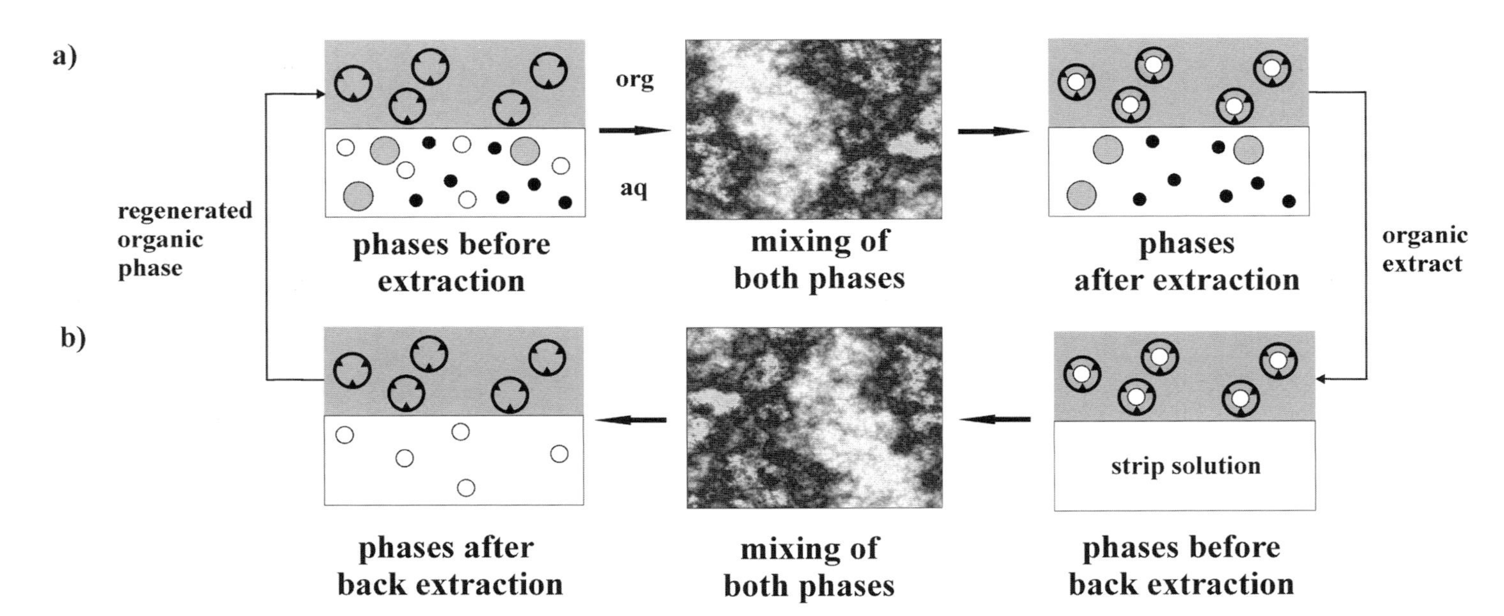

Fig. 4.1. General scheme of an extraction process showing (a) the extraction and (b) the back extraction step.

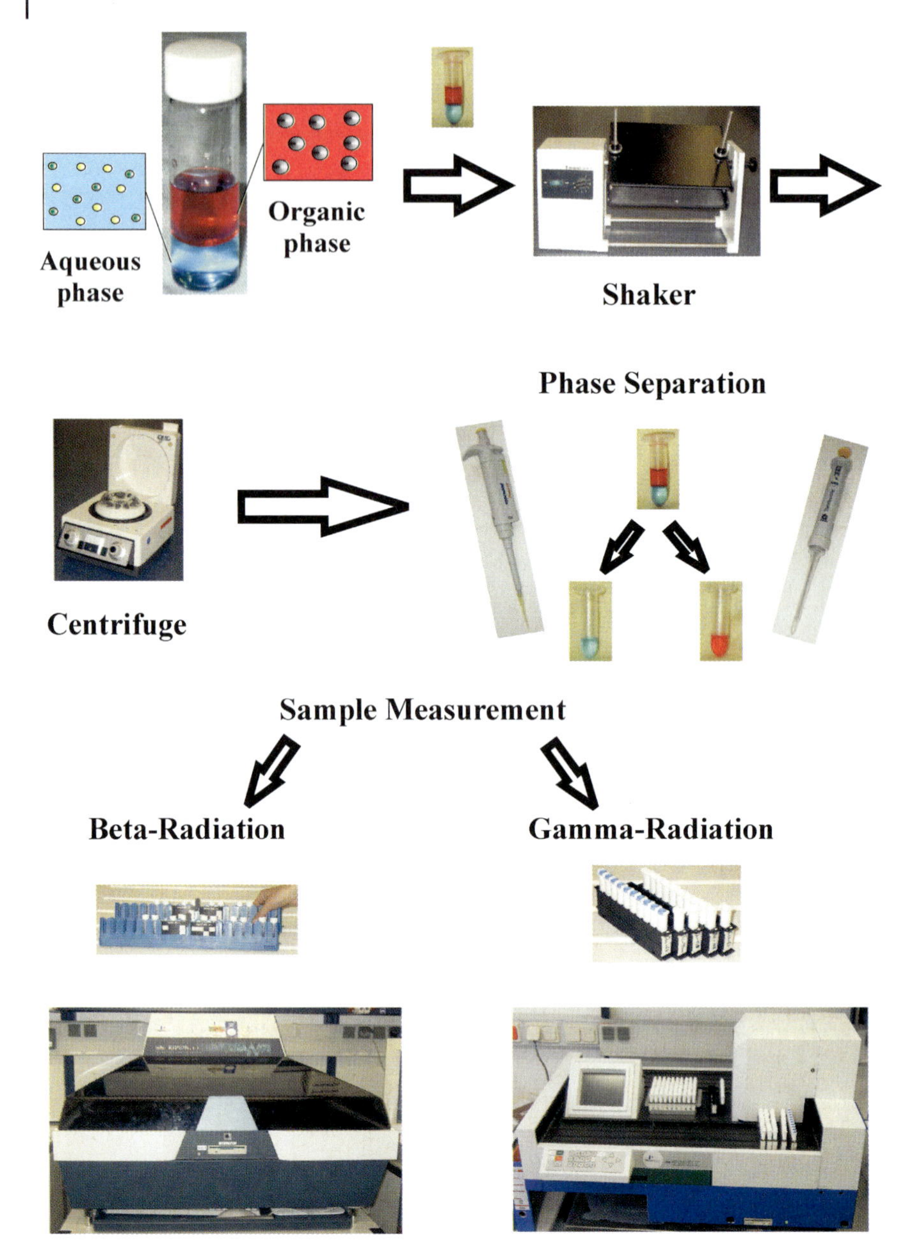

Fig. 4.2. Experimental procedure for extraction experiments using radiotracer technique.

mental conditions employed. D_{solute} represents the quotient of the analytical concentrations of the solute in the two phases and is defined by Eq. (4.1):

$$D_{solute} = \frac{[solute]_{(org)}}{[solute]_{(aq)}} \tag{4.1}$$

4.3
The Technical Process

Generally, reactive extraction processes represent efficient (and smart) technologies for the separation and concentration of ionic or molecular species in solution and are frequently used in both research and industry [16, 18, 19]. The industrial adaptation of the two-step extraction scheme discussed above leads to a closed extraction circuit of the type shown in Fig. 4.3.

The essential properties of a suitable extractant for an industrial application are especially related to its distinctive (selective) affinity for the substrate together with the need for reversible binding and a rapid reaction rate in both directions. Chemical stability and high lipophilicity of the complexing agent and the extracted complex are also requirements. Last but not least, the price of the extractant can be an important consideration concerning its use.

The currently applied extraction systems are mainly based on classical extractants, as for example "cation exchanging" acids or chelating agents, solvating ketones, ethers or esters for metal ion extraction and amines or quaternary ammonium salts for anion extraction [16, 18, 19].

Generally, a motivation for research in the area has been a clear need to develop more efficient and selective complexation reagents for various species for use in diverse applications. In many instances this reflects a need to separate very low concentrations of toxic and/or valuable species from large volumes of solution – often

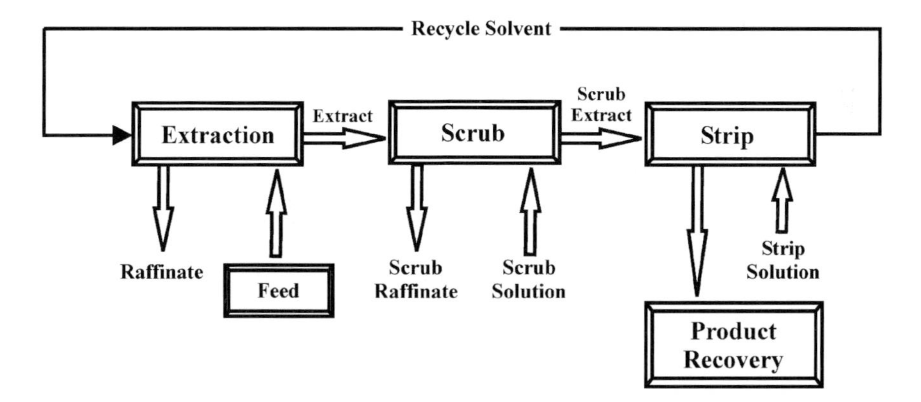

Fig. 4.3. Typical flowsheet of an industrial solvent extraction circuit.

present in difficult matrices. The use of supramolecular strategies offers considerable promise for solving such complicated problems as well as for controlling extraction selectivity [12–15].

4.4
The Extraction Equilibrium

Complete understanding of the extraction process in a heterogeneous two-phase extraction system is often inhibited by the changing nature and number of the individual interactions between the various species present in the two phases. Beside direct host–guest interaction, which can be controlled by the ligand design (at least in part), a key feature of importance is the role of the solvent. Solvation and desolvation processes of both host and guest can strongly influence the phase transfer of a given species. Hence the thermodynamics of an extraction reaction can be favored both enthalpically and/or entropically to various degrees that depend upon the conditions employed [3, 7].

Usually the description of the phase transfer behavior is presented in a simplified form based on the overall extraction equilibrium. This is illustrated schematically for a simple ion-pair extraction of a metal ion by a neutral macrocyclic extractant in Fig. 4.4.

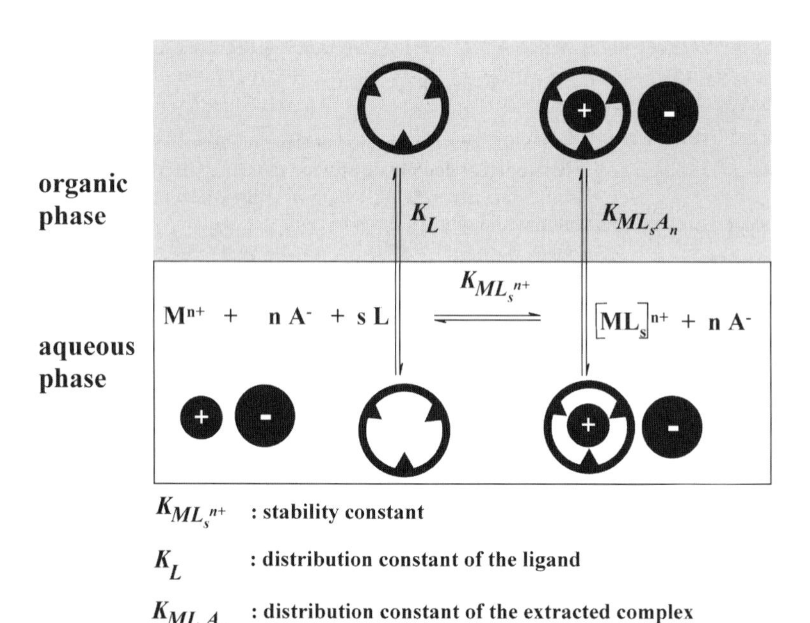

Fig. 4.4. Schematic representation of the extraction of cations by macrocycles.

The related extraction equilibrium is represented by Eq. (4.2) and the accompanying extraction constant K_{ex} by Eq. (4.3). The three individual equilibria for complex formation and partitioning of ligand and metal complex are given by:

$$M^{n+}{}_{(aq)} + nA^{-}{}_{(aq)} + sL_{(org)} \rightleftarrows ML_sA_{n(org)} \tag{4.2}$$

$$K_{ex} = K_{ML_S^{n+}} \cdot K_{ML_sA_n} \cdot K_L^{-S} \tag{4.3}$$

Equation (4.3) shows a direct proportionality of K_{ex} with the stability constant $K_{ML_S^{n+}}$ and the distribution constant of the extracted complex $K_{ML_SA_n}$. In contrast the distribution constant of the ligand K_L possesses a reverse dependence on K_{ex}.

Provided constant ionic strength of both phases is maintained (adjusted by salt or acid addition to the aqueous phase and ligand excess in the organic phase) the extraction constant can be formulated as given by Eq. (4.4):

$$K_{ex} = \frac{[ML_SA_n]_{(org)}}{[M^{n+}]_{(aq)} \cdot [A^-]^n_{(aq)} \cdot [L]^S_{(org)}} \tag{4.4}$$

Introducing the experimentally easily accessible distribution ratio D_M [(Eq. (4.1)] in Eq. (4.4) gives Eq. (4.5):

$$\log D_M = \log K_{ex} + n \log[A^-]_{(aq)} + s \log[L]_{(org)} \tag{4.5}$$

In the case when only one complex forming in the organic phase, it is possible to determine the stoichiometric coefficients n and s on the basis of two independent experimental data sets by a simple graphical slope analysis. This is illustrated in Fig. 4.5 showing the corresponding Log D_M – Log $c_{receptor}$ and Log D_M – Log $c_{picrate}$ diagrams for the extraction of barium picrate with a neutral hexameric amido-substituted *all*-homocalixarene **1**. The slope analysis clearly indicates the formation of a 1:1:2-complex of composition $(\mathbf{1} \subset \text{Ba}) \cdot \text{Pic}_{2(org)}$ (number of ligands bound in the extracted complex, $s = 1$; number of picrate anions bound in the complex, $n = 2$) [20].

However, because diverse other interactions in the aqueous and the organic phases are possible, the extraction equilibrium is often more complicated that may appear at first. In many cases the simultaneous formation of different complexes is observed. As a consequence a mathematical treatment of the extraction equilibrium is needed in such cases to obtain the stoichiometric coefficients. Computer programs using nonlinear regression methods allow the determination of the composition of the complexes extracted as well as the calculation of thermodynamic extraction constants [21–24]. In this way, one may also obtain further information about the presence of additional phenomena in the extraction system, such as ligand aggregation, oligomerization and/or solvation of the species extracted, and even about activity effects in the aqueous and organic phases.

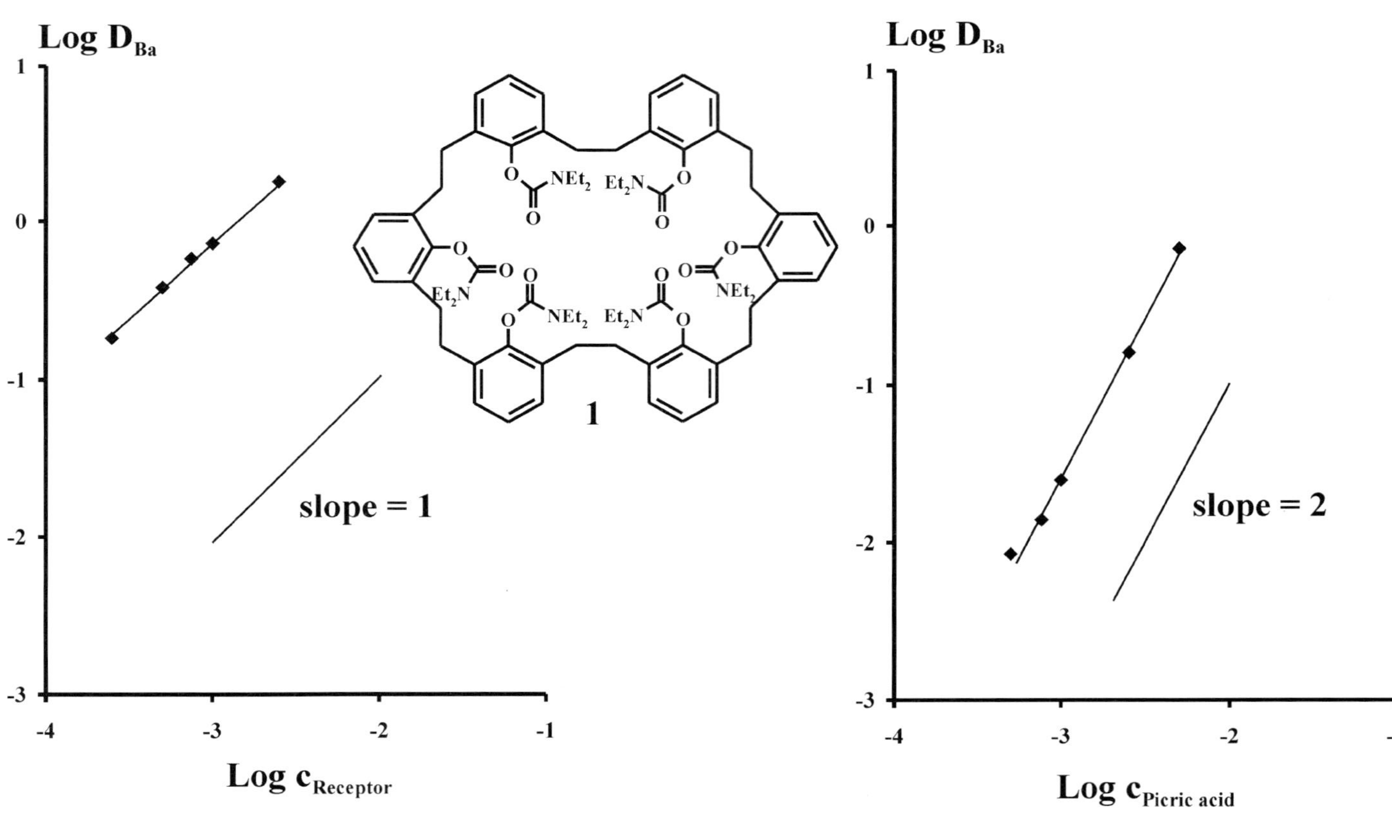

Fig. 4.5. Extraction of barium picrate with the all-homocalixarene **1** in $CHCl_3$.

4.5
Principles of Supramolecular Extraction

Nowadays the progress of supramolecular chemistry offers interesting possibilities to control selectivity and efficiency of an extraction process and to achieve a ligand system tailored for the specific binding and transport of a selected species. The basis for this is the application of the concept of molecular recognition. The key for a successful design of such a synthetic complexing system lies in the choice of appropriate binding sites. These may give rise to relatively strong coordinate bond formation or strong electrostatic interactions through to weaker hydrogen bonding and π-interactions – each being incorporated in an architecture which is complementary to a substrate of interest. Furthermore, a pronounced lipophilicity of the extracted complex is an important requirement for efficient transport from the aqueous into organic phase to occur.

Schematics illustrating these basic principles of supramolecular extraction of cations and anions are presented in Figs. 4.6 and 4.7. In all cases the supramolecular receptors are characterized by multifunctional binding modes. These are normally basic donor functions for cations or hydrogen bond donors for anions in combination with oppositely-charged entities and are incorporated in open-chain, macrocyclic or macrobicyclic architectures such that a complementary coordination pattern for the ion of interest is achieved [1–4]. Compared with the diversity of cation extraction systems so far reported, the possibilities for constructing anion systems are more limited [7–15]. This reflects the intrinsic nature of anions themselves; namely, their large size (and resulting lower charge density) together with their often strong hydration and, in particular instances, also their pronounced acid–base properties. The relative behavior of different anions often reflects the well-known Hofmeister series which ranks the various anions with respect to their "natural" lipophilicity order [25]. Considering the prime importance of the hydrophilicity/hydrophobicity balance for the phase transfer of anions, a frequent goal of supramolecular extraction studies is to design systems that will overcome the Hofmeister order.

Figure 4.6 gives an overview of the principles underlying single ion (cation or anion) extraction using neutral or charged ligands. For aqueous–organic phase transfer charge neutrality needs to be maintained. Thus a neutral host system, as present in the approaches **I**, **II** and **V**, inevitably requires co-transport of a counterion together with the charged supramolecular complex into the organic phase and efficient extraction of either a cation or an anion will depend strongly on the lipophilicity of this counterion. In contrast, this is not a requirement in the case of the cation- or anion-exchange approaches **IV**, **III** and **VI**, where the extraction is accompanied by a counter-transport of equally-charged ions, H^+ or X^-. The approach **IVb** illustrates a synergistic cation-exchange system based on pre-organization of an optimum metal ion extractant assembly using a self-organization process between an azacrown compound and a carboxylic acid. Examples **V** and **VI** are modifications of the approaches given by **I** and **II**, respectively, involving synergistic effects that are

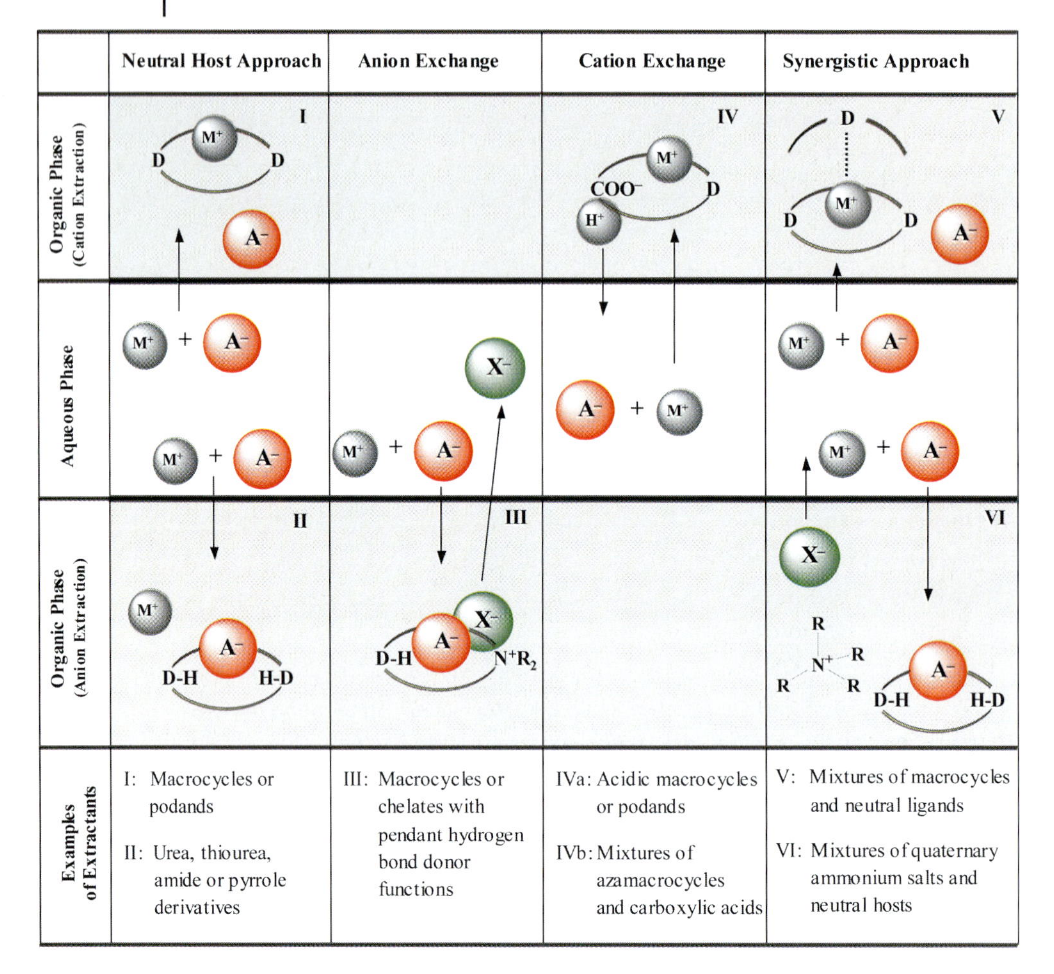

Fig. 4.6. Principles of supramolecular extraction of single ions.

generated through the introduction of a second complexing agent in the organic phase.

Figure 4.7 illustrates three supramolecular extraction principles that are concerned with the simultaneous extraction of cations and anions (**VII**, **VIII** and **IX**). Whereas in case of the dual-host approach **VII** two ligands are present to bind each ion type, while the approach given by **VIII** is based on the presence of two different binding sites in one heteroditopic ligand molecule. Here application of the principle illustrated by **VII** may require less synthetic effort because it is based on the use of single complexing agents that are often readily available. In contrast, **VIII** may lead to enhanced extraction selectivity and efficiency through the presence of direct electrostatic interaction (contact ion pairing) between the bound cation and

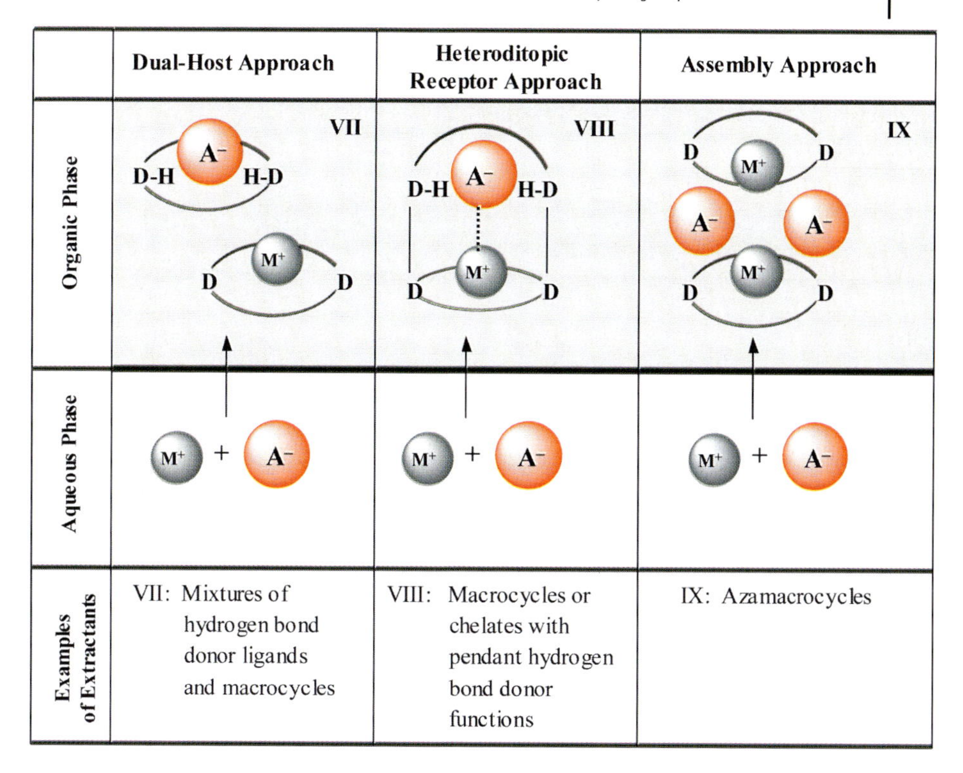

Fig. 4.7. Principles of simultaneous supramolecular extraction of cations and anions.

anion. Furthermore, specific self-organization processes involving defined metal binding sites may favor the effective phase transfer of formed assemblies incorporating metal salts as shown in the approach illustrated by **IX**.

4.6 Examples of Supramolecular Extraction

Crown compounds may be considered to be the first examples of supramolecular hosts that were able to extract metal ions selectively from aqueous into organic media. The advent of crowns triggered a revolution in liquid–liquid separations, particularly in analytical chemistry, because of their unique complexation behavior. Generally the tailoring of the macrocyclic cavity size allows the complexation and extraction of metal ions in a size-selective manner. For example, amongst the alkali metal ions 18-crown-6 (cavity radius: 1.33–1.43 Å) forms its strongest complex with potassium (ion radius: 1.38 Å) in water. This crown also yields the highest extraction of potassium amongst the alkali metal picrates, with the extractabilities of the 1:1:1 complexes falling in the order $K^+ > Rb^+ > Cs^+ > Na^+$ [26].

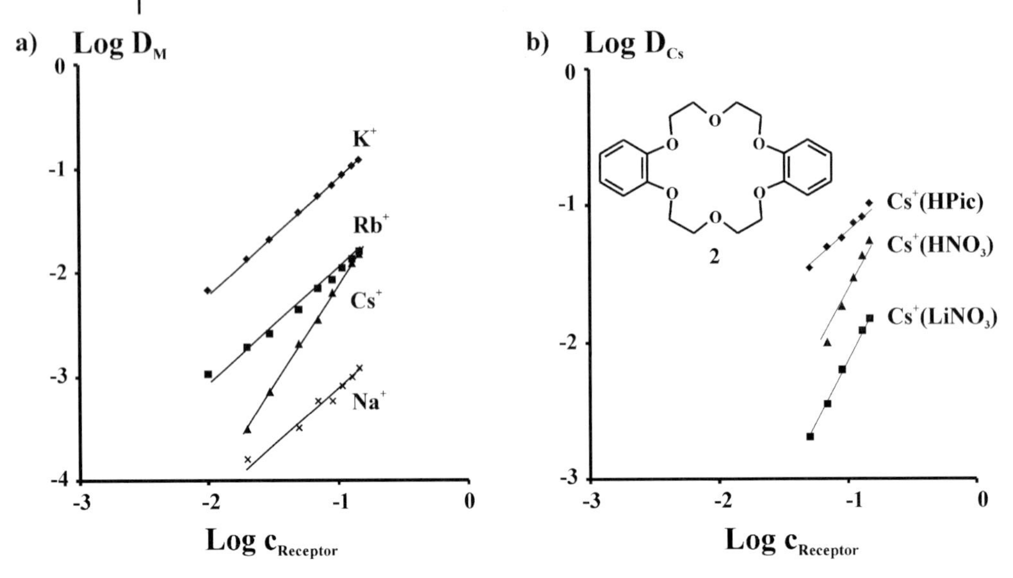

Fig. 4.8. Extraction of a) alkali metal nitrates and b) cesium picrate and nitrate with dibenzo-18-crown-6 (**2**) in $CHCl_3$.

It must be stressed that factors such as the hydration (or solvation) of the metal ion and anion effects on the extracted complex often make it difficult to predict the order of extractability for such systems. Such factors may even influence the stoichiometry of the extracted species. Thus, the simple "match of the metal to the whole concept" is only of limited utility. For example, potassium, rubidium and sodium nitrates are extracted in the presence of dibenzo-18-crown-6 (**2**) as 1:1:1 complexes. On the other hand, cesium forms a 1:2:1 sandwich complex with this crown (metal:crown:nitrate) in the organic phase and this affects the extraction order for the above metal ions, with the order being dependent on ligand concentration. In contrast, for picrate as the anion the composition of the extracted cesium complex is 1:1:1 (Fig. 4.8) [27].

Since the introduction of the crown ethers, strategies and methods have been developed that have resulted in the design and synthesis of an enormous variety (and complexity) of other supramolecular receptors that act as efficient and selective extraction reagents [1–4, 26]. For example, the cryptophane receptor **3**, featuring three *endo*-carboxylic acid groups and exhibiting a high state of pre-organization, is one such system that enforces 1:1 complex formation with (hard) alkali, alkaline earth and rare earth metal ions (see Fig. 4.9; only alkaline earth metal behavior shown). In this case molecular modeling calculations point to the octahedral arrangement of these metal ions inside the cavity bound by the carboxylate groups [28].

The use of ligands that are conformationally reinforced and have the appropriate number, nature and arrangement of their donor atoms for a particular metal ion,

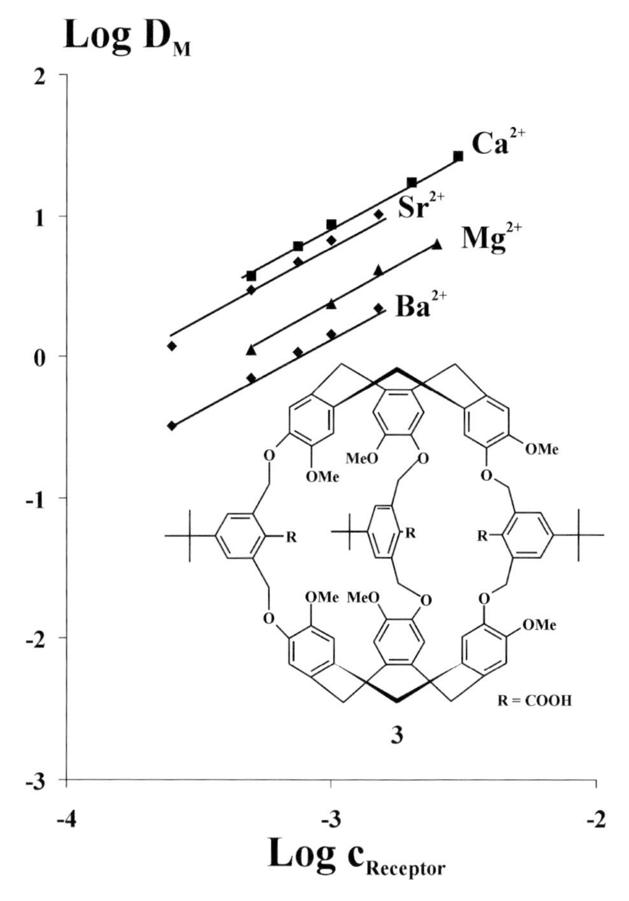

Fig. 4.9. Extraction of alkaline earth metal picrates with cryptophane **3** in $CHCl_3$.

may lead to peak selectivity in extraction. Three ligand examples of this concept are illustrated in Fig. 4.10.

The "crowned" calix[4]arene **4** bridged by an aliphatic chain of five carbon atoms spanning two opposite *para* positions shows exceptionally high Cs^+ extraction selectivity over sodium, calcium, strontium and barium [29]. However, the addition of the lipophilic picrate anion is necessary to transport the cesium ion into the organic phase. Using the calixarene biscrown **5** it is possible to extract the high-activity fission product ^{137}Cs with high selectivity from high-level nuclear waste solutions [30]. This will likely lead to its use being the first example of a large-scale application of a supramolecular extractant.

The introduction of intermediate and soft donor atoms like nitrogen and sulfur into a ligand architecture leads to a stronger preference for softer metal ions, such as the transition and precious metals. For example, the macrobicyclic tris(pyridine) cage **6** was shown to transport Ag^+ with high selectivity into an organic phase [31];

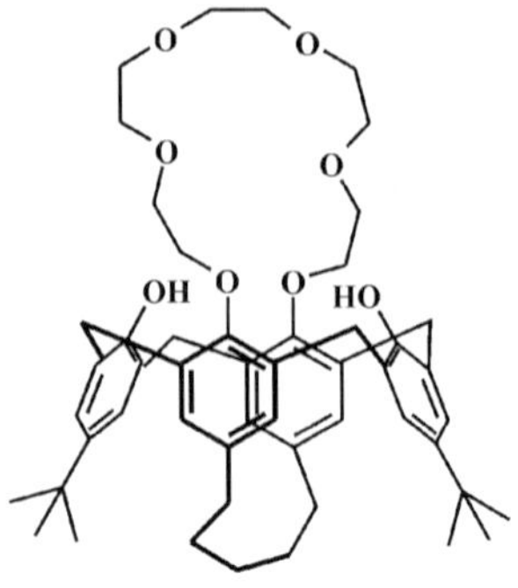

Fig. 4.10. Supramolecular extractants **4**, **5** and **6** showing peak selectivity in extraction.

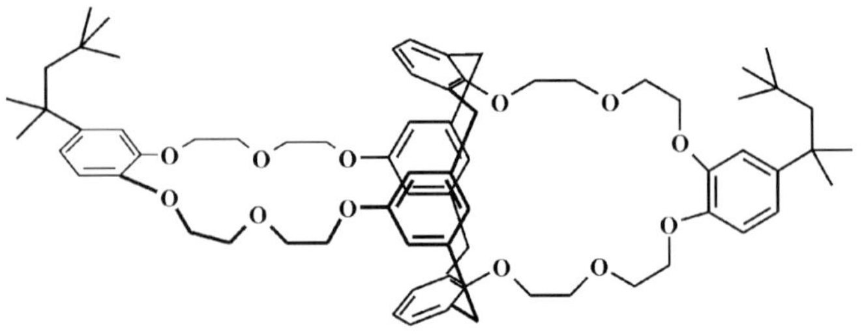

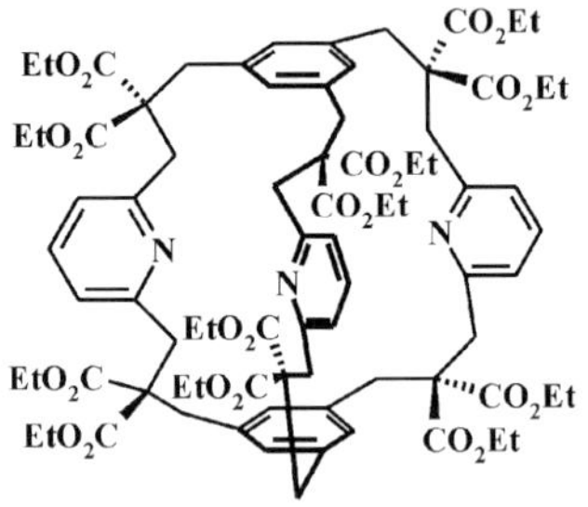

the univalent and bivalent metal ions Na^+, Tl^+, Cu^+, Ca^{2+}, Cu^{2+}, Zn^{2+} and Hg^{2+} were all retained in the aqueous phase.

Over many years there has been continuing strong interest in the chemistry of polyazacompounds of different structure (open-chain, macrocyclic or macrobicyclic) because such ligands are able to bind and extract both metal cations [32, 33] and anionic species [34, 35], depending on the pH of the aqueous solution.

In their protonated form polyamines are often attractive receptors for a variety of anions, as exemplified by many examples now known in nature, industry and model systems [34, 36]. The characteristic extraction properties of a range of such ligands toward halide, pertechnetate and perrhenate ions are strongly influenced by the ligand's protonation behavior and lipophilicity. Extraction therefore shows a pH-dependence, as well as being influenced by all manner of structural factors [35, 37, 38]. Some results for the perrhenate extraction by three different substituted tripodal tetraamines **7a–c** and two structure-related octaamino cryptands **7d**,**e** (Fig. 4.11) are shown in Fig. 4.12 [35, 37].

The open-chain tripodal ligands, **7a** and **7c**, yield remarkable extraction of ReO_4^- at pH 7.4. From among the amino cryptands investigated, only ligand **7d** gives similar extraction to that observed for **7a** and **7b**, whereas **7e** yields a comparable

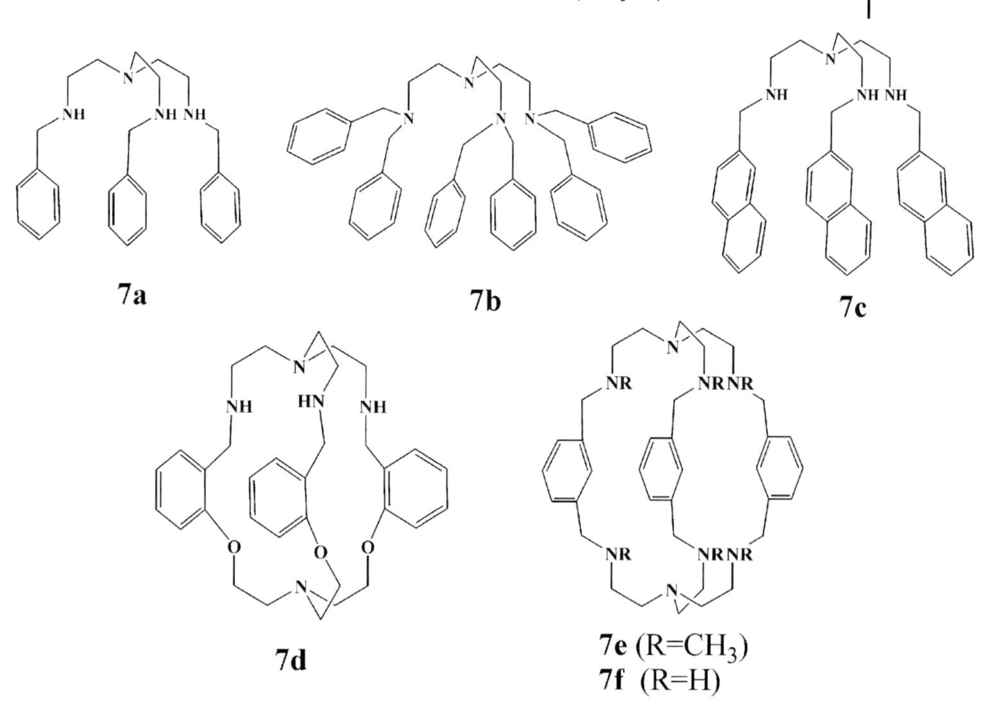

Fig. 4.11. Tripodal and macrobicyclic polyamines **7a–f** for anion extraction.

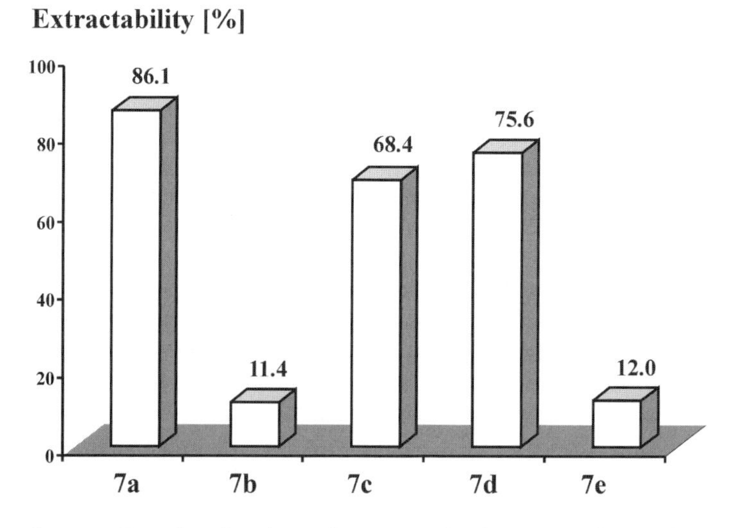

Fig. 4.12. Extraction of perrhenate by protonated polyamines **7a–f** in $CHCl_3$.

extraction to **7b**. These results are obviously a consequence of differences in the basicity of the secondary and tertiary amines and hence also reflect the nature of the resulting protonation equilibria. With respect to the observed exclusive transfer of mono- or diprotonated polyamine species into the organic phase, the use of ligands with a lower number of amine nitrogen groups are advantageous in view of their higher lipophilicity.

It is interesting to note that the structures of isolated crystalline anion complexes of the tripodal system **7a** are in each case characterized by a pronounced spreading of the three podand arms caused by repulsion of the positively-charged amine functions [38, 39]. This effect will thus tend to hinder anion chelation. The structure of $[H_4\mathbf{7a}]^{4+}\cdot(ReO_4)^-\cdot 3Cl^-$ shown in Fig. 4.13 [39] is characterized by a sandwich-like arrangement. The specific structural features of this complex reflect concerted (different) binding patterns of the cation with the two anions. The dimeric sandwich-like structure is characterized by strong charge-assisted N–H···O(Re) and weak C–H···O(Re) hydrogen bonds involving the large tetrahedral perrhenate. The presence of the latter aids optimization of the distances between the positively-charged ammonium ions. The smaller spherical chloride anions assist in the formation of a 2D network by participating in N–H···Cl and C–H···Cl hydrogen bonds. Finally additional aromatic edge-to-face C–H···π interactions between aromatic substituents of different layers lead to the observed 3D arrangement. Figure 4.13 also shows the structure of a perrhenate complex of the hexaprotonated octaamino cryptand **7e** that has one cavity included and five exo-cavity bound perrhenate anions [35]. Again the main interactions involve both strong and weak hydrogen bonds that link adjacent species to form an extended network. In contrast to the previous structure, this structure is stabilized by the additional binding of five water molecules. Provided a similar arrangement is maintained in solution, amongst other things the presence of the water molecules may influence the lipophilicity of the system resulting in a lower extraction efficiency. In the case of lowly-protonated octa-amino cryptands only complexes with exo-cavity bound anions could be identified. From the extraction and structural data it can be concluded that cavity inclusion is unfavorable for the phase transfer process of anions with such ligands, because inclusion normally requires a high protonation state of the ligand.

In other studies it was found that lipophilic urea-functionalized dendrimers of the poly(propylene amine) type **8** (such as shown in Fig. 4.14) are also strong extractants for the tetrahedral ions TcO_4^- and ReO_4^-. The extraction of the two anions increases with decreasing pH confirming the need for protonation of the amine functions in these molecules if efficient phase transfer is to take place [40].

An interesting example of pronounced anti-Hofmeister behavior was detected for sulfate extraction using a ditopic guanidinium compound of type **9b** [41, 42]. As shown in Fig. 4.15, **9b** extracts the strongly hydrophilic SO_4^{2-} ion with higher efficiency than Br^- or HPO_4^{2-}. The fact that the monotopic guanidinium derivative **9a** does not give comparable behavior points to **9b** yielding an optimum structural arrangement with the sulfate anion. The latter is in accord with the result of molecular modeling calculations of the structure of the corresponding 1:1 complex

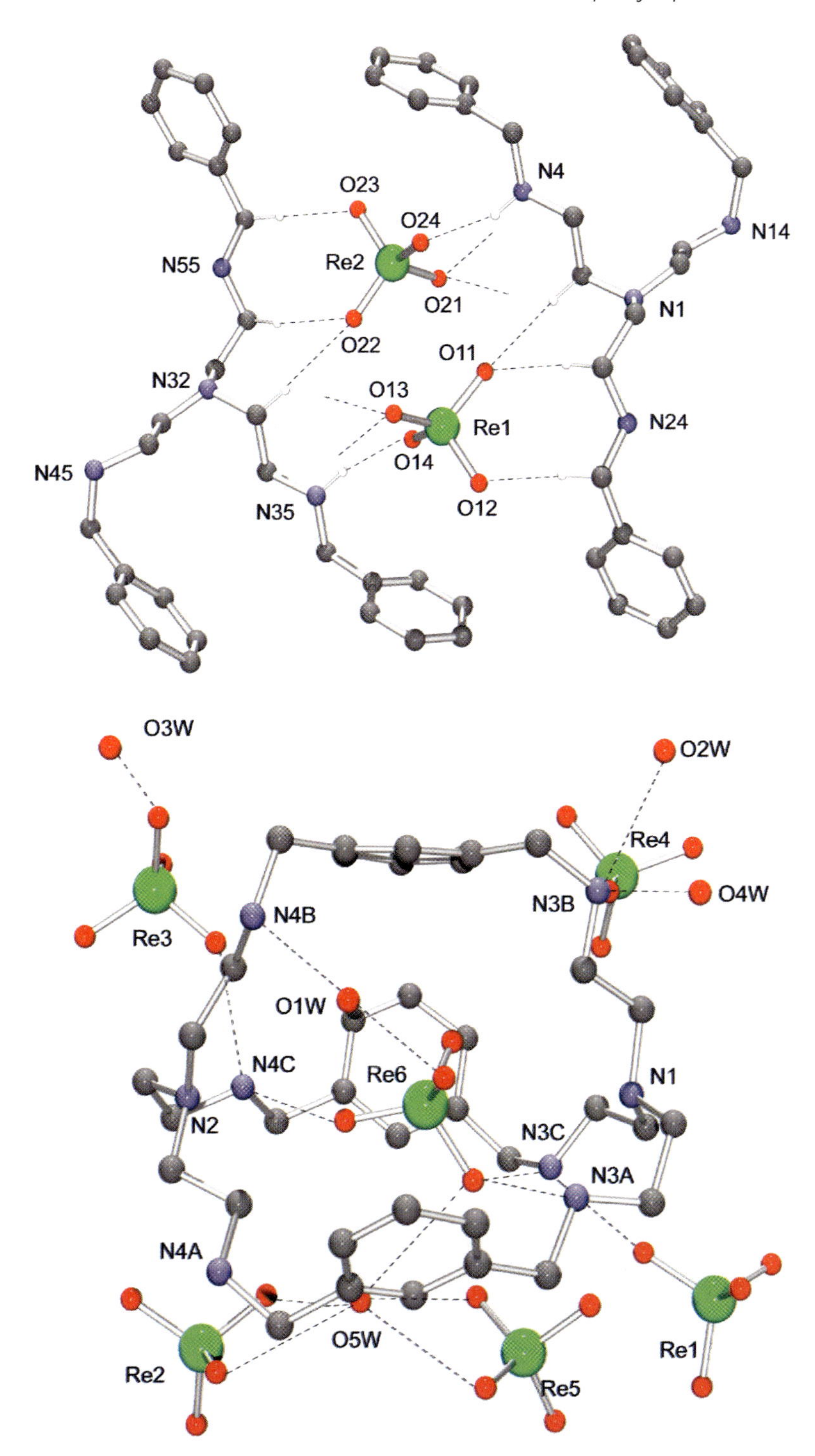

Fig. 4.13. Crystal structures of the perrhenate complexes $[H_4\mathbf{7a}]^{4+}\cdot(ReO_4)^-\cdot 3Cl^-$ and $[H_6\mathbf{7f} \subset ReO_4]^{5+}\cdot 5(ReO_4)^-\cdot 5H_2O$.

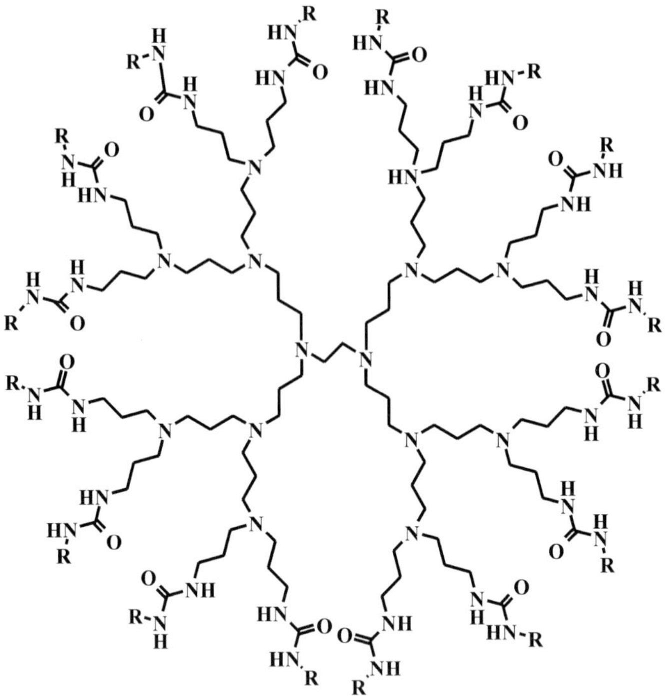

8 ($R = C_6H_{13}, C_{12}H_{25}$)

Fig. 4.14. Urea-functionalized dendrimer of the poly(propylene amine) type **8**.

[41] (Fig. 4.15) in which the anion is bound by the cooperative action of a salt bridge and at least two strong hydrogen bonds. Additional isothermal titration calorimetry measurements of sulfate binding by the above receptors in methanol attest to a dominating role of solvation during complex formation [43]. This result corroborates the proposal of possibly using solvation phenomena for the design of novel extraction systems [33]. As well as Br^-, I^- and SO_4^{2-}, the nucleotide anions AMP, ADP and ATP can also be extracted by the above bisguanidinium compound (with a preference for ATP) [42]. Furthermore **9a** also shows favorable phase transfer behavior for the 1:1 complex with pertechnetate [44].

Another interesting development is the use of bicyclic guanidinium receptors as heteroditopic systems for the enantioselective recognition of amino acids. In this case the guanidinium fragment was combined via a flexible linker with a crown ether moiety in order to promote the simultaneous binding of the carboxylate anion and the ammonium cation belonging to a zwitterionic amino acid [45].

Each of the extraction systems discussed so far in this section involve more or less fixed covalent binding sites that are present in a pre-organized architecture.

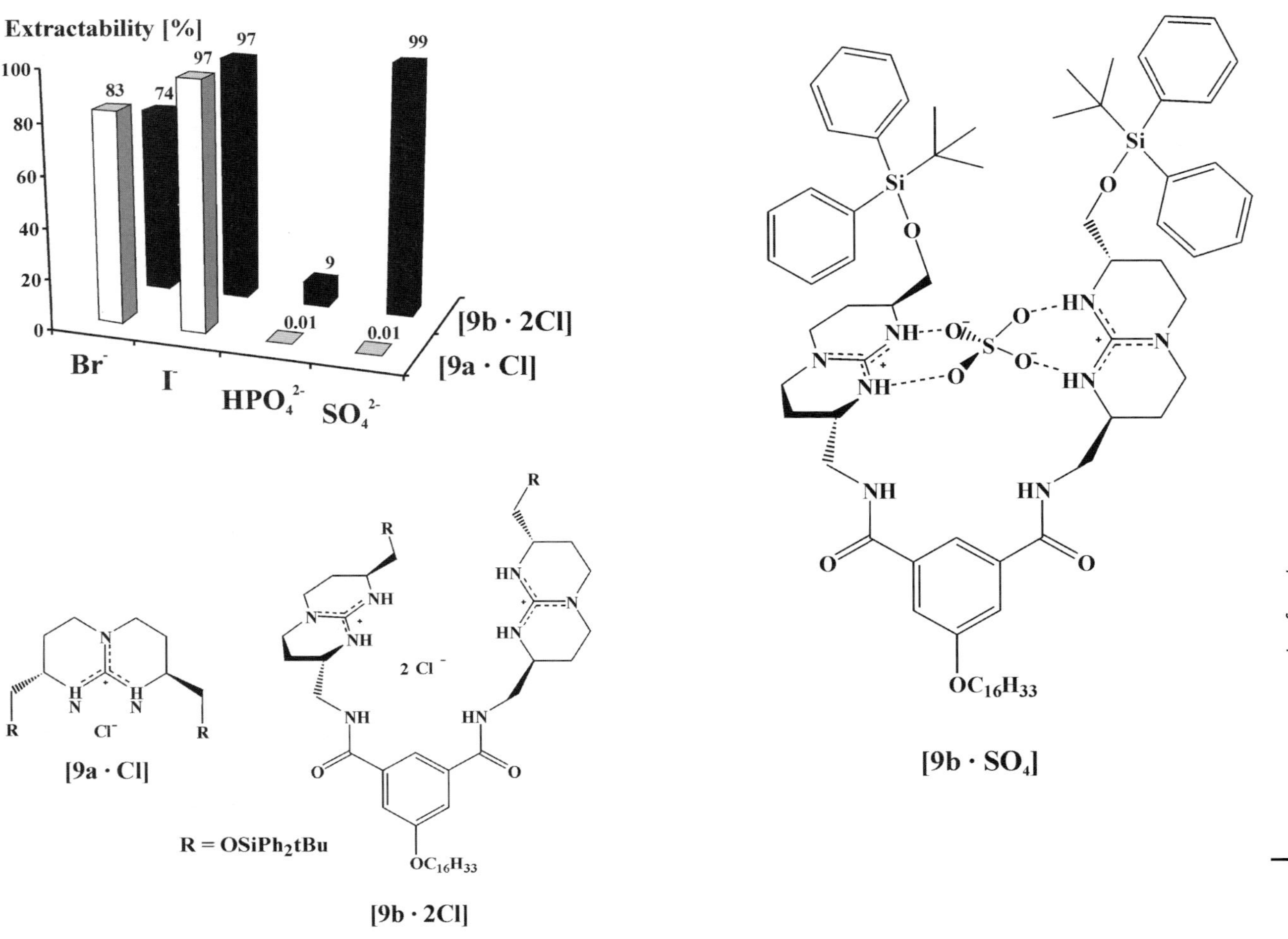

Fig. 4.15. Extraction of Br^-, I^-, HPO_4^{2-} and SO_4^{2-} by the guanidinium compounds [**9a**·Cl] and [**9b**·2Cl] in $CHCl_3$.

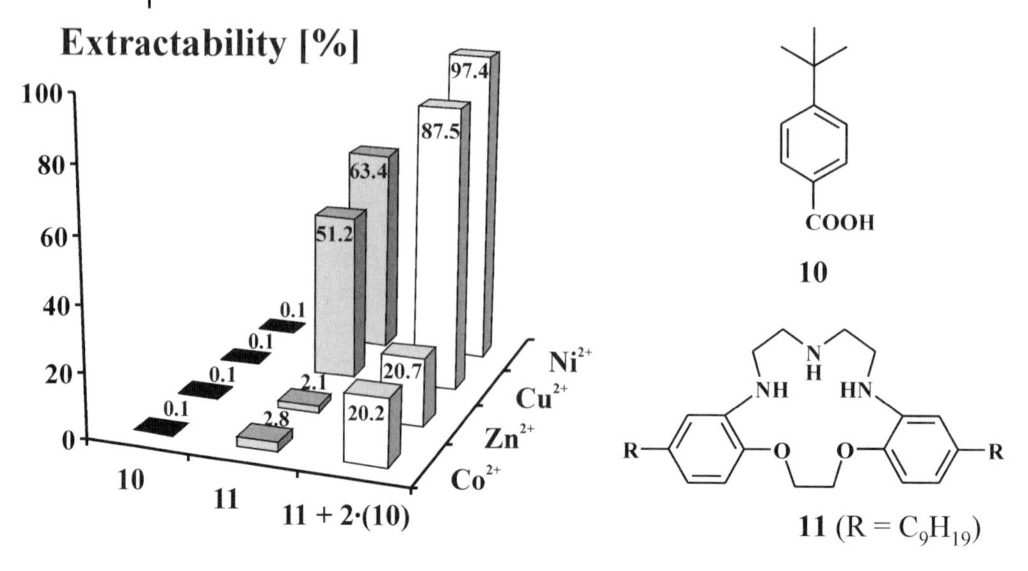

Fig. 4.16. Extraction of Co^{2+}, Zn^{2+}, Cu^{2+} and Ni^{2+} by the carboxylic acid **10**, the azacrown compound **11** and their 1:2 mixtures in $CHCl_3$.

The success of this approach over a wide range of examples for the binding and phase transfer of different species confirms its utility. Nevertheless, the preparation of the required ligand systems is often synthetically expensive. A more sophisticated supramolecular approach for the binding and extraction of either single ions or salts is based on self-organization processes employing comparatively simple components. This is illustrated by the results of extraction studies involving transition metal ions using a mixture of the carboxylic acid **10** and the azacrown compound **11** [46, 47]. As shown in Fig. 4.16, in comparison to the use of the individual components enhanced extraction of Co^{2+}, Zn^{2+}, Cu^{2+} and Ni^{2+} was observed when both extractants were used together. This synergistic effect can be attributed to a process (Fig. 4.17) starting with proton transfer from the acid to the nitrogens of the azacrown leading to formation of discrete assemblies with 1:1 and 2:1 composition (acid:crown). Such structures incorporate charge separated hydrogen bonds and the presence of water molecules in the hydrogen bonded network has also been identified in the solid state [48, 49]. On metal binding, the required number of protons necessary to maintain charge neutrality are exchanged for the metal ion.

Thus the assembly is pre-organized for metal ion uptake and no counter-anion transport into the organic phase is necessary. In other studies, it has also been demonstrated that another self-organization process involving azacrown compounds has allowed metal salt extraction against that counters the Hofmeister series for the anions involved [50].

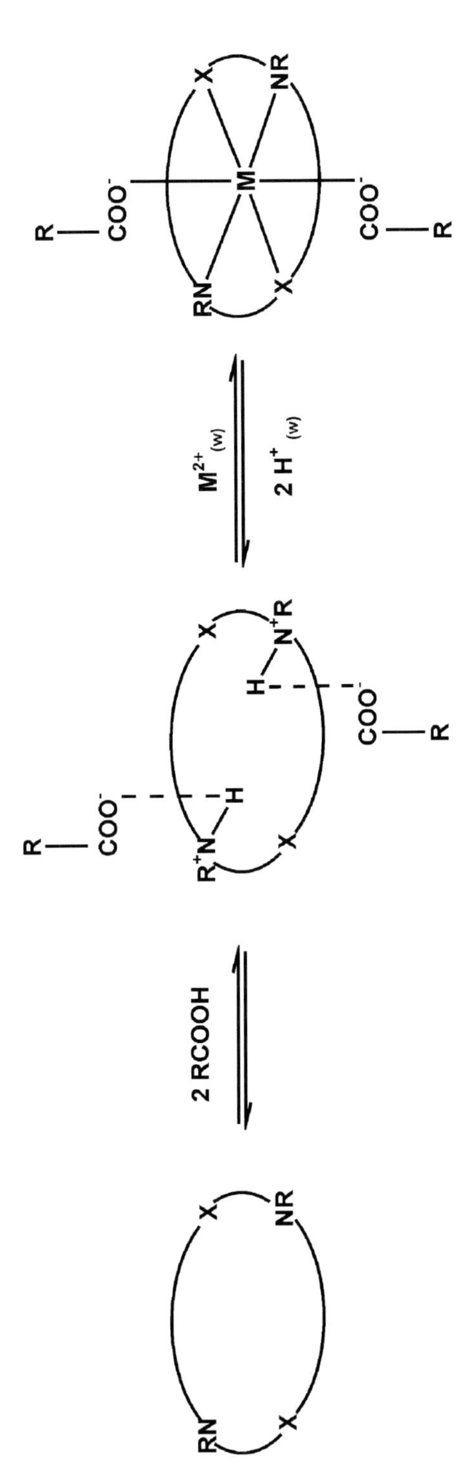

Fig. 4.17. General scheme for metal extraction on the basis of a 2:1 assembly between a carboxylic acid and an azacrown compound.

4.7
Conclusions and Future Perspectives

Recent advances in supramolecular chemistry have made available a wide range of synthetic supramolecular receptors for ionic and molecular species giving rise to a variety of interesting binding and structural motifs that occur both in solution and in the solid state. Extraction studies within the supramolecular realm continue to show strong potential for making significant contributions to both basic and applied research. At present, interest has been mainly focused on three main topics:

- the extensive characterization of novel receptor types
- the development of new separation systems for application in analysis, medicine, catalysis, synthesis and materials recycling as well as other aspects of environmental protection
- the modeling of biological recognition and transport processes based on new synthetic systems.

Because weak interactions are of key importance in supramolecular chemistry, the growing understanding of their role in supramolecular structures will also undoubtedly assist the understanding of the behavior of classical extraction systems, resulting in a greater control of extraction selectivity and efficiency. In complicated multicomponent systems, a range of interaction types usually contributes to species separation, with individual interactions frequently associated with typical supramolecular behavior. In this context, it seems especially appropriate to examine the role of water in the organic phase in terms of supramolecular principles, as well as the effect of the organic solvent employed on the selectivity and efficiency of phase transfer processes. In both cases much remains to be learned (and accomplished). Similarly, the effect of conformational change on host compounds as well as on their corresponding host–guest complexes in solution has so far been little investigated and is poorly understood. Furthermore, the influence of diverse phenomena, including aggregation, hydration and solvation effects on the extraction process also remains to be explored in more detail. An armoury of standard experimental techniques are available for this purpose while computational tools such as molecular dynamics simulations are also available to interpret and predict host–guest complexation in both aqueous and organic phases as well as the phase transfer behavior [51, 52].

Finally, the application of new extractants such as hyperbranched polymers [53] and the use of ionic liquids [54] are developments which tend to aid and promote supramolecular host–guest complexation and will undoubtedly further stimulate the development of more efficient and selective extraction systems.

Acknowledgements

We thank the Deutsche Forschungsgemeinschaft for financial support of our research on the topics of this review. B. A. expresses thanks to the Max-Buchner-

Stiftung for a research grant. The authors are especially grateful to P. K. Bharadwaj, Kanpur, V. Böhmer, Mainz, L. F. Lindoy, Sydney, J. Nelson, Belfast, F. P. Schmidtchen, Munich, E. Weber, Freiberg, and F. Vögtle, Bonn, for continued fruitful cooperation within the field of supramolecular binding and extraction as well as for stimulating discussions on the topics of this review.

References

1 *Comprehensive Supramolecular Chemistry*, J.L. Atwood, J.E. Davies, D.D. McNicol, F. Vögtle, Eds., Vol. 1–10, Elsevier, Oxford, **1996.**

2 J.W. Steed, J.L. Atwood, *Supramolecular Chemistry*, Wiley, Chichester, **2000.**

3 H.-J. Schneider, A. Yatsimirsky, *Principles and Methods in Supramolecular Chemistry*, Wiley, Chichester, **2000.**

4 *Encyclopedia of Supramolecular Chemistry*, J.L. Atwood, J.W. Steed, Eds., Marcel Dekker, New York, **2004.**

5 C.J. Pedersen, H.K. Frensdorff, *Angew. Chem. Int. Ed. Engl.* **1972**, *11*, 16–25.

6 S.S. Moore, T.L. Tarnowski, M. Newcomb, D.J. Cram, *J. Am. Chem. Soc.* **1977**, *99*, 6398–6405.

7 B.A. Moyer, *Complexation and transport*, in [1], Vol. 1, pp. 377–416.

8 A.T. Yordanov, D.M. Roundhill, *Coord. Chem. Rev.* **1998**, *170*, 93–124.

9 D.M. Roundhill, J.Y. Shen, *Phase transfer extraction of heavy metals*, in *Calixarenes 2001*, Z. Asfari, V. Böhmer, J. Harrowfield, J. Vicens, Eds., Kluwer Acad. Publ., Dordrecht, **2001**, pp. 407–420.

10 *Supramolecular Chemistry of Anions*, A. Bianchi, K. Bowman-James, E. Garcia-Espana, Eds., Wiley-VCH, New York, **1997.**

11 K. Gloe, H. Stephan, M. Grotjahn, *Chem. Eng. Technol.* **2003**, *26*, 1107–1117.

12 *Fundamentals and Applications of Anion Separations*, B.A. Moyer, R.P. Singh, Eds., Kluwer Acad./Plenum Publ., New York, **2004.**

13 B.A. Moyer, P.V. Bonnesen, R. Custelcean, L.H. Delmau, B.P. Hay, *Kem. Ind.* **2005**, *54*, 65–87.

14 S.G. Galbraith, P.A. Tasker, *Supramol. Chem.* **2005**, *17*, 191–207.

15 K. Gloe, B. Antonioli, K. Gloe, L.F. Lindoy, *Anion extraction*, in [4], online publ. **2006.**

16 *Solvent Extraction Principles and Practice*, J. Rydberg, M. Cox, C. Musikas, G.R. Choppin, Eds., Marcel Dekker, New York, **2004.**

17 K. Gloe, P. Mühl, *Isotopenpraxis* **1983**, *19*, 257–260.

18 *Handbook of Solvent Extraction*, T.C. Lo, M.H.I. Baird, C. Hanson, Eds., Krieger Publ. Comp., Malabar, **1991.**

19 E. Müller, R. Berger, E. Blass, D. Sluyts, *Liquid-liquid extraction, in Ullmann's Encyclopedia of Industrial Chemistry/Online Database*, Wiley-VCH, Weinheim, 2005. http://www.interscience.wiley.com/ullmanns

20 J. Schmitz, F. Vögtle, M. Nieger, K. Gloe, H. Stephan, O. Heitzsch, H.-J. Buschmann, *Supramol. Chem.* **1994**, *4*, 115–119.

21 M. Petrich, L. Beyer, K. Gloe, P. Mühl, *Anal. Chim Acta* **1990**, *228*, 229–234.

22 C.F. Baes, Jr., W.J. McDowell, S.A. Bryan, *Solv. Extr. Ion Exch.* **1987**, *5*, 1–27.

23 C.F. Baes, Jr., B.A. Moyer, G.N. Case, F.I. Case, *Sep. Sci. Technol.* **1990**, *25*, 1675–1688.

24 C.F. Baes, Jr., *Solv. Extr. Ion Exch.* **2001**, *19*, 193–213.

25 M.G. Cacace, E.M. Landau, J.J. Ramsden, *Quart. Rev. Biophys.* **1997**, 30, 241–277.

26 *Macrocyclic Compounds in Analytical Chemistry*, Yu.A. Zolotov, Ed., Wiley, New York, **1997.**

27 K. Gloe, P. Mühl, J. Beger, *Z. Chem.* **1988**, *28*, 1–14.
28 C.E.O. Roesky, E. Weber, T. Rambusch, H. Stephan, K. Gloe, M. Czugler, *Chem. Eur. J.* **2003**, *9*, 1104–1112.
29 H. Stephan, K. Gloe, E.F. Paulus, M. Saadioui, V. Böhmer, *Organic Lett.* **2000**, *2*, 839–841.
30 B.A. Moyer, J.F. Birdwell, P.V. Bonnesen, L.H. Delmau, *Use of macrocycles in nuclear-waste cleanup*, in *Macrocyclic Chemistry – Current Trends and Future Perspectives*, K. Gloe, Ed., Springer, Dordrecht, **2005**, pp. 383–405.
31 F. Vögtle, S. Ibach, M. Nieger, C. Chartroux, T. Krüger, H. Stephan, K. Gloe, *Chem. Commun.* **1997**, 1809–1810.
32 A.G. Blackman, *Polyhedron* **2005**, *24*, 1–39.
33 C. Chartroux, K. Wichmann, G. Goretzki, T. Rambusch, K. Gloe, U. Müller, W. Müller, F. Vögtle, *Ind. Eng. Chem. Res.* **2000**, *39*, 3616–36.
34 V. McKee, J. Nelson, R.M. Town, *Chem. Soc. Rev.* **2003**, *32*, 309–325.
35 D. Farrell, K. Gloe, K. Gloe, G. Goretzki, V. McKee, J. Nelson, M. Nienwenhuyzen, I. Pal, H. Stephan, R.M. Town, K. Wichmann, *Dalton Trans.* **2003**, 1961–1968.
36 J.M. Llindares, D. Powell, K. Bowman-James, *Coord. Chem. Rev.* **2003**, *240*, 57–75.
37 H. Stephan, K. Gloe, W. Kraus, H. Spies, B. Johannsen, K. Wichmann, G. Reck, D.K. Chand, P.K. Bharadwaj, U. Müller, W.M. Müller, F. Vögtle, in [12], pp. 151–161.
38 K. Wichmann, T. Söhnel, B. Antonioli, H. Heßke, M. Langer, M. Wenzel, K. Gloe, K. Gloe, J.R. Price, L.F. Lindoy, in *Solvent Extraction for Sustainable Development*, Proc. ISEC 2005, Beijing, Tsinghua University, **2005**, pp. 169–176.
39 B. Antonioli, K. Gloe, K. Gloe, G. Goretzki, M. Grotjahn, H. Heßke, M. Langer, L.F. Lindoy, A.M. Mills, T. Söhnel, *Z. Anorg. Allg. Chem.* **2004**, *630*, 998–1006.
40 H. Stephan, H. Spies, B. Johannsen, L. Klein, F. Vögtle, *Chem. Commun.* **1999**, 1875–1876.
41 K. Gloe, H. Stephan, T. Krüger, M. Czekalla, F.P. Schmidtchen, in *Value Adding Through Solvent Extraction*, Proc. ISEC'96, Melbourne, D.C. Shallcross, R. Paimin, L.M. Prvcic, Eds., The University of Melbourne, **1996**, Vol. 1, pp. 287–292.
42 H. Stephan, K. Gloe, P. Schiessl, F.P. Schmidtchen, *Supramol. Chem.* **1995**, *5*, 273–280.
43 M. Berger, F.P. Schmidtchen, *Angew. Chem. Int. Ed.* **1998**, *37*, 2694–2696.
44 H. Stephan, R. Berger, H. Spies, B. Johannsen, F.P. Schmidtchen, *J. Radioanal. Nucl. Chem.* **1999**, *242*, 399–403.
45 A. Metzger, K. Gloe, H. Stephan, F.P. Schmidtchen, *J. Org. Chem.* **1996**, *61*, 2051–2055.
46 V. Gasperov, K. Gloe, A.J. Leong, L.F. Lindoy, M.S. Mahinay, H. Stephan, P.A. Tasker, K. Wichmann, Proc. ISEC 2002, Cape Town, K.C. Sole, P.M. Cole, J.S. Preston, D.J. Robinson, Eds., South African Institute of Mining and Metallurgy, Johannesburg, **2002**, Vol. 1, pp. 353–359.
47 K.A. Byriel, V. Gasperov, K. Gloe, C.H.L. Kennard, A.J. Leong, L.F. Lindoy, M.S. Mahinay, H.T. Pham, P.A. Tasker, D. Thorp, P. Turner, *Dalton Trans.* **2003**, 3034–3040.
48 K.R. Adam, M. Antolovich, I.M. Atkinson, A.J. Leong, L.F. Lindoy, B.J. McCool, R.L. Davis, C.H.L. Kennard, P.A. Tasker, *Chem. Commun.* **1994**, 1539–1540.
49 K.R. Adam, I.M. Atkinson, R.L. Davies, L.F. Lindoy, M.S. Mahinay, B.J. McCool, B.W. Skelton, A.H. White, *Chem. Commun.* **1997**, 467–468.
50 V. Gasperov, S.G. Galbraith, L.F. Lindoy, B.R. Rumbel, B.W. Skelton, P.A. Tasker, A.H. White, *Dalton Trans.* **2005**, 139–145.
51 A. Varnek, G. Wipff, *Solv. Extr. Ion Exch.* **1999**, *17*, 1493–1505.

52 M. GROTJAHN, T. RAMBUSCH, K. GLOE, L.F. LINDOY, *Molecular modeling and related computational technique*, in [4], pp. 901–908.
53 K. GLOE, B. ANTONIOLI, K. GLOE, H. STEPHAN, in *Green Separation Processes*, A.M.C. AFONSO, J.P.S.G. CRESPO, Eds., Wiley-VCH, Weinheim, **2005**, pp. 304–322.
54 G.-T. WEI, Z. YANG, C.J. CHEN, *Anal. Chim. Acta*, **2003**, *488*, 183–192.

5
Mass Spectrometry and Gas Phase Chemistry of Supramolecules

Michael Kogej and Christoph A. Schalley

5.1
Introduction

Supramolecular chemistry has recently grown at a fascinating pace which would have been unthinkable without the appropriate methodological development. Supramolecular chemistry deals with noncovalent bonding between different building blocks of a complex or aggregate. The strengths of such individual weak interactions are usually of an order of magnitude comparable with those between the subunits and the molecule's environment. Often, it is only the cumulative or even cooperative action of several well-positioned weak interactions which brings a complex into existence in a competing environment.

Many researchers still consider mass spectrometry [1] to be a rather destructive method. The soft ionization methods – despite the revolution they caused in the life sciences – are often not appropriately highlighted in lecture courses on mass spectrometry, so prejudice has it that mass spectrometry almost unavoidably causes fragmentation even of covalent bonds. Consequently, the examination of noncovalent complexes is too often considered to be futile and successful only in some very special cases. However, the soft ionizaton methods developed in the 1980s reduce fragmentation to a minimum and even noncovalent, weakly-bound complexes can be ionized without complete destruction. Technically, the problem of intact ionization of weakly-bound complexes can be solved in many cases [2].

Consequently, a more fundamental question arises: Why should one apply mass spectrometry to supramolecules? What is the motivation and what is the added value of using this method together with other techniques that are maybe more commonly used in supramolecular chemistry? The present chapter elaborates on the hypothesis that the potential of mass spectrometry goes far beyond the analytical characterization of complexes with respect to their exact masses, charge states, stoichiometries, or purity. In fact, the information that can be gained is complementary to other methods such as NMR spectroscopy and includes structural aspects, reactivity, and even thermochemistry. Examination of supramolecules by mass spectrometry involves their transfer into the high vacuum of the mass spectrometer and thus implies that isolated particles are investigated. There is no

Analytical Methods in Supramolecular Chemistry. Edited by Christoph Schalley

ISBN: 978-3-527-31505-5

exchange of building blocks between gaseous supramolecules; fragmentation is irreversible, once it is induced for example by collisions or laser light. Our hypothesis is that a wealth of knowledge about supramolecules can be gained which cannot easily be obtained with other methods. Under the environment-free conditions inside a mass spectrometer, the intrinsic properties of ions can be evaluated. Comparison with the properties of the same species in solution allows us to analyze solvation effects. Since noncovalent bonds have strengths of similar magnitude to interactions with the solvent, one may expect that the comparison of intrinsic and solution-phase properties will reveal particularly pronounced effects for noncovalent complexes. Before discussing some examples, first some technical issues need closer inspection.

5.2 Instrumentation

In principle, a mass spectrometer consists of two segments: an ion source and a mass analyzer. Ion sources are required for the transfer of the sample from the condensed phase (i.e. from solution or the solid state) into the high vacuum as a

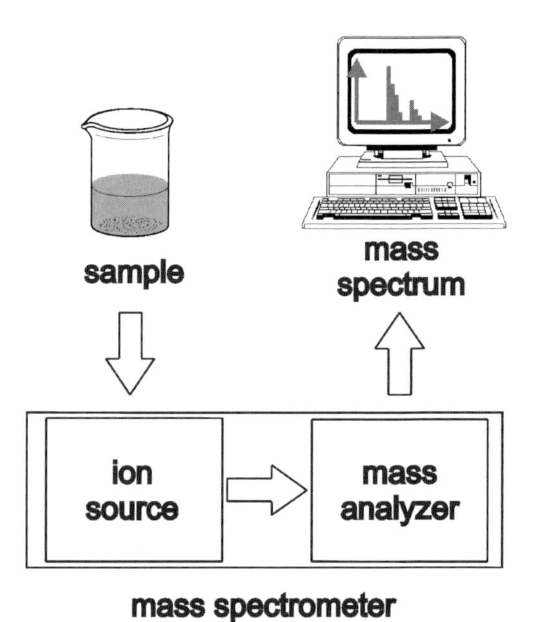

Fig. 5.1. Schematic representation of a mass spectrometer. Samples are typical for electrospray ionization. Depending on the ionization technique, the sample can be dissolved in a liquid solvent (ESI), cocrystallized in a matrix (MALDI) or dissolved in a liquid matrix (FAB).

charged species. A number of different ionization methods is available; we limit the discussion to those soft enough to accomplish the intact ionization of noncovalent species (Section 5.2.1). After leaving the ion source, the motion of the ions can be influenced by electric or magnetic fields inside the mass analyzer. Accordingly, the ion beam is resolved into its components according to the ions' masses or – to be more precise – the ions' mass-to-charge ratios (m/z). Several types of mass analyzers are available, and the most important ones will be presented in Section 5.2.2.

5.2.1 Ionization Techniques Suitable for Noncovalent Species

Table 5.1 summarizes the most common and often commercially available ionization methods together with brief comments regarding their utility for studies in supramolecular chemistry. Many of them are incompatible with noncovalent bonds because they are intrinsically harsh, and dissociate the weakly-bound ions of interest. In the following, we focus on those which can be used for the examination of noncovalent species.

5.2.1.1 Matrix-assisted Laser Desorption/Ionization (MALDI)

In MALDI [3], a laser is applied to vaporize and ionize the sample that consists of the analyte embedded in a crystalline matrix. In normal MALDI sample preparation, the matrix is used in large excess (e.g. matrix:analyte = 800:1). Both should be soluble in a common, easy to evaporate solvent in order to make their cocrystallization on the sample plate possible. The matrix is an integral part of this ionization technique. Samples which absorb at the wavelength of the laser beam (most commercial instruments use a N_2 laser with 337 nm) can also be analyzed by matrix-free laser desorption/ionization (LDI), but for all other samples, the matrix plays a threefold role: (i) It absorbs the laser light. Therefore, compounds such as 2,5-dihydroxy benzoic acid, dithranol, or 9-nitroanthracenen are typically used. The energy absorbed by the matrix is sufficient for vaporization and ionization of matrix and analyte molecules (Fig. 5.2). (ii) It can act as a proton donor, in particular, when organic acids are chosen. The matrix thus provides the charge and helps to generate ions. Nevertheless, photoionization, charge transfer, electron capture, self-protonation of the analyte, and its cationization for example with background sodium ions can occur [4], in particular in nonprotic matrices. The detailed mechanism of ion formation in MALDI is still a subject of continuous research and debate and we refer the reader to recent reviews on this topic for advanced reading [3]. (iii) Collisions between analyte and matrix molecules in the plume above the solid sample help to reduce the internal energy of the analyte ions so that fragmentation becomes less pronounced. Nevertheless, for many noncovalent ions, the conditions of the MALDI process are rather harsh. Intact ionization of weakly-bound complexes by MALDI is consequently often quite difficult.

In principle, the mass range of MALDI is almost unlimited and MALDI sources preferentially generate singly-charged ions so that broad charge distributions can by and large be avoided. The laser beam is applied in the form of short shots, so

Tab. 5.1. Common ionization methods, the underlying principles, and their utility for studies of non-covalent species.

Ionization method	Ionization principle	Utility for the examination of non-covalent species
Electron ionization (EI)	Radical-cation generation through collision with fast electrons $M + e^- \rightarrow M^{+\cdot} + 2\,e^-$	Need to vaporize sample before ionization limits mass range due to sample volatility to M < ca. 800 Da; high internal energies; high degree of fragmentation
Chemical ionization (CI)	Generation of protonated species from strong acids produced by EI of a suitable bath gas (H_2, CH_4...)	See EI; fragmentation less pronounced due to a lower internal energy of the ions controlled by the energetics of proton-transfer reaction
Atmospheric-pressure chemical ionization (APCI)	Analyte solution sprayed into a heated desolvation zone after being pneumatically nebulized; corona discharges produce primary ions which ionize the analyte (cf. CI)	Desolvation occurs in a heated vaporization chamber so that non-covalent adducts are usually destroyed easily
Fast atom bombardment (FAB)	Impulse transfer from fast atoms (Xe) or ions (Cs^+) through a liquid matrix transfers protonated samples into gas phase	Liquid matrices with high boiling points required, thus competition of polar, often protic matrices with non-covalent bonding, high degree of fragmentation; mass range usually limited to <5000–10000 Da
Field desorption (FD)	Desorption/ionization of ions in a high voltage (typically 8–12 kV)	Often low ionizaton efficiencies, some cluster ions have been observed, but the scope and limitations for supramolecules have not been fully examined
Electrospray ionization (ESI)	Generation of charged droplets, ion formation through shrinking – Coulomb explosion sequences and ion ejection from droplets	Ionization directly from sample solution possible; mass range virtually unlimited; low ion internal energies, thus no or very limited fragmentation levels
Matrix-assisted laser desorption/ionization (MALDI)	Sample in crystalline matrix heated by laser and ionized by attachment of H^+, Na^+, K^+, etc.	Higher fragmentation level than ESI, virtually no mass range limit, photochemical processes possible, when sample absorbs light at laser wavelength
Resonance-enhanced multi-photon ionization (REMPI)	Cold, neutral complexes from supersonic beams are photoionized after transfer to gas phase	Low internal ion energies due to cooling in supersonic jet expansion; radical cations generated

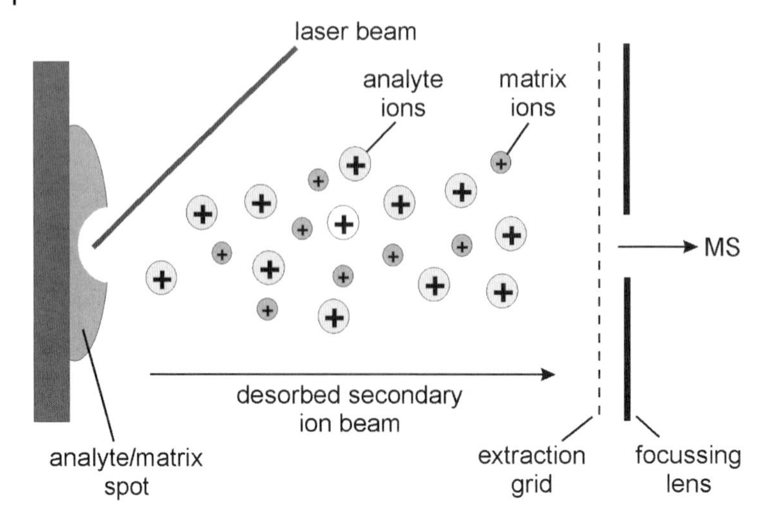

Fig. 5.2. Schematic representation of the ion generation in the MALDI process. A laser beam impacts on an analyte/matrix mixture and generates analyte and matrix ions.

MALDI sources do not continuously generate ions, but work in a pulsed manner. This feature matches perfectly the pulsed operation of time-of-flight (TOF) analyzers (see below). Consequently, MALDI–TOF mass spectrometers are quite common.

5.2.1.2 **Electrospray Ionization (ESI)**

Electrospray ionization [5] is a soft ionization technique that can be applied to transfer sample molecules from solution to the gas phase. Although there are different kinds of electrospray ionization methods like thermospray (TSP) [6] or electro-hydrodynamic ionization (EHI) [7], the principle of ESI as developed by Dole [8] is used in all these techniques. The analyte is dissolved in a volatile solvent and this solution is sprayed through a needle in an electrical field to create micrometer-sized charged droplets (Fig. 5.3). To help the evaporation of the solvent, the needle is embedded into a heatable stream of surrounding gas (normally N_2), the so-called nebulizing gas. Depending on the details of the instrument design, other gases than N_2 can be applied.

The solution containing the sample is pumped through the needle, at whose tip it forms the so-called Taylor cone due to the high voltage applied. From the tip of the Taylor cone, a jet is formed in which the droplets are generated with excess charge. Because of Coulomb repulsion, the droplets are driven away from each other. Upon desolvation, they shrink and undergo Coulomb explosions forming even smaller droplets. This process causes the generation of a fine spray. Two mechanisms for the formation of fully desolvated ions are discussed: in the so-called single-ion-in-droplet (SIDT) model, [9] the shrinking-explosion process con-

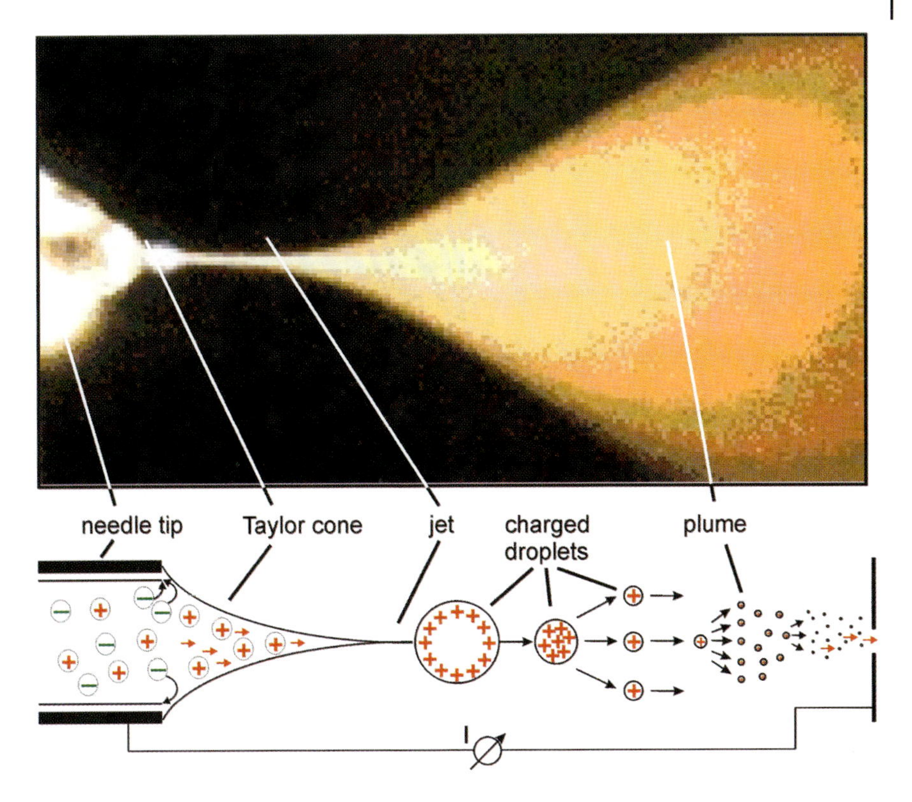

Fig. 5.3. A photograph of the spray generated in an electrospray ionization source (top) and a schematic view of the processes involved (bottom). From the Taylor cone, a jet is emitted and expands to a plume of micrometer-sized charged droplets from which finally desolvated ions are produced.

tinues until singly-charged droplets are formed which contain only a few solvent molecules so that the remaining ion can easily be desolvated completely. In contrast, the ion-evaporation model [10] assumes that single ions can be ejected from the droplet surface when charge repulsion becomes too strong. ESI is extremely useful in analysis of large, nonvolatile, chargeable molecules such as proteins [11] or nucleic acids [12, 13]. With these biopolymers, it preferentially forms a broad charge distribution of multiply-charged ions. This has the advantage that even very high molecular masses can be observed at much smaller m/z ratios which are easily accessible with most analyzers (see below). However, the disadvantage is that charge repulsion effects may destroy e.g. the tertiary structure of a protein under study.

A different, but related ionization method is coldspray ionization (CSI) [14]. The ionization occurs in a similar manner to ESI, with the exception that the source housing and drying gas are cooled in order to stabilize weakly-bound ions. The effect is often incomplete desolvation and distributions of ions with different numbers of attached solvent molecules are observed. Coldspray ionization has the advantage that supramolecules can be ionized without destruction and thus, the analytical data on molecular mass, charge state, isotope patterns and elemental composition is more easily obtained. For the gas-phase chemist, the solvent molecules, however, represent a disadvantage, because mass-selection of the completely desolvated ions becomes more difficult due to their usually low intensities.

5.2.1.3 Resonance-enhanced Multiphoton Ionization (REMPI)

The last ionization method to be discussed here is photoionization by resonance-enhanced multiphoton ionization (REMPI) [15]. Set-ups which are relevant for supramolecules use a supersonic jet expansion to vaporize the sample. The expansion leads to a strong cooling effect on the analyte molecules in the sample solution. Consequently, gaseous noncovalent complexes can easily be generated. However, they are still neutral, and thus, a laser beam is used to photoionize the complexes of interest through multiphoton absorption after their transfer into the gas phase. Consequently, this technique makes the mass spectrometrist more independent from ion generation problems. However, in certain cases, it might be a drawback that cation radicals are formed, which are usually more reactive because they contain an unpaired electron, rather than closed-shell cations, formed by protonation or cationization. The REMPI technique has not been applied very often to problems in supramolecular chemistry [16], but seems to be highly promising and thus is mentioned here, at least briefly.

5.2.1.4 Ionization of Noncovalent Species

Even with soft ionization methods at hand, it is not always easy to ionize weakly bound, noncovalent species without completely fragmenting them. For example, hydrogen-bonded aggregates often decompose in competing media such as the standard matrices used for MALDI or the common ESI spray solvents such as methanol. Protic media have the advantage of providing the protons necessary for ion generation. But, how should one examine hydrogen-bonded species in such a competitive environment? Hydrogen bonds become stronger in noncompeting solvents such as chloroform or dichloromethane and thus they are weak in protic media. But if we avoid protic solvents, where should the charge come from, if the supramolecular species of interest is neutral? The answer usually requires the experimenter's imagination and there is no general solution to this problem. Several ion labeling strategies have been developed, among them covalent modification with crown ether/alkali metal ion complexes [17], oxidation of a ferrocene center through addition of iodine before the electrospray process [18], Ag^+ coordination to aromatic or cyano groups present in the complex [19], attachment of a chloride ion [20], and the encapsulation of charged guests inside self-assembling capsular hosts [21]. With these "in-built" charges, ionization can also be achieved from noncompetitive media. These considerations are typical for hydrogen-bonded species, but certainly similar arguments apply to other noncovalent interactions. One may state that there is no general solution to the problem of ion generation and that it is a matter of the experimenter's chemical intuition to find the best, sometimes maybe quite unusual way.

5.2.2
Mass Analyzers

Mass analyzers use physical principles to separate the generated ions coming from the ionization source. Many physical principles have been tested for that purpose but only a few of them have become successful and are still enhanced to achieve better results in mass spectrometric analysis. The most common mass analyzers are listed in Table 5.2. Although in principle any source can be connected with any analyzer, some combinations are more common. For example, MALDI as a discontinuously working ion source can easily be coupled with time-of-flight analyzers, while ESI sources are often combined with quadrupoles and ion traps. Therefore, we restrict the discussion here to those analyzers which are typically available together with MALDI and ESI ion sources.

5.2.2.1 Quadrupole Instruments and Quadrupole Ion Traps

Quadrupole mass analyzers consist of four parallel metal rods which are either round-shaped or ideally have a hyperbolic section. Ion separation is realized by applying an alternating electrical field to the four rods with each pair of adjacent rods having the opposite signs of the potential. The potential Φ applied to two opposite rods is a superposition of a constant potential U and an alternating potential $V \cos \omega t$ (Eq. (5.1)). The other two opposite rods bear the potential $-\Phi$. ω describes the angular frequency and equals $2\pi\nu$ when ν is the frequency of the applied radio frequency field; t is the time.

$$\Phi = (U - V \cos \omega t) \tag{5.1}$$

Basically, ions entering the quadrupole move through it in a wavelike trajectory around the z-axis between the quadrupole's electrodes. This motion can be described by the so-called Mathieu differential equations (5.2) and (5.3).

$$\ddot{x} + \frac{2e}{mr^2}(U + V \cos \omega t)x = 0 \tag{5.2}$$

$$\ddot{y} - \frac{2e}{mr^2}(U + V \cos \omega t)y = 0 \tag{5.3}$$

with x, y providing the ions coordinates in x and y direction perpendicular to the long axis of the quadrupole (e: elemental charge, m: ion mass, $2r$: distance between opposing rods).

The solution of differential equations (5.2) and (5.3) is a specific pathway of predefined ions with a special m/z ratio inside the electric potential of the electrodes.

Tab. 5.2. Illustration of the common mass analyzers and their ion separation principles.

Type of analyzer	Ion separation principle	Comments	Accuracy/ ppm	m/z range	resolution
Sector field: (B, E, BE, EB...)[a]	Magnetic sector: Lorentz force; impulse selection; Electrostatic sector: kinetic energy selection	Limited resolution in MS/MS experiments[b], high-energy collisions; usually coupled to EI/CI, FAB	<5	10 000	30,000
Quadrupole (Q)	Superposition of a constant electric field (U) and a radiofrequency field (V) on four parallel metal rods, stable ion trajectories require appropriate settings of U and V, scanning voltages separates different m/z	Low cost instruments which can favorably be coupled to continuously operating ESI ion sources; MS/MS capable, if e.g. three quadrupols are used for mass selection, collisional activation and product ion scan	100	4000	4,000
Ion Trap (IT)	Similar to Q, traps have a ring electrode and two capping electrodes	Low cost instruments; collision gases can be subjected into the trap for MS/MS and beyond (MS^n)	100	4000	4,000
Time-of-flight (TOF)	All ions accelerated at the same time by a high voltage pulse, travel time corresponds to m/z	Pulsed analyzer typically used together with MALDI ion sources, not generally applicable for MS/MS	200	>300 000	8,000
Time-of-flight (reflectron)	Similar to TOF, but ions are reflected and thereby refocused by reflectron thus providing higher accurracy	TOF/TOF analyzers permitting MS/MS experiments available, growing number of applications	10	10 000	15,000
Ion-cyclotron resonance (ICR)	Cyclotron frequency of ions orbiting inside a cell within a superconducting magnet (4.7–12 T field strength) is measured	Costly equipment, highest accuracy and resolution, in principle unlimited MS^n capabilities (as long as ion intensity suffices), bimolecular reactions possible	<2	10 000	≫100,000

[a] B indicates a magnetic sector, E an electrostatic sector. The most common setups are double-focusing instruments with either a Nier–Johnson geometry (EB) or an inverse Nier-Johnson configuration (BE). Extended setups such as BEBE are available and can be used for up to MS^3 experiments.
[b] MS/MS, tandem MS and MS^n are expressions referring to experiments in which the ion of interest is mass-selected, then subjected to a gas-phase experiment, and finally examined with respect to the product ions formed in fragmentation reactions.

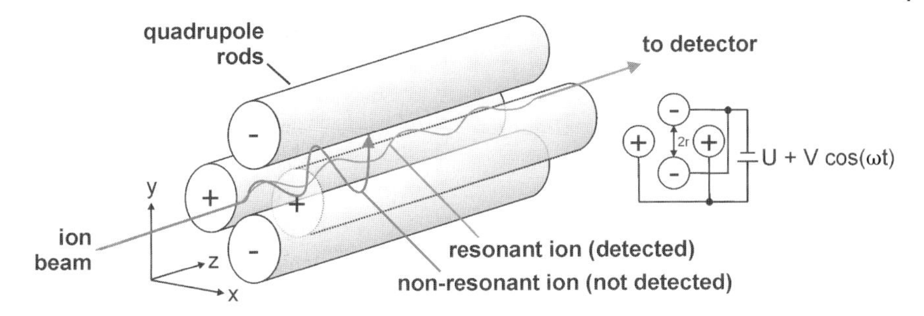

Fig. 5.4. Schematic view of a quadrupole system with stable (resonant ion) and unstable (non-resonant ion) ion trajectories.

Ions, whose m/z is not within the stable trajectories are neutralized when they touch one of the rods and thus do not leave the quadrupole into the detector. By varying the direct voltage and AC voltage fast scanning of a wide range of m/z values is possible (Fig. 5.4).

Quadrupole instruments are low in cost and thus are widespread in MS laboratories. The advantages are high transmission, a low weight, and compactness. They have low ion acceleration voltages and allow high scan speeds. In certain cases, the rather low mass accuracy and low resolution may however be somewhat problematic. Instruments with at least three linear quadrupoles in a row can be used for tandem mass spectrometric experiments. In these experiments, the first quadrupole is set to a constant m/z value in order to mass-select the ions of interest and to expel all unwanted ions from the ion flight path. The second quadrupole is operated as the collision cell, in which fragmentation of the ions can be induced by adding a collision gas. The product ions are detected by scanning the third quadrupole.

Quadrupole ion traps can be considered as a variant of linear quadrupoles. They bear a ring electrode and two end cap electrodes. The potential applied to these electrodes is again alternating, so that the ions again follow trajectories inside the cavity between the electrodes. One major advantage of ion trap instruments is their capability to perform MS^n experiments in the same trap, i.e. after mass selection (ejection of all unwanted ions from the trap), a collision gas is applied, the ions are accelerated inside the trap and fragment upon collision with the gas. Now either the products can be recorded by scanning the instrument settings for different m/z or another mass selection of one of the product ions can be done and so forth.

5.2.2.2 **Time-of-flight (TOF)**

When ions of different molecular masses are accelerated by the same voltage U, they all have the same kinetic energy E_{kin} (Eq. (5.4); z is the number of charges, v is the ion velocity). They differ however with respect to their velocities so that they need different time intervals for traveling the same distance. Heavier ions travel more slowly than lighter ones.

$$E_{kin} = zeU = \frac{1}{2}mv^2 \tag{5.4}$$

Rearrangement of Eq. (5.4) allows us to calculate the ion velocity v which corresponds to the quotient of the travel distance s and the flight time t (Eq. (5.5)). From the flight time, the m/z ratio can be calculated, if the acceleration voltage U and the path length s are known (Eq. (5.6)).

$$v = \frac{s}{t} = \sqrt{\frac{2zeU}{m}} \tag{5.5}$$

$$\frac{m}{z} = \frac{2eUt^2}{s^2} \tag{5.6}$$

Time-of-flight analyzers make use of this principle. One requirement is that the starting time must be known and the same for all ions within one package. This can be achieved by two rotating discs as shown in Fig. 5.5. Here, the time of flight of the ions, that pass the openings in the two rotating discs, can be detected because the rotational velocities of the two discs are known and also the distance between them so that the time intervals between two ion packages are known. The alternative is to apply a short high voltage pulse perpendicular to the ion flight path on which it entered the analyzer from the ion source. The ions at the deflector electrode simultaneously start to move on an almost perpendicular flight path and can be analyzed with respect to their flight times and thus m/z values. The pulsed manner of the TOF analyzer makes it ideally suited for MALDI ion sources which also operate in a pulsed way. Nevertheless, ESI–TOF mass spectrometers are also common.

One should distinguish between linear TOF analyzers and reflectron TOF analyzers. The reflectron in the latter instruments reflects the ions by electrostatic

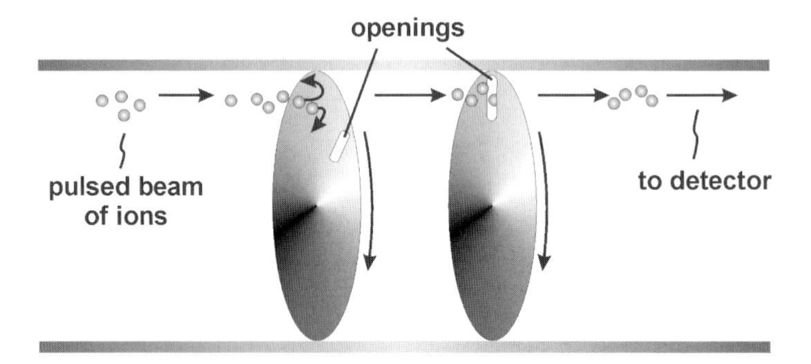

Fig. 5.5. Schematic view of a linear TOF analyzer. The time of flight of the generated ions that pass the openings in the rotating discs, is detected in the gap of the two discs.

potentials. This not only elongates the flight path and thus increases the ion separation in time, but also refocusses the ions. Consequently, reflectron TOF instruments usually achieve much higher mass accuracy and resolution than their linear counterparts. The price is a limitation in mass range, which at least in principle is not limited in linear TOF instruments. TOF instruments have high transmissions and permit extremely high scan rates, since each scan is obtained in the microsecond regime. Meanwhile, TOF/TOF instruments capable of tandem MS are commercially available.

5.2.2.3 **Ion Cyclotron Resonance (ICR)**

ICR instruments [22] employ the high magnetic fields of superconducting magnets (commercially available instruments are usually in the range of 4.7 to 9.4 T) to bring the ions on a small orbit within the analyzer cell. The Lorentz force is strong enough to prevent the ions from escaping the cell in the x and y directions perpendicular to the field axis z. After transfer of the ions into the ICR cell located inside the homogeneous magnetic field, they thus need to be trapped in z direction. Fig. 5.6 shows a schematic drawing of the cell. Two electrode plates, the trapping plates are charged with a positive voltage for cations (or a negative voltage for anions) in order to keep the ions inside the cell.

Equation (5.7) describes the radius r_m of the ions' orbits (B is the magnetic field strength, v the velocity of the ions on the circuit):

$$r_m = \frac{m\vec{v}}{ze\vec{B}} \tag{5.7}$$

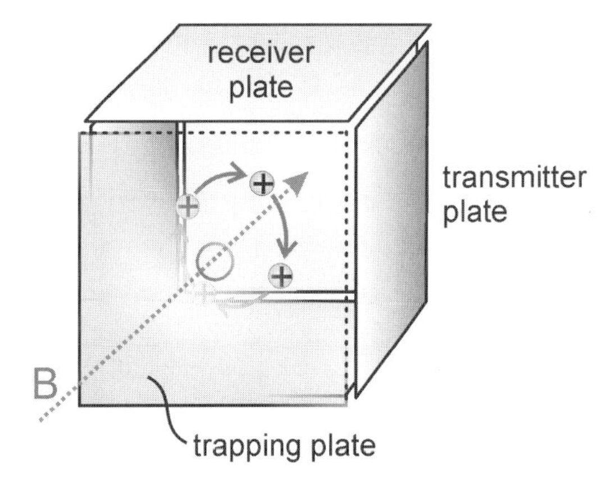

Fig. 5.6. Schematic view of an FT–ICR cell. Simply speaking, the cell consists of three pairs of oppositely positioned plates. The ions enter this cell along the magnetic field axis through the frontal trapping plate and are then trapped in the cell by charging the trapping plates. The transmitter plates are used to apply high-frequency pulses in order to manipulate the ions. Detection is done by measuring the image currents of orbiting ion packages on the receiver or detection plates.

Upon substitution of $\vec{v} = r_m\omega_c$ the angular frequency ω_c becomes:

$$\omega_c = \frac{ze\vec{B}}{m} \tag{5.8}$$

Consequently, the angular frequency is a function of the ion's mass, charge and also of the magnetic field, which is constant. Most importantly, it is independent from the ions' initial velocity and each m/z ratio corresponds to one particular cyclotron frequency. If a transverse high frequency alternating electrical field is applied to the transmitter plates shown in Fig. 5.6, the ions trapped inside the cell can be accelerated. The result of this acceleration is not a change of the cyclotron frequency, but the increase of the orbit radius. These changes in radius can be used for three different purposes: (i) If a package of ions is accelerated to a large radius, its detection is possible by measuring the frequency of the mirror current of the package at the detection plates. This frequency measurement can be done with extreme precision, so that ICR instruments provide highest resolutions and best mass accuracy. Since each ion package orbits with its unique cyclotron frequency, they can all be detected simultaneously as the transient free induction decay (FID; in analogy to NMR instruments), which is finally transformed into the mass spectrum by Fourier transformation of the time domain into the frequency domain. Quite rapid scan times are the result. (ii) By choosing the high frequency pulses appropriately, all ions except for a desired one can be accelerated to radii larger than the cell diameter. In this case, these undesired ions hit the cell walls and are neutralized. The result is the mass selection and isolation of a certain ion which then can be used for tandem MS experiments. In principle, FTICR instruments have no limits with respect to the number of MS^n steps. Practically, it is limited by the ion abundances. (iii) An ion of interest can be mass selected and after addition of a collision gas accelerated by a high frequency pulse. This makes the collisions harsher and permits to fragment ions in the gas phase. The experiment is called collision-induced decay (CID).

One of the most important features of FTICR instruments is their ability to store ions for a theoretically unlimited time (practically, there are limits caused by collisions with residual gas, but the times for which the ions can be stored are long enough to perform almost any desired experiment). As a consequence, this means that (i) not only unimolecular fragmentations, but also bimolecular reactions with neutral, volatile reactants added at pressures of ca. 10^{-8}–10^{-7} mbars through pulsed or leak valves can be studied. (ii) Reaction rate constants can be determined which for bimolecular reactions are often given as the fraction of productive collisions of the reaction partners relative to the collision rate. (iii) It is easy to couple FTICR instruments with, for example, infrared lasers to induce fragmentations by infrared multiphoton irradiation (IRMPD) for defined time intervals. Consequently, FTICR instruments have not only the advantage of high mass accuracy and extraordinary resolving power, they also offer the most versatile equipment for gas-phase chemistry beyond analytics. The major disadvantage is the cost of the equipment.

5.3 Particuliarities and Limitations of Mass Spectrometry

As for any other method, particular effects connected to the transition from solution to the gas phase and limitations arising from the mass spectrometric technique need to be considered for mass spectrometric experiments with supramolecules in order to avoid misinterpretations.

Let us first consider how noncovalent bonds are affected by the transition from condensed phase into the high vacuum of the mass spectrometer. Noncovalent forces range from very weak van der Waals forces in the order of sometimes less than a few kJ mol^{-1} up to the dative bond in metal coordination complexes, which often have strengths close to weak covalent bonds. In between are hydrogen bonds, π-donor π-acceptor interactions, cation-π forces, the hydrophobic effect, salt bridges, electrostatic attraction of ion pairs and others. They not only differ with respect to their strength, but also as far as directionality and thus geometry are concerned. Upon transition from solution to the gas phase, the environment changes dramatically. The change from a noncompetitive organic to a protic solvent often causes a complete change in binding selectivities, binding energies, or the stability of a large assembly. How can we then expect no changes to occur when transferring supramolecular species into the high vacuum of a mass spectrometer? More precisely, any interaction that competes with the solvent, for example a hydrogen bond in protic media, probably increases in strength upon evaporation. The same holds true for electrostatic attractions, which are weakened in solution by the dielectric constant of the solvent and by solvation with the dipoles of the solvent molecules. Instead, salt bridges, e.g. the carboxylate–guanine interaction in arginine-containing peptides, which are favorable in solution because the charges are stabilized by the solvent, may exist in the gas phase only under particularly beneficial circumstances. Other forces may become weaker in the gas phase. A special case is the hydrophobic effect. In water, nonpolar molecular surfaces tend to turn towards each other. However, this is not, because the attraction between these surfaces is strong, but rather because they minimize the unpolar surface which cannot be efficiently solvated by the surrounding water molecules. Thus, more hydrogen bonds between water molecules remain undisturbed, contributing their binding energy to the favor of the system as a whole. In the gas phase, the unpolar molecules after desolvation do not benefit from this effect any more and their binding energy will be much lower than that found in solution. A series of literature reports deal with the formation of cyclodextrin inclusion complexes of for example different drugs and use mass spectrometry as a means to provide evidence for their formation [23]. Sometimes, they are observed in the mass spectra, sometimes not, even if the complexes exist in solution in both cases. However, when thinking about it carefully, this is not surprising, because in solution it is mainly the hydrophobic effect which keeps the drug molecule inside the cyclodextrin cavity. In the gas phase, this effect is usually replaced by a proton bridge between the two molecules – but only for those guest molecules which have functional groups suitable for formation of these proton bridges. The result is the same in solution and the gas phase: a complex is formed. However, the reasons for which the complexes are held together could not be more different.

It cannot be overstated that treating mass spectrometric intensities as quantitatively correlating with solution concentrations is at least dangerous and a constant source of error and misinterpretation. Even, if many literature reports claim that ESI–MS provides a reliable picture of the situation in solution and cite a list of predecessors testifying in favor of this assumption, it is rather an exception than normality, if this hypothesis holds. It does not become any more accurate by repeating it over and over again. Factors such as the solvation energies influence the ESI response factors [24] which need to be determined for each single ion in order to obtain a quantitative correlation of concentration and intensitiy. There are many studies on crown ether–alkali ion complexes which even arrive at contradictory rankings of binding energies. A series of elegant, real gas-phase experiments [25] finally resolved the misunderstandings originating from them. These experiments clearly show that one either needs to carefully determine ESI response factors or to rely on experiments carried out truly in the gas phase after carefully mass-selecting the ion of interest. Quite recently, the serine octamer [26] was discovered (see below) as an amino acid cluster with a highly specific structure. It even exhibits a strong tendency for homochirality. Nevertheless, all attempts to detect this cluster in solution have failed so far. This example shows that in particular in the realm of weakly-bound species, there may be highly specific complexes which are not present in solution, but form upon electrospray ionization.

Unspecific binding is quite common and thus may become a source of error. It is often quite difficult to distinguish specific from unspecific binding and thus, one may be tempted to draw false conclusions for example on complex stoichiometry. Whenever possible, one should perform experiments aiming at a structural assignment of the ions or use suitable controls in order to exclude unspecific binding or a structural rearrangement during or after the ionization process.

Most weakly-bound species exhibit dynamic features in solution. For example, an equilibrium often exists between free host and guest on one side and the host–guest complex on the other. The equilibrium reflects thermodynamic stabilities. Often, the species interchange quickly and formation and dissociation of the host–guest complexes are fast processes. In the gas phase, there is no equilibrium situation, if the experiment is not deliberately and carefully set up to provide the necessary conditions. Upon decomposition of the complex ions in the gas phase, the two partners are irreversibly separated from each other. Consequently, the kinetic rather than thermodynamic stability determines the results of mass spectrometric experiments on noncovalent ions. Or in other words, the dynamics of the supramolecule may be significantly affected upon transfer into the high vacuum. This is not necessarily a disadvantage, because one gets insight into mechanistic aspects of the supramolecule's reactivity which are hard or even impossible to obtain from solution experiments, but one needs to take into consideration that the reactivity may be significantly altered.

When comparing gas-phase data with condensed-phase results, the influence of a charge cannot be neglected. Consequently, a comparison of e.g. neutral dimeric complexes in solution with proton-bridged dimers in the gas phase almost unavoidably leads to misinterpretation.

Finally, the ill-definition of temperature is a severe problem when attempting to collect thermochemical data in the gas phase. Isolated molecules or ions do not have a temperature, which is a macroscopic property and thus is only defined for an ensemble of molecules which are in thermal equilibrium with each other. Thus, the Boltzmann distribution of internal energies is not realized in the gas phase. In solution, all molecules reach thermal equilibrium by exchanging energy through collisions. In contrast, no such energy exchange is realized in the gas phase since collisions are avoided. Some ways to tackle this problem will be discussed below. While we are considering energetics, it should be mentioned that direct comparison of the ease of induction of fragmentation is not easily possible for ions with large size differences. Large ions have so many more degrees of vibrational freedom that they can store much more energy before fragmenting than smaller ions.

5.4 Beyond Analytical Characterization: Tandem MS Experiments for the Examination of the Gas-phase Chemistry of Supramolecules

Many researchers working in other fields than mass spectrometry use this method merely as a tool for the analytical characterization of the substances of interest. The identity of the compounds under study is of interest; maybe, if ionization methods such as EI are used, the fragments generated in the ion source provide some additional structural information. However, modern mass spectrometers are a whole laboratory in themselves whose potential to conduct ion chemistry in the gas phase is enormous. A large variety of experiments is available which allow the mass spectrometrist to examine any ion of interest – once it is successfully generated – in the high vacuum of the mass spectrometer under environment-free conditions. The central prerequisite for all of these experiments is mass selection. Usually, the ion of interest is not the only one generated upon ionization and thus, it is mandatory to get rid of all unwanted ions before doing the experiments on the desired ion. This is quite analogous to any synthetic step done in "real" labs: the reactants are first purified and prepared for the reaction, then the reaction is done, and afterwards, the products are separated and characterized. In the mass spectrometric version, these steps correspond to mass selection, fragmentation, and the product ion scan. It should be noted that different analyzers use different principles for mass selection which we do not describe in detail here, since they are based on the same principles used for ion separation and detection described above. The only change is that no scan is done, but the instrument is set to a single m/z – that of the ion of interest. In the following sections, we will describe briefly some of the experiments available nowadays for this kind of gas-phase chemistry.

5.4.1 Collision-induced Decay (CID)

The method most often applied to induce fragmentation is the use of collision gases – usually noble gases. For high-energy collisions within sector-field instruments, helium is the most prominent one, for the lower-energy collisions occurring in the FTICR cell, argon is typical. Other gases can however also be used. When mass-selected ions collide with a stationary collision gas, a part of their kinetic energy is converted into internal energy, i.e. rovibrational excitation. If the collision-energy is high enough, bonds can be broken or rearrangement reactions with subsequent fragmentations may occur. It is important to distinguish between low- and high-energy collisions. In high-energy collisions, rearrangements are more or less suppressed in favor of direct bond cleavages. These collision experiments provide quite direct information on atom connectivities in the ions under study. Low-energy collisions may give rise to a substantial amount of rearrangement reactions and the analysis in terms of atom connectivities is sometimes rather difficult. However, for supramolecular chemistry, the connectivities within the building blocks are known beforehand and their spatial arrangement in the complex is of major interest. In this case, the noncovalent bond energies are low enough that complex rearrangements within the covalent scaffold are not expected and this point does not play an important role. Nevertheless, one should be aware that there might be cases where it matters.

The pressure regime of the collision gas should be chosen to warrant that on average every ion undergoes only one collision (so-called single-collision conditions; in FTICR instruments, for example, 10^{-7} to 10^{-8} mbar is a good choice). But even if that holds true, the energy distribution within a population of ions after the collisions is broad. Consequently, in most cases, several competing fragmentation reactions are observed, although they differ with respect to their activation barriers. Nevertheless, thermodynamic data can be extracted from CID experiments, when the threshold energy is determined at which the fragmentation reaction begins to occur (so-called threshold-CID experiment; TCID) [27]. This technique requires a particular experimental set-up, which is not commercial today and thus only a few such instruments are available to date worldwide.

5.4.2 Infrared-multiphoton Dissociation (IRMPD)

Instead of collisional activation, the internal energy can also be increased by irradiating the mass-selected ions with an IR laser (usually a CO_2 laser operating at a wavelength of 10.6 μm). Absorption of a single photon does not provide sufficient energy to induce fragmentation reactions, but laser flux densities are high enough for multiphoton absorption to occur with subsequent fragmentation reactions. Thus, the experiment is called infrared multiphoton dissociation (IRMPD). One prerequisite of course is that the ions have at least one absorbing vibrational mode at the wavelength of the laser. This is the reason, why this technique works very well for larger ions with many coupled vibrations, but sometimes is problematic for very small species.

The IRMPD experiment has been developed into a quantitative tool for the estimation of activation barriers of fragmentation processes, the so-called FRAGMENT method [28]. The barriers correspond to binding energies for simple bond dissociation reactions, if the reverse reactions do not have significant barriers. Consequently, kinetic and thermodynamic data are accessible with this method despite the ill-definition of temperature. The experiment relies on the generation of a steady state of IR photon absorption and emission after a short induction period. In this case, thermal equilibrium is reached and internal energies (at least for large ions) follow Boltzmann distributions. Where this equilibrium lies and which actual temperature the ions thus have is not known, but when the laser flux density is increased, the steady state is shifted towards vibrationally more highly excited ions. Or expressed in a simplified way: the ion temperature depends on the laser flux density.

$$\ln k = Ae^{-E_A/RT} \tag{5.9}$$

Practically, the rate constants k for the dissociation of interest are determined at different laser flux densities P (given in W cm^{-2}), i.e. different ion temperatures. Fragmentations are unimolecular reactions, and thus, they follow the Arrhenius law (Eq. (5.9)). Since the actual temperature is unknown, the Arrhenius plot of $\ln k$ over $1/T$ needs to be modified in that $\ln k$ is plotted against $\ln P$. Linear relationsships are obtained. The activation energy E_A can be determined from the slope of the line. E_A contains the enthalpic contributions to the barrier. The entropic contributions remain unknown, because they are included in the pre-exponential factor A of the Arrhenius equation. This factor can only be determined from the intersection of the line with the ordinate and since the actual temperature is not known, there is no way to determine the intersection and the pre-exponential factor.

5.4.3 Blackbody Infrared Radiative Dissociation (BIRD)

The problem that the pre-exponential factor cannot be determined from quantitative IRMPD experiments can be circumvented by using an experiment which is called blackbody infrared radiative dissociation (BIRD). As in IRMPD experiments, BIRD is usually conducted in FTICR instruments. It does, however, require some special equipment, which is not always available. The BIRD experiment again is based on IR photon exchange with the ions in order to restore thermal equilibrium. The photon source, however, is different as compared to IRMPD. For the BIRD experiment, a heatable and temperature controlled FTICR cell is utilized which acts as the blackbody irradiation source. After a short induction period, the ions reach thermal equilibrium with the cell walls and thus their temperature is defined as the temperature of the cell wall. With the temperatures known, the rate constants of a gas-phase reaction can be determined at different temperatures and a simple Arrhenius plot will provide both E_A and A. Consequently, the enthalpic and entropic contributions to the activation barriers become accessible. In principle, even the activation enthalpy and activation entropy can be directly derived from the Eyring equation (Eq. (5.10)) by plotting $\ln(k/T)$ over $1/T$ (Eq. (5.11); k_B is the Boltzmann constant, h the Planck quantum, R the gas constant).

$$k = \frac{k_B T}{h} e^{-\Delta G^{\neq}/RT} = \frac{k_B T}{h} e^{-(\Delta H^{\neq} - T\Delta S^{\neq})/RT} \tag{5.10}$$

$$\ln \frac{k}{T} = \ln \frac{k_B}{h} + \frac{\Delta S^{\neq}}{R} - \frac{\Delta H^{\neq}}{RT} \tag{5.11}$$

5.4.4
Electron-capture Dissociation (ECD) and Electron Transfer Dissociation (ETD)

Electron-capture dissociation and electron-transfer dissociation are relatively new methods of inducing fragmentation reactions. Since electrons stemming from an electron-emitting cathode (ECD) or from an anion (ETD) react with cations, multiply-charged cations are required. Singly-charged cations will react with an electron to yield a neutral product which cannot be detected by mass spectrometry. This makes these methods highly interesting for proteomics, where protein and peptide ions usually are generated as multiply-charged species by electrospray ionization. In order to react efficiently with the cation, the electron must be slow. Once caught by the peptide, the resulting radical induces specific cleavages making the fragmentation pattern of the peptide/protein less complex and easier to interpret that those obtained by collisional activation or IRMPD. However, the two methods have not been utilized for supramolecules yet and we mention them only briefly here, because they may have an interesting potential for the future.

5.4.5
Bimolecular Reactions: H/D-exchange and Gas-phase Equilibria

In FTICR instruments, one can conduct bimolecular reactions between the mass-selected ion of interest and a neutral reactant. This is possible, because the ions are stored in the FTICR cell, so that they can react with a gaseous compound and the reaction can be followed over time. The rate constants are usually given as the fraction of the collision rate.

The major limitation of this experiment comes from the availability of suitable reactants, because they must be volatile enough to be evaporated into the ICR cell.

Sticky, polar molecules can cause practical problems, because they adsorb on the cell walls and the transfer lines and it takes quite a long time to get rid of them after the experiment. While this is normally no problem in the cell, which can be baked out over night, the transfer lines retain some amount of the reactants and do not have a bake out function. Consequently, there are limitations with respect to the reactants that can be used.

Nevertheless, two important types of experiments have been applied to supramolecules. The first one is the exchange of protons against deuterons by reaction of an ion with e.g. methanol-OD or deuterated ammonia. The kinetics of the exchange reaction can be followed, if one has a good model of which protons are exchangeable. The second experiment works with two different reactants present in the FTICR cell at the same pressure which thus compete in their reactions with the ions of interest. If one follows these reactions over time, an equilibrium situation is finally obtained. If the partial pressures of the two reactants are appropriately controlled (and this represents a nontrivial problem), one can deduce relative thermodynamic data, for example, the differences in binding energies for different host–guest complexes. It should be noted that this experiment works best if the two binding energies are close to each other. The smaller the difference in binding energies, the smaller the experimental error.

5.5 Selected Examples

In the following sections, the above methods will be illustrated with a couple of selected examples. The list cannot be comprehensive. In particular, mass spectrometric studies on the analytical characterization meanwhile number into the hundreds, while the other aspects described here are less prominent in the literature.

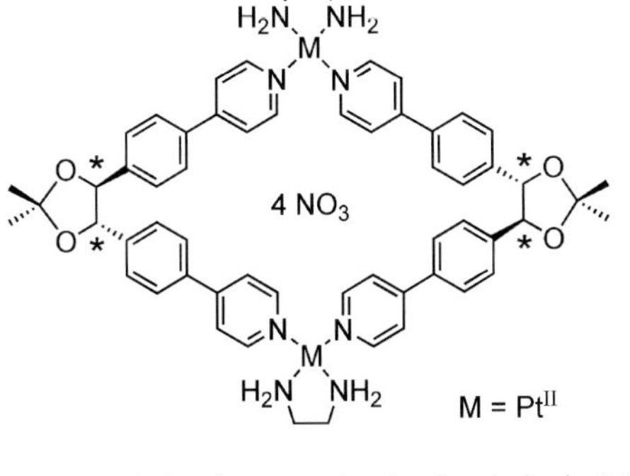

Fig. 5.7. A chiral supramolecular rhomb: both (R,R,R,R) and (S,S,S,S) enantiomers have been synthesized and examined with mass spectrometry.

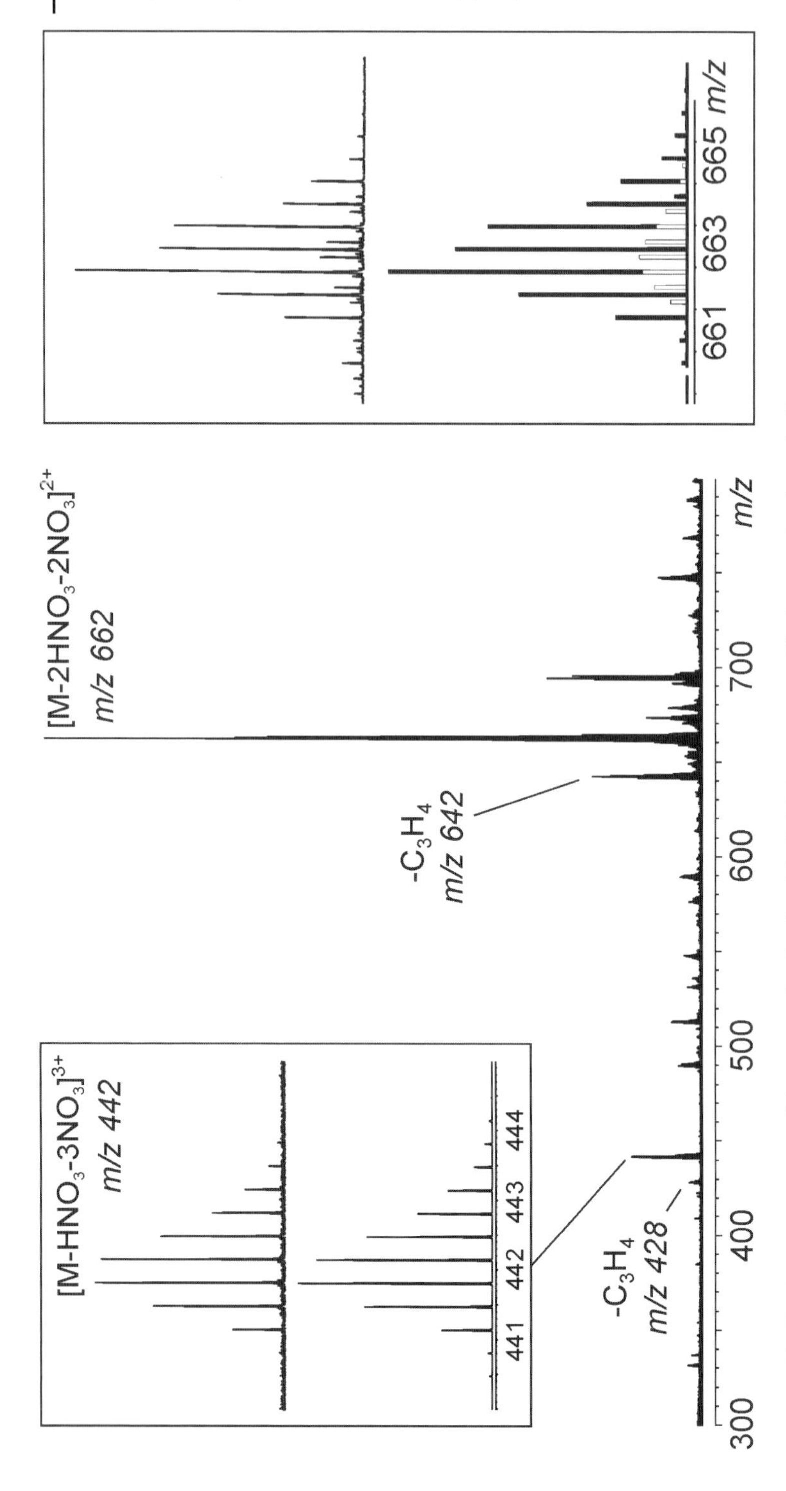

Fig. 5.8. ESI FTICR mass spectrum of the supramolecular rhomb's (S,S,S,S) enantiomer. The intense signal at *m/z* 662 represents the doubly-charged complex $[M{-}2\,HNO_3{-}2\,NO_3]^{2+}$. The insets show the experimental (top) and calculated (bottom) isotope patterns of the doubly-charged ion $[M{-}2\,HNO_3{-}2\,NO_3]^{2+}$ (right inset) and the triply-chargec ion $[M{-}HNO_3{-}3\,NO_3]^{3+}$ (left inset). The inset of the doubly-charged species at *m/z* 662 is superimposed by another isotope pattern of a triply-charged 3:3 complex, that is also included in the calculation.

5.5.1
Analytical Characterization: Exact Mass, Isotope Patterns, Charge State, Stoichiometry, Impurities

The chiral metal complex shown in Fig. 5.7 provides an excellent example for the information that can be retrieved from a simple mass spectrum. The two Pt(II) complexes at its corners mean that the complex is a tetracation. Ionization by ESI occurs through successive losses of nitrate counterions. Interestingly, no singly-charged ion for this complex is observed. So, when beginning the interpretation of the mass spectrum, one should always take a look at the isotope patterns. The distance between two adjacent signals equals $1/z$ (with z representing the number of charges). Consequently, the ion at m/z 442 is identified as trication, because the peak spacing is 0.33. This means that the actual mass of the corresponding ion is $m = (3 \times 442) = 1326$ Da which nicely corresponds to the mass of the complete rhomb (1575 Da) which has lost three NO_3^- counterions (62 Da each) and one HNO_3 molecule (63 Da; thus $1326 = 1575 - 63 - (3 \times 62)$ Da). From this mass, it is clear that the complex not only lost three nitrate counterions during the ionization, but also one HNO_3 molecule incorporating the last counterion. This information should be verified by determining the exact mass of the ion and by comparing the experimental isotope pattern with the calculated one (inset). Both data help to confirm the elemental composition.

Based on this information, we can assign the peak at m/z 662. When the ion is capable of losing counterions as HNO_3 molecules, one may expect that doubly-charged cations appear which not only lost two nitrate anions, but two additional HNO_3 molecules. The peak spacing in the isotope pattern should be 0.5 mass units. Indeed, the peak at m/z 662 corresponds to this doubly-charged ion. However, when the experimental isotope pattern is compared to the calculated one (which should basically be identical with the one calculated for the +3 charge state), quite significant deviations are observed. In particular, small signals arise between the large ones with peak spacings of 0.33 mass units. This indicates that a small contribution from a trication is superimposed on the dication and causes the deviations. Since both have the same m/z ratio, the trication must have the 1.5-fold mass of the complex under study. Consequently, a detailed isotope pattern analysis allows us to identify a second component in the mixture. Besides the 2:2 complex of ligand and corner, a 3:3 complex is formed in equilibrium with it.

Other signals are present in the spectrum indicating some impurity: the signal at m/z 642 corresponds to a doubly-charged ion in which one of the acetal groups has been cleaved. Whether this defect is a result of incomplete synthesis or of acetal cleavage in the spray solvent cannot be determined from the mass spectra alone.

This analysis can be extended to the other small peaks in the mass spectrum, and one will find some ion resulting from fragmentation of the complexes during ionization. We will not go into detail here, but rather conclude that the careful analysis of the exact mass and the isotope pattern provides information on the elemental composition and with it on the stoichiometry of the complex (e.g. 2:2

vs. 3:3). One can deduce the ionization mode and the charge state and gets at least a qualitative idea of the impurities in the sample. With respect to the stoichiometry, the warnings issued above on unspecific binding remain valid, of course.

5.5.2
Structural Characterization of Supramolecules

When discussing the contributions mass spectrometry can make for a structural assignment of supramolecules, we do not have in mind the connectivity of the atoms within the different building blocks. Rather, the "secondary" structure is meant: How are the different components positioned in the complex, how are they bound? Two examples will illustrate this: (i) Molecules with mechanical bonds (e.g. rotaxanes, catenanes, and knots) have special topologies and mass spectrometric experiments aiming at distinguishing different topologies are highly interesting not only from a fundamental point of view, but also for the synthetic chemist, who deals with these species. (ii) Encapsulation complexes which bear a guest molecule within a capsule held together by hydrogen bonding have attracted a lot of interest. In many cases, valuable evidence for encapsulation comes from mass spectrometry.

5.5.2.1 The Mechanical Bond: How to Distinguish Molecules with Respect to Their Topology

Figure 5.9 shows the structural formula of a rotaxane which can be synthesized by a hydrogen-bond mediated template effect [29]. The axle is mechanically trapped inside the cavity of the wheel by two sterically-demanding stopper groups. CID experiments can be used to provide definitive evidence for the threaded topology, because the structural alternative is a non-threaded, hydrogen-bonded complex of axle and wheel. It should be not too difficult to obtain structure-indicative spectra, because the complex dissociates by breaking a few hydrogen bonds, while the rotaxane can only decompose, when a covalent bond is broken. The rotaxane is easily

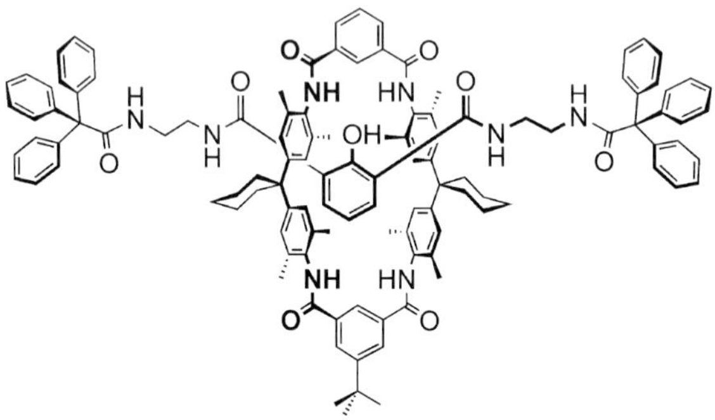

Fig. 5.9. Example for a rotaxane, which consists of an axle mechanically trapped inside the cavity of the macrocycle by large stopper groups attached to its two ends.

ionized by deprotonation of the central phenol group and can be observed as the corresponding anion in the negative mode. After ionization, the ion of interest is mass-selected. Subsequently, the ion is accelerated and collided with argon at a pressure of ca. 10^{-7} mbar. The non-intertwined axle-wheel complex can be independently generated from a 1:1 mixture of both components and then be subjected to the exact same experiment (same conditions, same parameters). Figure 5.10 shows the resulting CID mass spectra together with the analogous spectrum of the axle alone [30].

The CID mass spectrum of the rotaxane shows rather small fragments at m/z 243, m/z 505, m/z 518, which can unambiguously be assigned to products of covalent bond cleavages of the axle. This becomes clear from the same experiment performed with the axle in the absence of the wheel. Similar fragments are observed for the axle-wheel complex. The two CID spectra, however show two distinct differences: (i) The free, deprotonated axle is the major fragment in the CID spectrum of the complex, while it is virtually absent in the rotaxane CID spectrum. (ii) Both spectra are obtained under the same experimental conditons. While the rotaxane parent survives the experiment to an extent of ca. 85%, almost two thirds of the complex are destroyed. This points to less energy-demanding fragmentation for the complex. Both observations can be explained, when one considers that only four hydrogen bonds need to be broken in the complex in order to separate axle and wheel. The axle thus remains intact and the binding energy is low. From the rotaxane, the wheel can only be lost if the mechanical bond is released and this requires the cleavage of a covalent bond. Since bonds in the wheel are not easy to cleave, the axle is destroyed and only its fragments are observed. Covalent bond cleavage of course is more energy demanding.

While it is quite straightforward to distinguish these two species, the problem becomes more challenging when comparing the tetra- and octalactam macrocycles, catenanes and the trefoil knot shown in Fig. 5.11. For both, fragmentation of the octalactam macrocycle and of the catenane, covalent bonds need to be broken and one may ask whether mass spectrometry is still able to provide criteria for differentiating the two topologies.

Again, CID experiments provide the structural assignment. Ionization can be achieved by protonation in the positive mode [31] as well as deprotonation of an amide in the negative mode [30]. For both charge states, the result is similar: all monomacrocycles fragment through water losses originating from the amide bond, while other fragments are rather low in intensity and become clearly visible only at higher collision energies. Instead, the catenane first loses one wheel. The remaining macrocycle shows the same fragmentation reactions found for the tetralactam macrocycle. Mass spectrometry thus provides clear evidence for the intertwined topology of catenanes – an approach which has recently been used for other types of catenanes as well [32].

5.5.2.2 **Encapsulation of Guest Molecules in Self-assembling Capsules**

The encapsulation of guest molecules in self-assembling capsules is a topic of great interest in supramolecular chemistry. Quite often, it is rather difficult to provide

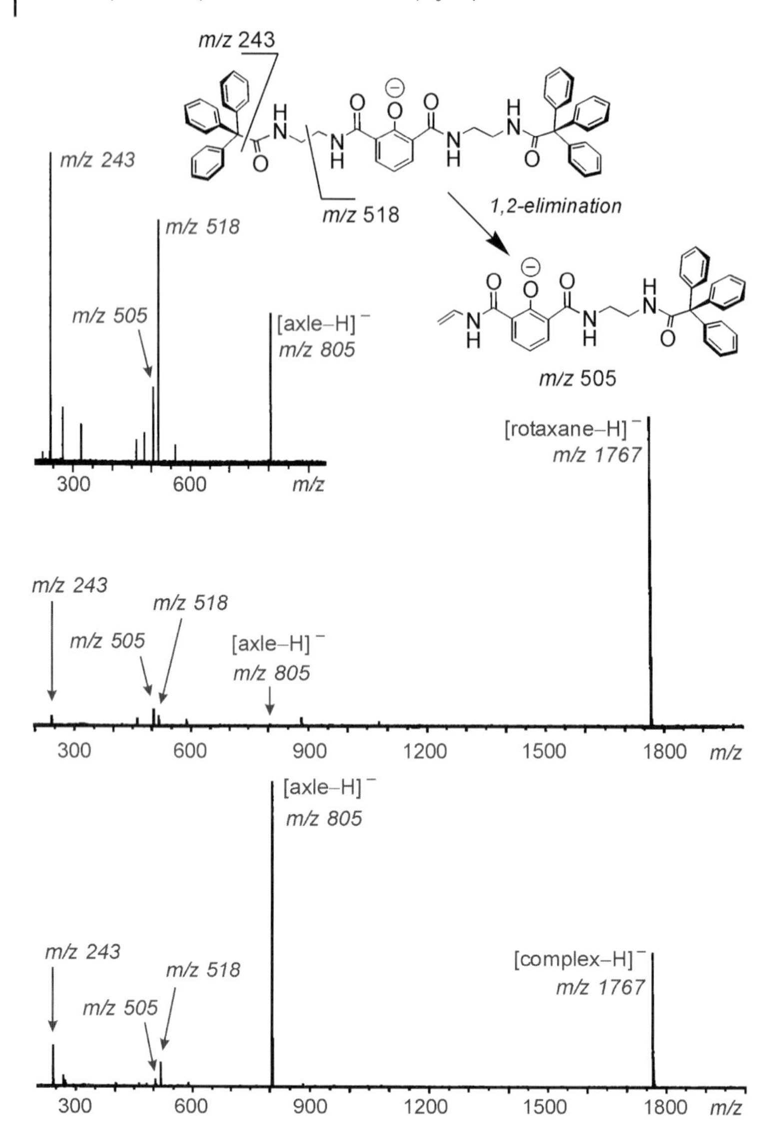

Fig. 5.10. CID experiments conducted with (top to bottom) the mass-selected axle, rotaxane and the non-intertwined hydrogen-bonded axle-wheel complex. The inset on top shows the fragmentations of the deprotonated axle.

clear evidence for the encapsulation inside the capsule's cavity – in particular when the capsule is a highly dynamic species [33]. NMR experiments may show two separate signals for the free and the encapsulated guest, provided that the guest exchange is slow enough. Complementary information from mass spectrometry is nevertheless very helpful. Mass spectrometric structural proof for encapsulation

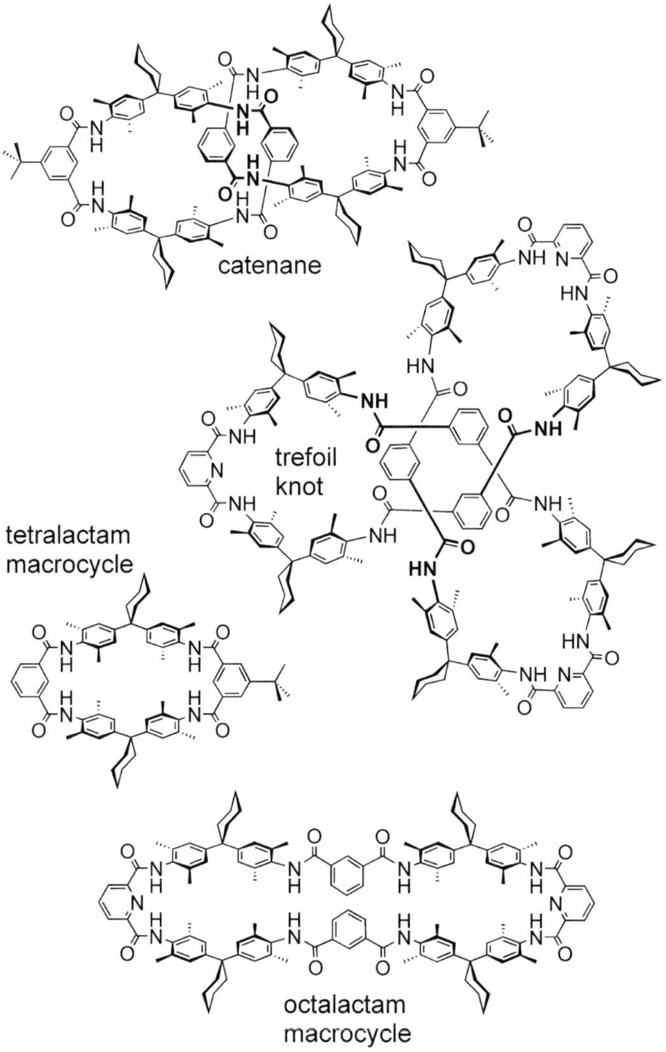

Fig. 5.11. Macrocycles, catenanes and knots of the amide type.

can be achieved by using charged guests which make ionization easy [21]. Competition experiments with different guest molecules can be used to address the size, shape and symmetry congruence of guest and capsule cavity. Heterocapsule formation can occur when two different capsules exchange their constituent subunits. This reaction can be used to address dynamic processes by mass spectrometry. Finally, tandem MS experiments may even provide some insight into the capsule's structure in the gas phase.

Figure 5.12 shows the Rebek softballs. Two C-shaped monomers can form a capsule self-assembled through a seam of hydrogen bonds. Inside the cavity, the

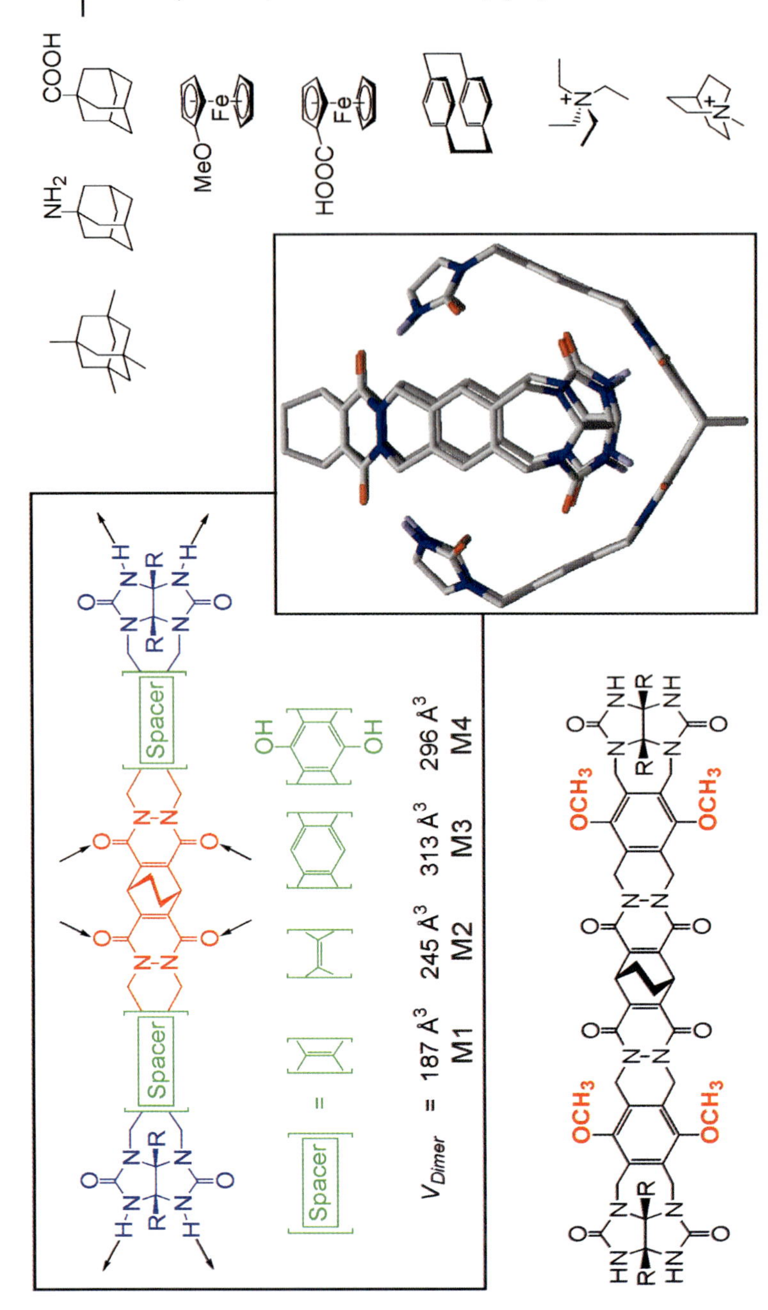

Fig. 5.12. Rebek's softballs – dimeric self-assembling capsules. The cavity volume can be tuned by the length of the spacer introduced between the center piece and glycoluril binding sites of the monomers. The methylated monomer (bottom) is not able to form dimeric capsules. Several hydrogen bonding donors and acceptors (arrows) that are responsible for dimerization, are implemented in each building block. A computer-generated model of an example of a dimeric capsule is shown in the box. Right: Typical guest molecules which can be encapsulated in the cavity.

capsules can accommodate guest molecules. For mass spectrometric characterization, electrospray ionization was used. Normally, methanol or water are used as the spray solvents in ESI MS. Since these solvents compete with the hydrogen bonds between the two capsule halves, a noncompetitive solvent needs to be used, e.g. chloroform or dichloromethane. In this case, a charge must be provided, which can elegantly be done by encapsulating cations such as tetraethylammonium. Finally, the counterion is important. Weakly coordinating counterions not only help to increase solubility of the guest salt, they neither interfere with the hydrogen-bonding pattern. This strategy is successful and Fig. 5.13 shows the mass spectra for four different softballs which all exhibit intense signals for the desired 2:1 complexes of capsule monomers and guest cation.

The desired structural information of these capsules is now elucidated in five steps: (i) Detecting the elemental composition and charge state for comparison with the calculated isotope pattern on the basis of natural abundances. (ii) Verification of the hydrogen-bonded nature of the assembly by addition of methanol to the sample that should disrupt the hydrogen bonds and thus make the signal of the intact capsule disappear. (iii) Competition experiments with guests of different sizes reveal whether a certain size selectivity governs guest binding. This is a good sign for encapsulation of the ion inside the cavity. (iv) A competition experiment with a 1:1 mixture of a softball monomer and the tetramethylated analog shown in Fig. 5.12 should give rise to signals only for the nonmethylated softball, because the methylated compound is known from solution phase studies (e.g. NMR) not to form capsules. This experiment provides evidence for recognition between the two capsule halves. (v) Further evidence for the hydrogen-bonded structure comes from exchange experiments, in which equimolar amounts of two different homodimers are mixed. Signals corresponding to the heterodimers reveal the exchange of monomers between the two capsules and thus support the dynamic nature of the capsules. All five experiments yielded the expected results and thus, mass spectrometry successfully provided insight into their capsular structure in solution.

Conclusive evidence for a capsular structure even in the gas phase after ionization is provided by collisional activation of capsule ions. In this experiment, covalent bond cleavages are observed to compete with loss of the guest, which consequently must be bound quite strongly (Fig. 5.14). Although the structural insight gained by these experiments is somewhat indirect, they suggest the capsule structure to be retained during ionization.

This approach succeeds not only for dimeric capsules, but can also be applied to pyrogallarenes and resorcinarenes such as those shown in Fig. 5.15, which self-assemble in the crystal [34] and in solution [35] to hexameric capsules. The assembly can be thought of as a cube with one bowl-shaped monomer on each face. The monomers are connected to each other by a complex network of hydrogen bonds and surround an interior volume of more than 1200 $Å^3$ which can be filled with guest molecules. If the monomers are dissolved in $CHCl_3$:acetone (2:1) without addition of a guest, clusters are observed with intensities decreasing with cluster size (Fig. 5.16a). This is expected, when unspecific binding occurs – an undesired effect which occurs quite often under soft ESI conditions. The addition of templating

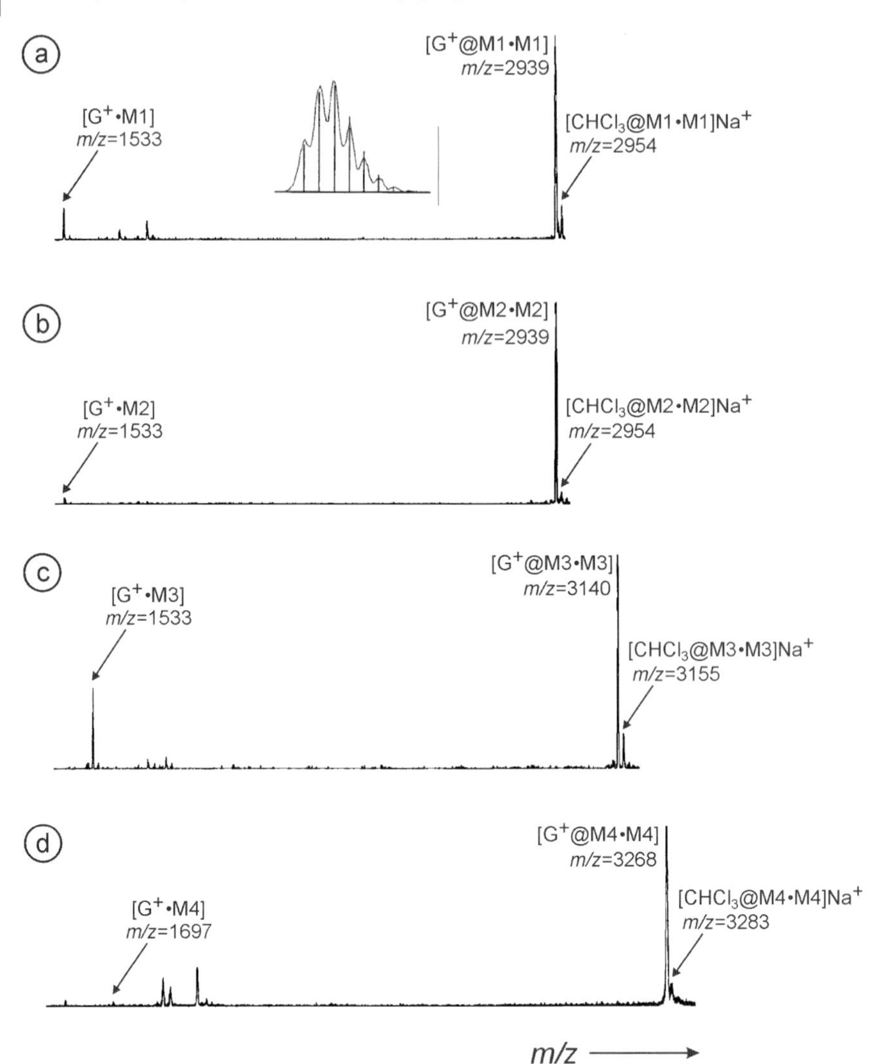

Fig. 5.13. ESI mass spectra of 2:1 mixtures of one of the four softball monomers shown in Fig. 5.12 (M1–M4) and tetraethylammonium as the guest cation (G^+).

cations changes the picture. Upon addition of tetramethyl ammonium salts, we observe the almost exclusive formation of dimers (Fig. 5.16b). In line with literature [36], these dimers can be assigned to small capsules which entrap one guest cation inside. Larger cations such as tetrabutyl ammonium change the size distribution (Fig. 5.16c), but do not specifically form hexamers. Only a guest such as tris-2,2′-bipyridine ruthenium(II), which is almost perfectly congruent with the geometry of the cavity in size, shape and symmetry (Fig. 5.15, right), produces a rather clean

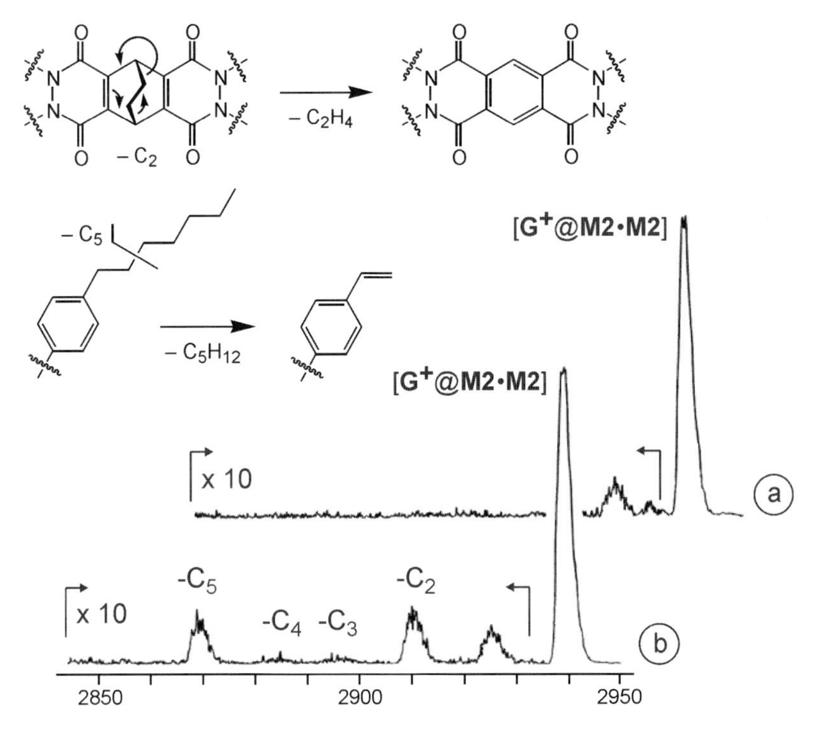

Fig. 5.14. Fragmentation processes during a CID experiment with a softball–tetraethylammonium complex. a) Mass spectrum of a sample solution without additional acceleration of the ions. b) The same experiment with an additional acceleration voltage of –100 V. The inset shows which fragmentation processes reasonably rationalize the observed product ion signals.

spectrum with the hexamer signal being the base peak (Fig. 5.16d) [37]. A control experiment with the tetramethyl resorcinarene in Fig. 5.15 clearly rules out unspecific hexamers: no signal for any hexamer is observed with this compound for which the formation of the hydrogen bond network is impossible. Finally, if two different resorcinarenes or a resorcinarene and a pyrogallarene are mixed directly before the mass spectrometric experiment, a completely statistical mixture of homo- and heterohexamers is observed indicating that the exchange of monomers is fast under these conditions.

The investigation of encapsulation processes is not restricted to hydrogen-bonded capsules. Also, metallosupramolecular encapsulation complexes can be studied. Figure 5.17 depicts an octa-anionic tetrahedron self-assembled around a templating tetraethylammonium cation. The mass spectrum is characterized by signals for multiply-charged ions formed by successively stripping away tetraethylammonium counterions. The presence of 12 Cd, 12 Br, and 12 Cl atoms in the whole structure, generates a very broad isotope pattern. Nevertheless, ESI–FTICR mass spectrome-

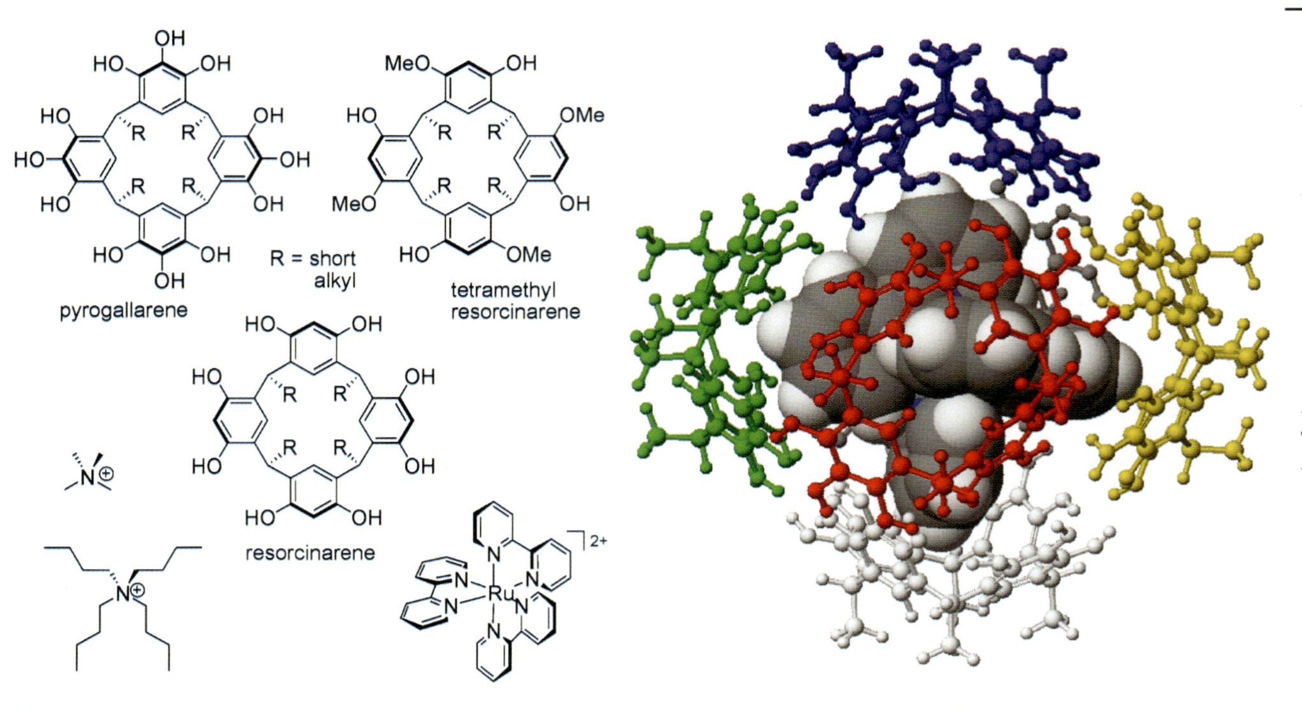

Fig. 5.15. Pyrogallarenes and resorcinarenes which form hexameric capsules through hydrogen bonding. The tetramethyl resorcinarene serves as a control compound. At the bottom, three different guest cations are shown, which carry charges to make the capsule MS-detectable. Right: Computer model of the $Ru(bpy)_3{}^{2+}$ guest encapsulated in the cavity of the hexamer. Each monomer is shown with a different color. It can nicely be seen that the Ru complex not only fits size-wise, but also with respect to its symmetry and shape.

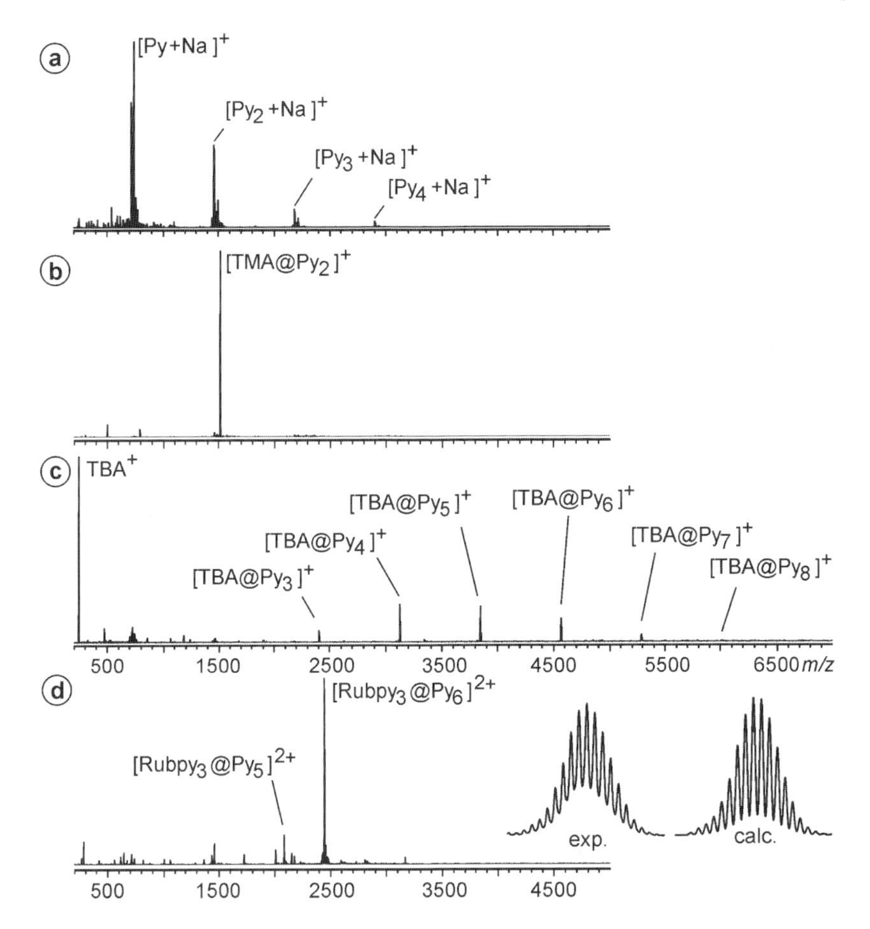

Fig. 5.16. (a) ESI–FTICR mass spectrum of pyrogallarene (Py) in $CHCl_3$:acetone (2:1). (b) ESI mass spectrum of the same solution after addition of 1 eq. of a tetramethyl ammonium (TMA) salt. (c) Distribution of clusters with tetrabutyl ammonium (TBA) as the guest cation. (d) ESI mass spectrum of the hexamer encapsulating $Ru(bpy)_3{}^{2+}$. The experimental and calculated isotope patterns nicely match.

try provides the necessary resolution even to analyze the pattern for the quadruply-charged anions.

A remarkable difference is detected between the mass spectra (Fig. 5.18) of tetrahedra containing $NEt_4{}^+$ or NEt_3H^+ ions as the templating guest [38]. When tetraethyl ammonium is encapsulated, the interior space is nicely filled. The mass spectra indicate that the signal for each charge state is accompanied by consecutive losses of NEt_4Cl or by the exchange of $NEt_4{}^+$ against H^+. The simulated pattern based on natural isotope abundances nicely agrees with the experimental pattern. This is no longer true, when triethyl ammonium acts as the guest. A calculation

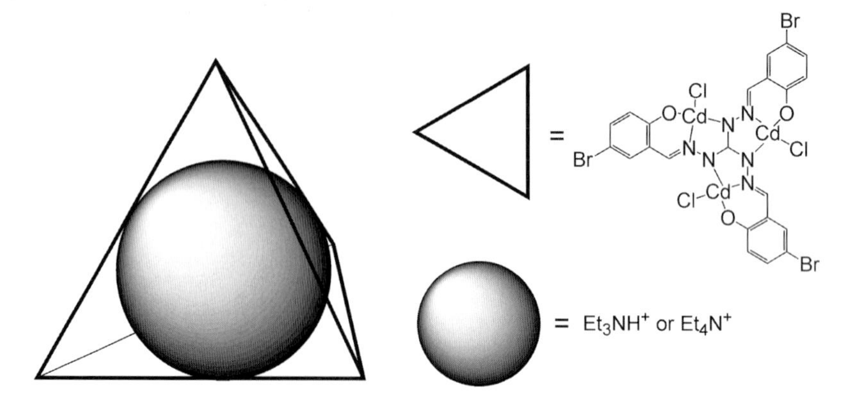

Fig. 5.17. Octaanionic tetrahedron with encapsulated templating guest ion (Et_3NH^+ or Et_4N^+). Each face consists of a triangular Cd complex. The faces are connected by a Cd–O–Cd–O four-membered ring binding motif and the surface is fully closed preventing any guest exchange because of the extended nature of the ligand with the Br atoms pointing towards the tetrahedron's tips.

based on an analogous sequence of ions does not match the experimental pattern closely. A detailed inspection revealed that the addition of a similar distribution of ions containing both a triethyl ammonium and one water molecule provided an excellent fit instead. The conclusion from these experiments is: the water molecule is coencapsulated and unable to escape on desolvation (which is complete in the case of the tetraethyl ammonium guest). It is thus tightly bound and must reside inside the cavity where, according to the crystal structure, it occupies the space of the missing forth ethyl group of the guest cation. Mass spectrometry thus strongly supports that the tetrahedron bears a fully closed surface, which even prevents the escape of the small water molecule from the cavity.

These examples show how structural data on different types of encapsulation complexes can be obtained by mass spectrometry. Again, the evidence is indirect, but helpful and complementary to that from other methods.

Fig. 5.18. Partial ESI–FTICR mass spectra of the tetrahedron. Left: Distribution of different ions generated by successive losses of NEt_4^+ counterions, exchange of the cation against protons, or NEt_4Cl losses. The calculated isotope patterns of the assigned ions almost perfectly fit the experimental ones. Right: Same for triply-charged tetrahedra encapsulating a Et_4NH^+ ion. The isotope patterns of a distribution analogous to the left one don't fit. If ions containing one water molecule are added, an excellent fit is achieved. According to the crystal structure, one water molecule is co-encapsulated with the triethyl ammonium ion.

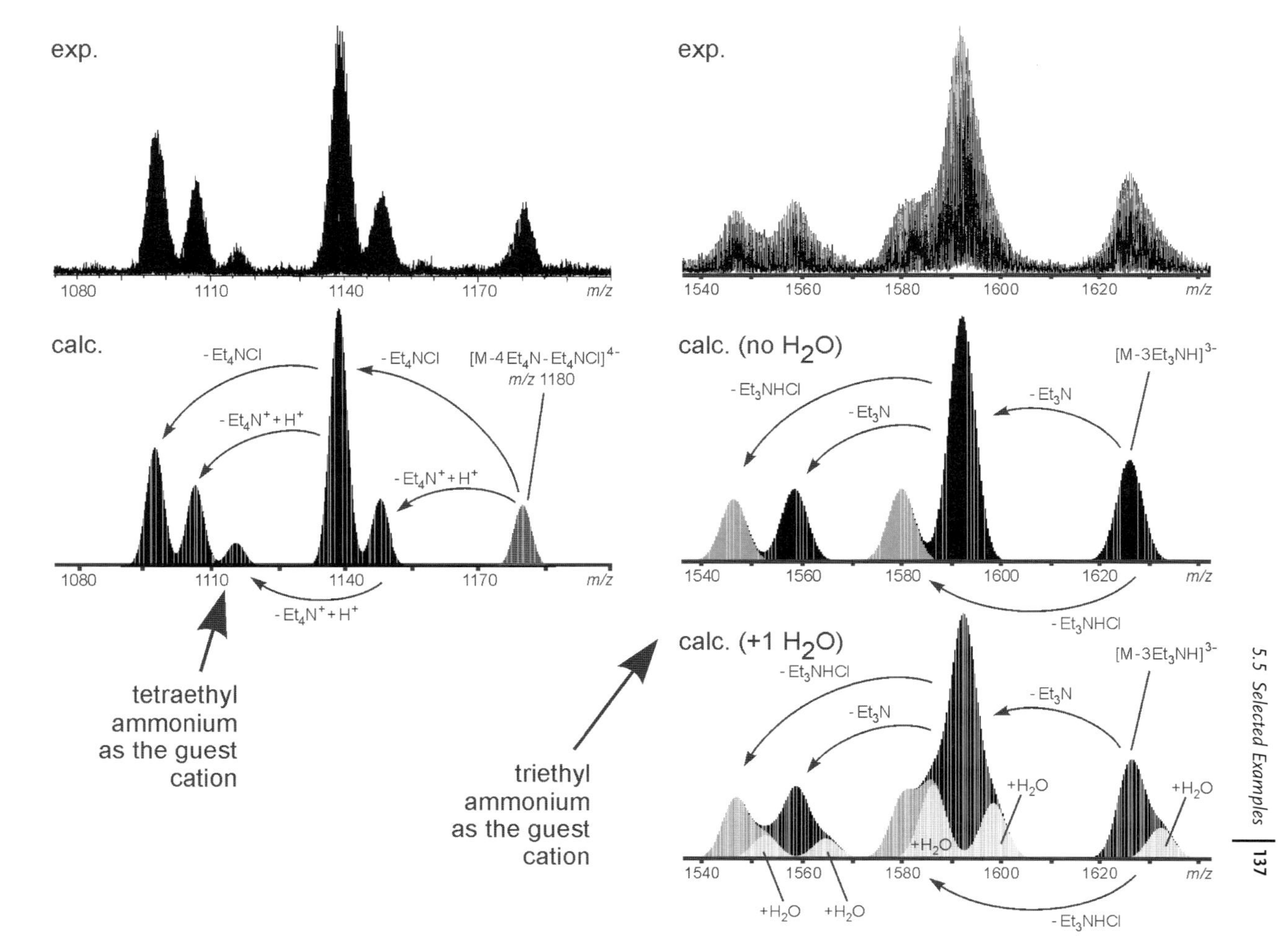

exp.
1080
1110
1140
1170
m/z
calc.
$-Et_4NCl$
$-Et_4N^+ + H^+$
$[M-4Et_4N-Et_4NCl]^{4-}$
m/z 1180
tetraethyl ammonium as the guest cation
exp.
1540
1560
1580
1600
1620
calc. (no H_2O)
$[M-3Et_3NH]^{3-}$
$-Et_3N$
$-Et_3NHCl$
calc. (+1 H_2O)
$+H_2O$
triethyl ammonium as the guest cation

5.5.3
Ion Mobility: A Zwitterionic Serine Octamer?

Quite remarkably, electrospray ionization of rather concentrated methanol solutions (e.g. 0.1 M) of serine yields an abundant ion corresponding to an octameric amino acid cluster [39]. Other cluster sizes are generated only with much lower intensity. The most intriguing feature of these ions is their strong tendency to form homochiral clusters. In a mixture of L-serine and isotopically-labeled [^{13}C]-D-serine, the homochiral octamer appears with an intensity higher than that expected from statistics by a factor of ca. 14, while the 4L/4D cluster is significantly less abundant than expected. In order to explain these features, several structures have been suggested in the literature, among them structures containing zwitterionic serine molecules and structures, where all serines existed in their non-zwitterionic form. One example for a zwitterionic structure is shown in Fig. 5.19. Both forms can be distinguished by their ion size. The strong attractive electrostatic forces within a cluster of zwitterionic serines, leads us to expect that the purely zwitterionic structure will be much more compact than non-zwitterionic ones. In this structure, all eight serine OH groups are involved in a cyclic seam of hydrogen bonds, each connected to a carboxylate oxygen of the next serine. If one changes the stereochemistry at one of the serines, this seam is disrupted.

Since the octamer has not been observed in solution, a mass spectrometric method is needed to distinguish between the two possibilities. With ion mobility experiments [40], such a method does indeed exist. In this experiment, mass-selected ions are subjected into a drift tube filled with helium. A low voltage pulls the ions through the tube. Collisions with helium atoms, which do not usually cause fragmentation because the collision energies are low, decelerate larger ions to a greater extent than smaller ions. The arrival time distribution determined at the end of the drift tube can be converted into the collision cross-section which correlates with the ion's size and shape. The collision cross-section can also be calculated with quite a high precision (uncertainty of experiment vs. theory is often less than 3%). For the serine octamer, cross-sections of 187 $Å^2$ and 191 $Å^2$ were obtained by experiment [41] – values much too low for non-zwitterionic structures. The zwitterionic structure shown in Fig. 5.19 however yields calculated values between 183 $Å^2$ and 191 $Å^2$ depending on the model used [42]. It consequently explains both, the compact size and the tendency towards homochiral clusters.

However, the picture became more differentiated recently, when H/D exchange experiments were performed in the gas phase with the serine octamer [43]. In these experiments, the ions of interest are mass selected and subsequently reacted with a deuterium source, e.g. ND_3 or CH_3OD. The kinetics of the isotope exchange can be followed by mass spectrometry since both isotopes differ in their atomic masses. In these experiments, the serine octamers turned out to consist of two non-interconverting structures, one of which exchanges protons against deuterons significantly faster than the other. The quickly exchanging population is less stable and fragments more easily. It is thus assigned to a non-zwitterionic form. One possible reason, why it has not been recognized in the ion mobility experiments, is its

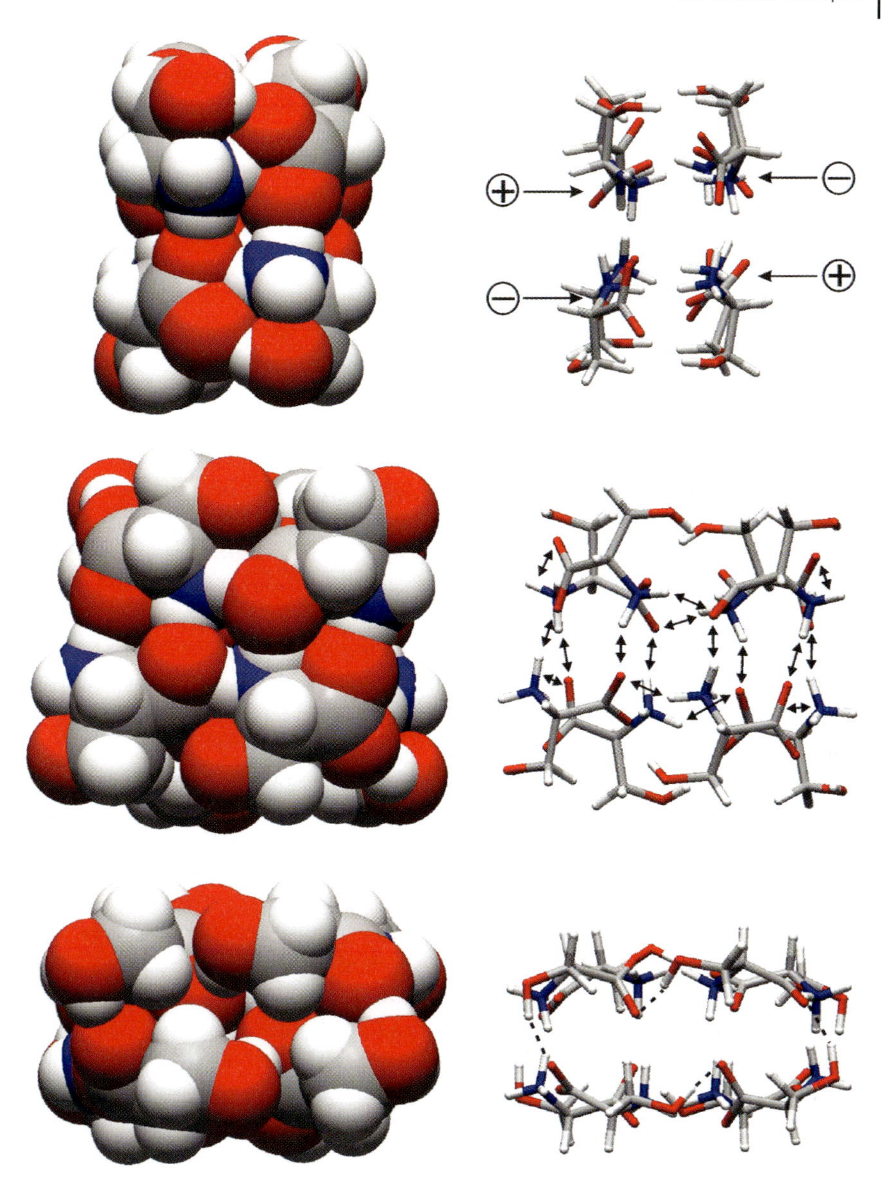

Fig. 5.19. A serine octamer cluster built from zwitterionic L-serine molecules. Left: Space filling representation; Right: favorable interactions (electrostatic attraction, hydrogen bonds) holding the cluster together. Bottom; View showing a cyclic array of hydrogen bonds connecting the serines' OH groups with carboxylates from adjacent serines. Changing the stereochemistry of only one serine will disconnect this array and thus destabilize heterochiral forms.

potential destruction upon submitting the ions into the drift tube. Ion mobility experiments which were used mainly for the examination of gas-phase conformations consequently show a great potential for elucidating details of supramolecular structures as well. However, the second population, which was initially overlooked, reminds us of the potential pitfalls of a purely mass spectrometric approach.

5.5.4 Mass Spectrometry for the Detection of Chirality

Mass spectrometry is an intrinsically achiral method. Consequently, we cannot expect it to distinguish two enantiomeric forms of a compound or supramolecular complex. However, diastereomers can have different properties and thus may well be distinguished by their mass spectra [44]. Two different methods have been developed for assessing the diastereomeric excess of host–guest complexes. Basically, the first one examines mixtures of one enantiomer of the host with one enantiomer of the guest and an achiral reference host [45]. In a second experiment, the other enantiomer of the guest is mixed with the same chiral and achiral hosts. From the relative peak intensities (RPI) of the two diastereomeric complexes as compared to the reference complex, the diastereoselectivity can be measured. Figure 5.20 shows a few examples for chiral crown ethers which have been used to investigate by mass spectrometry their ability to recognize chiral ammonium ions [45, 46].

One of the critical aspects of this approach is that two different experiments have to be performed between which the particular instrument conditions must be carefully kept constant in order not to affect the intensity ratios. This problem can be overcome by the enantiomer-labeled guest method [47]. It is based on the mass spectrometric examination of one enantiomer of the host with a pseudo-racemic mixture of the guest. In order to be able to detect both diastereomers separately, one enantiomer of the guest must be isotopically labeled, usually with deuterium. In the same experiment, both diastereotopic complexes are formed and their intensities can be compared directly. However, the stereochemical effect might additionally be superimposed by an unknown isotope effect. A way to separate stereochemical and isotope effects is to perform the same experiment with the second host enantiomer [48]. In one experiment both stereochemical and isotope effects disfavor the same complex and thus work in the same direction. In the other experiment, they partly cancel each other. If both experiments have been performed, one can use the two experimental values for the intensity ratios of both diasteromeric complexes to deconvolute both effects [49].

The diastereoselectivities observed in the MS experiments for some systems differ from those found in solution. Furthermore, some cases have been found, where different ionization methods, e.g. FAB *versus* ESI [50], gave rise to completely different diastereoselectivities. These findings point to the fact that the ionization procedure might alter the ratios of the species present in solution. Therefore, an approach using *true gas-phase* experiments would be advantageous. Several complexes of chiral ammonium ions with the chiral crown ethers in Fig. 5.20 have

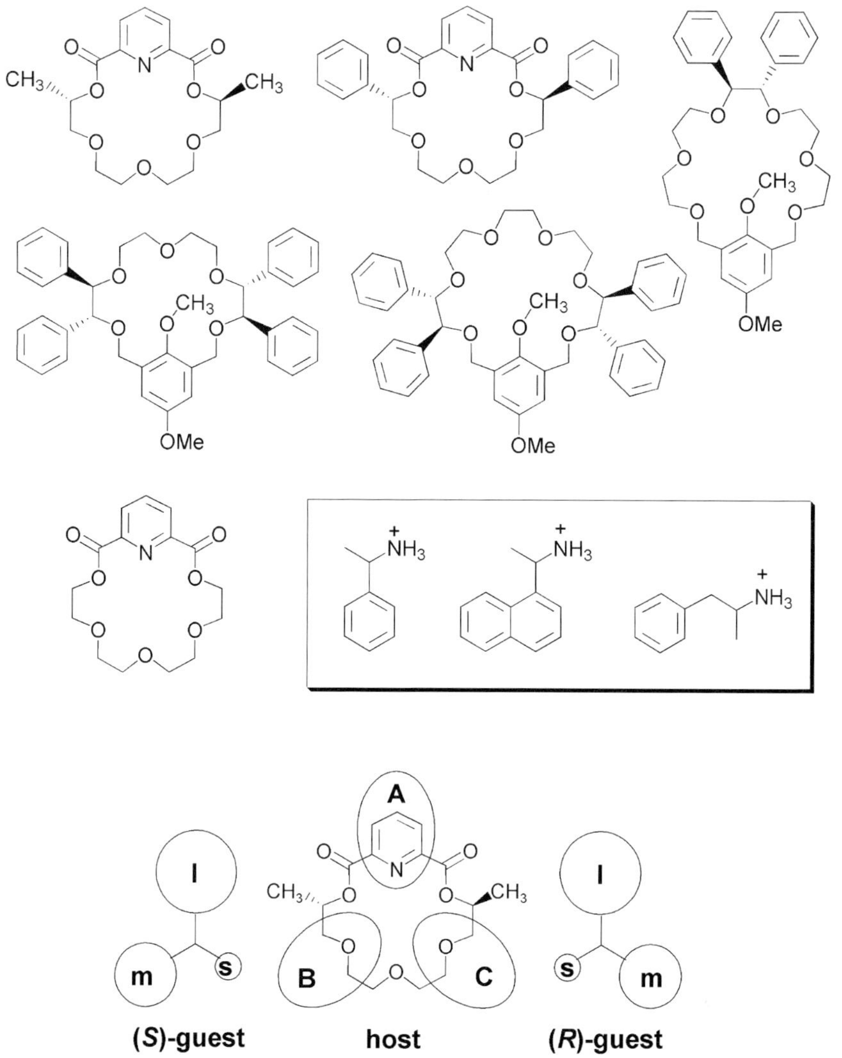

Fig. 5.20. Chiral crown ethers which have been used for chiral recognition studies with the ammonium ions in the inset. As a reference, the non-chiral crown at the center left was used. Bottom: Three-point model for chiral recognition of ammonium ions in crown ethers (l = large, m = medium, s = small substituents).

been studied with so-called cation transfer equilibrium experiments [51]. In this experiment, a crown ether–ammonium complex is generated in one of the two diastereomeric forms and mass-selected. The isolated complex ions are then reacted in the gas-phase with a pseudo-racemate of the neutral amine corresponding to the guest cation. The crown ether can be transferred together with the proton from one amine to another amine. This reaction is followed over time until an equilibrium is reached. The equilibrium distribution directly provides access to relative binding energies. Steric bulk and π–π interactions between the guest and the host contribute to the intrinsic stability difference of the two diasteromeric complexes. The steric interactions could be explained with a three-point model as depicted in Fig. 5.20 [51]. Consequently, with the right isotope labeling strategies, mass spectrometry provides access to the intrinsic energy differences between two diastereomeric complexes. As long as the intensity ratio of both diastereomers is in the range of ca. 0.2 to 4, this data is usually highly accurate because small energy differences translate into a quite large variation of the ratio of both diastereomers.

5.5.5 Reactivity Studies of Supramolecules in Solution

After the discussion of structural details such as the secondary structure of noncovalent complexes, the formation of (non-) zwitterionic clusters, and stereochemical features, reactivity will form the next topic in this chapter. In principle, mass spectrometry can provide two different kinds of reactivity data. It can merely serve as a highly sensitive detector for processes occurring in solution. By following the kinetics of for example ligand exchange reactions in solution, the dynamics of supramolecular systems can be addressed. However, if one considers time scales, the normal time scale for a mass spectrometric experiment is at least in the minutes regime. Consequently, the processes under study need to be quite slow. On the other hand, mass spectrometry directly provides access to the intrinsic reactivity of a supramolecule in the gas phase. While a number of different dynamic processes such as complex formation and dissociation may exist in solution, such processes do not occur in the gas phase. Consequently, quite different mechanistic aspects can be investigated in both experiments. We will thus discuss them separately starting with one example from solution here.

Carbonyl-substituted catechols (Fig. 5.21) react with Ti(IV) salts under slightly basic conditions to yield tris-catecholates in a first assembly step. In the presence of Li^+ ions (Na^+ and K^+ don't do the trick because their ion sizes are larger!), two of these tris-catecholate complexes form dimers bridged by three Li^+ ions [52]. The crystal structures of three representative examples only differing with respect to the

Fig. 5.21. Top: Self-assembly drives the formation of helical, homochiral dimeric titanium tris-catecholate complexes. Dimerization is only mediated by Li^+, while Na^+ and K^+ do not lead to comparable products. Bottom: Crystal structures of the dimers formed from the aldehyde (left, R = H), the ethyl ketone (centre, R = C_2H_5), and the methyl ester (right, R = OCH_3). Shown is a side view in space-filling representation and a ball-and-stick model with a view along the Ti–Ti axis.

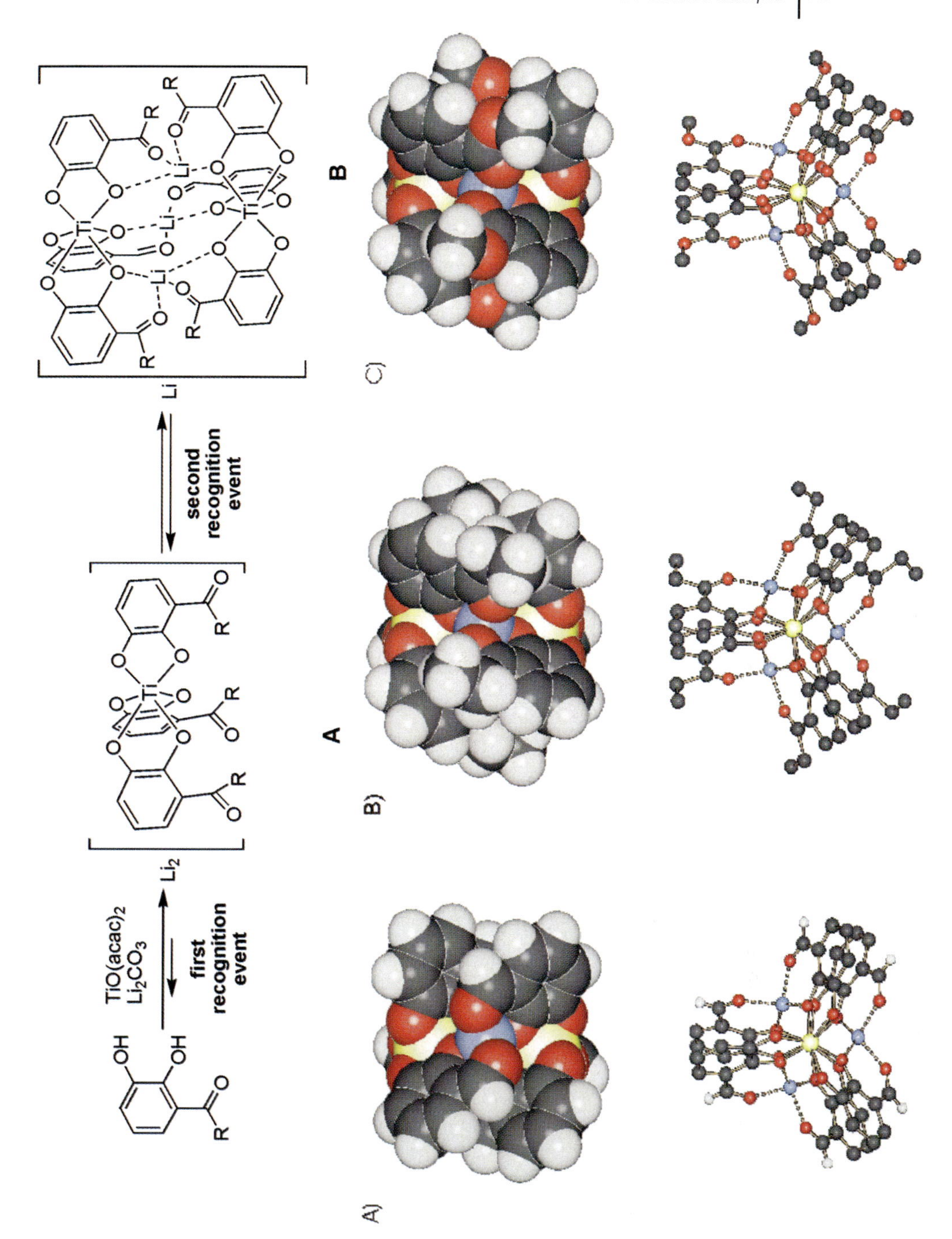

TiO(acac)$_2$
Li$_2$CO$_3$
first recognition event
Li$_2$
second recognition event
Li
A
B
A)
B)
C)

nature of the carbonyl substituent (aldehyde, ketone, or ester) are shown in Fig. 5.21.

The question to be answered by mass spectrometry is how ligand exchange proceeds in solution, when two different dimers are mixed with each other. By NMR methods, this is rather difficult to answer, because a large number of species, i.e. four different monomeric species and a multitude of different dimers (seven different ligand stoichiometries, some of which exist as mixtures of isomers), can be formed in the equilibrium. Mass spectrometry can however easily detect all different stoichiometries simultaneously as long as the two ligands differ in molecular weight. Different exchange processes are possible which lead to different expectations with respect to the mass spectra observed (Fig. 5.22). (i) Direct exchange of single ligands within the dimer without dissociation of the dimer into the two monomers would lead to a completely statistical distribution of all dimer stoichiometries. In the course of the reaction, a U-shaped distribution should appear at intermediate reaction times, because the first ligand must be exchanged, before the second one can be replaced. (ii) If no ligand exchange occurs at neither the dimer, nor the monomer, but dimers can dissociate into monomers, which reassemble to yield dimers, only three signals should appear – the two homodimers and the heterodimer, in which both monomers are intact with three identical li-

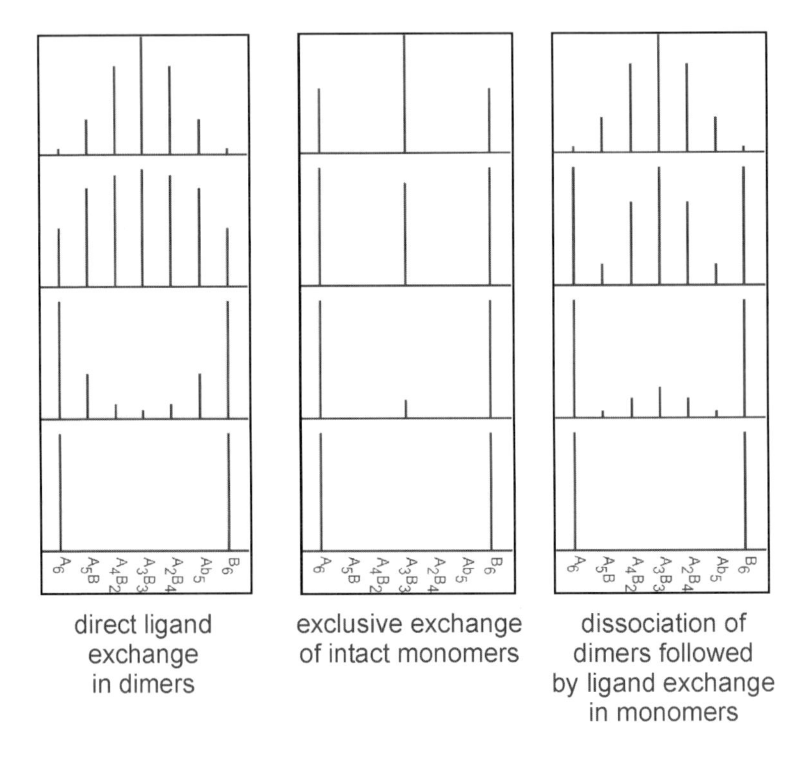

Fig. 5.22. Three different ligand exchange scenarios and their consequences for the mass spectrometric analysis.

gands. (iii) Finally, if ligands can exchange quickly on the monomer, but the exchange is slow within the dimer, the dimers must first dissociate, followed by ligand exchange in the monomer and reassembly to the dimeric species. If this scenario holds true, a statistical distribution of mixed forms grows between the two homodimers.

Figure 5.23 shows two series of mass spectra obtained from an equimolar mixture of two different lithium-bridged helicates with ketone-substituted catechols. Each series was measured in a particular solvent, i.e. tetrahydrofurane:methanol (1:1) and pure tetrahydrofurane (THF). In the mass spectra obtained immediately after mixing the two helicates (0 min), signals are observed exclusively for the two homodimers, while heterodimers are absent. While this is the same for both solvents, a first remarkable difference is found for the monomer signals (insets). While the monomers exchange ligands quickly in the presence of methanol (a statistical 1:3:3:1 mixture of the A_3, A_2B, AB_2, and B_3 monomers is found), the ligand exchange among monomers is slow in pure THF (only the two A_3 and B_3 monomers are observed). This finding is easily explained by the presence of methanol, a protic solvent which mediates catecholate exchange by protonating the catechol and offering methoxy ligands for stabilizing intermediates in which a catechol dissociates from the metal centre. A look at the last spectra in each series shows that finally an almost statistical mixture of all possible dimers (A_6, A_5B, A_4B_2, A_3B_3, A_2B_4, AB_5, B_6, some of which likely exist as mixtures of different positional isomers) is realized, although two very different reaction times are necessary depending on the solvent mixture.

Even more intriguing are the intermediate spectra. With these scenarios in mind, a look at the two series of spectra shows that no U-shaped distribution is generated in either of the solvent mixtures. Consequently, the exchange of individual ligands between dimers or between dimers and monomers does not occur. This is quite understandable, since the Li^+ ions coordinate the ligands within the dimer and slow down the ligand exchange process within the dimers. In pure THF, at first, an intense signal for the A_3B_3 dimer is found, while it takes much longer to generate the A_5B, A_4B_2, A_2B_4, and AB_5 dimers. This can only be explained with an exchange of complete monomers being faster than the exchange of individual ligands between monomers – in line with the observation that ligand exchange is slow for the monomer (see above). The latter reaction nevertheless occurs to some extent and finally gives rise to the statistical equilibrium mixture. In the presence of methanol, finally, ligand exchange proceeds via dissociation of the dimers into the monomers, a fast ligand exchange between monomers and reassociation of the monomers to yield statistical mixtures of dimers.

This example for mass-spectrometrically followed solution-phase reactivity clearly shows, how much information can be derived from some simple measurements. Qualitatively, it can be deduced which mechanisms contribute, which don't. Quantitatively, at least a ranking of the relative rates for the different processes is obtained. Finally, this example makes clear how large the influence of the solvent may be. The change of the mixture from pure THF to THF: methanol mixtures is not too drastical, but still, significant changes in reaction mechanism and rates are found.

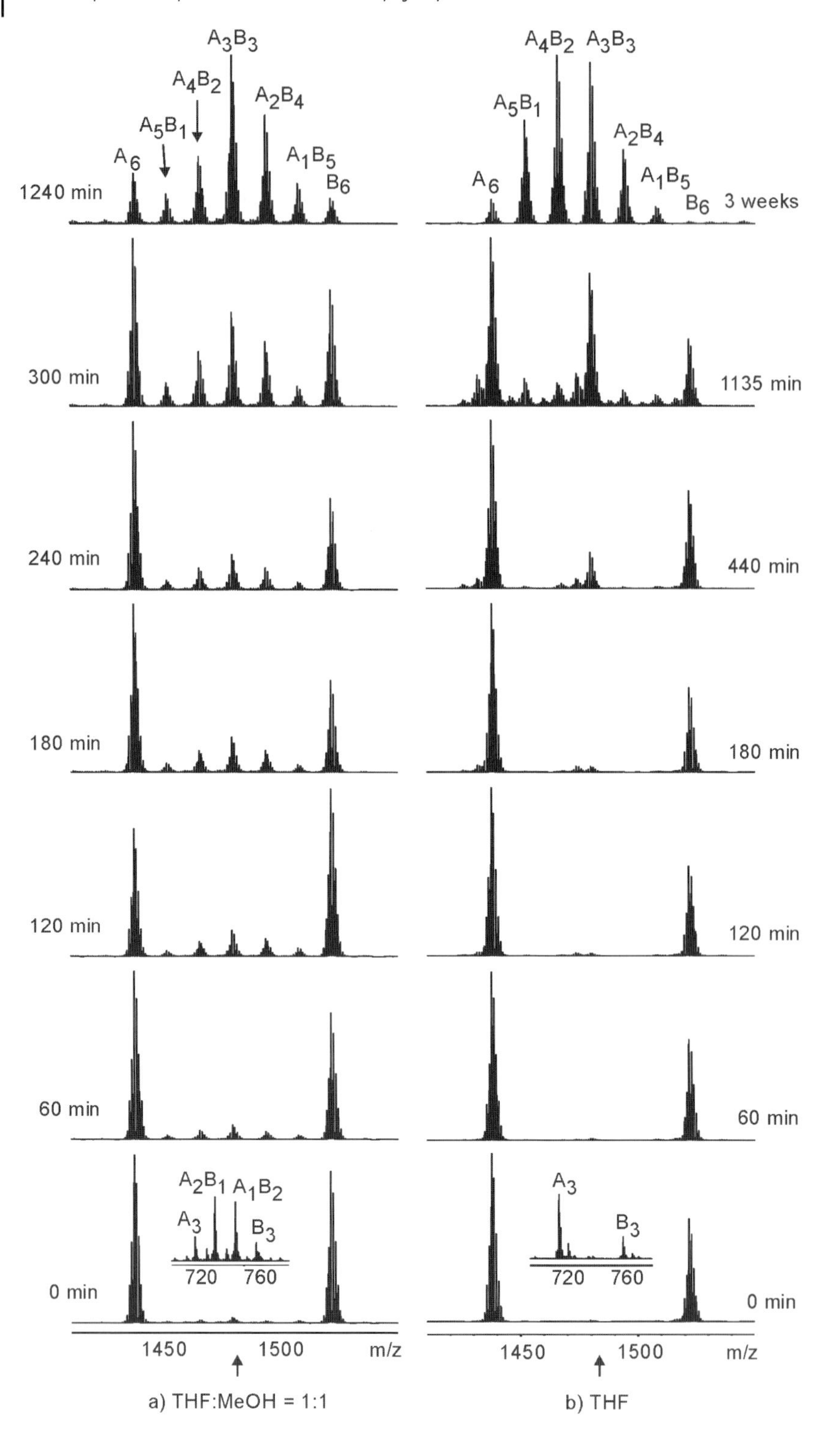

A3B3
A4B2
A2B4
A5B1
A6
A1B5
B6
1240 min
300 min
240 min
180 min
120 min
60 min
A2B1 A1B2
A3
B3
720 760
0 min
1450
1500
m/z
a) THF:MeOH = 1:1
A4B2 A3B3
A5B1
A2B4
A6
A1B5
B6
3 weeks
1135 min
440 min
180 min
120 min
60 min
A3
B3
720 760
0 min
1450
1500
m/z
b) THF

5.5.6
Reactivity in the Gas Phase: Isolated Species instead of Dynamic Interconverting Complexes

In this Section, two examples will be discussed which rely on IRMPD and CID experiments. It is important to keep in mind that ions in the gas phase are isolated particles, which do not undergo exchange processes among each other. Also, the neutral fragment formed upon dissociation does not come back; there is no equilibrium situation. This is the most important difference as compared to condensed-phase studies. The experiments described here thus provide a completely new view on these supramolecules and add new mechanistic insight into their reactivity which cannot be gained in solution where the equilibria present are complex.

5.5.6.1 Metallosupramolecular Squares: A Supramolecular Equivalent to Neighbor Group Assistance

If 4,4′-bipyridine is mixed with (dppp)Pd(II) or (dppp)Pt(II) triflates (dppp = bis-(diphenylphosphine) propane), the two reactants spontaneously self-assembly [53] to the metallo-supramolecular square shown in Fig. 5.24, because the weakly coordinating counterions leave two coordination sites open at the metal complex. By virtue of the dppp ligand, the coordination angle of 90° is fixed. It is rather difficult to ionize these squares as intact but completely desolvated entities, but under the

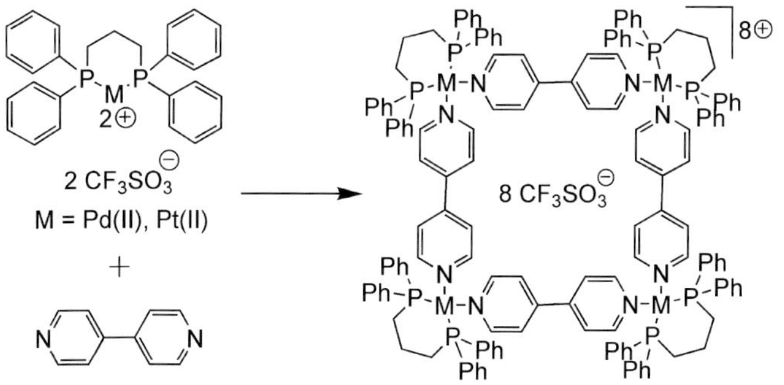

Fig. 5.24. A self-assembling square of the Stang-type.

Fig. 5.23. Ligand exchange between two different dimers of titanium(IV) tris-ketochatecholates. Left: Exchange followed in a 1:1 mixture of THF and methanol. Right: Exchange in pure THF. Insets in the bottom spectra: Monomer region showing fast ligand exchange between monomers in THF/MeOH and slow exchange in pure THF.

right conditions (low temperatures in the ion source, acetone as the solvent and a quite high 250-μM concentration), ionization is achieved through stripping away two or more counterions [54]. Since mass spectrometry measures the mass-to-charge (m/z) ratio of the ions under study, doubly-, triply-, quadruply- etc. charged species appear in the mass spectrum at half, a third, a quarter etc. of their real molecular weight. Consequently, a doubly-charged square may be superimposed by a singly-charged 2:2 complex of bipyridine and metal corner. This is indeed observed and can be analyzed by a closer look at the isotope pattern, where singly-charged ions have peak spacings of $\Delta m/z = 1$, while ions with n charges exhibit peak spacings of $\Delta m/z = 1/n$. Consequently, tandem MS experiments can easily be performed with squares in their +3 and +5 charge states, because these ions cannot contain any contributions from fragments.

If a tandem MS experiment is performed by carefully mass-selecting the triply-charged Pt(II) squares and irradiation of the parent ion with the IR laser (IRMPD), fragmentation of the squares is easily achieved. Interestingly, a doubly-charged 3:3 complex and a singly-charged 1:1 complex of corner and bipyridine are the primary products (Fig. 5.25). The complete absence of a doubly-charged 2:2 complex con-

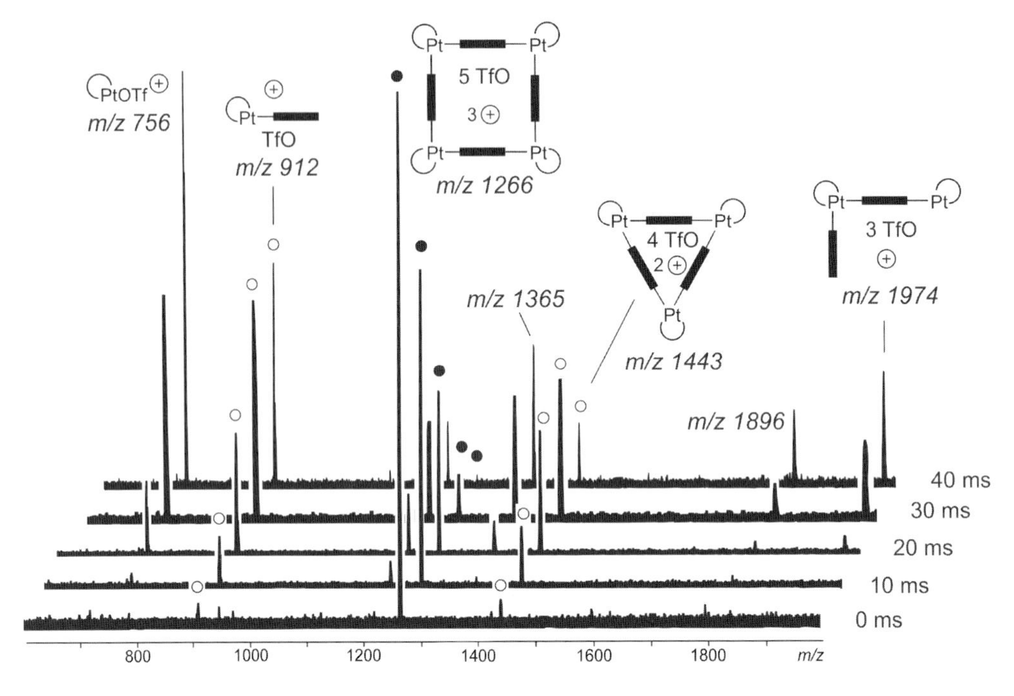

Fig. 5.25. IRMPD experiment with mass-selected triply-charged squares (full circles). At increasing irradiation times (right), the intensity of the parent ion decreases. Primary fragments are marked with open circles. They correspond to a doubly-charged 3:3 complex and a singly-charged 1:1 complex. Splitting into two half squares is not observed.

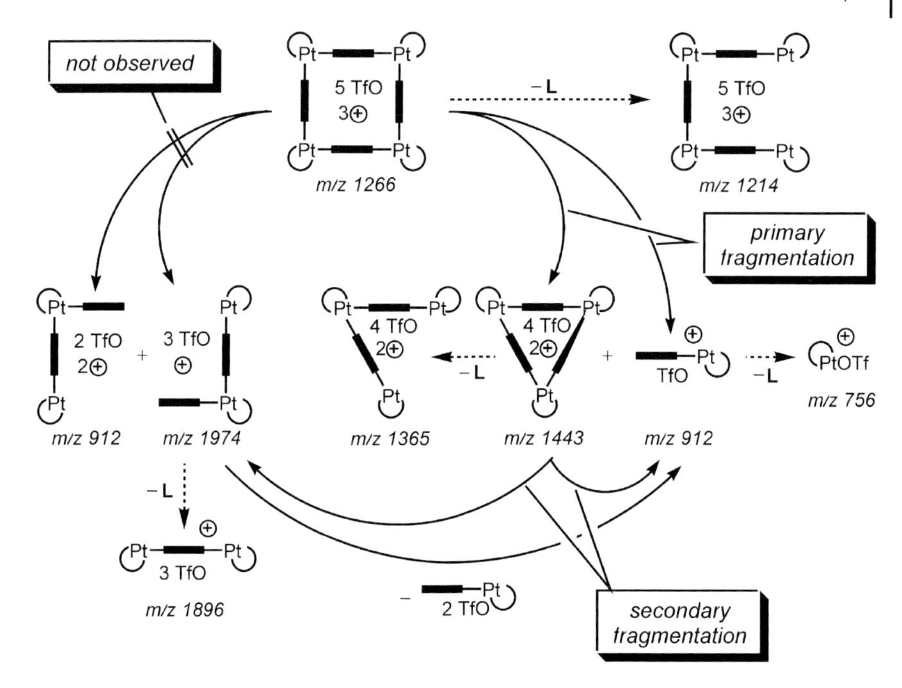

Fig. 5.26. Complete analysis of the fragmentation mechanisms observed for the triply-charged Pt(II) square.

firms that no fragmentation into two half squares occurs, although the same number of metal–nitrogen bonds need to be broken here. Instead, the singly-charged half square is a secondary fragmentation product formed from the 3:3 complex. These findings are summarized in Fig. 5.26.

One may ask, why only one of the two alternative reactions occurs, although both require the cleavage of two M–N bonds and thus should be expected to be similar in energy demand. One explanation involves a rear-attack mechanism. When irradiated by the IR laser, the ions' internal energy increases and finally, a first bond is opened. The result is a chain-like ion with a higher degree of conformational freedom. If the uncomplexed pyridine end of the chain attacks at the third metal centre (Fig. 5.27), cleavage of the second M–N bond benefits from the new M–N bond already forming during the reaction. Other conformations of the chain which would lead to two 2:2 fragments do not profit from a similar effect because that would cause too much strain. If this explanation holds true, two conclusions can be drawn: (i) Although not detected in the solution equilibrium by NMR experiments, triangles can form. However, they require the particular conditions of the gas phase to survive and be detected. (ii) If triangles can form, the strain within the cyclic structure must be lower than the binding energy of a Pt–N bond.

There is an alternative explanation for the experimental findings. Within a triply-charged square, charge repulsion must play a role. Upon fragmentation, the

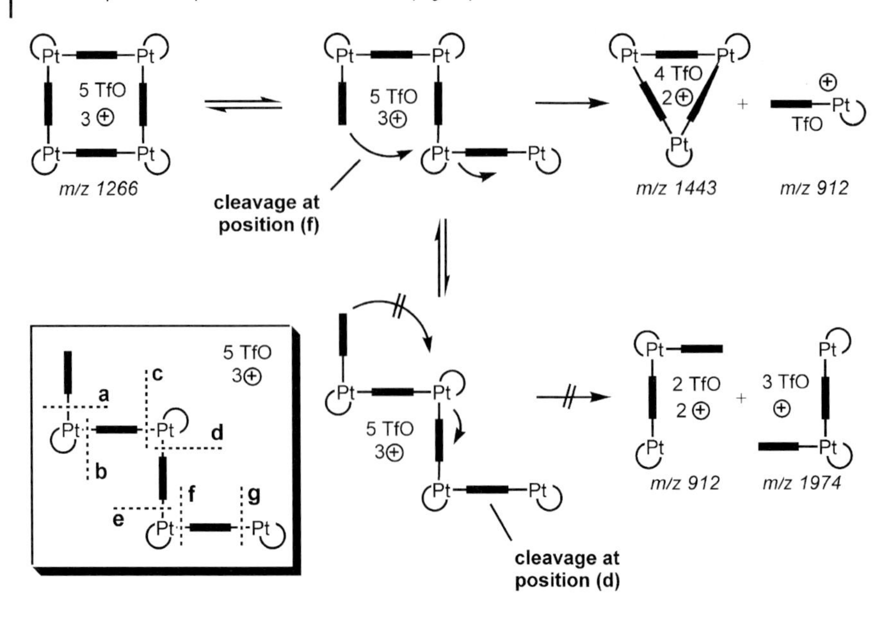

Fig. 5.27. Through a rear-attack mechanism, fragmentation into 3:3 and 1:1 complexes are energetically favored (top), while splitting into two half squares does not benefit from such a mechanism because of the strain imposed on the structure (bottom). The inset shows all possible cleavage sites a–g. Cleavage at position a is the only one which competes.

charges are distributed over the two fragments. This must happen in a non-symmetrical way, generating one singly and one doubly-charged fragment. Consequently, one may claim that an unsymmetric cleavage would provide better stabilization of the charges, when two charges are located on the larger and one charge remains on the smaller fragment. In order to distinguish both possibilities, the same experiments were performed for different charge states and all the fragments involved [55]. Only two of them are necessary to provide further insight into the mechanistic problem: the quadruply-charged triangle and the quintuply-charged square. The triangle in its +4 charge state undergoes fragmentation into different channels, distributing the charges over the two fragments in different ways. Instead, the fragmentation of the +5 square yields only two primary fragments: a singly-charged 1:1 complex and a 3:3 complex in its +4 charge state. Here, it would certainly be more favorable in terms of charge repulsion, if two charges would remain on one fragment and three on the other. These considerations nicely show the specificity of the fragmentation of the squares in contrast to those of the quadruply-charged triangle and thus rule out an explanation based solely on charge repulsion effects. Instead, we conclude that the supramolecular analog of neighbor group assistance is operative in the fragmentation of the squares.

5.5.6.2 A Surprising Dendritic Effect: Switching Fragmentation Mechanisms

Viologens are doubly-charged bipyridinium ions which form complexes with the molecular Klärner tweezer [56] shown in Fig. 5.28. Since the guest is a dication, the complexes can easily be transferred into the gas phase by ESI [57]. When one tries to ionize the guest cations alone, singly and doubly-charged clusters are observed with an appropriate amount of counterions balancing charge repulsion. There seems to be a stability trend for the naked dications. While under none of the tested ionization conditions, bare G0 viologen has been observed and we thus assume that it – if not intrisically unstable – may exist only as a short-lived metastable ion, the G1 viologen could be observed under very mild ESI conditions, while it was no problem at all to generate bare G2 viologen dications. In marked contrast, addition of the tweezer destroys all clusters and tweezer–viologen complexes are observed even for G0 as dications. Clearly, coordination by the tweezer stabilizes the dication and likely contributes to diminishing the charge repulsion effects through charge-transfer interactions.

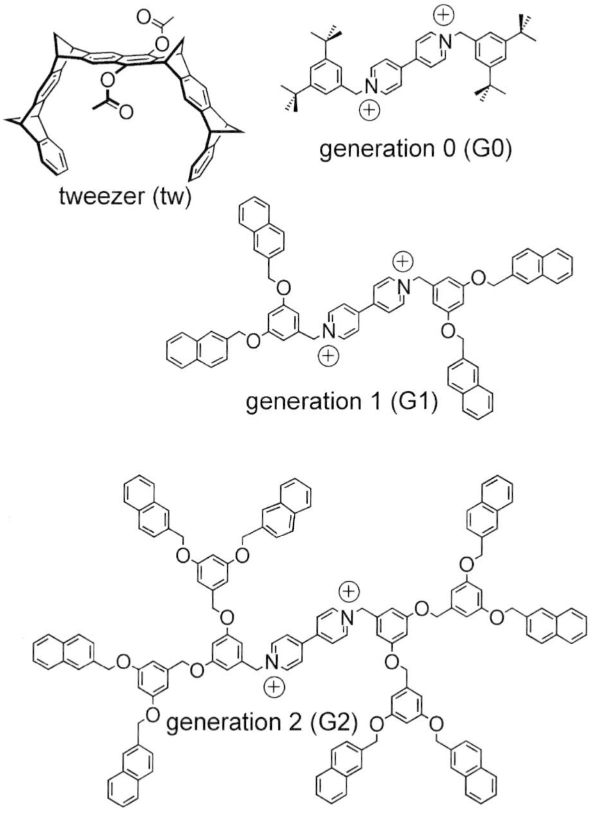

tweezer (tw)

generation 0 (G0)

generation 1 (G1)

generation 2 (G2)

Fig. 5.28. Dendritic viologens (G0–G2) and the Klärner tweezer which forms 1:1 complexes with the viologen guests.

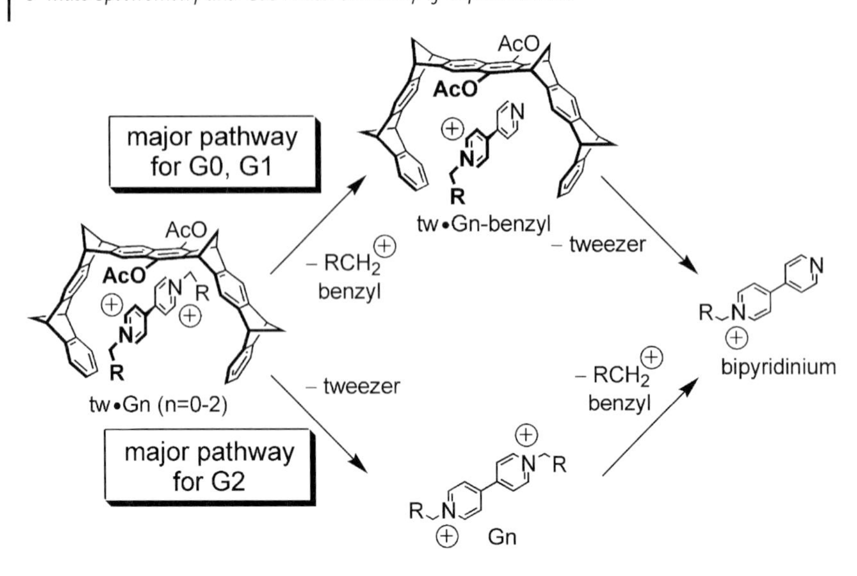

Fig. 5.29. Possible pathways for the fragmentation of the tweezer–viologen complexes. Depending on the dendron size, the mechanism switches between G1 and G2 from the upper to the lower pathway.

In a CID experiment with mass-selected tweezer–viologen dications, two possible reaction pathways exist, since two rather weak bonds are present: the noncovalent interactions between host and guest and the benzyl–nitrogen bond which upon cleavage generates a well-stabilized benzyl cation and suffers from charge repulsion. The supramolecular bond can be cleaved leading to the loss of the tweezer and formation of the bare dications. Based on the experience with the generation of these dications in the ESI source, one might expect that they undergo Coulomb expulsion to yield a singly-charged dendritic benzyl cation and a singly-charged bipyridinium cation bearing the second dendron (bottom pathway in Fig. 5.29). The alternative is initial cleavage of the weak benzyl–N bond producing a singly-charged dendritic benzyl cation. The tweezer may well have sufficient residual binding energy to the remaining bipyridinium monocation to remain present in the complex (upper pathway in Fig. 5.29). As a consecutive fragmentation, the tweezer may be lost, finally generating the same fragments as in the first pathway.

The CID spectra in Fig. 5.30 provide evidence that the mechanism switches from the upper pathway (realized for G0 and G1) to the bottom pathway (realized for G2) depending on the dendron size. In the CID spectrum of the tweezer complex with the G0 viologen guest dication, a fragment is observed at m/z 1059. It appears above the parent because the parent is doubly-charged and actually appears at half of its real molecular weight. This fragment corresponds to the loss of one di-*tert*-butyl benzyl cation from the complex. The tweezer remains bound to the singly-charged bipyridinium cation. A so-called double resonance experiment provides evidence that almost all of the bipyridinium cation at m/z 359 results from

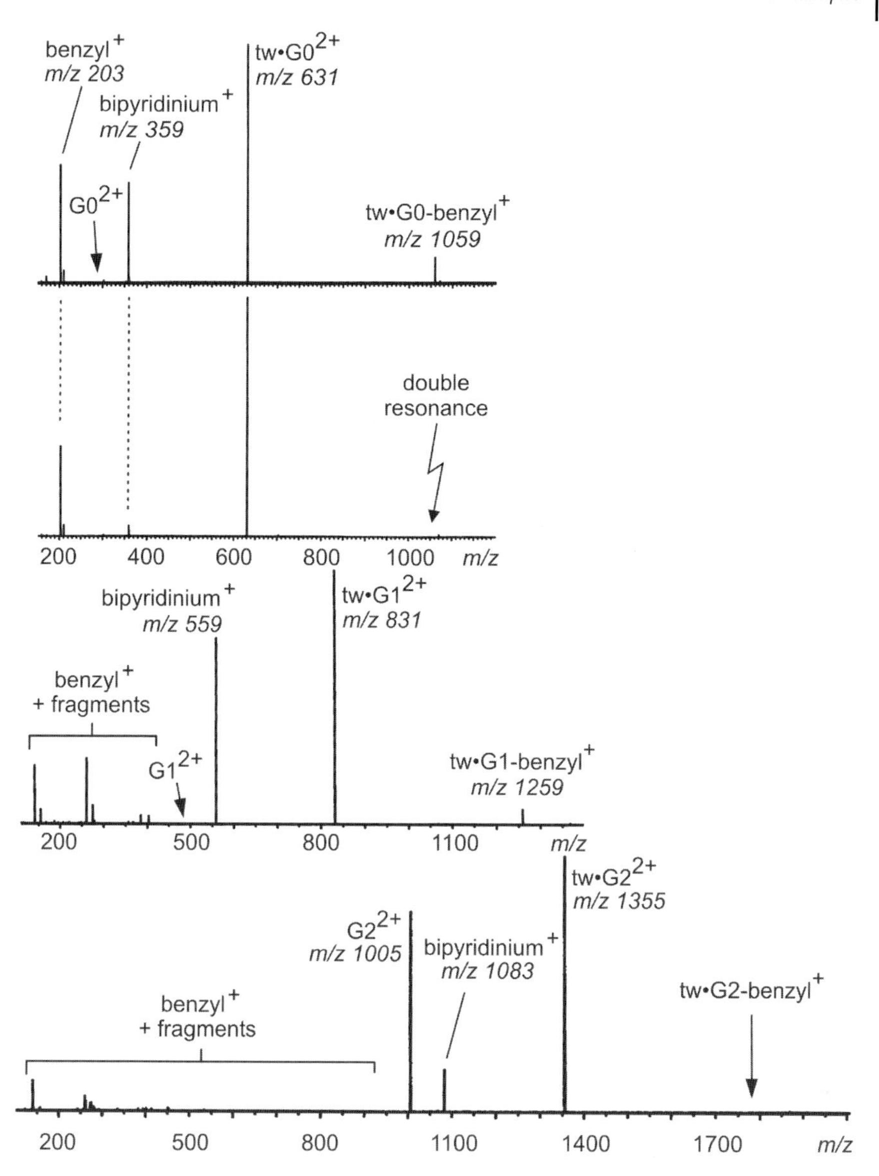

Fig. 5.30. CID mass spectra of mass-selected, dicationic tweezer–viologen complexes (G0: top two rows, G1 third row, G2 bottom row). The second trace shows a double-resonance experiment, which is a CID experiment during which the fragment at *m/z* 1059 was constantly ejected. All consecutive fragments are thus not formed and can easily be identified.

consecutive fragmentation of the complex at m/z 1059. In this experiment, the CID experiment is repeated und the exact same conditions with the exception that the intermediate at m/z 1059 is constantly removed and thus cannot produce any subsequent fragmentation products anymore. The intensity of the fragment at m/z 359 decreases significantly and thus identifies this ion as a product formed from the intermediate rather the parent.

For the tweezer complex with the G1 viologen, the overall picture is the same. However, the complex with the G2 viologen guest behaves very differently. The benzyl loss intermediate is not formed at all. Instead, the tweezer is lost generating the bare G2 dication at m/z 1005 which can then undergo further fragmentation by benzyl cation loss.

How can we understand this complete switch in mechanism which somehow parallels the stability differences of the bare dications encountered during the ionization? Molecular modeling suggests that the dendron arms can backfold to different extents. For G0, no backfolding is possible and the dication is not stabilized by internal solvation. For G1, the peripheral naphthalene groups can surround the dication and stabilize it through charge-transfer interactions. In the G2 viologen, such an interaction is even possible with the much more electron-rich dihydroxy benzyl branching units. Consequently, internal solvation stabilizes the dication with increasing dendron size. The tweezer competes with this interaction easily for the G0 and G1 viologens. The benzyl–N bond is the weakest in the complex and breaks first. However, the G2 viologen can self-stabilize to such an extent that the dication is stable. The internal stabilization also strengthens the benzyl–N bond and at the same time weakens the interaction with the tweezer. The interaction between tweezer and guest is now the weakest bond and thus in the CID experiment, the tweezer is lost preferentially.

This example shows gas-phase reactivity to be quite different from that in solution. Such a clear-cut dendritic effect cannot be expected to be found in solution, because counterions are present which compensate charge repulsion and thus waive this effect more or less completely. In summary, mass spectrometry adds a completely new view of the reactivity of supramolecular complexes.

5.5.7 Determining Thermochemical Data: The Influence of the Environment

5.5.7.1 Crown Ether – Alkali Complexes: Questioning the Best-fit Model

A qualitative understanding of a special issue, e.g. binding of cations to crown ethers, can be obtained by comparing two systems that do not differ much in structural and energetic features. Brodbelt et al. examined the binding attributes of alkali cations to crown ethers and their open-chained analogs with the so-called kinetic method [58]. This method examines the fragmentation pattern of a complex that consists of two different multidentate ligands (crown ethers or open-chained analogs) C_1 and C_2 that are attached to a central metal cation M^+. Collisional activation causes the complex fragments to appear and the relative abundances of the ions C_1M^+ and C_2M^+ can then be used to determine relative binding energies.

The results of these measurements are affinity ladders, in which a qualitative trend of binding affinities can be observed. Because of the temperature problem these values are only relative data and conversion into absolute data is not easily possible.

Another method is the measurement of equilibrium constants that can be performed in a FT–ICR mass spectrometer [59]. One example is an examination of a complex C_1M^+ consisting of crown ether C_1 and a metal cation M^+ that is reacted with a mixture of C_1 and a second neutral crown ether C_2 (both present with the same concentration) until the reaction has reached the equilibrium state. The intensity ratio of the two complexes C_1M^+ and C_2M^+ directly reflects the binding strengths of the crown ether to the metal cation.

The binding of alkali metal cations to crown ethers follows the best-fit model in solution, i.e. the cation which has the appropriate size for binding inside the crowns cavity will bind with the highest affinity (e.g. Na^+ in 15-crown-5, K^+ in 18-crown-6 etc.). Usually, these trends were observed for protic solvents such as methanol. In order to compare the binding energies in solution with gas phase data, Armentrout and his group collected a large number of gas-phase binding energies by threshold collision-induced fragmentation experiments. In Table 5.3, the gas-phase binding energies as determined by TCID experiments of monodentate dimethylether (DME), bidentate dimethoxy ethane (DXE), 12-crown-4, 15-crown-5, and

Tab. 5.3. Binding energies (in eV) of crown ethers and simple analogues (DME = dimethyl ether, DXE = dimethoxy ethane) to alkali metal ions. Each value refers to the binding energy of one ligand (e.g. 4 DME means the binding energy of the 4th DME ligand, when already three are present in the complex). For comparison, also the sums of the BDE's of two and four DME and two DXE ligands are given. All data taken from ref. [60].

Ligand	Li^+	Na^+	K^+	Rb^+	Cs^+
1 DME	1.71	0.95	0.76	0.64	0.59
2 DME	1.25	0.85	0.71	0.57	0.49
3 DME	0.92	0.72	0.59	0.38	0.41
4 DME	0.70	0.63	0.52	0.40[a]	0.37[a]
1 DXE	2.50	1.64	1.23	0.97	0.59
2 DXE	1.44	1.20	0.92	0.51	0.56
12-crown-4	3.85	2.61	1.96	0.96	0.88
15-crown-5	–	3.05	2.12	1.18	1.04
18-crown-6	–	3.07	2.43	1.98	1.74
1 + 2 DME	2.96	1.80	1.47	1.21	1.08
3 + 4 DME	1.62	1.35	1.11	0.89	0.78
1 – 4 DME	4.58	3.15	2.58	1.99	1.86
1 + 2 DXE	3.94	2.84	2.15	1.48	1.15

[a] estimated value.

18-crown-6 to the alkali metal cations $M = Li^+ - Cs^+$ are summarized [60]. In order to facilitate comparison, also the values for the sums of two DME, four DME, and two DXE ligands are included.

Several clear trends emerge: (i) As expected, the first ether ligand bound to the metal exhibits a higher bond dissociation energy than the second and so forth. (ii) The binding energy of any ligand L in the ML_n^+ complexes under study decreases from Li^+ to Cs^+, probably because the charge on Li^+ is more concentrated and has better orbital overlap, since Li^+ and the oxygen donor are in the same row of the periodic table. This trend has also been observed in a study employing the infrared multiphoton dissociation (IRMPD) technique [61]. (iii) The binding energy increases from 12-crown-4 to 18-crown-6 for all metal ions (for Li^+ this trend still needs to be confirmed, since no binding energies have been reported for the Li^+(15-crown-5) and Li^+(18-crown-6) complexes so far). The number of oxygen donors in the crown ether is thus much more important than the geometric fit of the cation inside the cavity of the ligand. (iv) Nevertheless, geometric properties play a secondary role. The binding energy of 15-crown-5 to Na^+, the preferred partner in solution, is the same as for 18-crown-6, although it bears only five instead of six oxygen donors. Seemingly, the better fit for 15-crown-5 compensates the additional binding energy from the sixth oxygen. The increase in binding energies from 15-crown-5 to 18-crown-6 is particularly drastic for Rb^+ and Cs^+, while the difference is much smaller between the two smaller crowns. This would not be expected, if there were no size argument, since the binding energies for each additional donor atom use to decrease with increasing number. Seemingly, the number of donor sites in the crown and the geometric fit both contribute to the total binding energy. (v) Finally, the sum of the binding energies for two DME ligands is larger than that of one bidentate DXE ligand. The same applies, if four DME groups are compared to two DXEs, and in the series of the tetradentate complexes $M(DME)_4^+$, $M(DXE)_2^+$, and M^+(12-crown-4), the crown ether always has the lowest total binding energy. This trend can be rationalized by the entropic (and enthalpic) costs of forming five membered cyclic $-M^+-O-CH_2-CH_2-O-$ subunits within the complexes. Conformations in the CH_2CH_2 backbone cannot freely interconvert as they can in the free ligand after dissociation and the optimal staggered geometry is not easily feasible. While the O–C–C–O dihedral angle in free dimethoxyethane is 74°, it is 48°, 55°, 59°, 61°, and 63° for $Li^+ - Cs^+$, respectively. Of course, the crown ethers suffer most from this effect.

The remaining question is why the "best fit" plays such a great role in polar solvents, while it is of minor importance in the gas phase. That brings us to the role of solvation, which can be analyzed based on these data. In methanol, for example, a crown ether competes with an appropriate number of solvent molecules for binding to the alkali metal ion. The intrinsic binding energies from the gas phase predict that the interaction with the solvent is somewhat, but probably not too much higher than that of the crown ether. However, two other effects come into play in solution: entropy and geometric fit. It is less favorable, for example, to bring together one Li^+ ion and four methanol molecules than to form a complex from just two components, Li^+ and 12-crown-6. This entropic effect is more or less the same for any alkali metal ion, independent of its size. Consequently, the geometric fit finally determines which ion binds best. In the gas phase however, there is no methanol replacement ligand competing with the crown. Consequently, the binding energies are governed by the number of oxygen donors, because there is no balanced interplay with the solvent. The entropy effects may play a role during ion formation, but once mass-selected, the reaction is always the same: loss of one ligand producing two particles from one, leveling out the entropy. Finally, geometric fit cannot overcompensate the binding energies determined by the number of donor atoms and thus only contributes a minor fraction of the binding energy. In conclusion, the major achievement resulting from the gas-phase data is a much more profound understanding of the role of solvation in cation binding by crown ethers.

5.5.7.2 BIRD: Arrhenius Kinetics of Oligonucleotide Strand Separation in the Gas Phase

The BIRD approach has been utilized to decide the question whether short oligonucleotide double strands still have a Watson–Crick base-paired structure. Schnier et al. [62] argued that one could conclude the Watson–Crick base pairing to be conserved in the gas phase, if the gas-phase barriers follow linearly the same trend as the melting enthalpies determined with UV methods in solution. The triply-charged double strands examined had four, six or seven base pairs and reveal features expected for a correct pairing pattern. A higher number of hydrogen bonds which relates to a higher G:C content leads to a higher activation energy. For example, palindromic $(TGCA)_2{}^{3-}$ has an activation barrier of 1.41 eV, while the $(CCGG)_2{}^{3-}$ double strand undergoes separation when a barrier of 1.51 eV can be surmounted. Also, mismatched complexes ($A_7{\cdot}A_7{}^{3-}$: 1.26 eV; $T_7{\cdot}T_7{}^{3-}$: 1.40 eV) melt at lower activation barriers than matched ones ($A_7{\cdot}T_7{}^{3-}$: 1.68 eV). When the gas-phase activation energy is plotted against melting enthalpies from solution, a nearly linear relationship is found. All these findings provide evidence for intact Watson–Crick base pairing in the gas phase. Molecular dynamics calculations agree with this conclusion as far as the hydrogen bonding is concerned. However, the effects of the three unsolvated negative charges on the backbone phosphates become clearly visible in the molecular dynamics simulation after ca. 100 fs. The double helix starts to disrupt. After ca. 400 fs, the helical structure is completely converted into a globular complex with maximized charge–charge distances, but still intact hydrogen bonding pattern.

5.6 Conclusions

The utility of mass spectrometry for supramolecular chemistry is quite often dramatically underestimated. The examples selected here hopefully make clear how far a detailed mass spectrometric analysis can go and how broad the spectrum of chemical problems is which can be addressed with this method. Nowadays, many researchers in supramolecular chemistry do not go beyond analytical characterization. Several reasons may exist: one of them certainly is the lack of easy access to the appropriate, admittedly costly equipment. Beyond analytics, mass spectrometry proves to provide structural evidence, which is often complementary to that coming from other methods. In part, this evidence overlaps and thus mass spectrometry is able to support other methods; in part, mass spectrometry adds new data on structure. The self-assembling hydrogen-bonded capsules represent an example nicely illustrating this: mass spectrometry provides evidence for a capsular structure in solution with the guest being located in the capsules cavity. This can easily be determined by NMR methods. However, several examples exist, where the formation of heterodimers or heterotetramers cannot be observed by NMR, because the monomers are too similar. In these cases, the hetero- and homocapsules differ in mass and thus the formation of heterocapsules is easily followed by mass spectrometry. The questions of structure determination even extend to subtle structural features such as stereochemical issues, gas-phase conformations, and the question whether a complex has a zwitterionic nature.

Mass spectrometers can also act as detectors for the dynamic processes going on in solution. Although the time frame is somewhat limited when fast processes are concerned, ligand exchange processes can be analyzed in terms of the details of the exchange mechanisms. As exemplarily shown by the noncovalent helicates discussed above, even the mechanistic differences induced by changes of the environment can be analyzed by mass spectrometry. However, the dynamic processes can be suppressed by examining the reactivity of supramolecular compounds in the gas phase, where the environment is absent and thus no reaction partners which may mediate dynamic processes are available. Here, supramolecular complexes reveal a completely different reactivity. For example, the metallosupramolecular squares fragment through a neighbor group assistance mechanism. The mechanistic change in the decomposition of the dendrimer host–guest complexes discussed above can only be analyzed in the gas phase, because of the presence of counterions in solution which stabilize the dications so that the solution-phase complexes are all well behaved. The major conclusion from this chapter is that the examination of reactions comes into reach in the gas phase, which are hard or impossible to analyze in solution. Consequently, mass spectrometry broadens our view on supramolecular reactivity and environment effects.

Finally, the role of solvation can be investigated by comparing thermochemical data from solution with that obtained from gas phase experiments. While crown ethers were believed to form those alkali cation complexes which have the best fit, gas phase studies tell us that this is true only when the environment allows for that. The second example of oligonucleotide double strands demonstrates convincingly that the Watson–Crick base pairing is conserved in the gas phase, but that charge repulsion effects and the missing hydrophobic effect cause the double helix structure to collapse. These studies lead to a much more profound understanding of the fundamental principles in supramolecular chemistry.

References and Notes

1 For a general introduction to mass spectrometry, we refer the reader to the two following excellent textbooks: a) E. de Hoffmann, V. Stroobant, *Mass Spectrometry*, Wiley, Chichester 2001; b) J. H. Gross, *Mass Spectrometry – A Textbook*, Springer, Heidelberg 2004.

2 Earlier reviews on the mass spectrometric analysis of supramolecular species: a) M. Vincenti, *J. Mass Spectrom.* **1995**, 30, 925; b) J. S. Brodbelt, D. V. Dearden, *Mass Spectrometry* in: *Comprehensive Supramolecular Chemistry*, Vol. 8, J. L. Atwood, J. E. D. Davies, D. D. MacNicol, F. Vögtle, J.-M. Lehn, J. A. Ripmeester (Eds.), Pergamon, Oxford, 1996, pp. 567; c) M. Przybylski, M. O. Glocker, *Angew. Chem.* **1996**, 108, 878; *Angew. Chem. Int. Ed.* **1996**, 35, 806; d) J. S. Brodbelt, *Int. J. Mass Spectrom.* **2000**, 200, 57; e) C. A. Schalley, *Int. J. Mass Spectrom.* **2000**, 194, 11; f) C. B. Lebrilla, *Acc. Chem. Res.* **2001**, 34, 653; g) C. A. Schalley, *Mass Spectrom. Rev.* **2001**, 20, 253.

3 Reviews on MALDI: a) R. Zenobi, R. Knochenmuss, *Mass Spectrom. Rev.* **1999**, 17, 337; b) C. Menzel, K. Dreisewerd, S. Berkenkamp, F. Hillenkamp, *Int. J. Mass Spectrom.* **2001**, 207, 73; c) K. Dreisewerd, S. Berkenkamp, A. Leisner, A. Rohlfing, C. Menzel, *Int. J. Mass Spectrom.* **2003**, 226, 189; d) K. Dreisewerd, *Chem. Rev.* **2003**, 103, 395; e) M. Karas, R. Krüger, *Chem. Rev.* **2003**, 103, 427; f) K. Tanaka, *Angew. Chem.* **2003**, 115, 3989; *Angew. Chem. Int. Ed.* **2003**, 42, 3989.

4 a) B. H. Wang, K. Dreisewerd, U. Bahr, M. Karas, F. Hillenkamp, *J. Am. Soc. Mass Spectrom.* **1993**, 4, 393; b) P. Juhasz, C. E. Costello, *Rapid Commun. Mass Spectrom.* **1993**, 7, 343; c) M. Glückmann, A. Pfenninger, R. Krüger, M. Thierolf, M. Karas, V. Horneffer, F. Hillenkamp, *Int. J. Mass Spectrom.* **2001**, 210/211, 121.

5 Reviews on ESI: a) J. B. Fenn, M. Mann, C. K. Meng, S. F. Wong, C. M. Whitehouse, *Mass Spectrom. Rev.* **1990**, 9, 37; b) P. Kebarle, L. Tang, *Anal. Chem.* **1993**, 65, 972A; c) S. J. Gaskell, *J. Mass Spectrom.* **1997**, 32, 677; d) J. B. Fenn, *Angew. Chem.* **2003**, 115, 3999; *Angew. Chem. Int. Ed.* **2003**, 42, 3871.

6 C. R. Blakley, J. J. Carmody, M. L. Vestal, *J. Am. Chem. Soc.* **1980**, 102, 5931.

7 C. A. Evans Jr., C. D. Hendricks, *Rev. Sci. Inst.* **1972**, 43, 1527.

8 a) M. Dole, L. L. Mack, R. L. Hines, R. C. Mobley, L. D. Ferguson, M. B. Alice, *Macromolecules* **1968**, 1, 96; b) M. Dole, L. L. Mack, R. L. Hines, R. C. Mobley, L. D. Ferguson, M. B. Alice, *J. Chem. Phys.* **1968**, 49, 2240.

9 a) M. Dole, L. L. Mack, R. L. Hines, R. C. Mobley, L. D. Ferguson, M. B. Alice, *J. Chem. Phys.* **1968**, 49, 2240; b) F. W. Röllgen, E. Bramer-Wegner, L. Buttering, *J. Phys. Colloq.* **1984**, 45, Suppl. 12, C9-297.

10 a) J. V. Iribarne, B. A. Thomson, *J. Chem. Phys.* **1976**, 64, 2287; b) B. A. Thomson, J. V. Iribarne, *J. Chem. Phys.* **1979**, 71, 4451.

11 A. T. Iavarone, J. C. Jurchen, E. R. Williams, *Anal. Chem.* **2001**, 73, 1455.

12 D. P. Little, R. A. Chorush, J. P. Speir, M. W. Senko, N. L. Kelleher, F. W. McLafferty, *J. Am. Chem. Soc.* **1994**, 116, 4893.

13 P. A. Limbach, P. F. Crain, J. A. McCloskey, *J. Am. Soc. Mass Spectrom.* **1995**, 6, 27.

14 S. Sakamoto, M. Fujita, K. Kim, K. Yamaguchi, *Tetrahedron* **2000**, 56, 955.

15 D. M. Lubman (Ed.), *Lasers and Mass Spectrometry*, Oxford University Press, New York 1990.

16 a) J. Taubitz, U. Lüning, J. Grotemeyer, *Chem. Commun.* **2004**, 2400. For a review on the use of REMPI to examine Van-der-Waals complexes, see: b) C. E. H. Dessent, K. Müller-Dethlefs, *Chem. Rev.* **2000**, 100, 3999.

17 K. C. Russell, E. Leize, A. Van Dorsselaer, J.-M. Lehn, *Angew. Chem.* **1995**, 107, 244–250; *Angew. Chem. Int. Ed.* **1995**, 34, 209.

18 M. Scherer, J. L. Sessler, M. Moini, A. Gebauer, V. Lynch, *Chem. Eur. J.* **1998**, 4, 152.

19 a) P. Timmerman, R. H. Vreekamp, R. Hulst, W. Verboom, D. N. Reinhoudt, K. Rissanen, K. A. Udachin, J. Ripmeester, *Chem. Eur. J.* **1997**, 3, 1823; b) K. A. Joliffe, M. Crego Calama, R. Fokkens, N. M. M. Nibbering, P. Timmerman, D. N. Reinhoudt, *Angew. Chem.* **1998**, 110, 1294; *Angew. Chem. Int. Ed.* **1998**, 37, 1247; c) P. Timmerman, K. A. Joliffe, M. Crego Calama, J.-L. Weidmann, L. J. Prins, F. Cardullo, B. H. M. Snellink-Ruël, R. H. Fokkens, N. M. M. Nibbering, S. Shinkai, *Chem. Eur. J.* **2000**, 6, 4104.

20 a) X. Cheng, Q. Gao, R. D. Smith, E. E. Simanek, M. Mammen, G. M. Whitesides, *Rap. Commun. Mass. Spectrom.* **1995**, 9, 312; b) X. Cheng, Q. Gao, R. D. Smith, E. E. Simanek, M. Mammen, G. M. Whitesides, *J. Org. Chem.* **1996**, 61, 2204.

21 a) C. A. Schalley, J. M. Rivera, T. Martín, J. Santamaría, G. Siuzdak, J. Rebek, Jr., *Eur. J. Org. Chem.* **1999**, 1325; b) C. A. Schalley, R. K. Castellano, M. S. Brody, D. M. Rudkevich, G. Siuzdak, J. Rebek, Jr., *J. Am. Chem Soc.* **1999**, 121, 4568; c) M. S. Brody, D. M. Rudkevich, C. A. Schalley, J. Rebek, Jr., *Angew. Chem.* **1999**, 111, 1738; *Angew. Chem. Int. Ed.* **1999**, 38, 1640; d) C. A. Schalley, T. Martín, U. Obst, J. Rebek, Jr., *J. Am. Chem. Soc.* **1999**, 121, 2133; (e) A. Lützen, A. R. Renslo, C. A. Schalley, B. M. O'Leary, J. Rebek, Jr., *J. Am. Chem. Soc.* **1999**, 121, 7455; f) B. M. O'Leary, T. Szabo, N. Svenstrup, C. A. Schalley, A. Lützen, J. Rebek, Jr., *J. Am. Chem. Soc.* **2001**, 123, 11519.

22 A. G. Marshall, C. L. Hendrickson, G. S. Jackson, *Mass Spectrom. Rev.* **1998**, 17, 1.

23 For a discussion of this problem, see: J. B. Cunniff, P. Vouros, *Rap. Commun. Mass Spectrom.* **1994**, 8, 715. For a review on cyclodextrins in the gas phase, see: C. B. Lebrilla, *Acc. Chem. Res.* **2001**, 34, 653.

24 E. Leize, A. Jaffrezic, A. Van Dorsselaer, *J. Mass Spectrom.* **1996**, 31, 537.

25 Review: J. S. Brodbelt, D. V. Dearden, *Mass Spectrometry* in: *Comprehensive Supramolecular Chemistry*, Vol. 8, J. E. D. Davies, J. A. Ripmeester (Eds.), Pergamon Press, Oxford 1996, p. 567.

26 Review: S. C. Nanita, R. G. Cooks, *Angew. Chem.* **2006**, 118, 568; *Angew. Chem. Int. Ed.* **2006**, 45, 554.

27 P. B. Armentrout, *Top. Curr. Chem.* **2003**, 225, 233.

28 a) M. A. Freitas, C. L. Hendrickson, A. G. Marshall, *Rap. Commun. Mass Spectrom.* **1999**, 13, 1639; b) M. A. Freitas, C. L. Hendrickson, A. G. Marshall, *J. Am. Chem. Soc.* **2000**, 122, 7768.

29 a) P. Ghosh, O. Mermagen, C. A. Schalley, *Chem. Commun.* **2002**, 2628; b) P. Ghosh, G. Federwisch, M. Kogej, C. A. Schalley, D. Haase, W. Saak, A. Lützen, R. Gschwind, *Org. Biomol. Chem.* **2005**, 3, 2691.

30 C. A. Schalley, P. Ghosh, M. Engeser, *Int. J. Mass Spectrom.* **2004**, 232–233, 249.

31 C. A. Schalley, J. Hoernschemeyer, X.-y. Li, G. Silva, P. Weis, *Int. J. Mass Spectrom.* **2003**, 228, 373.

32 a) M. Amman, A. Rang, C. A. Schalley, P. Bäuerle, *Eur. J. Org. Chem.* **2006**, 1940; b) M. Hutin, J. R. Nitschke, C. A. Schalley, G. Bernardinelli, *Chem. Eur. J.*, in press.

33 F. Hof, S. L. Craig, C. Nuckolls, J. Rebek, Jr., *Angew. Chem.* **2002**, 114, 1556; *Angew. Chem. Int. Ed.* **2002**, 41, 1488.

34 a) L. R. MacGillivray, J. L. Atwood, *Nature* **1997**, 389, 469; b) T. Gerkensmeier, W. Iwanek, C. Agena, R. Fröhlich, S. Kotila, C. Näther, J. Mattay, *Eur. J. Org. Chem.* **1999**, 2257.

35 See, for example: a) L. Avram, Y. Cohen, *J. Am. Chem. Soc.* **2002**, 124,

15148; b) A. Shivanyuk, J. Rebek, Jr., *J. Am. Chem. Soc.* **2003**, 125, 3432; c) L. Avram, Y. Cohen, *Org. Lett.* **2003**, 5, 3329; d) L. C. Palmer, J. Rebek, Jr., *Org. Lett.* **2005**, 7, 787; e) L. Avram, Y. Cohen, *Org. Lett.* **2006**, 8, 219.

36 a) H. Mansikkamäki, M. Nissinen, C. A. Schalley, K. Rissanen, *New. J. Chem.* **2003**, 27, 88; b) H. Mansikkamäki, C. A. Schalley, M. Nissinen, K. Rissanen, *New J. Chem.* **2005**, 29, 116.

37 N. K. Beyeh, M. Kogej, A. Åhman, K. Rissanen, C. A. Schalley, *Angew. Chem.*, **2006**, 118, 5339; *Angew. Chem. Int. Ed.* **2006**, 45, 5214.

38 I. Müller, D. Möller, C. A. Schalley, *Angew. Chem.* **2005**, 117, 485; *Angew. Chem. Int. Ed.* **2005**, 44, 480.

39 a) R. G. Cooks, D. Zhang, K. J. Koch, F. C. Gozzo, M. N. Eberlin, *Anal. Chem.* **2001**, 73, 3646; b) K. J. Koch, F. C. Gozzo, D. Zhang, M. N. Eberlin, R. G. Cooks, *Chem. Commun.* **2001**, 1854; c) K. J. Koch, F. C. Gozzo, S. C. Nanita, Z. Takats, M. N. Eberlin, R. G. Cooks, *Angew. Chem.* **2002**, 114, 1797; *Angew. Chem. Int. Ed.* **2002**, 41, 1721.

40 E. A. Mason, E. W. McDaniel, *Transport Properties of Ions in Gases*; Wiley, New York, 1988.

41 a) A. E. Counterman, D. E. Clemmer, *J. Phys. Chem. B* **2001**, 105, 8092; b) R. R. Julian, R. Hodyss, B. Kinnear, M. F. Jarrold, J. L. Beauchamp, *J. Phys. Chem. B* **2002**, 106, 1219.

42 C. A. Schalley, P. Weis, *Int. J. Mass Spectrom.* **2002**, 221, 9.

43 a) Z. Takats, S. C. Nanita, G. Schlosser, K. Vekey, R. G. Cooks, *Anal. Chem.* **2003**, 75, 6147; b) U. Mazurek, M. A. Farland, A. G. Marshall, C. Lifshitz, *Eur. J. Mass Spectrom.* **2004**, 10, 755.

44 J. S. Splitter, F. Turecek (Eds.), *Applications of Mass Spectrometry to Organic Stereochemistry*, VCH, Weinheim 1994.

45 a) M. Sawada, Y. Okumura, M. Shizuma, Y. Takai, Y. Hidaka, H. Yamada, T. Tanaka, T. Kaneda, K. Hirose, S. Misumi, S. Takahashi, *J. Am. Chem. Soc.* **1992**, 115, 7381; b) M. Sawada, Y. Okumura, H. Yamada, Y. Takai, S. Takahashi, T. Kaneda, K. Hirose, S. Misumi, *Org. Mass Spectrom.* **1993**, 28, 1525.

46 G. Pócsfalvi, M. Lipták, P. Huszthy, J. S. Bradshaw, R. M. Izatt, K. Vékey, *Anal. Chem.* **1996**, 68, 792.

47 Review: M. Sawada, *Mass Spectrom. Rev.* **1997**, 16, 73.

48 M. Sawada, Y. Takai, T. Kaneda, R. Arakawa, M. Okamoto, H. Doe, T. Matsuo, K. Naemura, K. Hirose, Y. Tobe, *Chem. Commun.* **1996**, 1735.

49 A simple formalism for the deconvolution of the stereochemical and isotope effects has been described: G. Hornung, C. A. Schalley, M. Dieterle, D. Schröder, H. Schwarz, *Chem. Eur. J.* **1997**, 3, 1866.

50 M. Sawada, Y. Takai, H. Yamada, J. Nishida, T. Kaneda, R. Arakawa, M. Okamoto, K. Hirose, T. Tanaka, K. Naemura, *J. Chem. Soc., Perkin Trans. 2* **1998**, 701.

51 a) D. V. Dearden, C. Dejsupa, Y. Liang, J. S. Bradshaw, R. M. Izatt, *J. Am. Chem. Soc.* **1997**, 119, 353; b) Y. Liang, J. S. Bradshaw, R. M. Izatt, R. M. Pope, D. V. Dearden, *Int. J. Mass Spectrom.* **1999**, 185–187, 977; c) I.-H. Chu, D. V. Dearden, J. S. Bradshaw, P. Huszthy, R. M. Izatt, *J. Am. Chem. Soc.* **1993**, 115, 4318.

52 a) M. Albrecht, S. Mirtschin, M. de Groot, I. Janser, J. Runsink, G. Raabe, M. Kogej, C. A. Schalley, R. Fröhlich, *J. Am. Chem. Soc.* **2005**, 127, 10371; M. Albrecht, M. Baumert, J. Klankermayer, M. Kogej, C. A. Schalley, R. Fröhlich, *J. Chem. Soc., Dalton Trans.* **2006**, 4395.

53 For reviews on self-assembled metallo-supramolecular complexes, see: a) D. L. Caulder, K. N. Raymond, *Acc. Chem. Res.* **1999**, 32, 975; b) S. Leininger, B. Olenyuk, P. J. Stang, *Chem. Rev.* **2000**, 100, 853; c) B. J. Holliday, C. A. Mirkin, *Angew. Chem.* **2001**, 113, 2076; *Angew. Chem. Int. Ed.* **2001**, 40, 2022; d) M. Fujita, K. Umemoto, M. Yoshizawa, N. Fujita, T. Kusukawa, K. Biradha, *Chem. Commun.* **2001**, 509.

54 C. A. Schalley, T. Müller, P. Linnartz, M. Witt, M. Schäfer, A. Lützen, *Chem. Eur. J.* **2002**, 8, 3538. For a review on coldspray ionization of metallo-supramolecular species, see ref. [14].

55 M. Engeser, A. Rang, M. Ferrer, A. Gutiérrez, C. A. Schalley, *Int. J. Mass Spectrom.* **2006**, 255–256, 185.

56 F.-G. Klärner, B. Kahlert, *Acc. Chem. Res.* **2003**, 36, 919.

57 C. A. Schalley, C. Verhaelen, F.-G. Klärner, U. Hahn, F. Vögtle, *Angew. Chem.* **2005**, 44, 477; *Angew. Chem. Int. Ed.* **2005**, 44, 477.

58 C.-C. Liou, J. S. Brodbelt, *J. Am. Soc. Mass Spectrom.* **1992**, 3, 543.

59 D. V. Dearden, *Physical Supramolecular Chemistry*, L. Echegoyen, A. E. Kaifer (Eds.), Kluwer, Dordrecht 1996.

60 a) M. B. More, E. D. Glendening, D. Ray, D. Feller, P. B. Armentrout, *J. Phys. Chem.* **1996**, 100, 1605; b) D. Ray, D. Feller, M. B. More, E. D. Glendening, P. B. Armentrout, *J. Phys. Chem.* **1996**, 100, 16116; c) M. B. More, D. Ray, P. B. Armentrout, *J. Phys. Chem. A* **1997**, 101, 831; d) M. B. More, D. Ray, P. B. Armentrout, *J. Phys. Chem. A* **1997**, 101, 4254; e) M. B. More, D. Ray, P. B. Armentrout, *J. Phys. Chem. A* **1997**, 101, 7007; f) M. B. More, D. Ray, P. B. Armentrout, *J. Am. Chem. Soc.* **1999**, 121, 417; g) P. B. Armentrout, *Int. J. Mass Spectrom.* **1999**, 193, 227.

61 D. M. Peiris, Y. Yang, R. Ramanathan, K. R. Williams, C. H. Watson, J. R. Eyler, *Int. J. Mass Spectrom.* **1996**, 157/158, 365.

62 P. D. Schnier, J. S. Klassen, E. F. Strittmatter, E. R. Williams, *J. Am. Chem. Soc.* **1998**, 120, 9605.

6
Diffusion NMR in Supramolecular Chemistry

Yoram Cohen, Liat Avram, Tamar Evan-Salem and Limor Frish

6.1
Introduction

Several different analytical methods are needed to determine the structure and dynamics and to map the intermolecular interactions that prevail in supramolecular systems in solution [1]. NMR is one of the most powerful of these methods [1d–e]. The conventional NMR parameters that are used to characterize the structure and dynamics of supramolecular systems in solution are chemical shifts, spin–spin coupling, relaxation times, NOEs (Nuclear Overhauser Effect) and the correlation thereof. We shall demonstrate that the diffusion coefficient, which is currently underused, should be added to this arsenal of NMR parameters when characterizing supramolecular systems in solution.

Despite the fact that it was realized quite early that molecular diffusion can be measured by NMR methods [2], the most practical pulse sequence for measuring diffusion with NMR was introduced by Stejskal and Tanner only in 1965 [3a]. Since then, and until the 1990s, there was a steady increase in the use of diffusion NMR. However, this increase was relatively limited since the required hardware and software were not available on commercial NMR spectrometers [3b–c]. Since the beginning of the 1990s we have witnessed a dramatic increase in the use of gradients in all areas of NMR [4]. In the mid-1990s, with the advent of high-resolution gradient-enhanced spectroscopy [4, 5] and the improvement in gradient performance, such gradient sets became conventional accessories of standard high-resolution NMR spectrometers. These gradient-containing high-resolution probes enable simultaneous determination of the diffusion coefficients for the entire sets of peaks in a high-resolution spectrum NMR with high sensitivity and accuracy. Indeed, in the last decade there has been an increase in the use of diffusion NMR in many fields [6], however, this technique is, in our opinion, still underused by supramolecular chemists – a situation that may change in the near future.

In the present chapter, after a brief description of the notion of diffusion and the basics of diffusion NMR, we briefly survey a few technical issues concerning diffusion NMR. Readers interested in more detailed descriptions of the theory of diffusion NMR can consult the many excellent reviews published in recent years [5–7].

Analytical Methods in Supramolecular Chemistry. Edited by Christoph Schalley

ISBN: 978-3-527-31505-5

We shall then describe selected applications of diffusion NMR that will form the main body of the present chapter. Thereafter, we shall provide a brief discussion on limitations of diffusion NMR and the effect of chemical exchange on diffusion NMR. We conclude with a short summary and outlook.

6.2 Concepts of Molecular Diffusion

Self-diffusion is the random translational motion of ensembles of particles (molecules or ions) originating from their thermal energy. It is well known that diffusion, which is closely related to the molecular size of the diffusing species, is given by the Einstein–Smoluchowski equation, Eq. (6.1) [8]:

$$D = k_b T/f \tag{6.1}$$

where k_b, T and f are the Boltzmann constant, the absolute temperature and the so-called hydrodynamic frictional coefficient, respectively. For a hard sphere in a continuous medium of viscosity η, f is given by the Stokes' equation;

$$f = 6\pi\eta r_s \tag{6.2}$$

In Eq. (6.2), r_s is the hydrodynamic radius, often called the Stokes' radius. Combining equations (6.1) and (6.2) leads to the familiar Stokes–Einstein equation, Eq. (6.3), which implies that the hydrodynamic radius of a molecular species can be extracted from NMR diffusion measurements [9].

$$D = k_b T/6\pi\eta r_s \tag{6.3}$$

It should be noted, however, that for molecular species of geometries other than spheres, different approximations are needed to describe the hydrodynamic frictional coefficient, f [8]. Since there is a relationship between r_s and the effective molecular size and shape of the molecular species, it is clear that diffusion coefficients are sensitive to the structural properties and aggregation modes of these species. Therefore, the diffusion coefficient can be used to probe binding phenomena and intermolecular interactions [10, 11].

6.3 Measuring Diffusion with NMR

6.3.1 The Basic Pulse Sequence

The basis for diffusion measurements is the fact that magnetic field gradients can be used to indirectly label the position of spins through their Larmor frequency

(ω_0). This is done by applying an external linear gradient of the magnetic field (G), which is described by:

$$G = \frac{\partial Bz}{\partial x}\hat{i} + \frac{\partial Bz}{\partial y}\hat{j} + \frac{\partial Bz}{\partial z}\hat{k} \tag{6.4}$$

where $\hat{i}$, $\hat{j}$ and $\hat{k}$ are the unit vectors in the x, y and z directions, respectively. Thus, the total external magnetic field at position r ($B(r)$) is given by:

$$B(r) = B_0 + G \cdot r \tag{6.5}$$

where B_0 is the external static magnetic field. Under these conditions, the effective Larmor frequency, ($\omega(r)$), depends on the spatial position as shown in Eq. (6.6)

$$\omega(r) = \gamma B(r) \tag{6.6}$$

where γ is the gyromagnetic ratio. If the linear gradient G is applied over a defined time period, δ, the acquired phase angle depends linearly on $B(r)$. Under these conditions, assuming that only a Z gradient is applied (G_z), the acquired phase shift is position dependent and is given by a phase angle of:

$$\Phi(z) = -\gamma B(z)\delta \tag{6.7}$$

From the above, it is clear that the magnetic field gradient can be used to label the position of the spins. The most common NMR pulse sequence for measuring diffusion is the pulsed gradient spin echo (PGSE) sequence [3a], which is a modification of the Hahn spin-echo pulse sequence [2]. In this sequence, two identical gradient pulses are inserted, one into each τ period of the spin echo sequence (Fig. 6.1) [7a]. The PGSE sequence and a schematic representation of its effect on the magnetization of an ensemble of spins are shown in Fig. 6.1. At the beginning of the experiment, the net magnetization is oriented along the z-axis, meaning that the spins are in thermal equilibrium. Then, a 90° rf pulse is applied, tipping the magnetization to the x–y plane. At a time point t_1, a pulse gradient, the duration and magnitude of which are δ and G, respectively, is applied. As a result, at the end of the first τ period, just before the 180° rf pulse, each spin experiences a phase shift according to Eq. (6.8):

$$\Phi i(\tau) = \gamma \mathrm{B}_0 \tau + \gamma \mathrm{G} \int_{t_1}^{t_1+\delta} z_i(t)\, dt \tag{6.8}$$

In Eq. (6.8) the first term is the phase shift due to the static magnetic field and the second term is the phase shift due to the magnetic gradient pulse applied along the z-direction. The next step is applying a 180° rf pulse, which inverts the sign of the precession and the phase as depicted in Fig. 6.1. At time $t_1 + \Delta$ a second identical pulse gradient is applied which cancels out the induced phase shifts,

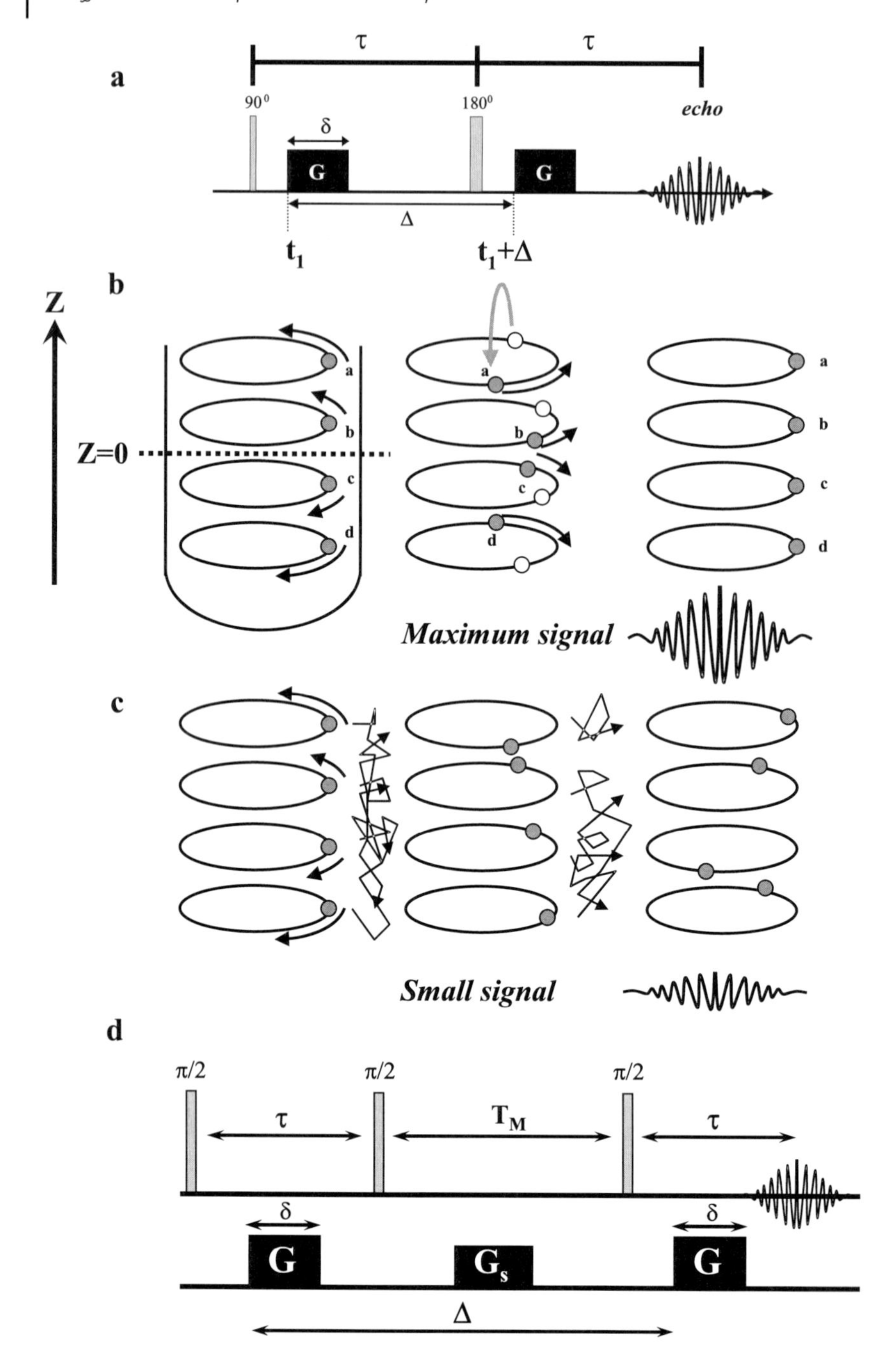

Fig. 6.1. (a) The PGSE pulse sequence [3a] and the effect of the absence (b) and presence (c) of diffusion on the phase shift and signal intensity in a PGSE experiment, and (d) the stimulated-echo (STE) diffusion sequence [12]. In the STE diffusion sequence shown in (d) the spoiling gradient (G_s) is optional. (Adapted with permission from ref. [10]. Copyright 2005 Wiley-VCH.)

as shown in Fig. 6.1b. However, it should be noted that this description is adequate for spins which do not undergo any translational motion along the z-direction, i.e. there is no diffusion during the Δ time interval. In this situation, an echo of maximal intensity is obtained. In the presence of diffusion (Fig. 6.1c), however, the phase shift of each spin due to the first and the second τ periods is different in magnitude. This occurs since, in the presence of diffusion, each species is located at a different position along the z-axis at times t_1 and $t_1 + \Delta$ and hence each species is situated in a different magnetic field, meaning that its Larmor frequencies ($\omega(r)$) at the two time points are different. Thus, in the presence of diffusion, the second gradient pulse is not able to completely refocus the magnetization. One important advantage of the PGSE experiment is that if the experiment is preformed while keeping τ constant, it is possible to separate the T_2 relaxation and the diffusion contributions. Hence, after normalizing out the attenuation due to T_2 relaxation, only the attenuation due the diffusion remains. Stejskal and Tanner [3a] showed that, for a single free diffusing component, the signal intensity is given, in the case of rectangular pulse gradients, by

$$\begin{aligned} I_{(2\tau, G)} &= I_{(0,0)} \exp\left(\frac{-2\tau}{T_2}\right) \exp(-\gamma^2 G^2 \delta^2 (\Delta - \delta/3) D) \\ &= I_{(2\tau, 0)} \exp(-\gamma^2 G^2 \delta^2 (\Delta - \delta/3) D) \end{aligned} \tag{6.9}$$

Hence:

$$\ln\left(\frac{I_{(2\tau, G)}}{I_{(2\tau, 0)}}\right) = -\gamma^2 G^2 \delta^2 (\Delta - \delta/3) D = -bD \tag{6.10}$$

where $I_{(0,0)}$, $I_{(2\tau,0)}$ and $I_{(2\tau,G)}$ are the signal intensity immediately after the first 90° rf pulse and after 2τ in the absence and presence of gradient pulses, respectively, G is the pulsed gradient strength, Δ and δ are the time separation between the pulsed gradients and their duration, respectively (Fig. 6.1), and D is the diffusion coefficient. The product $\gamma^2 G^2 \delta^2 (\Delta - \delta/3)$ is often abbreviated as the b value and represents the *diffusing weighting*. Thus, for an isotropic solution, a plot of $\ln(I_{(2\tau,G)}/I_{(2\tau,0)})$ vs. b should give a straight line, the slope of which is equal to $-D$. In principle, any of the parameters, δ, Δ and G, can be varied to affect signal attenuation and hence measure D. However, technical factors and the relaxation characteristics of the sample may limit our choice. The term $(\Delta - \delta/3)$ is generally referred to as the *diffusion time* [3b].

Figure 6.2 shows the experimental signal decay, shown as a stackplot, for a PGSE experiment, in which the magnetic field gradient strength was incremented from 0 to approximately 30 G cm^{-1} in ten steps while δ and Δ were kept constant, for molecules diffusing with diffusion coefficients of 1.81×10^{-5} and 0.33×10^{-5} cm^2 s^{-1}, respectively along with the normalized signal decays ($\ln(I_{(2\tau,G)}/I_{(2\tau,0)})$) as a function of the diffusion weighting (b values). It should be noted that $\ln(I_{(2\tau,G)}/I_{(2\tau,0)})$ is generally abbreviated as $\ln(I/I_0)$.

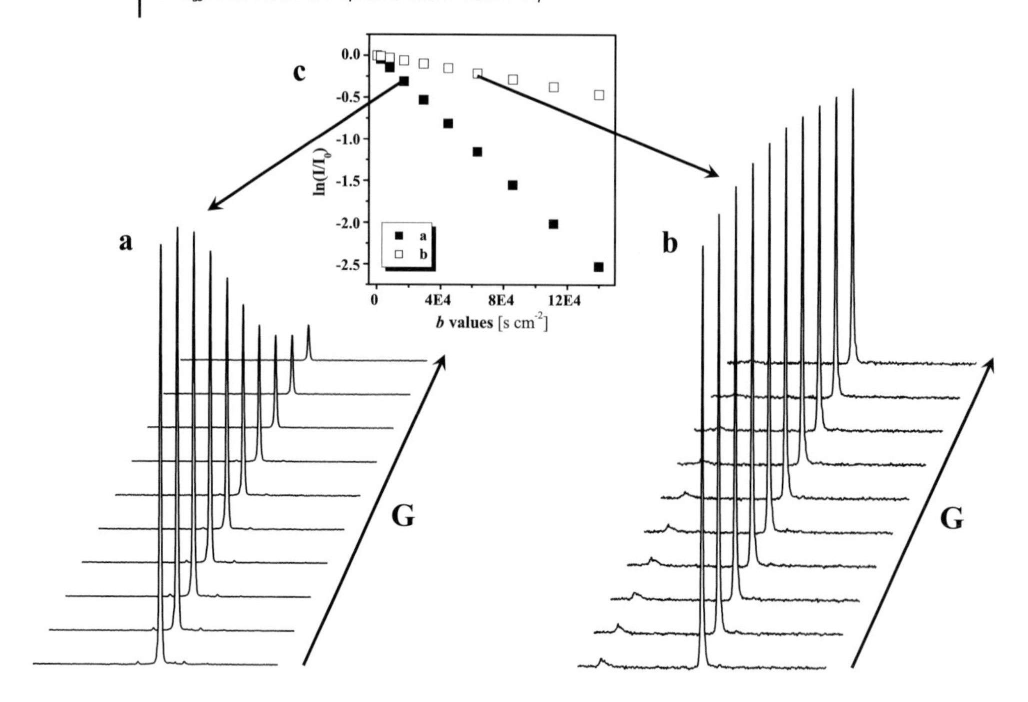

Fig. 6.2. Signal decays for a compound having diffusion coefficients of (a) $D = 1.81 \times 10^{-5}$ cm^2 s^{-1}, (b) $D = 0.33 \times 10^{-5}$ cm^2 s^{-1}, and (c) the corresponding graphical analysis of the data in (a) and (b). (Reproduced with permission from ref. [10]. Copyright 2005 Wiley-VCH.)

6.3.2
The Stimulated Echo (STE) Diffusion Sequence

The stimulated echo (STE) diffusion pulse sequence is depicted in Fig. 6.1d [12]. The signal intensity of the STE diffusion experiment, for rectangular pulse gradients, is given by Eq. (6.11),

$$\begin{aligned} I_{(TM+2\tau,\,G)} &= (I_{(0,0)}/2)\exp\left[\left(\frac{-2\tau}{T_2}\right) - \left(\frac{T_M}{T_1}\right)\right]\exp(-\gamma^2 G^2 \delta^2 (\Delta - \delta/3) D) \\ &= I_{(TM+2\tau,\,0)}\exp(-\gamma^2 G^2 \delta^2 (\Delta - \delta/3) D) \end{aligned} \tag{6.11}$$

From Eq. (6.11) it is clear that, again, we can separate the effects of relaxation and diffusion on the signal decay. In this sequence, signal decay, during the mixing time (T_M), depends on T_1-relaxation. T_2-relaxation is operative only during 2τ, thus allowing us to increase the diffusion time and hence the diffusion weighting without increasing τ. Therefore, the main advantage of the STE diffusion sequence over the PGSE sequence is that the STE diffusion sequence is more suitable for obtaining diffusion spectra of systems characterized by short T_2. This is an advantage since, in many systems, T_1 is significantly longer than T_2. The STE diffusion sequence allows the increase of the diffusion weighting by "paying" in T_1 and not T_2 relaxation time. Longer diffusion times are generally required for measuring low diffusion coefficients and in situations where the diffusion coefficients may be dependent on the diffusion times. The main disadvantage of STE-based sequences versus spin echo-based sequences is the loss of a factor of two in the echo intensity. However, for $T_M \cong 2\tau$, the STE/SE signal ratio is about 5 when $T_1 \cong 5T_2$. For $T_M \cong 8\tau$, for example, this ratio is more than 10 when $T_1 \cong 5T_2$. It should be noted that the equations shown above describe the case of single quantum transition where the coherence order p is 1. For multi-quantum transitions of coherence p, the signal decay due to diffusion in Eqs. (6.9)–(6.11) should be multiplied by p^2.

6.3.3
Technical Issues in Diffusion NMR

The gradient system used is the most important hardware component for diffusion NMR [7]. First, one has to determine its strength and quality. The determination of the gradient strength can, in principle, be obtained by simply measuring the one-dimensional profile of a sample with known dimensions. However, the most practical approach is to measure the diffusion coefficients of several solvents for which the self-diffusion coefficients are well known. It is recommended to measure three different solvents which have different diffusion coefficients. The diffusion coefficient of each solvent should be measured with several combinations of parameters (echo time, diffusion time and gradient length) and the normalized signal, i.e. I/I_0 should decay to about 10% of its original value. This will allow determining the actual strength of the pulsed gradient produced by the specific gradient system.

The most important issue affecting the accuracy of NMR diffusion measurements is the quality of the gradient pulses generated by the gradient system. In this context, the most serious problem is the eddy currents generated as a result of the pulse gradients. Eddy currents result from the rapid changes in pulse gradients and are proportional to their strength and the rising and falling times of the gradient pulses, i.e. dG/dt. The magnetic field associated with these eddy currents can interfere with the main magnetic field applied. If such a magnetic field expands to the second τ period of the PGSE experiment, for example, there will be a residual phase shift, even for spins which do not undergo any motion and wiggles will be observed in the spectra. One method to minimize eddy currents is to use shielded gradient coils. In fact, most modern gradient systems do use active shielded gradient coils. This reduces the eddy currents considerably, to about 1% of their original values [14]. Nevertheless, for acquiring high quality diffusion NMR data, particularly for systems characterized by short T_2 and low diffusion coefficients, one needs to reduce the eddy currents even further.

A quick evaluation of the performance of the gradient system in use can be obtained using the pulse gradient shown in Fig. 6.3a. This sequence is, in fact, the

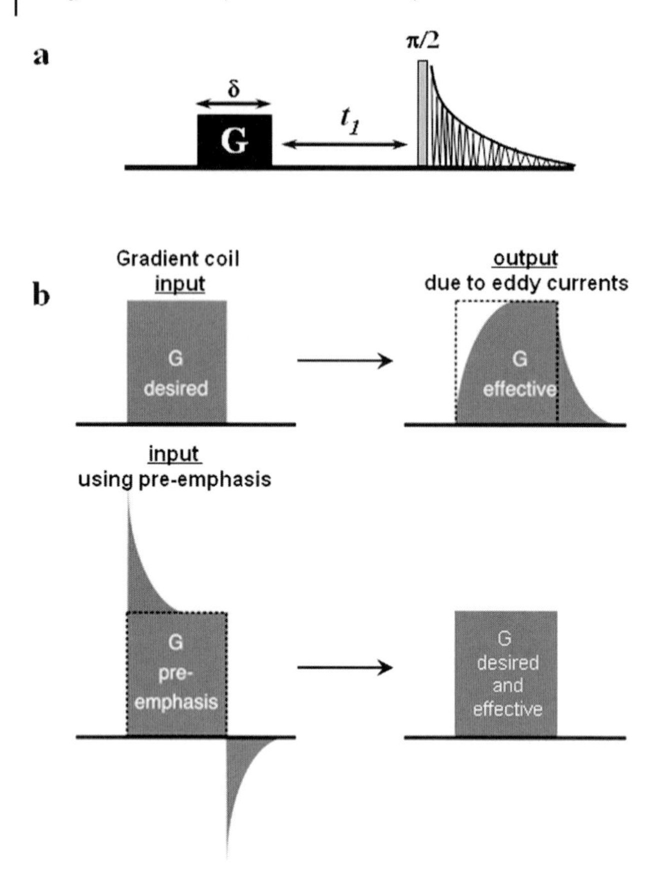

Fig. 6.3. (a) The pulse sequence used to evaluate gradient performance and for pre-emphasis. (b) The principle of pre-emphasis.

sequence used for pre-emphasis of the gradient systems. The logic behind it is quite simple. Here, one has to use a gradient pulse, which is stronger and longer than the maximal gradient pulses generally used. In the first experiment, a very long t_1 of about 1 s should be used where eddy currents should have no effect on the signal. Then, the phase of the obtained signal is corrected and followed in subsequent experiments in which t_1 is shortened. The aim of the experiment is to determine the minimal t_1 for which there is no effect on the phase and intensity of the observed signal. High quality gradient systems show no effect on the signal for t_1s in the order of several μs when typical gradient pulses of about 30 G cm^{-1} of several ms length are used. This pulse sequence can also be used to pre-emphasize the gradient coils. The principle of the pre-emphasis procedure is shown in Fig. 6.3b. Input of an ideal waveform into the gradient coil does not result in such an ideal gradient shape, because of the generation of eddy currents. Therefore, small extra corrections, which overdrive the leading and tailing edges of the gradients, are used to partially compensate for this effect. Although pre-emphasis further improves the performance of the gradient systems, there is sometime a need for more advanced pulse sequences which were devised to even further reduce eddy current effects. Two such sequences are the longitudinal eddy current delay (LED) sequence [15a] and the bi-polar LED (BPLED) sequence [15b] which will be described in the Section 6.3.4 after we briefly discuss the effect of temperature gradients and convection on diffusion NMR.

NMR diffusion experiments are sensitive to temperature instability and temperature gradients which in turn may induce convection in the sample. Temperature gradients along the z-direction (in probes equipped with a single gradient coil, the gradient is generally along the z-direction) are likely to occur since temperature regulation in NMR probes is generally achieved by flowing gas through the probe. The flow and temperature are regulated by a remote feedback detector and heater. Convection is a flow phenomenon and therefore should produce a coherent phase shift. Since the resulting convections have a distribution of velocities, the result is often a reduction in signal intensity. Convection artifacts are more apparent when longer diffusion times are used and in general they are much more severe when diffusion is measured in low viscosity solvents, such as most of the common organic solvents. There are several possibilities to alleviate the problem. First, one can use shorter samples but this implies more difficult shimming and smaller signal-to-noise ratios. Another simple alternative, which from experience works quite well, is the use of a thermally isolated sample. Indeed it is recommended to perform our NMR diffusion measurements in 4-mm NMR tubes inserted in a 5-mm NMR tube. Therefore, NMR diffusion measurements should be attempted only after the probe has reached thermal equilibrium. If, after all these precautions, convection artifacts are still apparent, one can always turn to flow-compensated sequences such as the double stimulated echo sequence introduced by Jerschow and Müller [16].

6.3.4
The LED and BPLED Sequences

One of the best alternatives to reduce the effects of eddy currents in NMR diffusion measurements is to use the LED and the BPLED sequences shown in Figs. 6.4a and 6.4b, respectively. From Fig. 6.4a, it is clear that the difference between the LED sequence and the STE diffusion sequence is the addition of two 90° pulses separated by a delay (t_e) at the end of the STE diffusion sequence. As a result of the fourth 90° rf pulse, the magnetization is again stored in the longitudinal direction while the eddy currents can decay during an eddy current settling period, t_e. Then, the fifth 90° rf pulse rotates the magnetization into the x–y plane where it can be detected.

The BPLED sequence shown in Fig. 6.4b is a modification of the LED sequence shown in Fig. 6.4a. In this sequence, each gradient pulse (G), which has a duration of δ is replaced in the LED sequence by two pulses of different polarity (G and $-G$) separated by a 180° rf pulse with a duration of $\delta/2$. The G-180-($-G$) sequence causes the eddy currents to be cancelled out, while the diffusion gradients accumu-

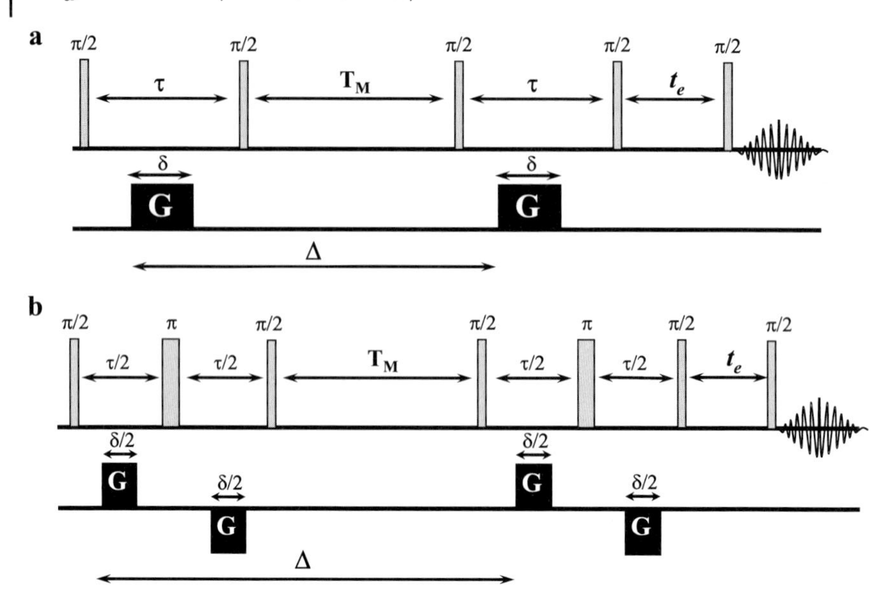

Fig. 6.4. (a) The LED pulse sequence [15a], and (b) the bi-polar-LED (BPLED) pulse sequence [15b].

late. There, there are two main advantages to using the BPLED over the LED experiment. First, eddy currents are reduced to a minimum, and secondly, the effective gradient output is doubled, making it more suitable for measuring diffusion in systems characterized by slow diffusion. It should be noted that the BPLED sequence is in fact much less sensitive to the background gradient, however this is less of a problem when diffusion NMR is applied to homogenous solutions in high-resolution probes.

The four sequences shown in Figs. 6.1 and 6.4 are by far the most important and widespread pulse sequences for measuring diffusion in solution. The sequences shown in Fig. 6.4 are the ones generally used in diffusion ordered spectroscopy (DOSY, see below and ref. [17]) and are less prone to eddy current artifacts. However, to decrease the sensitivity of all these sequences to eddy currents even further, one can use them with shaped gradient pulses as opposed to rectangular gradient pulses. Since the induction of eddy currents is proportional to dG/dt, the rise and fall times of the gradient pulses, it is recommended to use shaped pulsed gradients. The most popular shaped gradient pulses are the sine-shaped ones, for which the gradient effect is scaled-down by a factor of $2/\pi$. Here, the effective diffusion time is $(\Delta - \delta/4)$, and not $(\Delta - \delta/3)$ as in the case of rectangular pulse gradients. However, since in most diffusion NMR experiments $\Delta \gg \delta$, this has only a minor effect. Readers who are interested in more technical issues including those related to diffusion hardware, are encouraged to consults several of the reviews that have been published in recent years [5–7].

6.3.5
DOSY – Diffusion Ordered Spectroscopy

One of the most important developments in the field of diffusion NMR in the last decade seems to be the introduction of diffusion-ordered spectroscopy (DOSY) by Johnson Jr. [15, 17]. DOSY would appear to be one of the main reasons for the recent increase in popularity of diffusion NMR among non-expert users. This method uses the inverse Laplace transform (ILT), and provides a 2-D-map (like those in 2-D NMR) in which one axis is the chemical shift while the other is that of the diffusion coefficient. Figure 6.5 shows a simulated DOSY spectrum of four components having different diffusion coefficients.

The basic idea behind the DOSY concept is similar to the one behind multidimensional NMR. In 2-D NMR, a modulation in the phase or signal intensity with respect to a known time increment is recovered by inverse FT. In a DOSY experiment, the diffusion coefficient is recovered from the signal decay as a function of a diffusion increment by an ILT. In fact, the approximate ILT of the signal amplitude with respect to q^2, where q is defined as $\gamma g\delta f(t)$, yields the second dimension of the spectrum which correlates the chemical shift with the diffusion coefficient. Therefore, it was termed diffusion ordered spectroscopy (DOSY). However, unlike the FT of the time domain signal that yields a unique solution, ILT does not yield a unique solution. Therefore, several software packages were developed to overcome this problem. Readers interested in more details concerning the DOSY techniques can consult a recent extensive review on the subject [17].

One of the most important features of DOSY is its ability to separate signals of compounds within a mixture, based on their diffusion coefficients which, in fact, reflect their size and shape, thus providing a means for "virtual separation". Here, the information is spread over the entire plan thus also simplifying peaks assignment. However, it was commented that for separating peaks of different compounds, which happen to have the same chemical shift, a large difference in the diffusion coefficients (a ratio of about 3) is needed for DOSY to provide the accurate numbers. An important fact to remember is that diffusion sequences, in fact, act as filters and can therefore be imbedded or coupled to nearly any NMR sequence, including most of the multidimensional NMR sequences.

In 3-D-DOSY experiments, a diffusion dimension is added to the conventional 2-D experiment. The 2-D NMR spectrum spreads the NMR signals of the same species over an entire 2-D plane, while the spreading of the different species on a

Fig. 6.5. (a) A schematic representation of a 2-D-DOSY spectrum showing four different species characterized by four different diffusion coefficients, (b) the DOSY–COSY pulse sequence [18a] and (c) a schematic representation of 3D-DOSY data obtained from a pulse sequence such as the one shown in (b). (Reproduced with permission from ref. [10]. Copyright 2005 Wiley-VCH.)

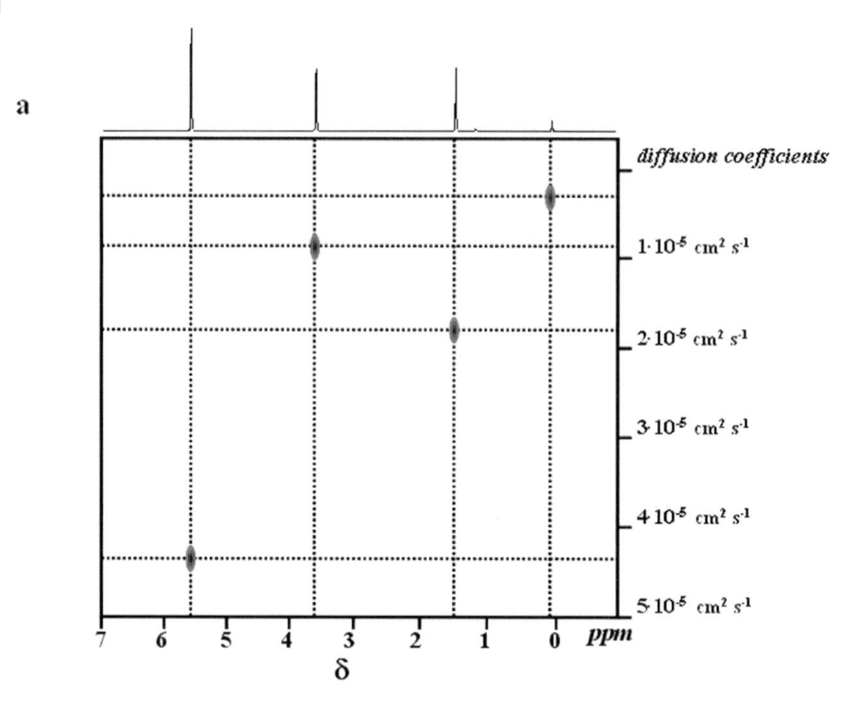
a
diffusion coefficients
1·10-5 cm2 s-1
2·10-5 cm2 s-1
3·10-5 cm2 s-1
4·10-5 cm2 s-1
5·10-5 cm2 s-1
7 6 5 4 3 2 1 0 ppm
δ

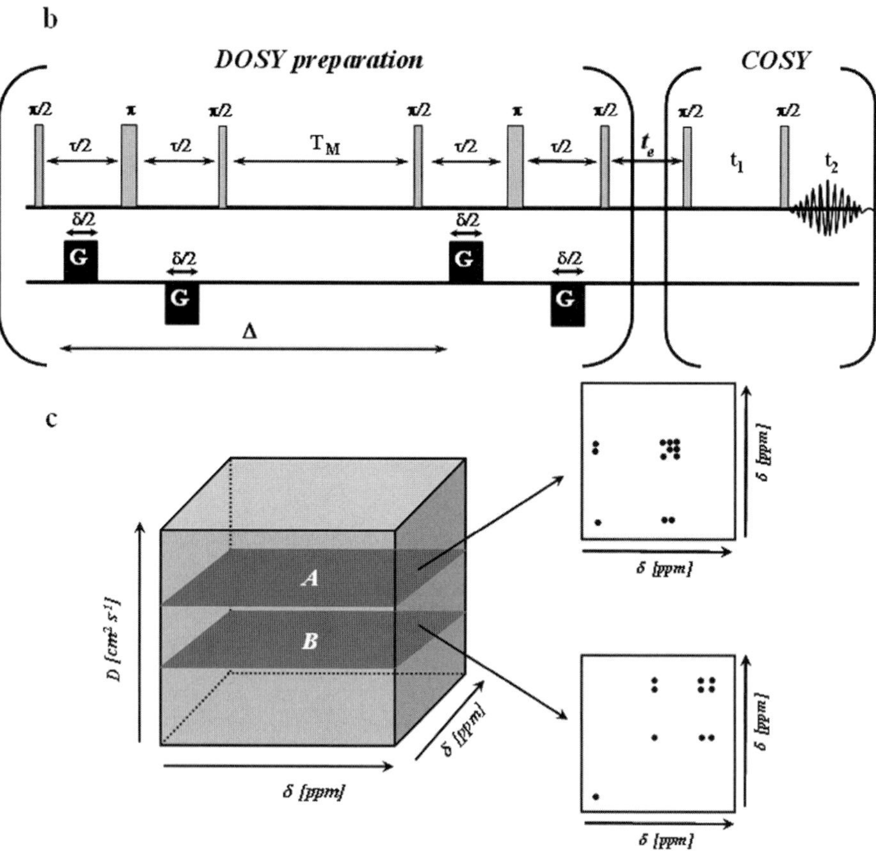
b
DOSY preparation
COSY
π/2
π
τ/2
T_M
t_e
t1
t2
δ/2
G
Δ
c
A
B
D [cm2 s-1]
δ [ppm]

third axis is performed using their diffusion coefficients. Figure 6.5b shows the DOSY–COSY pulse sequence along with the results expected from such an experiment [18a]. This specific pulse sequence is obtained by linking the BPLED preparation sequence with a COSY sequence, where at the end of the eddy currents delay, t_e, the fifth rf pulse which rotates the magnetization to the xy detection plane also serves as the first pulse of the COSY sequence. The results of such an experiment are shown in Fig. 6.5c. This figure, showing a schematic representation of the results of a 3-D-DOSY–COSY sequence, also depicts the ability of this sequence to "virtually separate" different components. The various compounds are spectroscopically separated according to their different diffusion coefficients. Thus, one can find in each plane, at different altitudes on the diffusion axis of the cube, the 2-D spectrum of a different species. Figure 6.5c shows two schematic COSY spectra of two compounds which differ in their diffusion coefficients. Other 3-D-DOSY pulse sequences were subsequently designed using the same rationale [18]. However, the main disadvantage of 3-D-DOSY sequences, in general, is the fact that their collection is time consuming. For a more comprehensive discussion of DOSY, readers are encouraged to consult the recent review of Johnson Jr. on the topic [17].

6.4 Applications of Diffusion NMR in Supramolecular Chemistry: Selected Examples

To motivate the reader, inexperienced with diffusion NMR, to use these techniques to study supramolecular systems, we chose to provide selected examples which emphasize the questions and the type of information that can be obtained from such NMR diffusion measurements.

6.4.1 Binding and Association Constants

Diffusion NMR can be very useful in determining association constants in many supramolecular systems [10, 19]. Association constants, K_a, are an important measure of the strength of intermolecular interactions in supramolecular systems. Indeed, since the early days of supramolecular chemistry, a huge number of such constants were reported using several different analytical methods [20]. Despite the fact that NMR had become an important tool for studying association constants and the fact that the possibility of using diffusion NMR for the determination of association constants was proposed decades ago [21], the first relatively isolated example of such an application on a host–guest was only reported in 1983 [22]. During the 1990s, scattered studies were reported in which diffusion NMR was used to determine association constants. However, even currently, this parameter is still underused relative to studies based on NMR chemical shifts. This is despite the fact that there are cases and systems in which the diffusion coefficient is a superior parameter to chemical shifts for determining association constants. Even when diffusion NMR provides the same information as conventional NMR chemical shift titrations, it is important to attempt to use it, since it is rewarding to obtain similar results from two independent parameters using the same experimental set-up which are obtained under strictly the same experimental conditions, like concentration for example.

NMR diffusion coefficients, being an NMR observable, can be used in a very similar way as chemical shifts to determine the stoichiometry of the complexes and their association constants. For a simple 1:1 host–guest complex, for example, in the case of slow exchange, on the NMR time scale, the association constant can be determined by simple integration of the peaks of a solution of known concentrations. In these cases, NMR diffusion measurements can only indicate the formation of the complex, but cannot provide the K_a values. However, in the case of fast exchange, the values of the association constants can be extracted using diffusion NMR by first calculating the bound fraction.

The rationale behind the extraction of the bound fraction from NMR diffusion measurements is simple [10, 21–23]. The host and guest have their own diffusion coefficients in the free unbound states that are a reflection of their effective molecular weights and shapes. However, when a strong complex is formed, the complex components diffuse as a single molecular entity and should have the same diffusion coefficient. In the case of no interaction, the diffusion coefficients of the host and guest will remain as they were in their free states. For any other case, assuming fast exchange on the NMR timescale, the observed diffusion coefficient (D_{obs}) should be a weighted average of the free and bound diffusion coefficients (D_{free} and D_{bound}, respectively) and can therefore be used to calculate the bound fraction, (X), in the same way that chemical shifts are used for this purpose as shown in Eq. (6.12),

$$D_{obs} = XD_{bound} + (1 - X)D_{free} \tag{6.12}$$

Therefore, in principle, the same methodologies used to obtain K_as from changes in chemical shift titration experiments [10, 19] can be used to obtain association constants from diffusion NMR. The main advantage of using diffusion NMR here is that, in many cases, a complete titration to find D_{bound} for the guest is not a necessity. This is true in cases where there is a large difference between the molecular weight of the host and the guest. In such cases, one can, *a priori*, predict that D_{bound} of the guest will be very similar to D_{free} of the much larger host [23, 10]. In principle, the determination of association constants using NMR diffusion measurements has the inherent advantages and limitations associated with an NMR-based method [10, 19, 23]. Consequently, this method is less prone to misinterpretation resulting from minor impurities than techniques based on other spectroscopic methods. However, diffusion NMR is only suitable for measuring, in a direct way, association constants in the range of 10–10^5 M^{-1}. An important advantage of using diffusion NMR to extract association constants, as compared with NMR chemical shift titrations for example, is the elimination of one of the main possible sources of error in such techniques, namely confusing acid–base

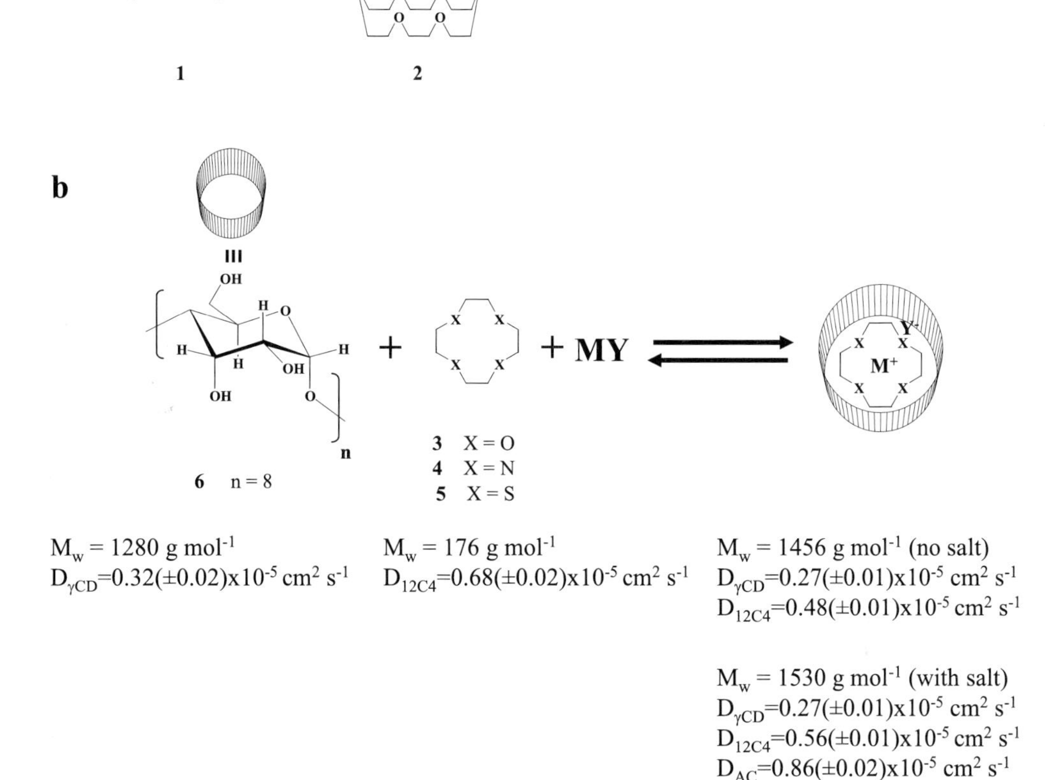

Fig. 6.6. Schematic representation of the formation of the γ-CD:macrocycle complexes along with representative molecular weight (M_w) and respective diffusion coefficients (D). More numerical values can be found in Table 6.1 [25a].

chemistry, i.e., chemical shift changes originating from proton transfer, with binding processes [23]. This is because proton transfer, which generally affects chemical shifts, has only a marginal effect on the measured diffusion coefficient. This enables easy extraction of the association constant between methyl ammonium chloride (**1**) and [2.2.2]cryptand (**2**), where partial protonation of **2** occurred [23] (Fig. 6.6a). As the diffusion coefficient of a certain molecular species depends on its effective molecular size, which changes with any intermolecular interactions, diffusion NMR is the only method in which the value expected for the bound guest, in a host–guest system, can be predicted, *a priori*, thus reducing the need for a complete titration to find this value. The main limitation of the diffusion coefficient as a parameter for the determination of K_a is in systems where the interacting species happen to have a similar diffusion coefficient in their free states. In such systems, the method is much less sensitive or even impractical.

In multicomponent systems, diffusion NMR may be an even more attractive alternative for probing complexation and calculating association constants. Such systems may be systems in which the guest itself can be used as a host for yet another guest. A system of this kind, which was originally described by Vögtle and Müller [24], was recently studied in solutions by diffusion NMR [25]. There, the complexation of 12-crown-4 (**3**), cyclen (**4**) and 1,4,7,10-tetrathiacyclododecane (**5**) with γ-CD (**6**) was studied in the absence and presence of salts [25]. In these systems, where ^{1}H-NMR chemical shift changes were found to be very small, the difference in the molecular weight of the different macrocycles and that of γ-CD (**6**) enabled accurate determination of the association constants. Figure 6.6 shows the studied equilibrium along with the molecular weights and the diffusion coefficients extracted from diffusion NMR for the **3**/**6** system, both before and after the addition of lithium acetate. Table 6.1 provides the diffusion coefficients of these components in

Tab. 6.1. Diffusion coefficients (D) of the free components, γ-CD:macrocycle, macrocycle:salt and the three component systems in D_2O at 298 K and calculated associated constants (K_a).[a] [25]

System[b]	D_6 [10^{-5} cm^2 s^{-1}]	$D_{macrocycle}$ [10^{-5} cm^2 s^{-1}]	D_{OAc^-} [10^{-5} cm^2 s^{-1}]	K_a [M^{-1}]
3+6+LiOAc	0.27 ± 0.01	0.56 ± 0.01	0.86 ± 0.02	11
3+6	0.27 ± 0.01	0.48 ± 0.01	–	187
3	–	0.68 ± 0.02	–	
6	0.32 ± 0.02	–	–	
LiOAc	–	–	1.02 ± 0.01	
3+LiOAc	–	0.60 ± 0.01	0.90 ± 0.01	40
4+6+LiOAc	0.30 ± 0.01	0.53 ± 0.01	0.96 ± 0.01	19
4+6	0.29 ± 0.01	0.42 ± 0.01	–	165
4	–	0.60 ± 0.01	–	
6	0.32 ± 0.02	–	–	
LiOAc	–	–	1.09 ± 0.01	
4+LiOAc	–	0.58 ± 0.01	0.96 ± 0.01	29
5+6+LiOAc	0.19 ± 0.02	0.36 ± 0.02	0.82 ± 0.01	21
5+6	0.19 ± 0.01	0.34 ± 0.02	–	69
5	–	0.41 ± 0.02	–	
6	0.24 ± 0.02	–	–	
LiOAc	–	–	0.87 ± 0.02	
5+LiOAc	–	0.39 ± 0.01	0.84 ± 0.01	10

[a] All experiments were performed three times and the reported values are means ± standard deviation.
[b] The 12C4 and the 12N4 systems were measured in a D_2O solution having a pD of 7.6 while the 12S4 was measured in DMSO-d_6.

the unbound state and that of the two and three component complexes along with the extracted association constants. It was found that the presence of alkali salts decreased the association between the macrocycles and the γ-CD and that the pH had practically no effect on the extracted association constants [25a]. Because diffusion NMR is relatively unaffected by acid–base equilibrium, the authors could easily determine the association constants at different pHs and concluded that hydrogen bonding is not the major factor responsible for complexation of these specific γ-CD complexes. In these systems, the changes in the chemical shifts were rather small and both the cation and the γ-CD had some effect on the chemical shift of the macrocycles making the extraction of association constants from this parameter difficult in general and hardly practical in the three-component cases. This example emphasizes the advantage of using diffusion coefficients to map the interaction between more than two species simultaneously. In this early study the measured diffusion coefficients for the γ-CD were slightly higher than more recent measurements reported very recently [25b, 26].

Pseudorotaxanes obtained from the interaction of α-CD (**7**) with diaminoalkanes and their salts (**8**) [26a, b] were also studied by diffusion NMR (Fig. 6.7) [27]. In this study, where the motivation, *inter alia*, was to evaluate the effect of protonation on the stability of the pseudorotaxanes, there was a need to determine the association constants of the different diaminoalkanes to α-CD both before and after protonation. Since the chemical shift changes upon the formation of these pseudorotaxanes were relatively small, and in order to avoid attributing chemical shift changes because of protonation to binding phenomena in the case of the protonated amines, diffusion measurements were used rather than NMR chemical shift titrations [27]. Representative diffusion coefficients in these systems along with the log K_a values are shown in Fig. 6.7 and Table 6.2. Table 6.2 shows that, even in the last entry, describing the formation of the pseudorotaxanes of α-CD with 1,12 diaminododecane (**8e**), the longest diaminoalkane used, there was only a small effect on the diffusion coefficient of **7** upon binding. The changes in the diffusion coefficients of **7** and **8e**, as a function of the **8e**/**7** ratio, are depicted graphically in Fig. 6.7b. A dependency was found between the length of the diaminoalkanes and the association constants to **7**. It was found that protonation considerably reduced the stability of the pseudorotaxanes with the shorter diaminoalkanes. Only with the longest diaminoalkane studied, i.e. (**8e**), were the same association constants found for the diaminoalkane and its disalts form [27].

As previously stated, one of the main advantages of using diffusion NMR to obtain association constants of host–guest systems, over chemical shifts for example, is the fact that D_{bound} for the guest in usually not very different from D_{free} of the host since, in most host–guest systems, the host is much larger than the guest. This may, in some cases, eliminate the need for a complete titration in order to obtain D_{bound} of the guest. It is clear that the larger the difference in the size of the interacting molecules, the better this approximation. Recently, Cameron and Fielding tested this assumption, both theoretically and experimentally [28], by evaluating the effect of concentration of the diffusion coefficients of β-CD (**9**), cyclohexylacetic acid (**10**) and cholic acid (**11**) (Fig. 6.8a). It was found that when β-CD

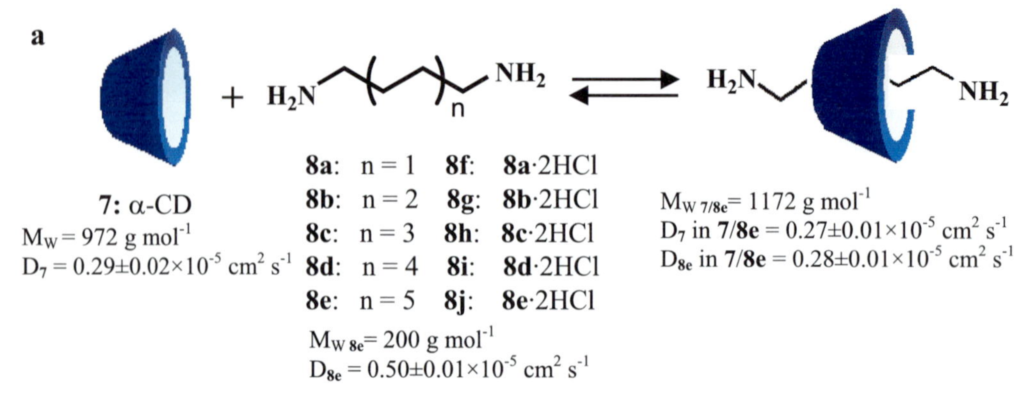

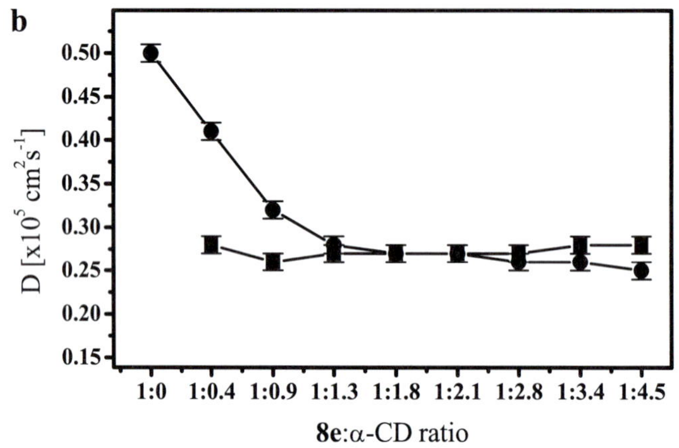

Fig. 6.7. (a) Schematic representation of the formation of the pseudorotaxanes of α-CD (**7**) and the diaminoalkane and their respective salts along with a few representative M_w and diffusion coefficients (D) and (b) the effect of the **8e**/**7** ratio on the diffusion coefficient of α-CD (**7**) (▪) and **8e** (•). Further numerical values can be found in Table 6.2 [27]. (Reproduced in part with permission from ref. [27]. Copyright 2002 American Chemical Society.)

(M_W of 1135 g mol^{-1}) (**9**) forms a complex with cyclohexylacetic acid (**10**) (M_W of 136 g mol^{-1}) (Fig. 6.8b), for which the association constant K_a is in the order of 1800 $\pm$ 100, there is a small gradual reduction in the diffusion coefficient of β-CD upon the addition of **10**. In this case, however, these changes were in the order of the error bars of the diffusion measurements. When the guest was cholic acid (**11**) (M_W of 420 g mol^{-1}) (Fig. 6.8c) then much more significant changes in D_{host} were found following the addition of **11**, implying that in this case $D_{complex}$ differs from D_{host}, emphasizing the need for a complete titration, or at least several points, to obtain the binding curve in this complex. These results clearly indicate that single point binding experiments using diffusion NMR need to be viewed with caution,

Tab. 6.2. Diffusion coefficients (D) of α-CD (7), diamines (**8a–8e**) and their mixtures before and after protonation, and the calculated log K_a in D_2O at 298 K.[a] [26]

System[b]	D_{guest} [$\times 10^{-5}$ cm^2 s^{-1}]	D_{CD} [$\times 10^{-5}$ cm^2 s^{-1}]	D_{water} [$\times 10^{-5}$ cm^2 s^{-1}]	log K_a [M^{-1}]
α-CD		0.30 ± 0.01	1.96 ± 0.01	/
8a	0.76 ± 0.01	/	1.97 ± 0.01	/
8a:α-CD (1:2.9)	0.71 ± 0.01	0.29 ± 0.01	1.97 ± 0.01	1.18 ± 0.14
8b	0.65 ± 0.01	/	1.96 ± 0.01	/
8b:α-CD (1:5.1)	0.49 ± 0.02	0.30 ± 0.01	1.96 ± 0.01	1.81 ± 0.07
8g:α-CD (1:1.3)	0.63 ± 0.01	0.30 ± 0.01	1.95 ± 0.01	/
8c	0.60 ± 0.01	/	1.96 ± 0.01	/
8c:α-CD (1:0.8)	0.47 ± 0.01	0.29 ± 0.01	1.95 ± 0.01	2.83 ± 0.21
8h:α-CD (1:1.1)	0.52 ± 0.01	0.29 ± 0.01	1.96 ± 0.01	1.87 ± 0.15
8d	0.55 ± 0.01	/	1.96 ± 0.01	/
8d:α-CD (1:0.7)	0.38 ± 0.01	0.29 ± 0.01	1.96 ± 0.01	4.13 ± 0.30
8i:α-CD (1:1.7)	0.33 ± 0.01	0.29 ± 0.01	1.96 ± 0.01	3.34 ± 0.19
8e	0.50 ± 0.01	/	1.94 ± 0.01	/
8e:α-CD (1:0.9)	0.32 ± 0.01	0.26 ± 0.01	1.94 ± 0.01	±3.85 ± 0.27
8j:α-CD (1:0.9)	0.34 ± 0.01	0.29 ± 0.01	1.94 ± 0.01	3.83 ± 0.21

[a] All experiments were performed at least three times and the reported values are means ± standard deviation.
[b] The stoichiometric ratios of the mixtures are given in parentheses.

in cases where the different partners of the host–guest systems do not differ much in size.

The determination of association constants, using diffusion NMR, is based on the fact that there is a fast exchange between the free and bound states of the guest. However, there are important applications of diffusion measurements in the case of slow exchange. Such an example is the study of molecular capsules and the encapsulation phenomenon, as will be outlined in the next section.

6.4.2 Encapsulation and Molecular Capsules

Molecular capsules [29, 30] are molecules that can accommodate smaller molecules in their cavities. Therefore, diffusion coefficients which reflect the "effective" molecular size of the investigated molecular species should be a sensitive tool for

a

9 n=7 10 11

b

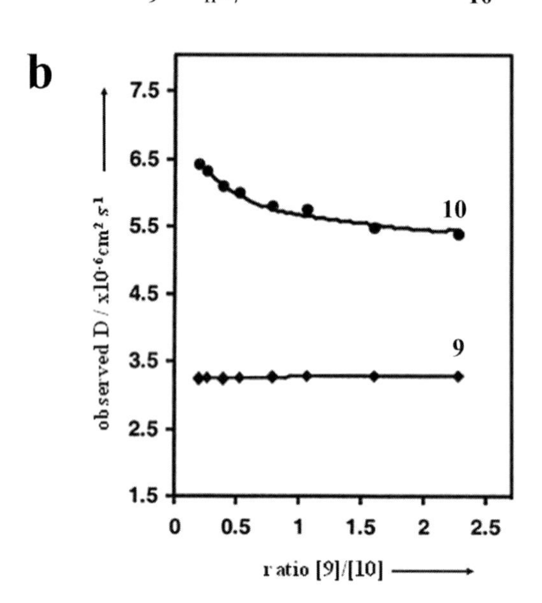
7.5
6.5
5.5
4.5
3.5
2.5
1.5
observed D / x10-6cm2 s-1
10
9
0
0.5
1
1.5
2
2.5
ratio [9]/[10]

c

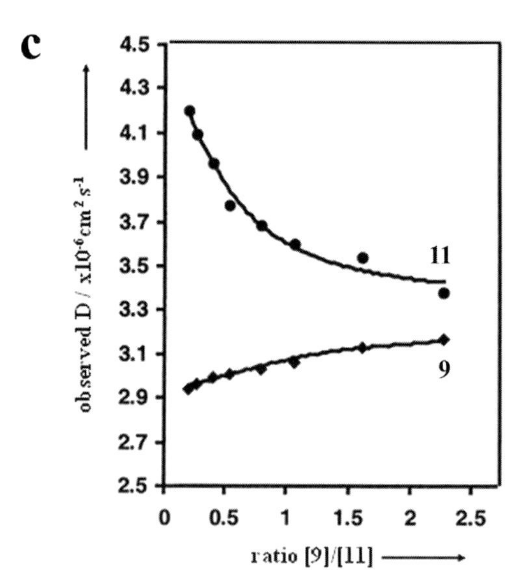
4.5
4.3
4.1
3.9
3.7
3.5
3.3
3.1
2.9
2.7
2.5
observed D / x10-6cm2 s-1
11
9
0
0.5
1
1.5
2
2.5
ratio [9]/[11]

detecting and probing encapsulation. The relatively large differences between the capsule and the encapsulated guest, implies that the guest should have a much lower diffusion coefficient in the capsule than in its free state. Moreover, the encapsulated molecules, if they are in slow exchange with their bulk population, should have the same diffusion coefficient as the capsule itself since the capsule and the encapsulated molecules diffuse as a single molecular entity. Noncovalent capsules are relatively labile systems and are therefore more difficult to probe in solution. Indeed, NMR diffusion measurements can be expected to, and have played, an important role in the corroboration of the formation of noncovalent molecular capsules [31–33].

The first application of diffusion NMR to corroborate the encapsulation of benzene in the tetraureacalix[4]arene dimer (**12**•**12**, Fig. 6.9) was published in 1999 [31a]. When this dimer was prepared in a mixture of C_6H_6 and C_6D_6, a new peak at about 4.4 ppm was observed in the ^1H NMR spectrum. This peak, that was tentatively assigned to an encapsulated benzene molecule, was found to have a much lower diffusion coefficient than bulk "free" benzene (at 7.17 ppm). Indeed, the peak at ~4.4 ppm was found to have the same diffusion coefficient as the peaks representing the dimer as shown in Fig. 6.9b. This figure depicts the signal decay as a function of the diffusion gradient strength (G), and the diffusion coefficients extracted therefrom for the peak at 4.4 ppm along with that of the peak of bulk benzene (7.17 ppm) and one representative peak of the dimer (at 1.95 ppm); it demonstrates how easy it is to reach the conclusion that the entire capsule diffuses as a single entity [31a]. To corroborate this conclusion, a titration of the **12**•**12** solution with DMSO, which should disaggregate the dimer, was performed. Although DMSO increases the viscosity of the solution and a decrease in the diffusion coefficients would be anticipated, an increase in the diffusion coefficient was observed. The most plausible explanation for this observation is that the DMSO indeed disintegrates the hydrogen-bond aggregates [31a].

The power and simplicity of DOSY in probing intermolecular interactions in general and encapsulation in particular, is shown in Fig. 6.10. Upon addition of the cobaltocenium cation (**13**) to the $C_2D_4Cl_2$ solution of the dimer **12**•**12**, an extra peak at about 2.8 ppm was observed beyond that of **13**. When ferrocene (**14**) was added to the solution, no additional high-field peak was observed. The DOSY spectrum, in Fig. 6.10b, of the $C_2D_4Cl_2$ solution of dimer **12**•**12** in the presence of **13** and **14** clearly demonstrates how trivial it is to map the mutual interactions that prevail in the solution between the different molecular species [31c]. Only the peak

◄

Fig. 6.8. (a) The structure of β-CD (**9**), cyclohexylacetic acid (**10**) and cholic acid (**11**), and the effect of the (b) **9**/**10** ratio and (c) **9**/**11** on their diffusion coefficients. Both the experimental values and the fitting of the data to a theoretical model are shown. For more details see ref. [28]. (Reproduced in part with permission from ref. [28]. Copyright 2001 American Chemical Society.)

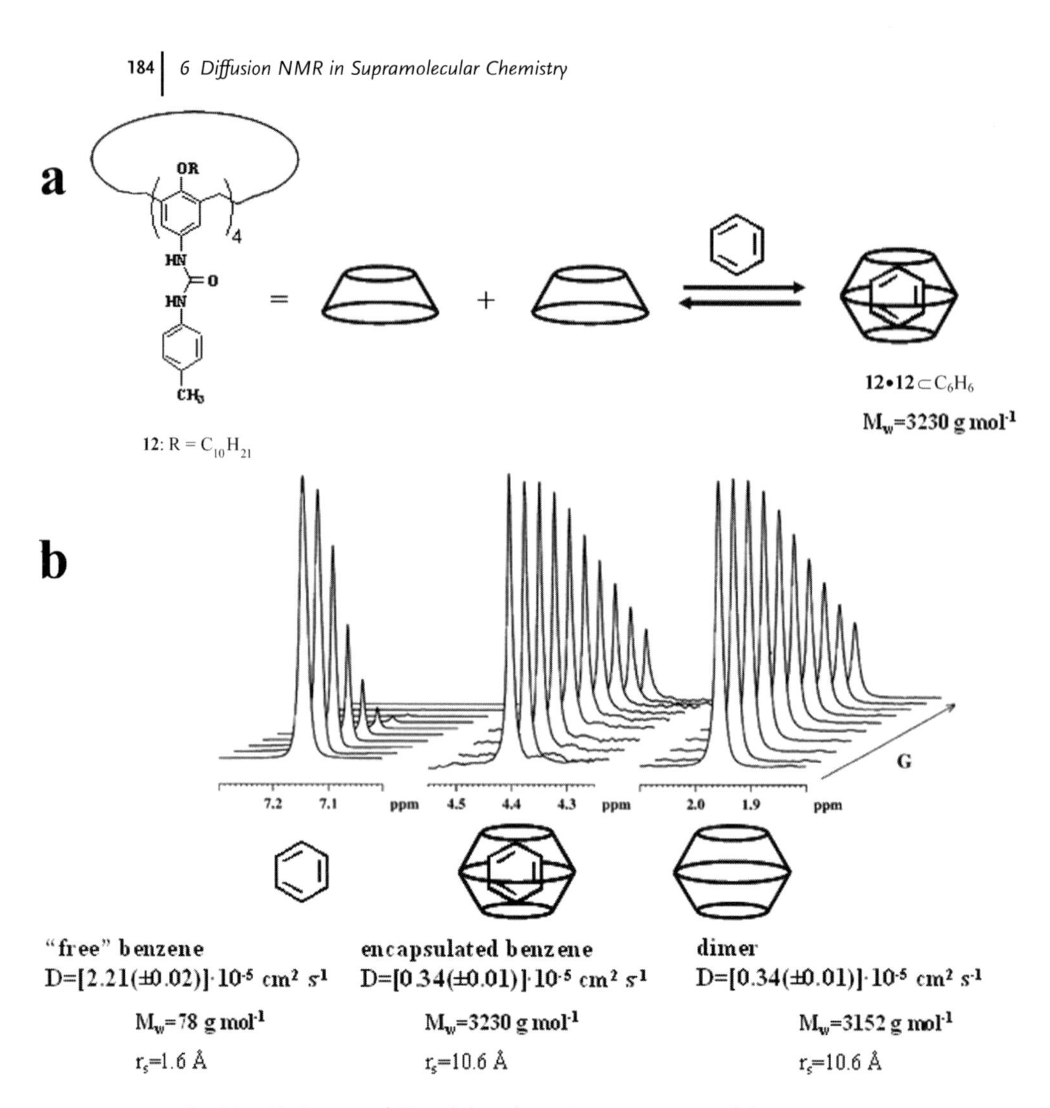

Fig. 6.9. (a) Compound **12** and the schematic representation of the formation of a dimeric capsule of **12** and (b) the signal decay as a function of the pulsed gradient strength (G) for "free" (at 7.17 ppm) and encapsulated benzene (at 4.4 ppm) and one representative peak of **12·12** (at 1.95 ppm) along with their molecular weights and calculated radii (r_s) [31a]. (Adapted from ref. [10].)

at 2.72 ppm has the same diffusion coefficient as all the other peaks of the dimer, thus corroborating the assignment of this peak to the encapsulated cobaltocenium cation. The same methodology was used to demonstrate that, in this type of dimeric capsules, the tropylium cation has a much higher affinity toward the capsule cavity than benzene [31b].

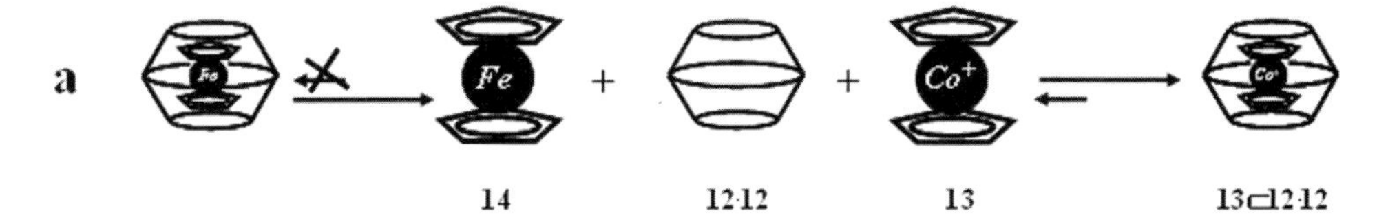

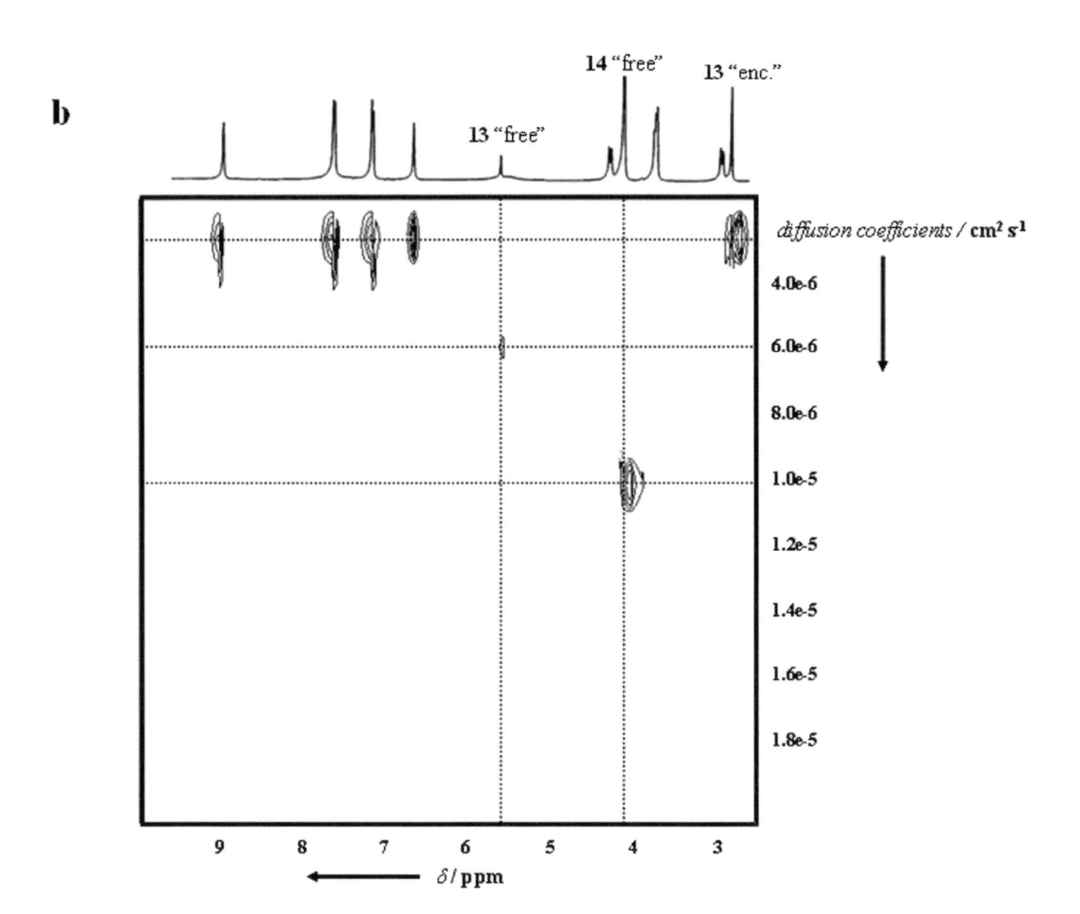

Fig. 6.10. (a) Schematic representation of the reaction of **12·12** with **13** and **14** and (b) the DOSY spectrum (400 MHz, 298 K) of a $C_2D_4Cl_2$ solution of **12·12** in the presence of **13** and **14**, showing that the peak at 2.72 ppm, suspected to be the encapsulated cobaltocenium cation, has the same diffusion coefficient as all the peaks of **12·12**. The peaks of the free ferrocene and cobaltocenium cation at 4.06 and 5.61 ppm, respectively, were found to have much higher diffusion coefficients, as expected (data from ref. [31c]). (Reproduced with permission from ref. [10]. Copyright 2005 Wiley-VCH.)

The real power of diffusion NMR, however, is manifested in multicomponent supramolecular systems and even more so in supramolecular systems that conserve the symmetry and spectroscopic characteristics of their building blocks. In such supramolecular systems, diffusion NMR can be crucial for accurate determination of the structures that prevail in solution, as demonstrated for the hexameric capsule of resorcin[4]arenes (**15**) and pyrogallol[4]arenes (**16**) (Fig. 6.11) [32–33]. The structures of these systems are shown in Fig. 6.11a. These systems, first characterized by the Atwood and Mattay groups, were found to form, in the solid state, a $[(\mathbf{15a})_6(H_2O)_8]$ and $(\mathbf{16b})_6$ type hexamers, respectively as shown in Fig. 6.11b [34]. Subsequently, it was shown that **15c** forms a hexamer in the presence of suitable guests in water saturated $CDCl_3$ solutions [35]. Tetrahexylammonium bromide (**17**) (see Fig. 6.11c, upper spectrum) and later Bu_4SbBr (**18**) were found to be ideal guests for the hexamer of **15c** and were therefore claimed to induce the formation of these hexamers. However, when the diffusion coefficient of **15c** was measured in the absence and presence of **17** in the $CDCl_3$ solutions, the same low diffusion coefficients were measured for the peak of **15c** in both solutions (see Fig. 6.11c). The only plausible explanation for these results is that the same aggregation mode prevails in the two solutions and it was therefore suggested that **15c** self-assembles spontaneously, in wet $CDCl_3$ solutions, into a hexameric capsule [32a]. Indeed, when **15c** was subsequently dissolved in $CHCl_3$, new high-field signals appeared at about 4.8–5.1 ppm. These peaks, tentatively assigned as encapsulated $CHCl_3$ molecules when measured with a STE diffusion pulse sequence, were found to have the same low diffusion coefficient as that of the resorcinarene peaks, thus further supporting the assignment of these signals to encapsulated chloroform molecules [32a]. Moreover, the addition of **17** to this chloroform solution resulted in the disappearance of these signals. In titration experiments, in which the change in the diffusion coefficient of the resorcinarene peaks was followed as a function of the addition of DMSO, an increase in the diffusion coefficient of the **15c** peaks was found upon addition of DMSO. Very similar titration curves were found for the hexamer in the presence and absence of **17**. All these results strongly support the conclusion that the process followed is indeed the disruption of the higher aggregates, namely the hexamers, into their lower aggregates [32a].

Prompted by its crystal structure [34a] and the fact that some of these aggregates were formed in water-saturated chloroform solutions, the role of the water molecules in the formation of the hexameric capsules of **15c** and **16c** was also studied. Diffusion NMR was selected since the chemical shift of water protons in the presence of acidic compounds is very sensitive to the amount of water, the concentration of the acidic material, the exchange rate and solvent pH and is therefore not a predictable parameter. In addition, since water molecules are very small relative to the resorcinarene and pyrogallolarene hexamers, the water diffusion coefficient was anticipated to be a rather sensitive parameter to follow water interaction with these supramolecular structures. Indeed, when the water diffusion coefficient was followed as a function of the **15c**/H_2O ratio, a decrease in the diffusion coefficient of the water peak was observed with the increase of the **15c**/H_2O ratio, as shown in Fig. 6.12. Fig. 6.12a–d shows the signal decay of the water (Fig. 6.12a, c) and the

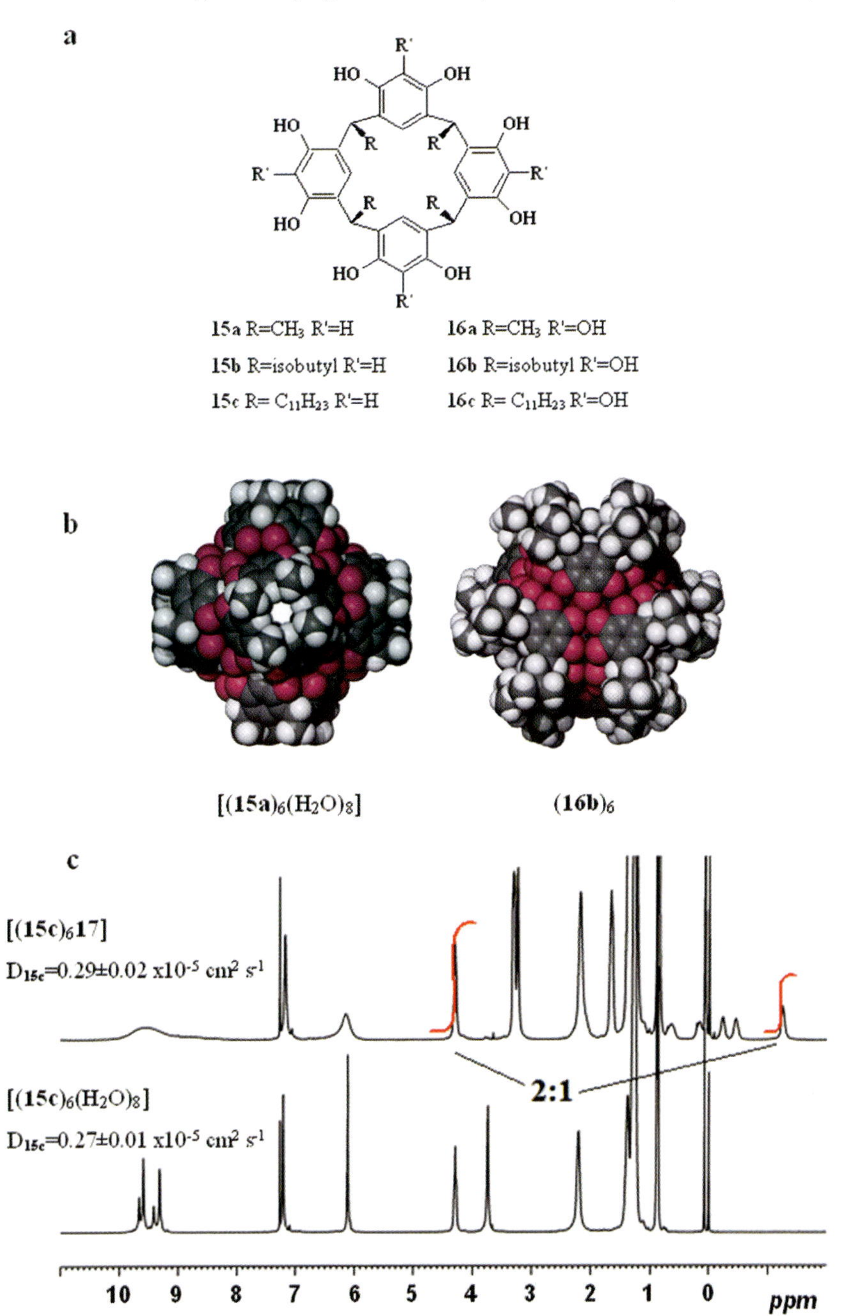

Fig. 6.11. (a) The structure of compounds **15a–c** and **16a–c**, (b) the crystal structure of $[(\mathbf{15a})_6(H_2O)_8]$ and $(\mathbf{16b})_6$ [34] and (c) (top) the ^1H NMR spectrum of the hexameric capsule of **15c** encapsulating **17** and (bottom) of **15c** in a $CDCl_3$ solution along with the diffusion coefficient of **15c** in each solution [32a].

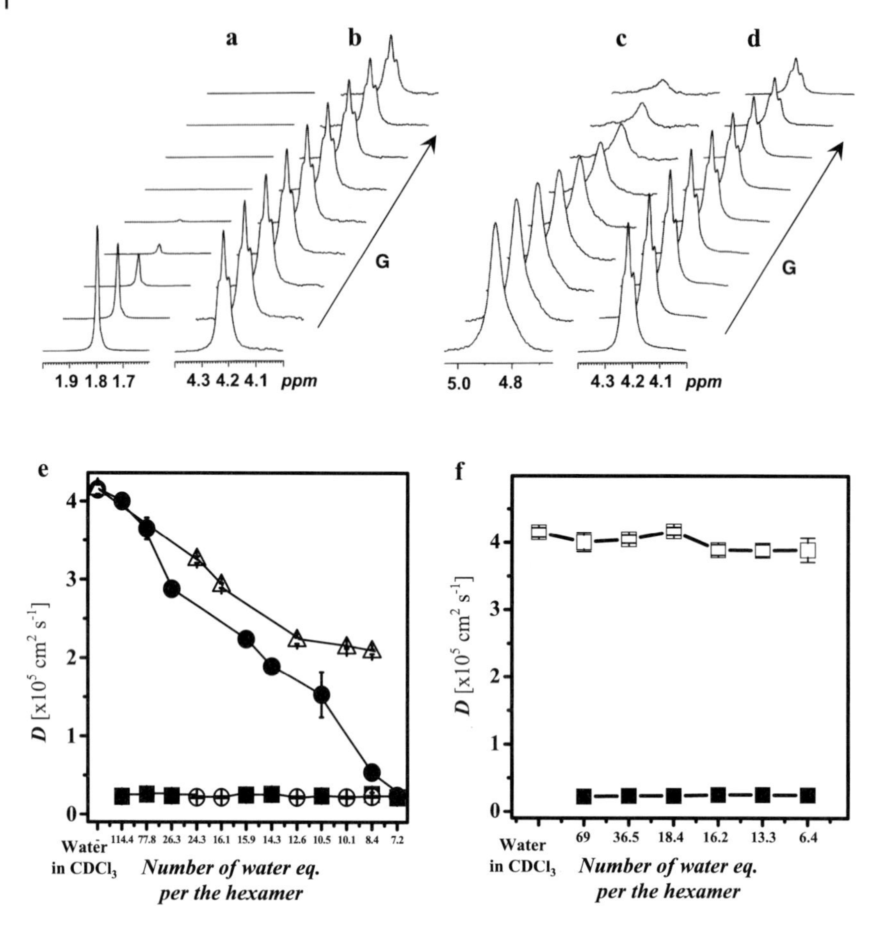

Fig. 6.12. Signal decay for the water peak (a, c) and one representative peak of the hexameric capsule of **15c** (b, d) when the **15c**:water ratios are 6:114 (a, b) and 6:8.4 (c, d) and changes in the diffusion coefficients of (e) the water (●) and **15c** (■) before and after addition of **17** (water △, **15c** ○) and for (f) water □ and **16c** ■ in $CDCl_3$ solutions [32b, c, 33a].

macrocycle (Fig. 6.12b, d) peaks as a function of the gradient strength (G) in the two different **15c**/H_2O ratios. Clearly, when the **15c**/H_2O ratio was changed from 6/114 to 6/8.4 a dramatic change in the signal decay of the water peak was observed. Figure 6.12e shows the extracted diffusion coefficient as a function of the **15c**/H_2O ratio. This figure shows that the water diffusion coefficient is dramatically affected by the **15c**/H_2O ratio. When the relative amount of H_2O is decreased there is a decrease in the extracted diffusion coefficient of water (Fig. 6.12e). Interestingly, even when the **15c**/H_2O ratio was 6:8.4 the diffusion coefficient for the water peak was nearly twice that of the hexamer. However, when that ratio was

Tab. 6.3. Diffusion coefficients of **15c**, **16c** and water in $CDCl_3$ at 298 K as a function of the macrocycle/water ratios.[a] [32b, 32c, 33a]

System[b]	Ratio macrocycle/water	Diffusion coefficients [$\times 10^{-5}$ cm^2 s^{-1}]		
		15c 1.3 ppm	16c 1.3 ppm	water
Water saturated $CDCl_3$				4.15 ± 0.06
15c 3.1 mM Water saturated $CDCl_3$	6:114.4	0.23 ± 0.01		4.00 ± 0.06
15c 5.6 mM Water saturated $CDCl_3$	6:77.8	0.26 ± 0.01		3.65 ± 0.14
15c 32 mM Water saturated $CDCl_3$	6:26.3	0.24 ± 0.01		2.88 ± 0.02
15c 32 mM Commercial $CDCl_3$	6:15.9	0.25 ± 0.01		2.24 ± 0.02
15c 26.7 mM Water saturated $CDCl_3$	6:14.3	0.26 ± 0.01		1.89 ± 0.02
15c 31.6 mM $CDCl_3$ (ampule)	6:10.5	0.24 ± 0.01		1.53 ± 0.29
15c 27.4 mM Commercial $CDCl_3$	6:8.4	0.26 ± 0.01		0.54 ± 0.01
15c 30.6 mM $CDCl_3$ (ampule)	6:7.2	0.22 ± 0.01		0.24 ± 0.03
15c/THABr 1:1 43.9 mM dry $CDCl_3$+H_2O	6:24.3	0.22 ± 0.01		3.25 ± 0.04
15c/THABr 1:1 43.9 mM dry $CDCl_3$+H_2O	6:16.1	0.22 ± 0.01		2.92 ± 0.03
15c/THABr 1:1 28.7 mM $CDCl_3$	6:12.6	0.22 ± 0.01		2.22 ± 0.04
15c/THABr 1:1 43.9 mM dry $CDCl_3$+H_2O	6:10.1	0.22 ± 0.01		2.13 ± 0.01
15c/THABr 1:1 43.9 mM dry $CDCl_3$+H_2O	6:8.4	0.23 ± 0.01		2.08 ± 0.03
15c/THABr 1:1 28.7 mM $CDCl_3$	6:4.1	0.20 ± 0.01		1.26 ± 0.02
16c 5.1 mM Water saturated $CDCl_3$	6:69		0.23 ± 0.01	4.00 ± 0.13
16c 24.8 mM Water saturated $CDCl_3$	6:18.4		0.23 ± 0.01	4.17 ± 0.05

Tab. 6.3 *(continued)*

System[b]	Ratio macrocycle/water	Diffusion coefficients [$\times 10^{-5}$ cm^2 s^{-1}]		
		15c 1.3 ppm	16c 1.3 ppm	water
16c 3 mM Commercial $CDCl_3$	6:36.9		0.23 ± 0.01	4.05 ± 0.05
16c 6.5 mM Commercial $CDCl_3$	6:16.2		0.25 ± 0.01	3.89 ± 0.07
16c 28.3 mM Commercial $CDCl_3$	6:6.4		0.24 ± 0.01	3.89 ± 0.16

[a] All experiments were performed three times and the reported values are means $\pm$ standard deviation.
[b] Both the macrocycle concentration and source of the $CDCl_3$ are given.

less than 6:8, i.e. 6:7.2, the diffusion coefficient of the water peak was found to be equal to that of the hexamer (Fig. 6.12e, Table 6.3). The only plausible explanation for this observation is that the eight water molecules interact with the six molecules of **15c**, thus suggesting that a hexamer of the $[(\mathbf{15c})_6(H_2O)_8]$-type prevails in solution [32b], as in the solid state [34a, c]. These results demonstrate that diffusion NMR enabled us to probe unequivocally, and in a simple way, the interaction between the fourteen components in this supramolecular system. Because of the high sensitivity of the diffusion coefficient of the water molecules, the same experiments were used to evaluate the role of the water molecules in the case of the **15c** hexamer that encapsulates **17** and for the hexamer of **16c**. For the hexamer of **16c** the **16c**/H_2O ratio had practically no effect on the diffusion coefficient of the water molecules, suggesting the formation of a $(\mathbf{16c})_6$-type hexamer (Fig. 6.12f, Table 6.3). Interestingly, for the hexamer of **15c** encapsulating **17**, the **15c**/H_2O ratio had only a limited effect on the diffusion coefficient of the water molecules. Even for a **15c**/H_2O ratio of less than 6:8 the diffusion coefficient of the water molecule was several times that of the hexamer (Fig. 6.12e), indicating that the water molecules can exchange with the hydroxyl group of the hexamer, however they are not part of the hexameric capsule [32c]. The NMR diffusion measurements seem to indicate that the water molecules leave the supramolecular structure when **17** replaces the encapsulated $CDCl_3$ molecules in the hexamer of **15c**. It should be noted, however, that diffusion measurements cannot distinguish between encapsulated water molecules and water molecules which are incorporated in the hexamer's seam of hydrogen bonds since, in both situations, in the case of slow exchange, the water diffusion coefficient should be equal to the diffusion coefficient of the hexamer. However, because of the fast exchange of H_2O with bulk water and the fact that $[(\mathbf{15c})_6(H_2O)_8]$-type capsules were observed in the solid state [34a], the more plau-

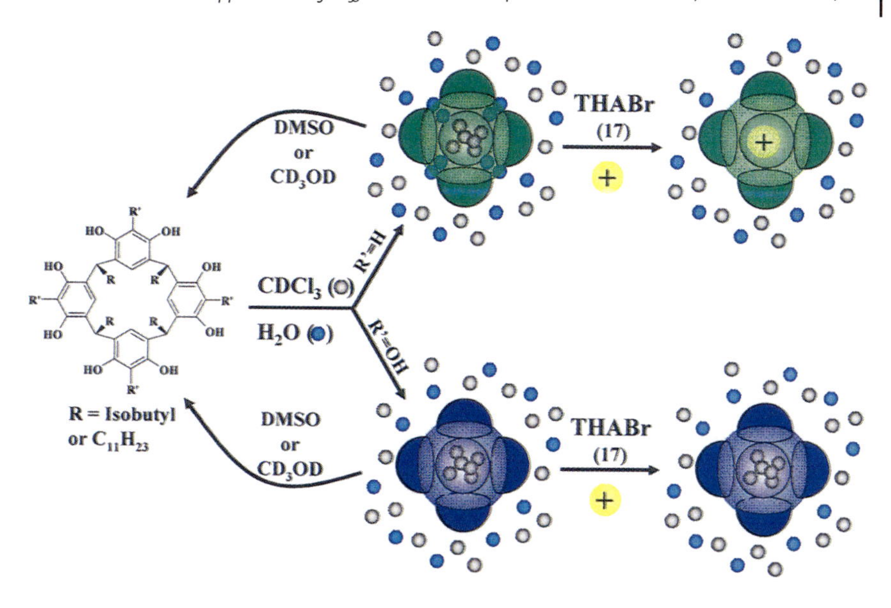

Fig. 6.13. Schematic representation of some of the different molecular interactions and aggregation modes identified by diffusion NMR in these giant capsules [32–33].

sible explanation seems to be that the eight water molecules, which have the same diffusion coefficient as the hexamer, are part of the supramolecular structure in the solution. The maps of the molecular interactions that prevail in these systems that were obtained from these diffusion NMR studies are summarized graphically in Fig. 6.13 [32–33]. Kaifer et al. recently used diffusion NMR in conjunction with electrochemistry to characterize the interaction of the hexameric capsules of **15c** and **16c** with **13** and **14** [36]. Interestingly, here again, like in the case of the dimeric capsule **12·12** [31c], they found that only **13** is encapsulated in the hexamers although it is clear that sterically both **13** and **14** can easily be accommodated in the cavity of these hexamers.

Self-recognition [37] in the self-assembly process of the resorcin[4]arene and pyrogallol[4]arene hexamers was also studied using diffusion NMR [33b]. Here again the chemical shift was not informative. However, the diffusion coefficient was found to be a very good parameter for following the systems, as shown in Fig. 6.14. For evaluation of the self-recognition in the self-assembly processes of these hexameric capsules, more derivatives that differ in their molecular weights and hence in their diffusion coefficients, i.e. **15b** and **16b** were prepared. When mixtures of **15b,c**, **16b,c** and **15b**, **16c** were prepared, nearly no changes in chemical shift could be observed and the ^{1}H NMR spectra of the mixture were a superposition of the spectra of the individual compounds, as shown in Figs. 6.14a and 6.14c for the **15b,c** and **15b**, **16c** mixtures, respectively. However, the diffusion results were quite different for the different mixtures studied. It was found that for the **15b,c** (see for example Fig. 6.14b) and **16b,c** mixtures, two different diffusion

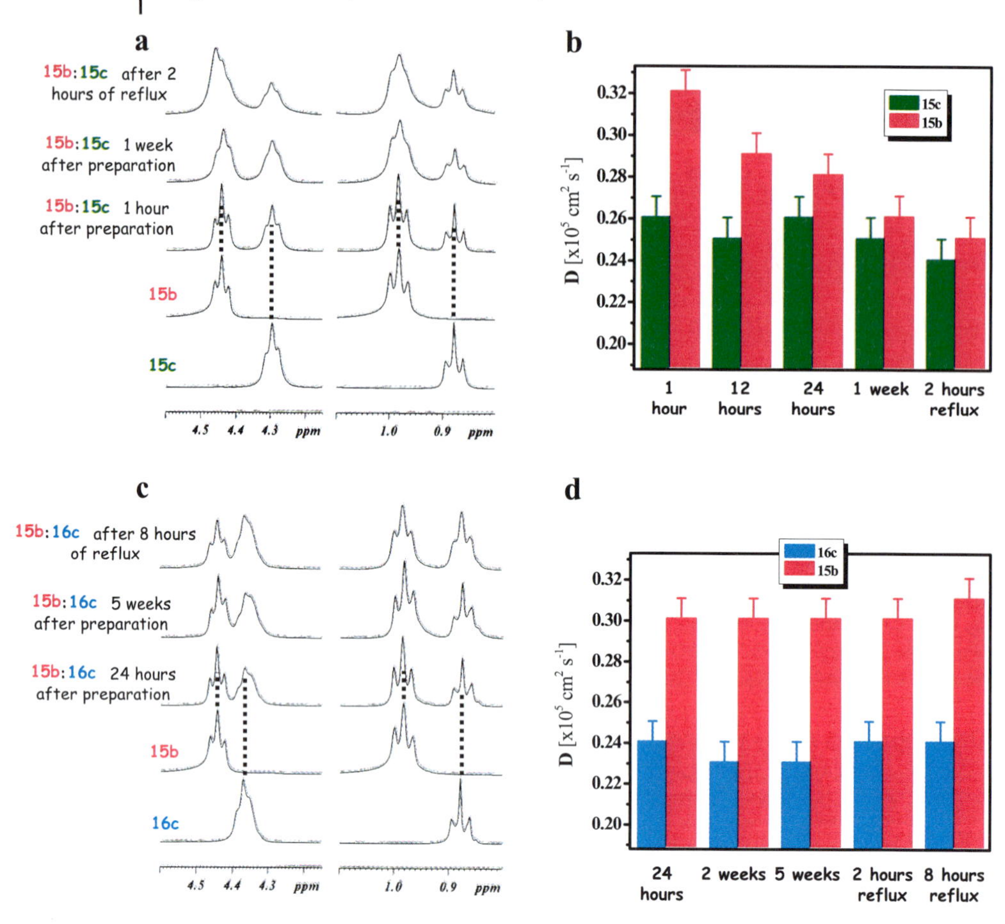

Fig. 6.14. Sections of ^{1}H NMR spectra for (a) **15c** and **15b** and their mixture and (b) the diffusion coefficient of **15b** and **15c** in $CDCl_3$ as a function of time that elapsed since the preparation of the mixture; (c) and (d) show the same type of data for the mixture of **15c** and **16b** [33b].

coefficients were observed in each mixture, immediately after their preparation, and they were found to equilibrate with time. These results were not observed for the mixtures of **15b**, **16c** or **15c**, **16b**. There the diffusion coefficient remained different for weeks. These diffusion results, summarized in Fig. 6.14, clearly demonstrate that *within* the macrocycle type the self-assembly process proceeds without self-recognition but *across* macrocycle types, the self-assembly proceeds with complete self-recognition. Such conclusions could not have been obtained from analyzing the chemical shifts but were completely conclusive when diffusion coefficients were used as a parameter to follow the process [33b]. Although the difference between **15b** and **15c** or **16b** and **16c** is only about 30% the diffusion results were nonetheless conclusive. It should be noted, however, that, in this case, diffusion does not have the resolution required to distinguish between the different hetero-hexamers formed. All these diffusion experiments demonstrate the power of diffusion NMR in characterizing multicomponent noncovalent molecular capsules.

Indeed, very recently, several examples for the use of diffusion NMR for studying other capsules were reported by different groups [38].

6.4.3 Molecular Size, Shape and Self-aggregation

The Stokes' radius of a molecule is related to its partial specific volume ($\bar{v}$) and its molecular weight (M) by Eq. (6.13):

$$r_s = \sqrt[3]{\frac{3M\bar{v}}{4\pi N}} \tag{6.13}$$

Substitution of this equation into Eq. (6.2) and then into Eq. (6.1) when k_b is replaced with R/N where R is the gas constant and N is Avogadro's number affords Eq. (6.14):

$$D = \frac{RT}{\eta 2^{1/3} 3^{4/3} \pi^{4/3} N^{2/3} v^{-1/3} M^{1/3}} \tag{6.14}$$

This equation shows the reciprocal cubic root dependency of the diffusion coefficient on the molecular weight. This implies that, under identical conditions for two molecules diffusing with diffusion coefficients of D_1 and D_2 and with molecular weights of M_1 and M_2 the following relation holds:

$$\frac{D_1}{D_2} = \sqrt[3]{\frac{M_2}{M_1}} \tag{6.15}$$

This equation, which is correct for spherical molecules, implies that one should expect a decrease of about 26% in the diffusion coefficient when the molecular weight is doubled. For linear molecules, however, larger decrease in the diffusion coefficient of about 40% should be observed when the molecular weight is doubled. Therefore, diffusion can be used to determine the size and shape of molecules by just weighing them. This was demonstrated on several single (SR), double (DR) and tetra-rosettes (TR) originally prepared by the Reinhoudt group [39], which were characterized by diffusion NMR [40]. Diffusion could easily distinguish between the SR ($\mathbf{19}_3 \cdot \mathbf{20}_3$), DR ($\mathbf{21}_3 \cdot \mathbf{20}_6$) and TR ($\mathbf{22}_3 \cdot \mathbf{20}_{12}$) as shown in Fig. 6.15. For example, Fig. 6.15b clearly shows that the signal decay, as a function of the diffusion weighting (b), for the three types of rosettes is very different. In addition, relatively good agreement was found between the Stokes' diameters, extracted from the NMR diffusion measurements and those obtained from gas-

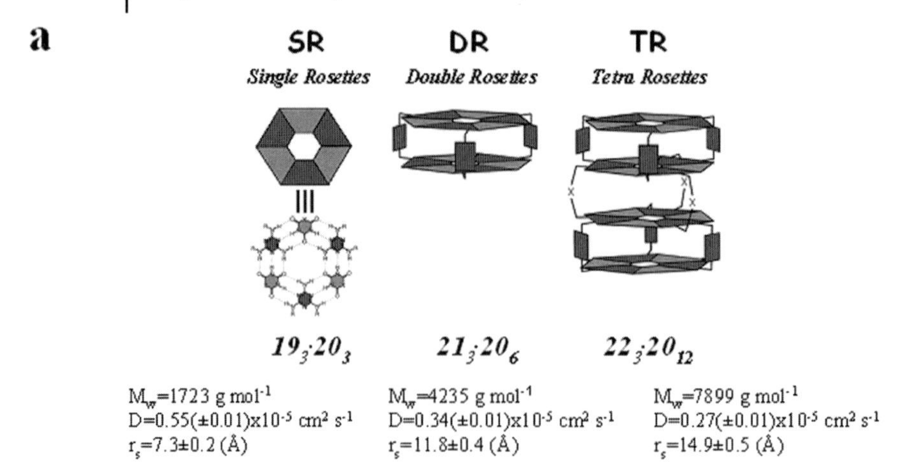

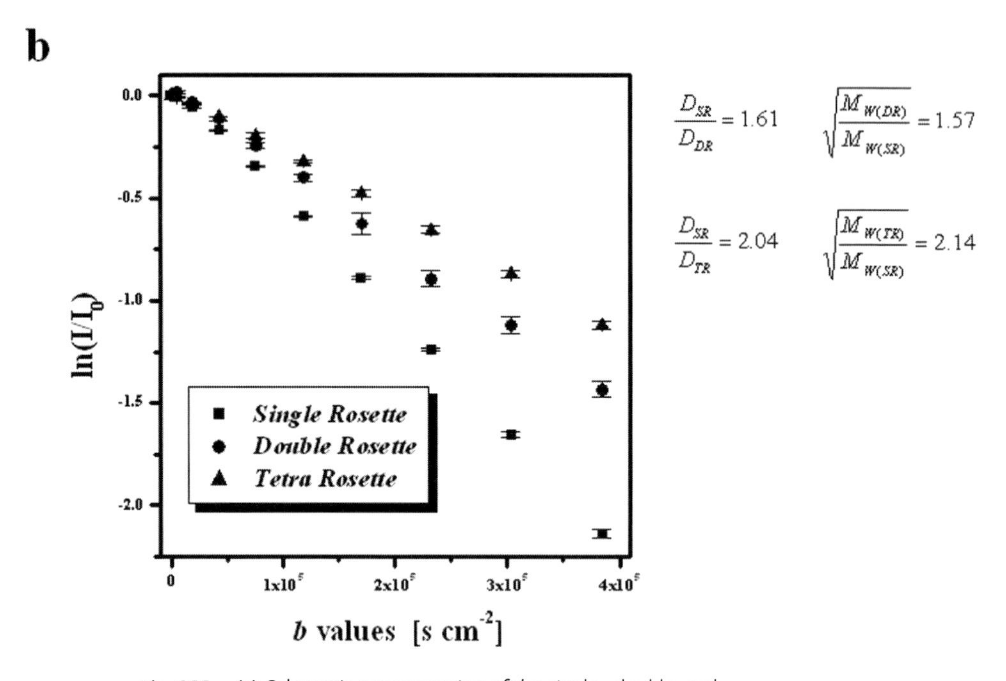

Fig. 6.15. (a) Schematic representation of the single-, double- and tetra-rosettes (ST, DR and TR, respectively) along with their molecular weights, diffusion coefficients and Stokes radii (r_s) and (b) the signal decay ($\ln I/I_0$) for three representative rosettes along with the ratios of their diffusion coefficients (D) and molecular weights (M_w) [40]. (Adapted with permission from ref. [40]. Copyright 2000 Royal Society of Chemistry.)

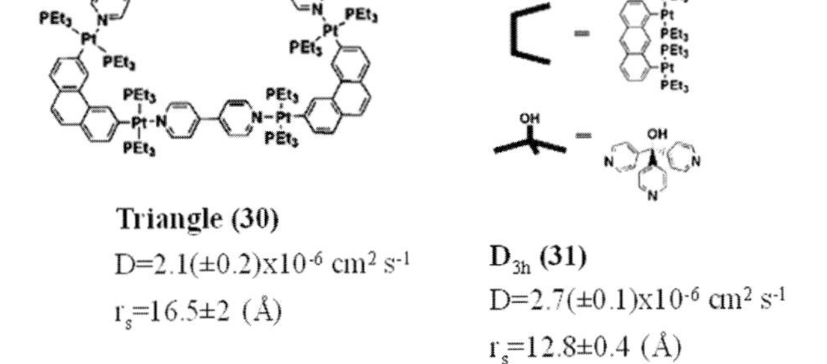

Fig. 6.16. Schematic representation of a selected coordinative supramolecular systems structure prepared by the Stang group, along with their diffusion coefficients, molecular weights and calculated Stokes radii (r_S) [41, 42].

phase-minimized structures [40]. More importantly, by measuring the diffusion coefficient of the SR and DR by addition of their components it was possible to determine which of the rosettes are kinetically labile. It was found that the SR and only some of the DRs, are kinetically labile on the NMR timescale under the given conditions (2 mM sample in $CDCl_3$, 298 K, 500 MHz). This information was difficult to obtain from conventional NMR of these systems.

The Stang group used the PGSE to corroborate the formation of their spectacular dodecahedra **26** and **27** shown in Fig. 6.16a, constructed from 50 components [41]. The diffusion coefficients of the two dodecahedra, **26** and **27**, the molecular weights of which were calculated to be 41656 and 61955 Daltons, were found to be $0.18 \pm 0.05 \times 10^{-5}$ cm^2 s^{-1} and $0.13 \pm 0.06 \times 10^{-5}$ cm^2 s^{-1}, respectively in the nonviscous acetone–dichloromethane solutions. Simulation of motion with such diffusion coefficients through that medium enabled them to extract diameters of 5.2 nm and 7.5 nm for **26** and **27**, respectively, values which are in good agreement with the estimated diameters of 5.5 nm and 7.5 nm, respectively, as obtained from calculations [41]. Figure 6.16b depicts diffusion data of several complexes, recently prepared by the Stang group [42]. They found good agreement between the r_s calculated from the diffusion data and the molecular dimensions obtained from X-ray crystallography. The authors used this comparison to corroborate the fact that systems **28–31** are indeed intact supramolecular assemblies in solution [42].

Larive and coworkers recently used NMR diffusion measurements to characterize a series of ligands (**32–36**) and rhenium complexes that were used as building units in the construction of molecular squares that could not be characterized by mass spectrometry (Fig. 6.17, [43]). There, good correlation was found between the diffusion coefficient and the reciprocal of the estimated Stokes' radii ($1/r_s$), as shown in Fig. 6.17c. The authors also concluded, based on the NMR diffusion measurements, that the complexity of some of the spectra are intrinsic characteristics of the supramolecular systems prepared, rather than contamination from low molecular species. The data extracted from these NMR diffusion experiments is presented in Table 6.4.

Recently, the Kaifer group, whose work combines supramolecular electrochemistry with solid spectroscopic characterization of their systems, studied the interaction of cucurbit[7]uril (**CB7**, **37**) with ferrocene-containing dendrimers up to three generations (**FG1** (**38**), **FG2** (**39**), and **FG3** (**40**); see Fig. 6.18). They found, *inter alia*, that, at pH 7, the diffusion coefficients of the functionalized dendrimers are smaller than the corresponding diffusion coefficients measured at pH 2 which, according to the authors, should be attributed to the fact that, under acidic condi-

→

Fig. 6.17. (a) The structure of compounds **32–36**, (b) their rhenium corner and square complexes and (c) the correlation between their diffusion coefficients and the reciprocal of the estimated Stokes radii ($1/r_s$) for ligands (●), corners (▲), and squares (■) [43]. (Adapted with permission from ref. [43]. Copyright 2002 American Chemical Society.)

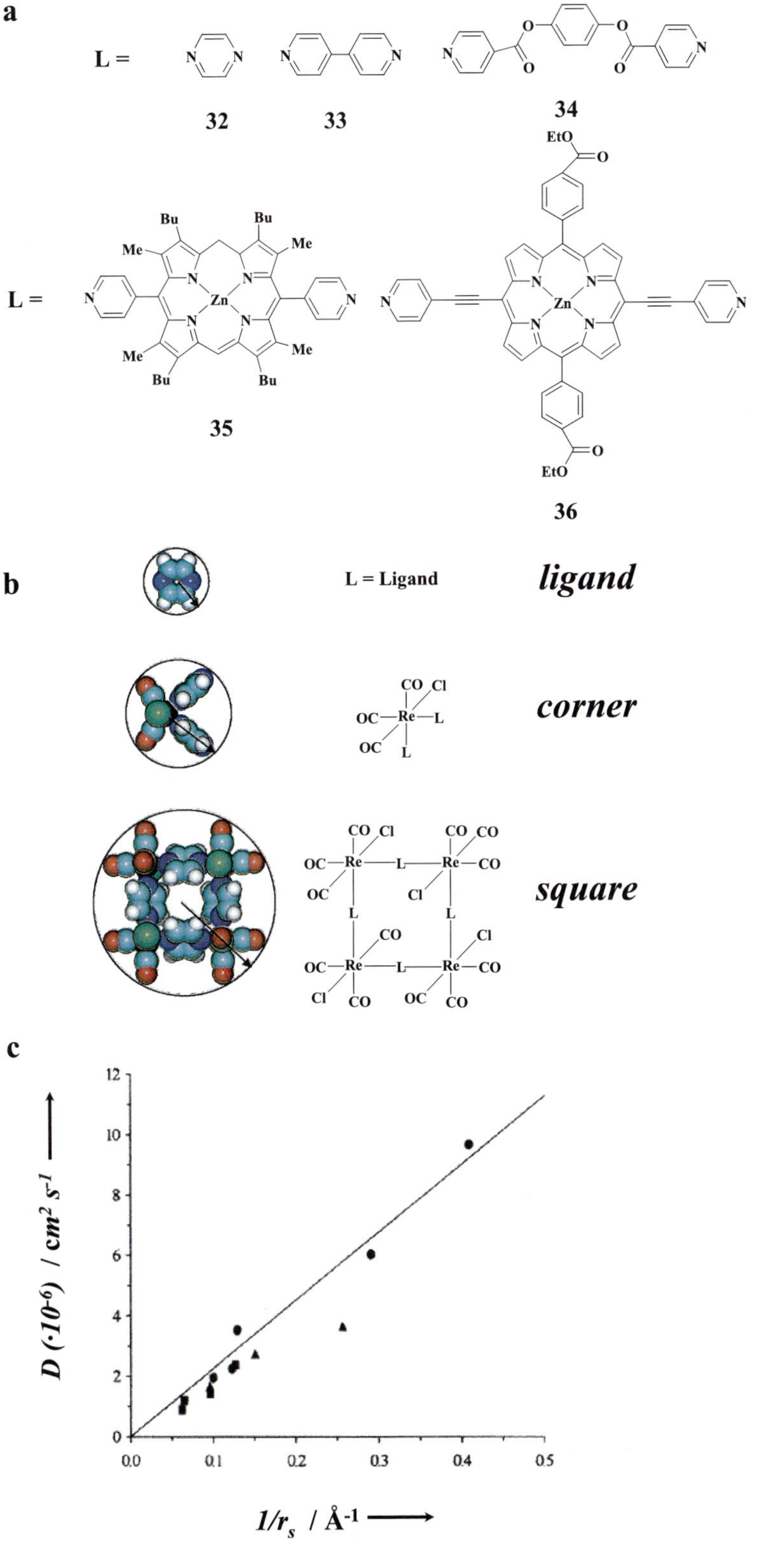

a

L =

32 33 34

L =

35 36

b

L = Ligand

ligand

corner

square

c

$D\ (\cdot 10^{-6})\ /\ cm^2\ s^{-1}$

$1/r_s\ /\ Å^{-1}$

Tab. 6.4. Diffusion coefficients (D) of the ligands **32–36** and their rhenium complexes. [43]

Molecule	D_{32} [$\times 10^{-6}$ cm^2 s^{-1}]	D_{33} [$\times 10^{-6}$ cm^2 s^{-1}]	D_{34} [$\times 10^{-6}$ cm^2 s^{-1}]	D_{35} [$\times 10^{-6}$ cm^2 s^{-1}]	D_{36} [$\times 10^{-6}$ cm^2 s^{-1}]
ligand	9.68 ± 0.09	6.03 ± 0.04	3.53 ± 0.04	2.24 ± 0.02	1.94 ± 0.06
corner	3.62 ± 0.08	2.69 ± 0.02		1.61 ± 0.03	
square	2.37 ± 0.11	1.42 ± 0.05	0.87 ± 0.02	1.20 ± 0.04	1.04 ± 0.03

tions, the carboxylic end-groups are protonated and therefore the dendrimers have more collapsed structures [44a]. At pH 7, the carboxylic end-groups are charged and therefore repulsion forces keep the dendrimer branches in the extended form. This is an example of the power of diffusion NMR to report on the shape of molecules. It should be noted that such an observation was previously reported for other Newkome-type dendrimers [44b].

Another interesting application on some lipophilic guanosine systems was recently described by the Davis group [45]. Guanosine nucleosides and their lipophilic models attracted much interest because they have tremendous biological importance and an ability to form peculiar supramolecular systems by self-aggregation, in time, through the mediation of cations and/or anions [46]. For example, **41a** was found to form the hexadecamer **42** following the addition of potassium picrate, while lipophilic isoguanosine **41b** was claimed to form a mixture of an octamer and a hexadecamer (**43**, **44**, respectively) after the addition of potassium dinitrophenolate (DNP) as shown in Fig. 6.19. This figure shows the structure of the monomers, the aggregates formed, and the signal decays as a function of the diffusion weighting along with the extracted diffusion coefficients [45]. When studying the diffusion characteristics of the systems formed by **41a**, the authors used compound **45** as an internal reference. Compound **45** was selected as a reference since it is unable to self-aggregate and interact with **41a**. This mixture (**45**, **41a** and **42**) was then used as a test case for diffusion NMR. Based on crystallographic data, they estimated the molecular volume of **42** and **41a** to be 12140 Å^3 and 586 Å^3, respectively. Using the hard sphere model, they predicted the D_{42}/D_{41a} to be 0.36. The experimental ratios were indeed found to be 0.35 and 0.39 for D_{42}/D_{45} and D_{42}/D_{41a}, respectively. Since there was good agreement between the predicted value and the experimental value for D_{42}/D_{45}, the deviation of the D_{42}/D_{41a} value was thought to originate from self-association of **41a** in the CD_3CN solution [45]. Indeed D_{41a}/D_{45} was found to be 0.88 in CD_3CN, but only 0.96 in [D_6]DMSO where self-aggregation of **41a** is marginal. This is a good example of the benefit obtained by using an appropriate internal standard in NMR diffusion measurements [45]. However, it should be noted that when it is difficult to find the appropriate internal standard and self-aggregation is of concern, one can

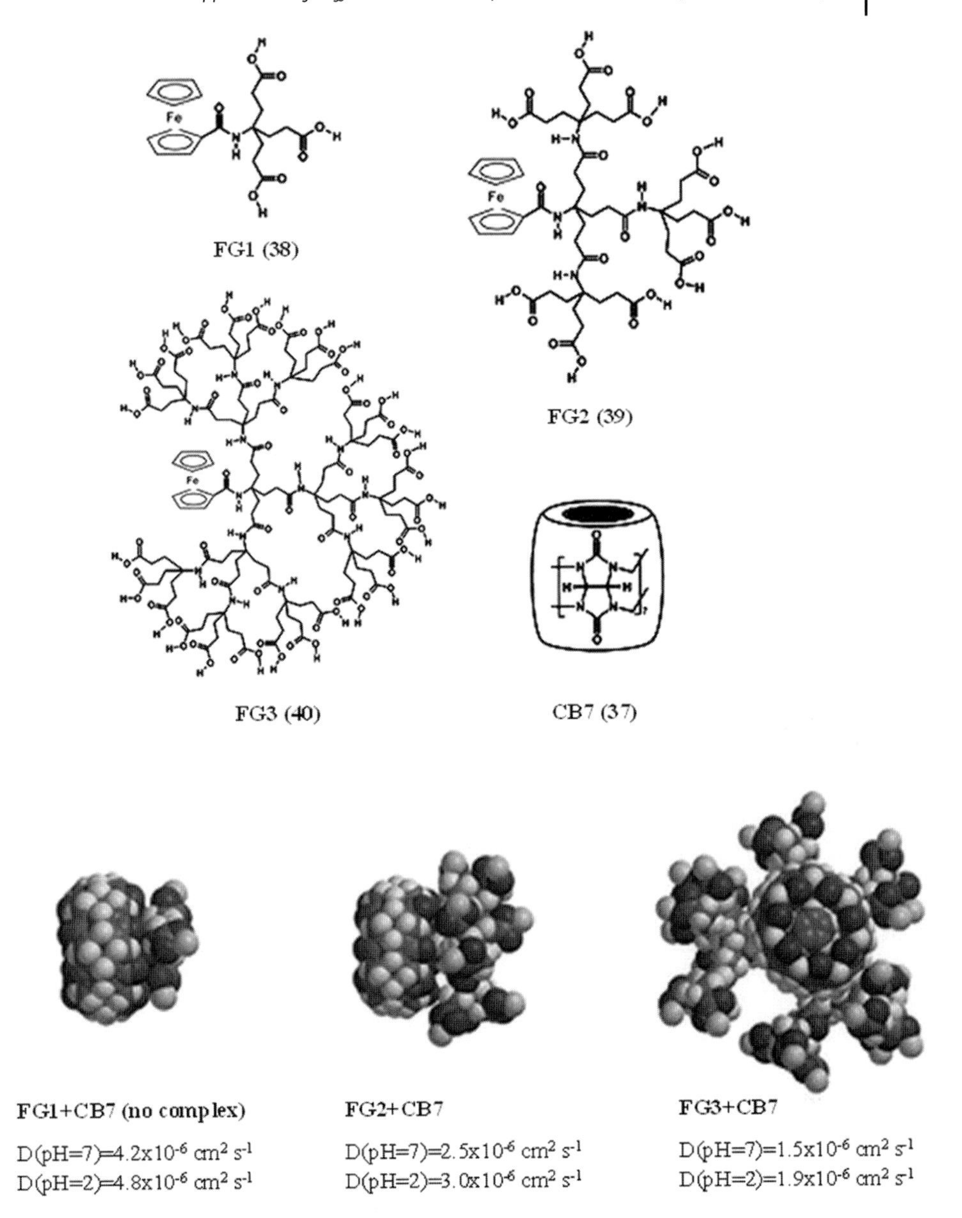

Fig. 6.18. The structure of CB[7] and the first, second and third generation ferrocene-dendrimers (**FG1**, **FG2** and **FG3**, respectively) and a schematic representation of their complexes along with their diffusion coefficients at different pH [44a]. (Adapted with permission from ref. [44a]. Copyright 2005 Royal Society of Chemistry.)

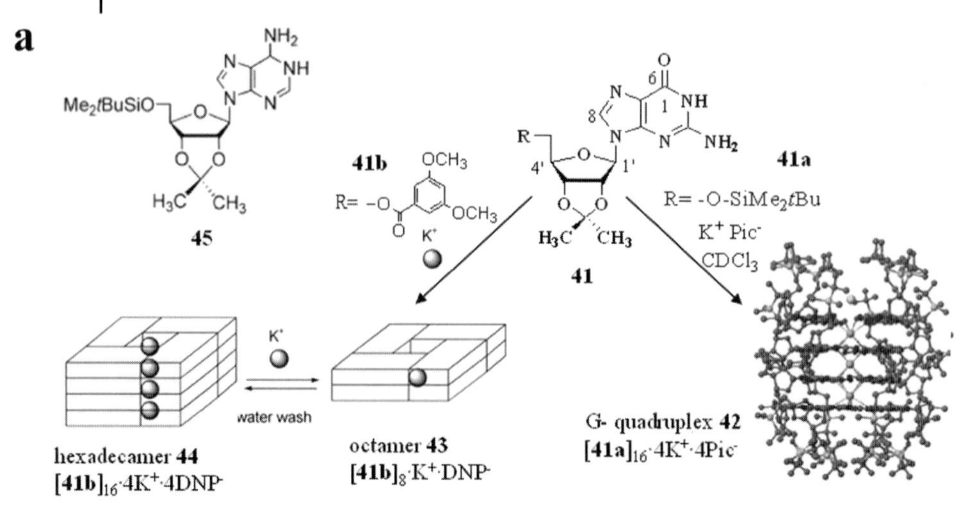

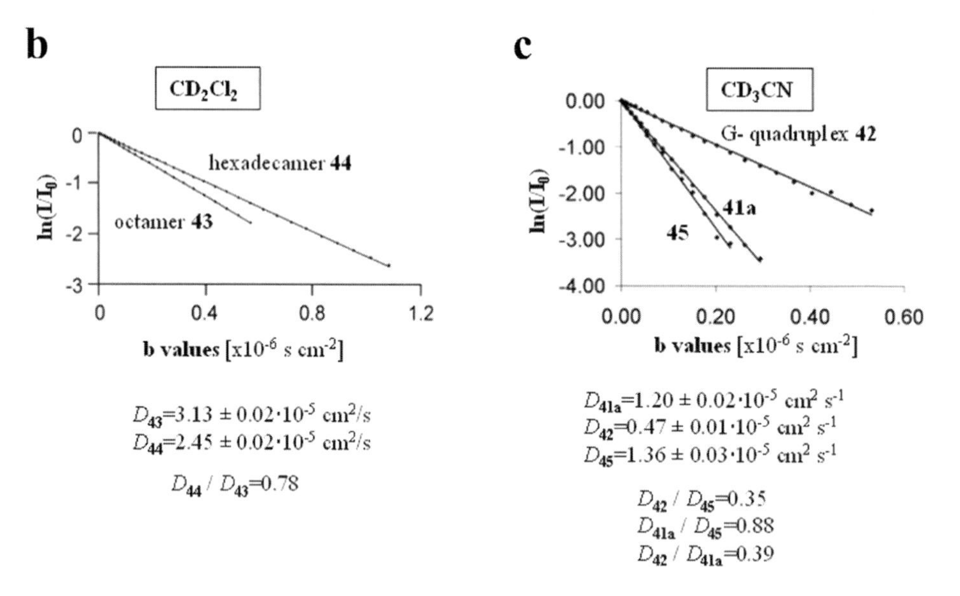

Fig. 6.19. (a) The structures of lipophilic isoguanosine **41a** and **41b** and the respective supramolecular systems which they form under different conditions, the structure of the reference compound **45**, and the signal decay as a function of the diffusing weighting factor *b* of (b) octamer (**43**) and hexadecamer (**44**) in CD_2Cl_2 and (c) **41a**, **45** and **42** in CD_3CN along with the diffusion coefficients and molecular weights and their ratios [45].

measure the diffusion coefficient of the sample following dilution. An increase in the diffusion coefficient when the sample is diluted can serve as an indication for self-aggregation provided that the viscosity of the sample is not changed significantly during the dilution process. In this study, only the free **41a** and the corresponding hexadecamer **42** were found, indicating that this self-assembly process occurs via a cooperative mechanism.

When **41b** was used to extract KDNP from water to CD_2Cl_2 a complex was obtained that was attributed to the octamer **43**. However, when solid–liquid extraction was performed, a different spectrum was obtained. Here again, diffusion NMR came to the rescue. When the diffusion coefficients of the two species were measured, a ratio of 0.78 was found experimentally which agrees very well with the expected theoretical D_{16mer}/D_{8mer} ratio of 0.79. Consequently, compound **44** could be assigned as the hexadecamer of **41b** while **43** was assigned as the octamer [45].

DOSY spectra were recently used to assist in the identification of some spectacular supramolecular structures prepared by the Kim and Fujita groups (Figs. 6.20 and 6.21). The pentameric necklace **48**, for which the structure could not be confirmed by X-ray crystallography when prepared, was found to have a hydrodynamic volume which is 8.7 times larger than that of CB[8] (**47**) alone, thus corroborating

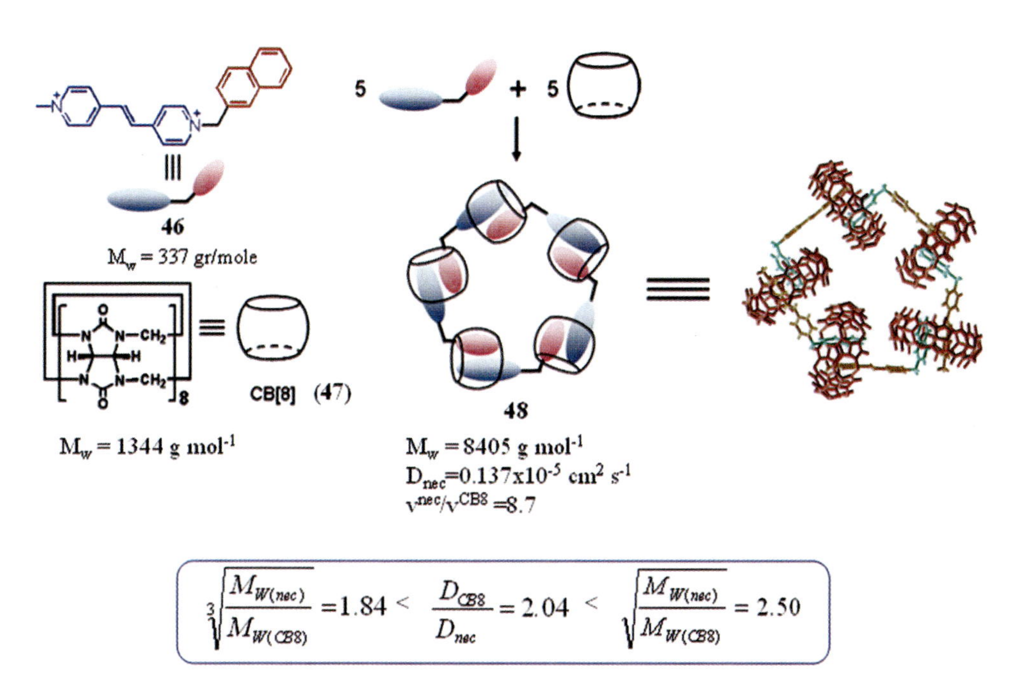

Fig. 6.20. Schematic representation of the formation of the supramolecular necklace **48** and the molecular weights, diffusion coefficients and volumes of **48** and its components [47]. (Adapted with permission from ref. [47]. Copyright 2004 American Chemical Society.)

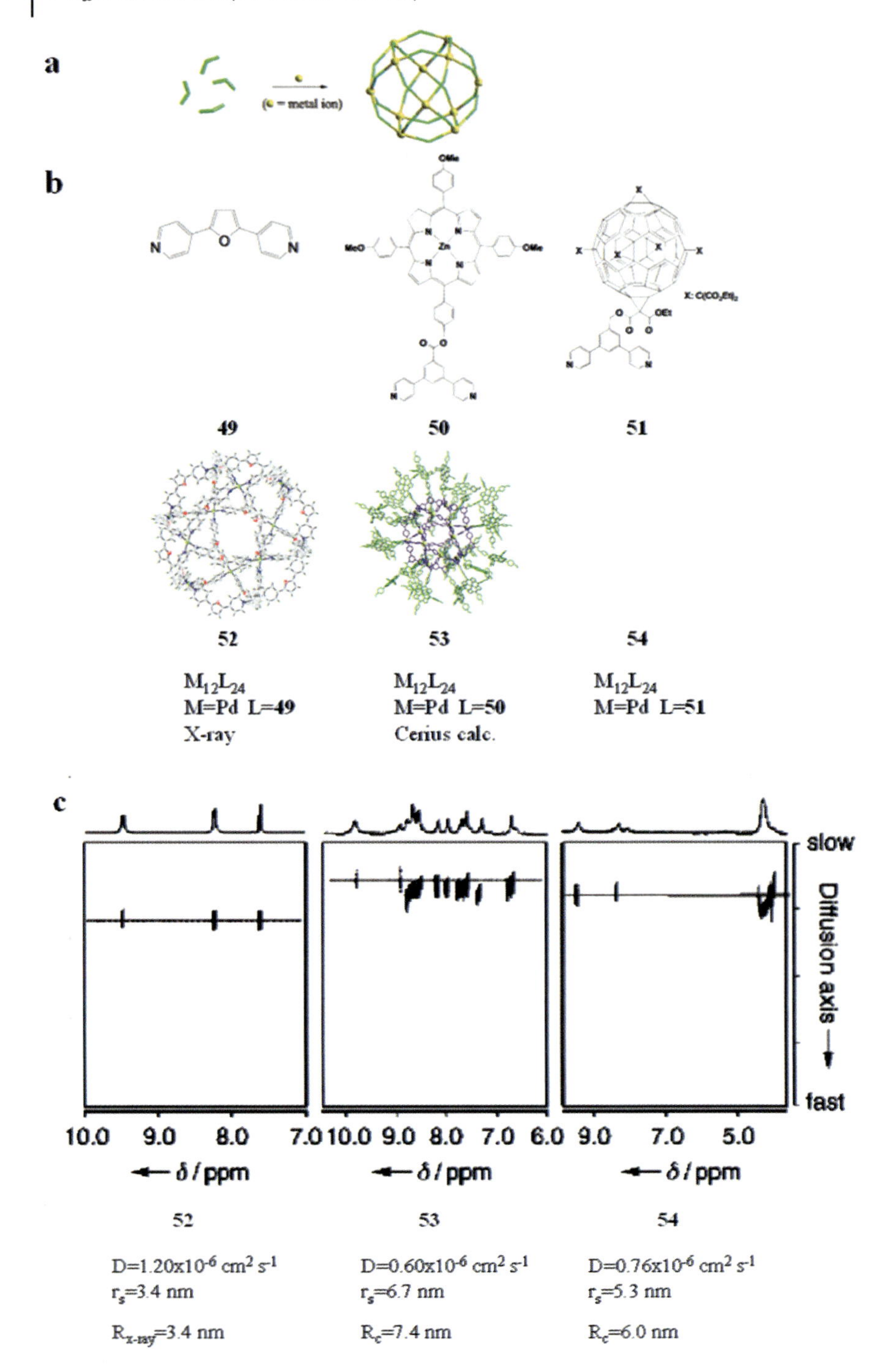

a
(● = metal ion)
b
49
50
51
52
53
54
M12L24
M=Pd L=49
X-ray
M12L24
M=Pd L=50
Cerius calc.
M12L24
M=Pd L=51
c
slow
Diffusion axis
fast
δ / ppm
D=1.20x10-6 cm2 s-1
rs=3.4 nm
Rx-ray=3.4 nm
D=0.60x10-6 cm2 s-1
rs=6.7 nm
Rc=7.4 nm
D=0.76x10-6 cm2 s-1
rs=5.3 nm
Rc=6.0 nm

the formation of the large necklace. The DOSY spectrum clearly showed that all peaks of the two components had the same low diffusion coefficient [47].

Fujita's group recently described the preparation of spherical assemblies from 36 components having a general formula of $M_{12}L_{24}$ where M was Pd^{+2} (Fig. 6.21). When compounds **49–51** were reacted with $Pd(NO_3)_2$ in $[D_6]$DMSO, the 36 component systems **52**, **53** and **54**, respectively were obtained as a single product [48]. Compound **52** was characterized by X-ray crystallography while the structures of **53** and **54** were estimated using Cerius calculations [48]. DOSY experiments performed on these solutions corroborated the formation of a large single supramolecular system in each case (Fig. 6.21c). Indeed, the hydrodynamic radii (r_s) extracted from the measured diffusion coefficients were 3.4, 6.7 and 5.3 nm for **52**, **53** and **54**, respectively. The first value is in excellent agreement with the value obtained from the X-ray structure of **52** (3.4 nm) and the others are within acceptable agreement with the values obtained from the Cerius program for **53** and **54** which were found to be 7.4 and 6.0 nm, respectively [48].

6.4.4
Diffusion as a Filter: Virtual Separation and Ligand Screening

Diffusion is, in fact, a filter, which filters compounds according to their effective molecular weight and shape, and, as such, can be combined with any NMR sequence. The fact that diffusion is a filter enables us to use diffusion for solvent suppression, virtual separation and screening for potential ligands for a given receptor [49]. Since solvents are generally small molecules relative to solutes, one can use diffusion weighting as a preferable means of eliminating the signal of the solvent. This filtering-off of solvent peaks is more efficient if the solutes are large molecules with M_W of around 10 000 Daltons or more.

As shown in Fig. 6.5, diffusion in general and DOSY spectra in particular can be used for "virtual separation" of compounds. DOSY provides 2-D maps in which one axis is that of the chemical shift and the other is that of the diffusion coefficient. Although the 1-D ^{1}H NMR spectrum shown in Fig. 6.22a is a superposition of the spectra of all the compounds in the mixture, each horizontal line in the diffusion coefficient axis in Fig. 6.22b represents only one compound. Here compounds of very different molecular weight were used and peak separation on the diffusion axis was complete leaving only one compound in each horizontal line in the 2-D DOSY spectrum. However, Figs. 6.22c and 6.22d show the same kind of

Fig. 6.21. (a) Schematic representation of the formation of the 36 component supramolecular spheres, (b) the ligands (**49**, **50** and **51**) used to construct these spheres, along with the X-ray structure of **52** and the calculated structure of **53**, and (c) the DOSY spectra of **52**, and **53** and **54** along with their diffusion coefficients, the Stokes radii (r_s) and the X-ray and calculated radii of these compounds [48]. (Adapted with permission from ref. [48]. Copyright 2004 Wiley-VCH.)

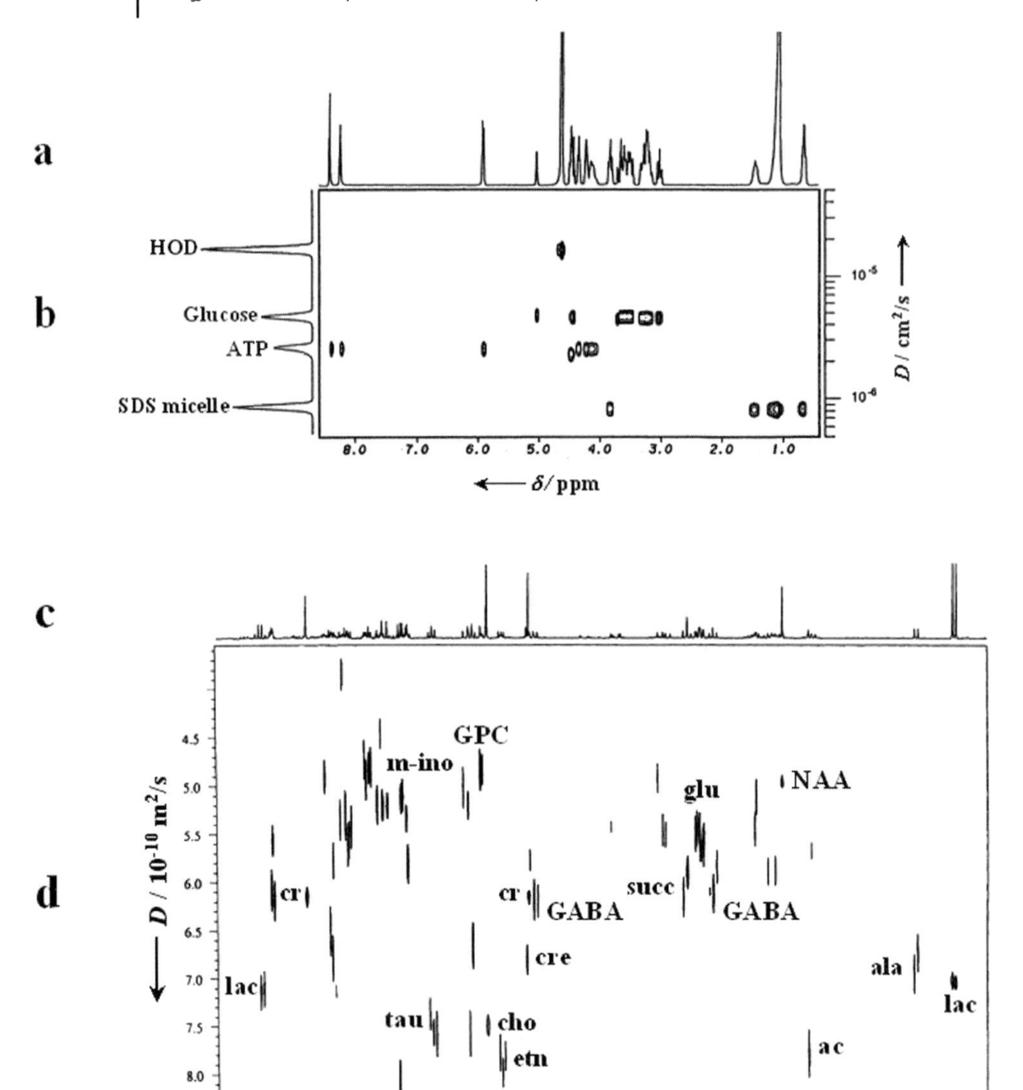

Fig. 6.22. The (a) 1-D and (b) 2-D-DOSY spectrum of a mixture containing D_2O, glucose, ATP and SDS micelles [50]. (Adapted with permission from ref. [50]. Copyright 1993 American Chemical Society). (c) 1-D and (d) 2D-DOSY spectrum of the perchloric acid extract of a gerbil brain in D_2O. Selected assignments are: ac = acetate; ala = alanine; cho = choline; cr = creatinine; etn = ethanolamine; GABA = γ-aminobutyric acid; glu = glutamine; GPC = glycerophospho-choline; lac = lactate; m-ino = myo-inositol; NAA = N-acetylaspartate; succ = succinate and tau = taurine [51]. (Adapted with permission from ref. [51]. Copyright 1997 Elsevier.)

data for a mixture containing many metabolites, some of which differ only marginally in their molecular weight. Here again, the 1-D spectrum in Fig. 6.22c is that of the entire mixture and each horizontal line in Fig. 6.22d represents only part of the metabolites in the mixture. Figures 6.22b and 6.22d indeed represent "virtual separations" of the compounds. Here, spectra of each component (or part of them) are obtained separately without the need to actually physically separate them, by chromatography for example. With the current technology, one can now separate compounds which differ by less than a factor of 2 in there molecular weights. Although the DOSY presentation is very attractive if one uses single peak analysis of the signal decay, one can still differentiate between molecular species that differ by only 20–30% in their molecular weight without resorting the DOSY procedure. This can be done with the aid of high quality gradient system if the molecular weights of the investigated molecules are in the range of hundreds to a few thousand Daltons. It should be noted that overlapping signals may present a real problem when mixtures of large numbers of compounds or complex compounds are present in the mixture. One way to partially alleviate this problem is to couple the DOSY filter with a 2-D-NMR sequence where overlapping signals are less of a problem. This results in 3-D sequences, referred to as 3-D-DOSY sequences. Since diffusion is a filter that can easily be coupled to nearly any 2-D-NMR sequence, many 3-D-DOSY sequences have been developed with relative ease. However, there are only a limited number of applications of these techniques in real chemical systems. The principle of a 3-D-DOSY in virtual separation is schematically outlined in Fig. 6.5b. Here, the result of a 3-D-DOSY–COSY sequence is presented, where COSY maps of each compound in the mixture are separated on the diffusion axis and appear on a separate plan.

Like other parameters, diffusion NMR can be used to screen for potential ligands for a given receptor or host. The diffusion coefficient of a small molecule changes upon addition of a large molecule if they interact. Therefore, diffusion coefficients can be used to screen for the interactions between small molecules and specific receptors, allowing a distinction to be made between ligands that do not interact with a host and those which interact with the receptor or host molecule. One example for such an experiment, which is based on the single-point approach developed by Shapiro and coworkers, is that depicted in Fig. 6.23 [52]. These researchers first measured the 1-D ^{1}H spectrum of the mixture of the eight potential ligands (compounds **55–62**, see Fig. 6.23a) and hydroquinine 9-phenanthryl ether (**63**) (Fig. 6.23b). They then found the experimental conditions (gradient strength and duration) needed to eliminate the spectrum of the mixture of the eight tentative ligands in the absence of **63** (Fig. 6.23c). Subsequently, they repeated the same 1-D ^{1}H diffusion spectrum after addition of **63** to the mixture, and they indeed observed, in addition to the signals of **63**, further signals that originate from compounds **55** and **56** that interact with the receptor molecule **63** (Fig. 6.23d) [52]. Only the compounds which interact with the receptor, experience a decrease in their diffusion coefficients and, hence, their signals reappear under the previously selected PFG conditions.

The advantages of using this simple scheme of diffusion NMR as a screening method for potential ligands for specific receptors are the following: The measured

a

55 M_w = 172 g mol^{-1} **56** M_w = 193 g mol^{-1} **57** M_w = 162 g mol^{-1}

58 M_w = 116 g mol^{-1} **59** M_w = 102 g mol^{-1} **60** M_w = 102 g mol^{-1} **61** M_w = 158 g mol^{-1}

62 M_w = 172 g mol^{-1} **63** M_w = 504 g mol^{-1}

b

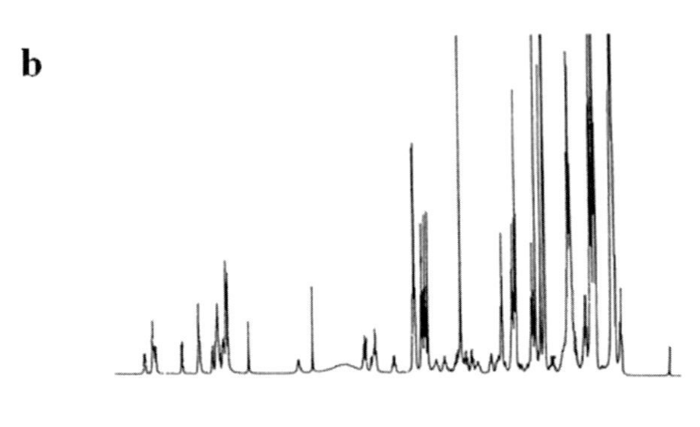

c

d

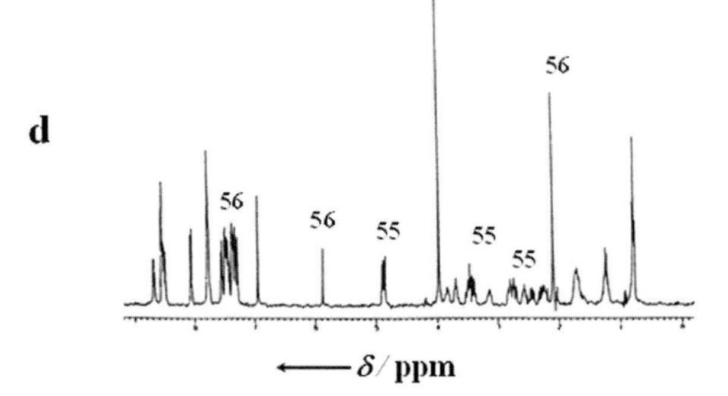

parameter, i.e. the diffusion coefficient, is intuitively related to the followed phenomenon, i.e., one has to expect a decrease in the diffusion coefficient for the compounds that interact with the receptor, as one indeed observes. This is in contrast with other NMR parameters such as chemical shifts and relaxation, where the magnitude or even the direction of the changes cannot be predicted *a priori*. Additionally, the method can be performed in a few minutes for a mixture of several ligands when their concentrations are in the mM range. A further important advantage of screening by diffusion NMR is the fact that the method is applicable to all receptors, even to systems of unknown ligands and receptors or hosts. This is in clear contrast with the main competing NMR approach for ligand screening [53] on which we will not elaborate here. Other approaches based on NMR diffusion measurements for ligand screening have also been reported [10, 54].

6.4.5
From Organometallics to Supercharged Supramolecular Systems

Diffusion NMR can be, and has been used to study ion–pairing phenomena [55] and in the last decade (mostly since the year 2000) diffusion NMR has become more popular among organometallic chemists [56]. Organometallic compounds and complexes that tend to self-aggregate are relatively labile and often involved in association–dissociation reactions when ligands enter or leave the complexes. Generally, the labile ligand is small relative to the entire complex and their association dissociation can be followed relatively easily by diffusion NMR. Indeed, in the last seven years, Pregosin and others have used diffusion NMR to study many organometallic complexes [56]. Although some organometallic complexes can only remotely be classified as supramolecular systems, many of them clearly are. Since the subject has recently been reviewed [11a, 57], we will focus here on charged polycycles that form supercharged supramolecular systems.

In 1995, the diffusion coefficients of a series of polycyclics and their respective dianions as well as corannulene (**64**) and its tetraanion ($\mathbf{64}^{4-}$) were measured in $[D_8]$THF solutions [58a]. Upon the formation of the charged species, a large decrease in the diffusion coefficients was observed and, partially, assigned to the increased solvation of the charged systems in the $[D_8]$THF solutions. However, some of the systems showed too large a decrease of their diffusion coefficients upon

Fig. 6.23. (a) The structure of DL-isocitric lactone (**55**), (S)-(+)-O-acetylmandelic acid (**56**), DL-*N*-acetylhomocysteine thiolactone (**57**), (±)-sec-butyl acetate (**58**), propyl acetate (**59**), isopropyl butyrate (**60**), ethyl butyrylacetate (**61**), butyl levulinate (**62**), hydroquinine 9-phenanthryl ether (**63**), (b) the 1-D ^{1}H NMR spectrum of the eight compounds **55**–**62**, (c) 1-D pulse gradient ^{1}H NMR spectrum of the mixture without **63** using the LED sequence, and (d) 1-D pulse gradient ^{1}H NMR spectrum of the nine-component mixture using the same experimental conditions as in (c). Peaks arising from compounds **55** and **56** are labeled. All other peaks are from compound **63** [52]. (Reproduced with permission from ref. [52]. Copyright 1997 American Chemical Society.)

reduction. One of the most dramatic decreases in the diffusion coefficient upon reduction was found for the **64**/$\mathbf{64}^{4-}$ pair and it was suggested that this decrease reflects the dimerization of the $\mathbf{64}^{4-}$, which was indeed proven when two different derivatives of corannulene were reduced and evidence for the formation of the heterodimer of $\mathbf{64}^{4-}$ was found [58b].

However, one of the most intriguing examples in this field, which also demonstrates the power of diffusion NMR to characterize supramolecular systems obtained by self-aggregation, is the assignment of the different species that prevail in solution following the reduction of 2,5,8,11-tetra-*tert*-butylcycloocta[1,2,3,4-def-,5,6,7,8]bis-biphenylene (**65**) to its respective tetraanions [59]. In this sample, different species were observed and only diffusion NMR provided a proof, in conjunction with 2-D NOESY, that the different molecular species are indeed different helically-stacked anionic aggregates of $\mathbf{65}^{4-}$. When NMR diffusion measurements were performed on the obtained solution, four different diffusion coefficients were found for the mixture. These coefficients that were assigned to the monomer, dimer, trimer and tetramer of $\mathbf{65}^{4-}$ (Fig. 6.24). Based on Eq. (6.16):

$$D = \frac{k_B T}{6\pi\eta\left(3v_0 \cdot \dfrac{n}{4\pi}\right)^{1/3} F} \tag{6.16}$$

where all the parameters have their above meanings and F is the Perrin factor for each aggregate. The authors plotted the diffusion coefficient against $1/(n^{1/3}F)$ and obtained a straight line ($R^2 = 0.98$) from the slope of which the effective monomer

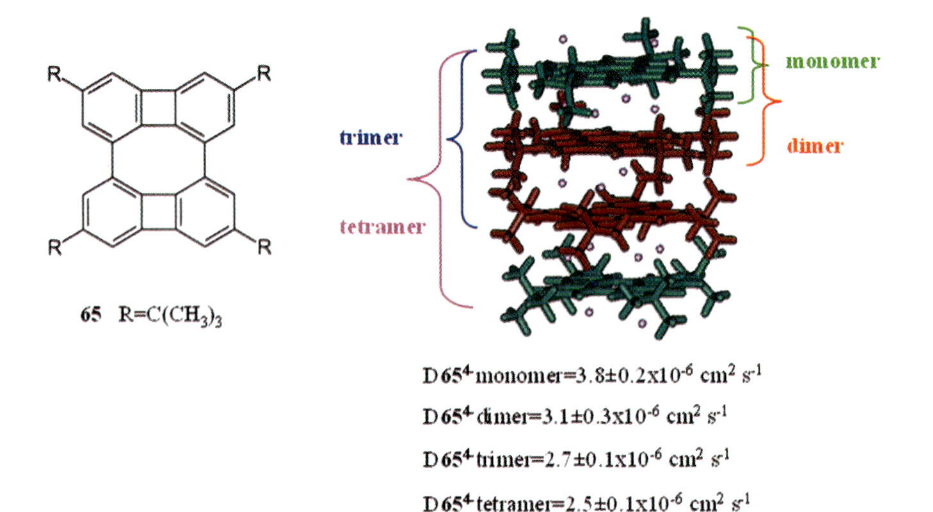

Fig. 6.24. The structure of compound **65** along with the computerized calculated structure of the monomer, dimer, trimer, and tetramer of $\mathbf{65}^{4-}$ along with the corresponding experimental diffusion coefficients of these systems in $[D_8]$THF at 258 K [59].

volume (v_0) was calculated to be 520 Å^3. This value was said to perfectly match the MNDO-based estimation of the monomer's effective dimension. With the aid of diffusion NMR, the authors provided evidence for the formation of a tetramer composed of four layers of **65**$^{4-}$ unit. This is a supercharged supramolecular system that contains sixteen negative charges which are probably formed, through the mediation of the cations in the system [59]. Very recently, some additional supercharged supramolecular systems were studied with diffusion NMR [60].

6.5 Advantages and Limitations of Diffusion NMR

Diffusion NMR, using the current technology, provides a fast and accurate means for measuring the diffusion coefficient of entire sets of peaks simultaneously. The strength of the conventional gradient systems in commercial NMR spectrometers can produce up to 50 G cm^{-1} which enables one to determine the diffusion coefficient of systems of molecular weight of up to 10 000 Daltons relatively easily. The main advantage of using diffusion coefficients as a probe to intermolecular interactions as compared with other NMR parameters is that it is the only NMR parameter which is intuitively connected to the process followed. In addition, it is the only parameter which directly reflects the size and the shape of molecular species.

It should be noted, however, that for spherical molecules for example, the change in the diffusion coefficient is only proportional to the cubic root of the effective molecular weight. Diffusion measurements are much easier to perform on systems that have long T_2 relaxation times. Since diffusion sequences are based on spin-echo or stimulated-echo, they can be performed only on systems where T_2 is not too short. This means that, for large molecules that generally have low diffusion coefficients and are characterized by short T_2 values, accurate determination of their diffusion coefficients is more difficult. In addition, for accurate determination of diffusion coefficients from NMR, well calibrated high quality gradient systems are required. Current technology and the available sequences already enable quite satisfying results on conventional supramolecular systems. Special care should also be given to temperature stabilization and the prevention of convections. In addition, it should be noted that in order to map intermolecular interactions with high sensitivity using diffusion NMR, the interacting species should have different molecular weights. The interactions between two different species that happen to have the same molecular weight are difficult to probe if the two systems do not interact strongly. In addition, one should bear in mind that diffusion coefficients reflect the "effective" size and molecular weight of the observed system and that different systems may happen to have the same diffusion coefficients. The diffusion coefficient may represent the size of the molecular systems and its solvation. For example solvation may be more important for multiply charged systems. Therefore, diffusion coefficients of systems where solvation is changed are more difficult to compare. For example, one should take care not to compare the diffusion coefficients of neutral and charged systems without paying attention to the possible effect of solvation. With the right approach, solvation can indeed be studied by diffusion NMR. One of the main limitations of diffusion NMR is that it does not provide local structural information on the investigated molecular systems. Therefore, when studying supramolecular systems, diffusion NMR should be combined with other NMR methods that provide local structural information. Since NOE and NOESY are NMR methods for probing intermolecular interactions which provide also structural information, it seems that the combination of these major NMR techniques with diffusion should be beneficial. Indeed, during the last year, two extensive review articles were published emphasizing the value of this combination of NMR methodologies when studying intermolecular interactions [11a, b]. There, the focus was organometallic systems.

Another factor that may further complicate the extraction of the "true" diffusion coefficient is the existence of exchange. Therefore, in the following section we will discuss briefly the effect of exchange on diffusion NMR.

6.6
Diffusion NMR and Chemical Exchange

In the same way that chemical exchange between two different chemical sites may be followed by several other NMR techniques, diffusion NMR can be used to probe the exchange between sites that differ in their hydrodynamic characteristics. Exchange rates are generally defined by their rates relative to the time scale of the NMR techniques used. These exchange processes can be defined as slow exchange when the individual sites can be distinguished or fast exchange where only a time-averaged site is observed. In diffusion NMR, fast and slow exchanges are defined relative to the diffusion time. In the two extreme cases (fast and slow exchange) the extraction of the real diffusion coefficient is relatively simple. In the case of fast exchange there is only one diffusing component and the data can be analyzed using Eqs. (6.10) and (6.11). In the case of slow exchange, the signal decay caused by diffusion can be analyzed using Eq. (6.17):

$$\frac{I}{I_0} = \sum_{i=1}^{n} A_i \exp\left[-\gamma^2 g^2 \delta^2 \left(\Delta - \frac{\delta}{3}\right) D_i\right] \tag{6.17}$$

In this equation, A_i is the molar fraction having a diffusion coefficient D_i. The more difficult cases to analyze are those of intermediate exchange. Under these conditions, a two site exchange model incorporating the mean lifetime of each site, the diffusion weighting, and the diffusion time as shown in equations 6.18a–f should be used to simulate the signal decay and extract the real diffusion coefficient values [61]. The mathematical model is the following: there are two sites, namely A and B. In each site there are molecules which have different diffusion characteristics. D_A and D_B are the diffusion coefficients of the two populations in the two sites, in the absence of exchange. P_A and P_B are the fractions of the two populations in each site in the absence of exchange and τ_A and τ_B represent the average lifetime of each population in each site. When $\tau \to \infty$ there is no exchange since each population is at its "own" site. Thus, this model uses D_A, D_B, P_A, P_B, τ_A, τ_B and K (Eq. 6.18a) in order to calculate D_1, D_2, P_1 and P_2 which are the diffusion coefficients and the two populations' fractions obtained as a result of the exchange, respectively. This calculation is performed according to Eqs. (6.18b–e).

$$K = \gamma^2 g^2 \delta^2 \tag{6.18a}$$

$$D_1 = \frac{1}{2}\left\{ D_A + D_B + \frac{1}{K}\left(\frac{1}{\tau_a} + \frac{1}{\tau_B}\right) - \sqrt{\left[D_B - D_A + \frac{1}{K}\left(\frac{1}{\tau_B} - \frac{1}{\tau_A}\right)\right]^2 + \frac{4}{K^2}\tau_A\tau_B} \right\} \tag{6.18b}$$

$$D_2 = \frac{1}{2}\left\{ D_A + D_B + \frac{1}{K}\left(\frac{1}{\tau_a} + \frac{1}{\tau_B}\right) + \sqrt{\left[D_B - D_A + \frac{1}{K}\left(\frac{1}{\tau_B} - \frac{1}{\tau_A}\right)\right]^2 + \frac{4}{K^2}\tau_A\tau_B} \right\} \tag{6.18c}$$

$$P_2 = \frac{1}{D_2 - D_1}(P_A D_A + P_B D_B - D_1) \tag{6.18d}$$

$$P_1 = 1 - P_2 \tag{6.18e}$$

After calculating D_1, D_2, P_1 and P_2 according to Eqs. (6.18a–e), the signal decay (R) is simulated using Eq. (6.18f):

$$R = P_1 \exp\left(-KD_1\left(\Delta - \frac{\delta}{3}\right)\right) + P_2 \exp\left(-KD_2\left(\Delta - \frac{\delta}{3}\right)\right) \tag{6.18f}$$

The effect of the diffusion time on the simulated signal decay for two exchanging sites with different diffusion coefficients is presented in Fig. 6.25 [61].

In the present discussion we wish to describe a different response of exchanging peaks when measured with the PGSE or the STE diffusion sequences as compared with the results obtained when diffusion is measured with the more recently published LED and BPLED sequences generally used in DOSY [62]. Fig. 6.26a–d shows the signal decay as a function of gradient strength (G) of the water peak (Fig. 6.26a, b) and one of the peaks of the macrocycle **15c** (Fig. 6.26c, d) in the $CDCl_3$ solution of $[(\mathbf{15c})_6(H_2O)_8]$ obtained using BPLED (Fig. 6.26a, c) or PGSE (Fig. 6.26b, d) sequences. This figure shows that the signal decay of the peak of **15c** in the two sequences is exactly the same. However, the signal decay of the water peak is very different when collected with the two diffusion sequences. The graphs shown in Fig. 6.26e and f clearly demonstrates that this is a general phenomenon. The peaks of **15c** show a mono-exponential signal decay which is exactly the same for all four sequences used. However, for the water peak, the signal decay in the PGSE and STE diffusion is mono-exponential and bi-exponential when collected with the LED and BPLED sequences, implying the existence of a fast and a slow

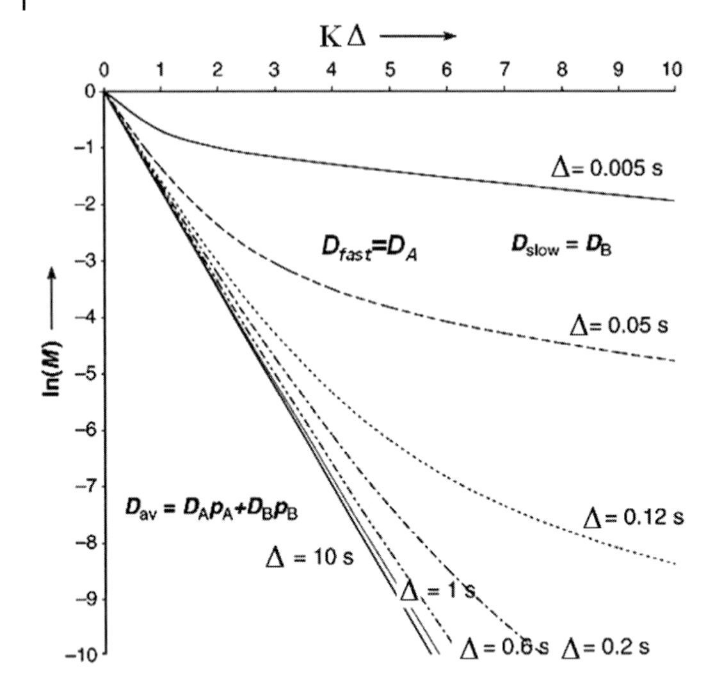

Fig. 6.25. Simulation of the effect of the diffusion time on the normalized signal decay for the two site exchange model as a function of the diffusion time (Δ) in a stimulated echo-based diffusion sequence. (Reproduced with permission from ref. [11c]. Copyright 2005 Wiley-VCH.)

diffusing component [62]. The question that arises is: "What is the difference between the two peaks that resulted in a different response to the four different diffusion sequences used?" In addition, it is strange and annoying that two different diffusion responses are obtained when different sequences are used to characterize the same system. In the chloroform solution of $[(\mathbf{15c})_6(H_2O)_8]$ there is only one water peak representing two different water pools which differ considerably in their diffusion coefficients. This means that, in fact, the two modern diffusion sequences i.e. LED and BPLED produce an artifact in the water peak which

Fig. 6.26. Signal decay of the water peak (a, b) and one representative peak of **15c** (c, d) as a function of the gradient strength (G) in the $CDCl_3$ solution of $[(\mathbf{15c})_6(H_2O)_8]$ collected with a BPLED (a, c) and spin echo diffusion (b, d) sequence along with the normalized signal decay ($\ln I/I_0$) as a function of the b-values taken with the sequences shown in Figs. 6.1 and 6.4 for (e) the water peak and (f) one of the peaks of **15c** [62]. (Reproduced with permission from ref. [62]. Copyright 2005 American Chemical Society.)

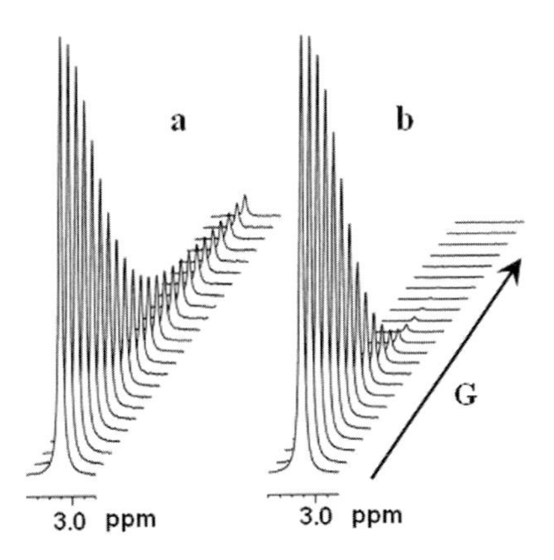
a
b
G
3.0 ppm
3.0 ppm

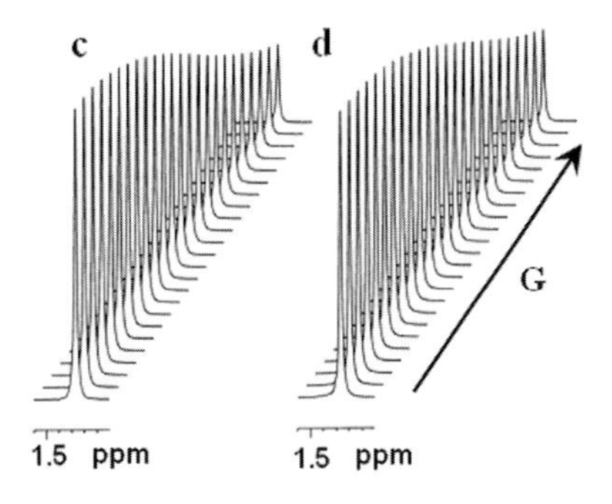
c
d
G
1.5 ppm
1.5 ppm

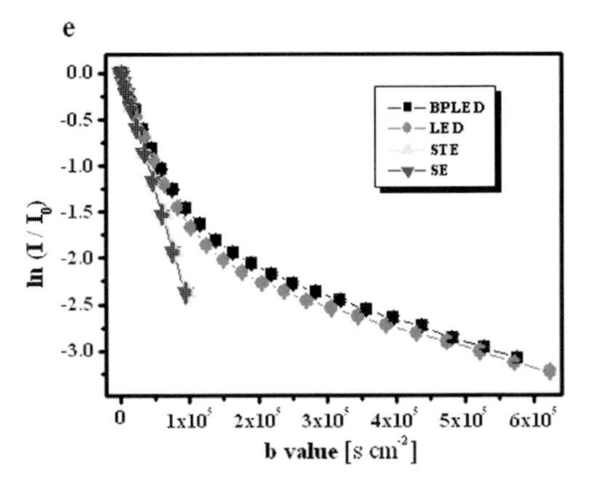
e
BPLED
LED
STE
SE
0.0
-0.5
-1.0
-1.5
-2.0
-2.5
-3.0
ln (I / I0)
0
1x10^5
2x10^5
3x10^5
4x10^5
5x10^5
6x10^5
b value [s cm^-2]

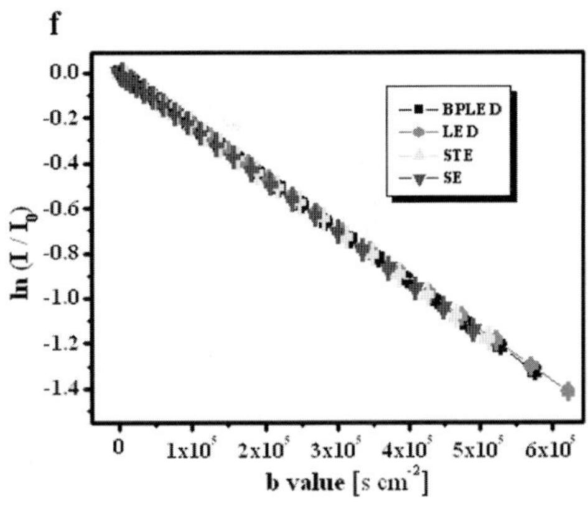
f
BPLED
LED
STE
SE
0.0
-0.2
-0.4
-0.6
-0.8
-1.0
-1.2
-1.4
ln (I / I0)
0
1x10^5
2x10^5
3x10^5
4x10^5
5x10^5
6x10^5
b value [s cm^-2]

exchanges between the two diffusion sites. When the different parameters were changed it was found that the spurious slow diffusing component of water observed in the LED and the BPLED originates from the t_e time period. This is demonstrated in Fig. 6.27 which shows the signal decay of the water peak collected with the BPLED diffusion sequence with different t_e values. It was found that the

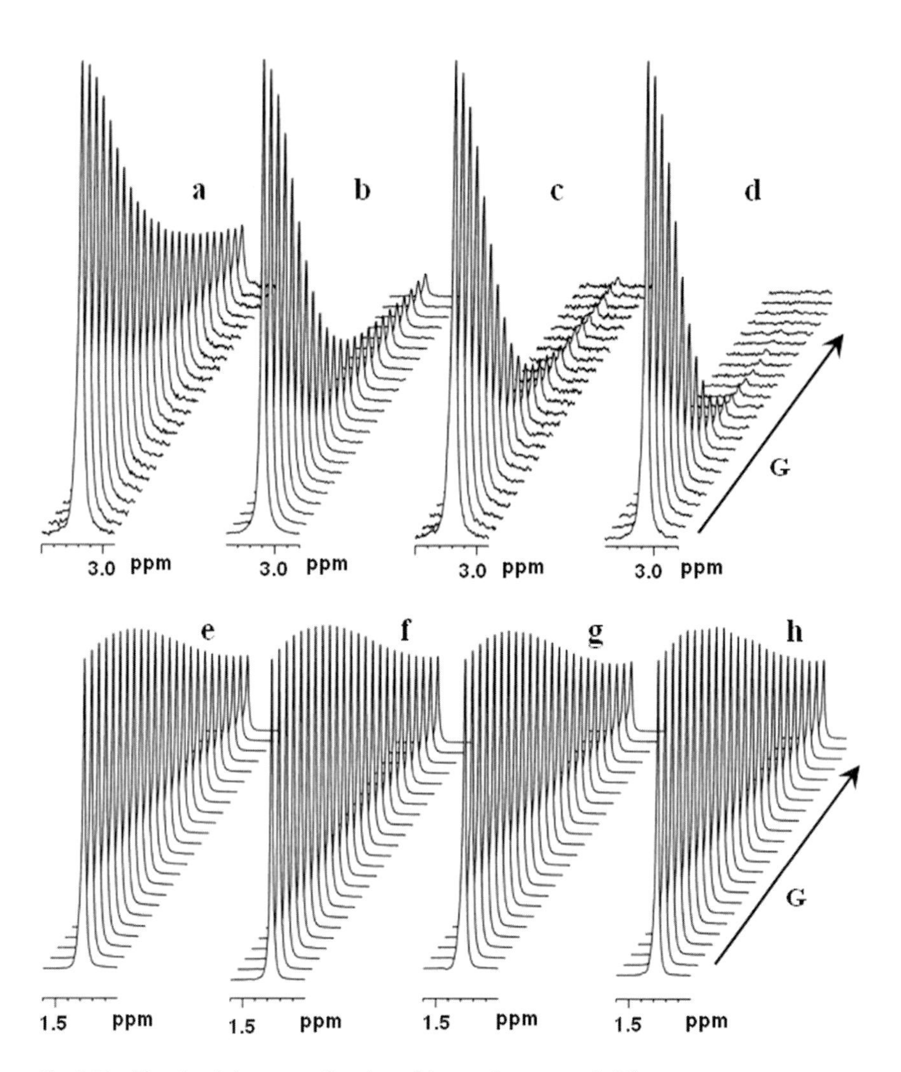

Fig. 6.27. The signal decay as a function of the gradient strength (G) of the water peak (a–d) and one representative peak of **15c** (e–h) in the $CDCl_3$ solution of $[(\mathbf{15c})_6(H_2O)_8]$ collected with a BPLED sequence when the eddy-current delay time (t_e) was (a, e) 150 (b, f) 50 (c, g) 14 and (d, h) 4 ms, respectively [62]. (Reproduced with permission from ref. [62]. Copyright 2005 American Chemical Society.)

longer the t_e the larger the additional spurious slow diffusion component is. We shall not elaborate on the origin of this observation here but it is clear that one should exercise care when using DOSY packages, which routinely use the LED and BPLED sequences, when studying systems with exchange. More details can be found in reference [62].

6.7 Summary and Outlook

We have demonstrated that high-resolution diffusion NMR is a very simple, flexible and accurate means for simultaneously obtaining diffusion coefficients in solution on ensembles of peaks, using standard NMR technology, thus allowing the addition of the diffusion coefficient into the set of NMR parameters used to characterize supramolecular systems in solution. As the diffusion coefficient reflects the "effective" size and shape of the different molecular components in solution, it is an excellent tool for probing intermolecular interactions and, as such, is a valuable tool for determining the structure of supramolecular systems. Diffusion coefficients can be used to determine association constants, probe aggregation, encapsulation, hydration and ion-pairing and provide a means for estimating the mutual interaction in many bodied systems. Diffusion NMR can be applied, in principle, to all systems which have NMR spectra. It is more important in multicomponent systems especially those which are formed by self-aggregation where the supramolecular system has the same symmetry as its monomer.

For such supramolecular systems, diffusion NMR may be extremely important for the determination of the structure and dynamics of the system. Diffusion is also extremely important in determining the association constant in systems, which on their formation induce a small change in the NMR spectra, and in systems in which chemical shift changes occur for reasons other than complexation, for example when protonation takes place. Diffusion NMR can also easily be used to probe the kinetic stability of multicomponent systems just by monitoring the effect of a small excess of one of the components on the diffusion coefficient of the supramolecular system.

In addition, diffusion NMR can be used for "virtual separation" of a mixture or library of compounds, of modest size, and can provide an efficient and general means, which imposes no requirement on the investigated system, for screening for lead compounds and potential ligands and even for unknown receptors. An important feature of diffusion NMR is that the measured parameter, i.e. the diffusion coefficient, is intuitively related to many of the phenomena studied in a much more direct and intuitive way than many of the more conventional and heavily used NMR parameters i.e. chemical shifts and relaxation times. In addition, diffusion is, in fact, a filter which can relatively easily be coupled to nearly any known NMR sequence. Even with current technology, diffusion measurement can be performed on a wealth of supramolecular systems. As it is anticipated that gradient technology will improve with time, and since we already have a large number of diffusion sequences and commercial packages, we speculate that diffusion NMR will be increasingly used by supramolecular chemists in the near future.

Until recently, most of the applications of diffusion NMR in the field of supramolecular chemistry concentrated on such systems in solutions, but it seems that this approach should be equally useful for studying the interaction of molecules with surfaces. We believe that the diffusion coefficient, as measured by diffusion NMR, should be added to the arsenal of NMR parameters used to routinely characterize supramolecular systems. The triumph of diffusion NMR will be when it becomes as transparent as other NMR methods.

References and Notes

1 a) J.-M. Lehn, *Supramolecular Chemistry Concepts and Perspectives*, VCH Weinheim, **1995**; b) J. W. Steed, J. L. Atwood, *Supramolecular Chemistry*. J. Wiley & Sons, New York, **2000**. c) *Comprehensive Supramolecular Chemistry, Vol 1-10* (Eds.: J. L. Atwood, J. E. D. Davis, D. D. MacNicol, F. Vögtle, J.-M. Lehn), Pergamon, Oxford, **1996**. d) M. Pons, O. Millet, *Prog. NMR Spectrosc.* **2001**, *38*, 267–324. e) L. Fielding, *Tetrahedron*, **2000**, *56*, 6151–6170.

2 E. L. Hahn, *Phys. Rev.* **1950**, *80*, 580–594.

3 a) O. E. Stejskal, J. E. Tanner, *J. Chem. Phys.* **1965**, *42*, 288–292. b) For early reviews on the application of diffusion NMR in chemical systems see: P. Stilbs, *Prog. NMR Spectrosc.* **1987**, *17*, 1–45. c) J. Kärger, H. Pfeifer, W. Heink, *Adv. Magn. Reson.* **1988**, *12*, 1–88.

4 a) T. Parella, *Magn. Reson. Chem.* **1996**, *34*, 329–347. b) T. Parella, *Magn. Reson. Chem.* **1998**, *36*, 467–495. c) D. D. Stark, W. G. Bardley, *Magnetic Resonance Imaging*, 2nd Ed. Mosby Company, **1992**; d) *Methods of Magnetic Resonance Imaging and Spectroscopy*, (Ed.: I. R. Young), John Wiley, Chichester, **2000**.

5 W. S. Price in *Annual Reports on NMR spectroscopy*, (Ed.: G. A. Webb), Academic, London, **1996**, pp. 51–142.

6 a) J. Kärger, D. M. Ruthven, *Diffusion in Zeolites and Other Microporous Solids*, Wiley, New York, **1992**; b) T. Nose, *Ann. Rep. NMR Spectrosc.* **1993**, *27*, 218–253. c) J. Kärger, H. Pfeifer, *NMR and Catalysis*, (Eds.: A. Pines, A. Bell), Dekker, New York, **1994**; d) P. T. Callaghan, A. Coy in *NMR Probes and Molecular Dynamics*, (Ed.: R. Tycko), Kluwer, Dordrecht, **1993**, pp. 490–523. e) O. Söderman, P. Stilbs, *Prog. NMR Spectrosc.* **1994**, *26*, 445–482. f) G. Lindblom, G. Oradd, *Prog. NMR Spectrosc.* **1994**, *26*, 483–515. g) A. R. Waldeck, P. W. Kuchel, A. J. Lennon, B. E. Chapman, *Prog. NMR Spectrosc.* **1997**, *30*, 39–68. h) G. Morris, (Ed.), *Magn. Reson. Chem.* **2002**, *40*, S15–S152. i) Y. Cohen, M. Neeman, (Eds.), *Israel J. Chem.* **2003**, *43*, 1–163.

7 a) W. S. Price, *Concepts Magn. Reson.* **1997**, *9*, 299–336. b) W. S. Price, *Concepts Magn. Reson.* **1998**, *10*, 197–237. c) B. Antalek, *Concepts Magn. Reson.* **2002**, *14*, 225–258.

8 a) J. Crank, The Mathematics of Diffusion, 2nd Edition, Claxendon Press, Oxford, **1975**; b) E. L. Cussler, Diffusion: Mass Transfer in Fluid Systems, Cambridge University Press, Cambridge 1984.

9 A. Einstein, *Ann. Phys.* **1906**, *19*, 298–306.

10 For a recent more comprehensive review on the use of diffusion NMR in the context of supramolecular and combinatorial chemistry see: Y. Cohen, L. Avram, L. Frish, *Angew.*

Chem. Int. Ed. **2005**, *44*, 520–554; *Angew. Chem.* **2005**, *117*, 524–560.

11 a) P. S. Pregosin, P. G. A. Kumar, I. Fernandez, *Chem. Rev.* **2005**, *105*, 2977–2998. b) T. Brand, E. J. Cabrita, S. Berger, *Prog. NMR Spectrosc.* **2005**, *46*, 159–196. c) A. Dehner, H. Kessler, *ChemBioChem.* **2005**, *6*, 1550–1565.

12 J. E. Tanner, *J. Chem. Phys.* **1970**, *52*, 2523–2526.

13 a) P. Mansfield, B. Chapman, *J. Phys. E* **1986**, *19*, 540–545. b) P. Mansfield, B. Chapman, *J. Magn. Reson.* **1986**, *66*, 573–576.

14 M. Burl, I. R. Young, in *Encyclopedia of Nuclear Magnetic Resonance, Vol. 3* (Eds. D. M. Grannt, R. K. Harris), Wiley, New York, **1996**, p. 1841–1850.

15 a) S. J. Gibbs, C. S. Johnson, Jr., *J. Magn. Reson.* **1991**, *93*, 395–402. b) D. Wu, A. Chen, C. S. Johnson, Jr., *J. Magn. Reson. A* **1995**, *115*, 123–126.

16 A. Jerschow, N. Müller, *J. Magn. Reson.* **1997**, *125*, 372–375.

17 C. S. Johnson, Jr., *Prog. NMR Spectrosc.* **1999**, *34*, 203–256.

18 a) Wu, A. Chen, C. S. Johnson, Jr., *J. Magn. Reson. Ser. A* **1996**, *123*, 215–218. b) E. Gozansky and D. G. Gorenstein, *J. Magn. Reson. Ser B*, **1996**, *111*, 94–96. c) D. Wu, A. Chen and C. S. Johnson, Jr., *J. Magn. Reson. Ser B*, **1996**, *121*, 88–91. d) A. Jerschow, N. Müller, *J. Magn. Reson. Ser. A*, **1996**, *123*, 222–225.

19 L. H. Lucas, C. K. Larive, *Concepts Magn. Reson.* **2004**, *20A*, 24–41.

20 For some selected examples of compilation of association constants see: R. M. Izatt, J. S. Bradshaw, S. A. Nielsen, J. D. Lamb, J. J. Christensen, *Chem. Rev.* **1985**, *85*, 271–339; K. E. Krakowiak, J. S. Bradshaw, D. J. Zamecka-Krakowiak, *Chem. Rev.* **1989**, *89*, 929–972; R. M. Izatt, K. Pawlak, J. S. Bradshaw, *Chem. Rev.* **1991**, *91*, 1721–2085.

21 B. Lévay, *J. Chem. Phys.* **1973**, *77*, 2118–2121.

22 R. Rydén, J. Carlfors, P. Stilbs, *J. Incl. Phenomena* **1983**, *1*, 159–167.

23 O. Mayzel, Y. Cohen, *J. Chem. Soc. Chem. Commun.* **1994**, 1901–1902.

24 F. Vögtle, W. M. Muller, *Angew. Chem.* **1979**, *91*, 676–677; *Angew. Chem. Int. Ed. Engl.* **1979**, *18*, 623–624.

25 a) A. Gafni, Y. Cohen, *J. Org. Chem.* **1997**, *62*, 120–125. b) S. Simova, S. Berger, *J. Incl. Phenom. Macrocyclic Chem.* **2005**, *53*, 163–170.

26 a) R. Isnin, C. Salam, A. E. Kaifer, *J. Org. Chem.* **1991**, *56*, 35–41. b) A. Nepogodiev, J. F. Stoddart, *Chem. Rev.* **1998**, *98*, 1959–1976.

27 L. Avram, Y. Cohen, *J. Org. Chem.* **2002**, *67*, 2639–2644.

28 K. S. Cameron, L. Fielding, *J. Org. Chem.* **2001**, *66*, 6891–6895.

29 Covalent molecular capsules: a) R. G. Chapman, J. C. Sherman, *Tetrahedron* **1997**, *53*, 15911–15945. b) A. Jasat, J. C. Sherman, *Chem. Rev.* **1999**, *99*, 931–967.

30 Non-covalent molecular capsules: a) M. M. Conn, J. Rebek. Jr., *Chem. Rev.* **1997**, *97*, 1647–1668; b) M. Fujita, K. Umemoto, M. Yoshizawa, N. Fujita, T. Kusukawa, K. Biradha, *Chem. Commun.* **2001**, 509–518; c) F. Hof, S. L. Craig, C. Nuckolls, J. Rebek, Jr., *Angew. Chem.* **2002**, *114*, 1556–1578; *Angew. Chem. Int. Ed.* **2002**, *41*, 1488–1508.

31 Diffusion on dimeric capsules: a) L. Frish, S. E. Matthews, V. Böhmer, Y. Cohen, *J. Chem. Soc. Perkin Trans. 2*, **1999**, 669–671. b) L. Frish, M. O. Vysotsky, S. E. Matthews, V. Böhmer, Y. Cohen, *J. Chem. Soc. Perkin Trans. 2*, **2002**, 88–93; c) L. Frish, M. O. Vysotsky, V. Böhmer, Y. Cohen, *Org. Biomol. Chem.* **2003**, *1*, 2011–2014.

32 Diffusion on hexameric capsules: a) L. Avram, Y. Cohen, *J. Am. Chem. Soc.* **2002**, *124*, 15148–15149. b) L. Avram, Y. Cohen, *Org. Lett.* **2002**, *4*, 4365–4368; c) L. Avram, Y. Cohen, *Org. Lett.* **2003**, *5*, 1099–1102.

33 Diffusion on hexameric capsules: a) L. Avram, Y. Cohen, *Org. Lett.* **2003**, *5*, 3329–3332; b) L. Avram, Y. Cohen, *J. Am. Chem. Soc.* **2004**, *126*, 11556–11563.

34 a) L. R. MacGillivray, J. L. Atwood, *Nature* **1997**, *389*, 469–472. b) T. Gerkensmeier, W. Iwanek, C. Agena, R. Fröhlich, S. Kotila, C. Näther, J. Mattay, *Eur. J. Org. Chem.* **1999**, 2257–2262. c) J. L. Atwood, L. J. Barbour, A. Jerga, *Proc. Natl. Acad. Sci. USA* **2002**, *99*, 4837–4841.

35 a) A. Shivanyuk, J. Rebek, Jr., *Proc. Natl. Acad. Sci. USA* **2001**, *98*, 7662–7665; b) A. Shivanyuk, J. Rebek, Jr., *Chem. Commun.* **2001**, 2424–2425.

36 I. Philip, A. E. Kaifer, *J. Org. Chem.* **2005**, *70*, 1558–1564.

37 a) R. Krämer, J.-M. Lehn, A. Marquis-Rigault, *Proc. Natl. Acad. Sci. USA.* **1993**, *90*, 5394–5398. b) C. Piguet, *J. Incl. Phenom. Macro. Chem.* **1999**, *34*, 361–391. c) M. Shaul, Y. Cohen, *J. Org. Chem.* **1999**, *64*, 9358–9364. d) M. Greenwald, D. Wessely, E. Katz, I. Willner, Y. Cohen, *J. Org. Chem.* **2000**, *65*, 1050–1058.

38 a) D. Zuccaccia, L. Pirondini, R. Pinalli, E. Dalcanale, A. Macchioni, *J. Am. Chem. Soc.* **2005**, *127*, 7025–7032. b) M. Alajarin, A. Pastor, R.-A. Orenes, E. Martinez-Viviente, P. S. Pregosin, *Chem. Eur. J.* **2006**, *12*, 877–886. c) C. Schmuck, T. Rehm, F. Gröhn, K. Klein, F. Reinhold, *J. Am. Chem. Soc.* **2006**, *128*, 1430–1431. d) X. Liu, Y. Liu, G. Li, R. Warmuth, *Angew. Chem. Int. Ed.* **2006**, *45*, 901–904; *Angew. Chem.* **2006**, *118*, 915–918. e) N. Kuhnert, A. Le-Gresley, *Org.&Biomol. Chem.* **2005**, *3*, 2175–2182.

39 a) R. H. Vreekamp, J. P. M. van Duynhoven, M. Hubert, W. Verboom, D. N. Reinhoudt, *Angew. Chem. Int. Ed. Engl.* **1996**, *35*, 1215–1218; *Angew. Chem.* **1996**, *108*, 1306–1309; b) P. Timmerman, R. H. Vreekamp, R. Hulst, W. Verboom, D. N. Reinhoudt, K. Rissanen, K. A. Udachin, J. Ripmeester, *Chem. Eur. J.* **1997**, *3*, 1823–1832; c) K. A. Jolliffe, P. Timmerman, D. N. Reinhoudt, *Angew. Chem. Int. Ed. Engl.* **1999**, *38*, 933–937. *Angew. Chem.* **1999**, *111*, 983–986.

40 P. Timmerman, J. L. Weidmann, K. A. Jolliffe, L. J. Prins, D. N. Reinhoudt, S. Shinkai, L. Frish, Y. Cohen, *J. Chem. Soc. Perkin Trans. 2*, **2000**, 2077–2089.

41 B. Olenyuk, M. D. Levin, J. A. Whiteford, J. E. Shield, P. J. Stang, *J. Am. Chem. Soc.* **1999**, *121*, 10434–10435.

42 T. Megyes, H. Jude, T. Grosz, I. Bako, T. Radnai, G. Tarkanyi, G. Palinkas, P. J. Stang, *J. Am. Chem. Soc.* **2005**, *127*, 10731–10738.

43 W. H. Otto, M. H. Keefe, K. E. Splan, J. T. Hupp, C. K. Larive, *Inorg. Chem.* **2002**, *41*, 6172–6174.

44 a) D. Sobransingh, A. E. Kaifer, *Chem. Commun.* **2005**, 5071–5073. b) G. R. Newkome, J. K. Young, G. R. Baker, R. L. Potter, L. Audoly, D. Cooper, C. D. Weis, K. Morris, C. S. Johnson, Jr., *Macromolecules* **1993**, *26*, 2394–2396.

45 M. S. Kaucher, Y.-F. Lam, S. Pieraccini, G. Gottarelli, J. T. Davis, *Chem. Eur. J.* **2005**, *11*, 164–173.

46 G. P. Spada, G. Gottarelli, Synlett 2004, 596–602. b) J. T. Davis, *Angew. Chem. Int. Ed. Engl.* **2004**, *43*, 668–698; *Angew. Chem.* **2004**, *116*, 684–716.

47 Y. H. Ko, K. Kim, J.-K. Kang, H. Chun, J. W. Lee, S. Sakamoto, K. Yamaguchi, J. Fettinger, K. Kim, *J. Am. Chem. Soc.* **2004**, *126*, 1932–1933.

48 M. Tominaga, K. Suzuki, M. Kawano, T. Kusukawa, T. Ozeki, S. Sakamoto, K. Yamguchi, M. Fujita, *Angew. Chem. Int. Ed. Engl.* **2004**, *43*, 5621–5625; *Angew. Chem.* **2004**, *116*, 5739–5743.

49 a) P. C. M. van Zijl, C. T. W. Moonen, *J. Magn. Reson.* **1990**, *87*, 18–25. b) J. C. Lindon, M. Liu, J. K. Nicholson, *Rev. Anal. Chem.* **1999**, *18*, 23–66. c) A. Chen, M. J. Shapiro, *Anal. Chem.* **1999**, *71*, 699A–675A.

50 D. Wu, A. Chen, C. S. Johnson, Jr., *J. Am. Chem. Soc.* **1993**, *115*, 4291–4299.

51 G. A. Morris, H. Barjat, in *Methods for structure Elucidation by High-*

Resolution NMR, (Eds. Gy. Batta, K. E. Kövér, Cs. Szántay, Jr.), Elsevier, Amsterdam, **1997**, pp. 209–226.

52 M. Lin, M. J. Shapiro, J. R. Wareing, *J. Am. Chem. Soc.* **1997**, *119*, 5249–5250.

53 a) P. J. Hajduk, R. P. Meadows, S. W. Fesik, *Q. Rev. Biophys.* **1999**, *32*, 211–240; b) M. van Dongen, J. Weight, J. Uppenberg, J. Schultz, M. Wikström, *Drug Discovery Today* **2002**, 7, 471–478. c) S. B. Shuker, P. J. Hajduk, R. P. Meadows, S. W. Fesik, *Science* **1996**, *274*, 1531–1534. d) P. J. Hajduk, R. P. Meadows, S. W. Fesik, *Science* **1997**, *278*, 497–499.

54 a) P. J. Hajduk, E. T. Olejniczak, S. W. Fesik, *J. Am. Chem. Soc.* **1997**, *119*, 12257–12261. b) R. C. Anderson, M. Lin, M. J. Shapiro, *J. Comb. Chem.* **1999**, *1*, 69–72.

55 a) S. S. Pochapsky, H. Mo, T. C. Pochapsky, *J. Chem. Soc. Chem. Commun.* **1995**, 2513–2514. b) H. Mo, T. C. Pochapsky, *J. Phys. Chem. B* **1997**, *101*, 4485–4486.

56 a) S. Beck, A. Geyer, H.-H. Brintzinger, *Chem. Commun.* **1999**, 2477–2478. b) N. G. Stahl, C. Zuccaccia, T. R. Jensen, T. J. Marks, *J. Am. Chem. Soc.* **2003**, *125*, 5256–5257. c) C. Zuccaccia, G. Bellachioma, G. Cardaci, A. Macchioni, *Organometallics* **2000**, *19*, 4663–4665. d) M. Valentini, P. S. Pregosin, H. Rüegger, *Organometallics* **2000**, *19*, 2551–2555. e) M. Valentini, P. S. Pregosin, H. Rüegger, *J. Chem. Soc. Dalton Trans.* **2000**, 4507–4510. f) A. Pichota, P. S. Pregosin, M. Valentini, M. Wörle, D. Seebach, *Angew. Chem. Int. Ed. Engl.* **2000**, *39*, 153–156; *Angew. Chem.* **2000**, *112*, 157–160. g) N. E. Schlorer, E. J. Cabrita, S. Berger, *Angew. Chem.* **2002**, *114*, 114–116; *Angew. Chem. Int. Ed. Engl.* **2002**, *41*, 107–109. h) C. Zuccaccia, N. G. Stahl, A. Macchioni, M.-C. Chen, J. A. Roberts, T. J. Marks, *J. Am. Chem. Soc.* **2004**, *126*, 1148–1164. i) I. Fernandez, F. Breher, P. S. Pregosin, Z. Fei, P. J. Dyson, *Inorg. Chem.* **2005**, *44*, 7616–7623.

57 M. Valentini, H. Rüegger, P. S. Pregosin, *Helv. Chim. Acta.* **2001**, *84*, 2833–2853.

58 a) Y. Cohen, A. Ayalon, *Angew. Chem.* **1995**, *107*, 888–890; *Angew. Chem. Int. Ed. Engl.* **1995**, *34*, 816–818. b) A. Ayalon, A. Sygula, P.-C. Cheng, M. Rabinovitz, P. W. Rabideau, L. T. Scott, *Science* **1994**, *265*, 1065–1067.

59 R. Shenhar, H. Wang, R. E. Hoffman, L. Frish, L. Avram, I. Willner, A. Rajca, M. Rabinovitz, *J. Am. Chem. Soc.* **2002**, *124*, 4685–4692.

60 a) I. Aprahamian, D. H. Preda, M. Bancu, A. P. Belanger, T. Sheradsky, L. T. Scott, M. Rabinovitz, *J. Org. Chem.* **2006**, *71*, 290–298. b) I. Aprahamian, D. Eisenberg, R. E. Hoffman, T. Sternfeld, Y. Matsuo, E. A. Jackson, E. Nakamura, L. T. Scott, T. Sheradsky, M. Rabinovitz, *J. Am. Chem. Soc.* **2005**, *127*, 9581–9587.

61 a) J. Andrasko, *Biochim. Biophys. Acta* **1976**, *428*, 304–311. b) C. S. Johnson, Jr., *J. Magn. Reson. Ser. A* **1993**, *102*, 214–218. b) E. J. Cabrita, S. Berger, P. Bräuder, J. Kärger, *J. Magn. Reson.* **2002**, *157*, 124–131. d) G. J. Stanisz, *Israel, J. Chem.* **2003**, *43*, 33–44.

62 L. Avram, Y. Cohen, *J. Am. Chem. Soc.* **2005**, *127*, 5714–5719.

7
Photophysics and Photochemistry of Supramolecular Systems

Bernard Valeur, Mário Nuno Berberan-Santos and Monique M. Martin

7.1
Introduction

This chapter is devoted to photoresponsive supramolecular systems undergoing in most cases light-induced processes such as electron transfer, charge transfer, energy transfer, excimer formation, photoisomerization, etc. Such systems are conceived [1]:

- for ion and molecule recognition using molecular luminescent sensors [2]
- to perform specific functions triggered by light absorption, e.g. in supramolecular devices [3]
- to mimick light-induced biological processes, e.g. artificial photosynthesis [4].

The aim of this chapter is to show how conventional spectrophotometry and spectrofluorometry, time-resolved fluorometry or anisotropy, and transient absorption spectroscopy can be used to characterize, chemically and photochemically, supramolecular systems containing chromophores or fluorophores, and to follow in real time their photoresponse.

In the first section, steady-state spectroscopy is used to determine the stoichiometry and association constants of molecular ensembles, emphasize the changes due to light irradiation and provide information on the existence of photoinduced processes. Investigation of the dynamics of photoinduced processes, i.e. the determination of the rate constants for these processes, is best done with time-resolved techniques aiming at determining the temporal evolution of absorbance or fluorescence intensity (or anisotropy). The principles of these techniques (pulse fluorometry, phase-modulation fluorometry, transient absorption spectroscopy) will be described, and in each case pertinent examples of applications in the fields of supramolecular photophysics and photochemistry will be presented.

Analytical Methods in Supramolecular Chemistry. Edited by Christoph Schalley

ISBN: 978-3-527-31505-5

7.2 Spectrophotometry and Spectrofluorometry

Absorption and emission spectra provide basic information (molar absorption coefficients, luminescence quantum yields), but also their changes upon association between two species can be used to determine the stoichiometry and stability constant of host–guest complexes. Moreover, evidence for the existence of photo-induced processes can be simply obtained in some cases from the fluorescence spectra.

7.2.1 Determination of the Stoichiometry and Association Constant of Supramolecular Complexes from Spectrophotometric or Spectrofluorometric Titrations

Let us consider a molecular entity L containing one or several chromophores or fluorophores that undergo changes in absorbance or fluorescence intensity upon association with a species M (ion or molecule). We examine first the simplest case where a 1:1 complex is formed:

$$M + L \rightleftarrows ML$$

The stability of the complex ML is characterized by the equilibrium constant for which various terms are employed: *stability constant*, *binding constant*, *association constant*, *affinity constant* (K_s or β) or *dissociation constant* (K_d, reciprocal of K_s). From the thermodynamic point of view, the true equilibrium constant (that depends only on temperature) must be written with activities. But we will consider only dilute solutions so that the activity coefficients can be approximated to 1 (reference state: solute at infinite dilution). Activities can then be replaced by molar fractions (dimensionless quantities), but in solution they are generally replaced by molar concentrations:

$$K_s = \frac{[ML]}{[M][L]} \tag{7.1}$$

However, *the true equilibrium constant is in fact a dimensionless quantity* (according to the classical relation $-\Delta G^0 = RT \ln K_s$). All molar concentrations in the expression of K_s should thus be interpreted as molar concentrations relative to a standard state of 1 mol dm^{-3}: that is, they are the numerical values of the molar concentrations [5]. If the solution is not dilute enough, the equilibrium constants can still be written with concentrations but they must be considered as *apparent stability constants*.

When a second complex of stoichiometry 2:1 (M:L) is formed, the following equilibria should be considered, assuming stepwise binding

$$M + L \rightleftarrows ML$$

$$ML + M \rightleftarrows M_2L$$

The *stepwise binding constants* are defined as

$$K_{11} = \frac{[ML]}{[M][L]} \tag{7.2}$$

$$K_{21} = \frac{[M_2L]}{[ML][L]} \tag{7.3}$$

It is possible to write the formation of the 2:1 complex directly from L and M as follows

$$2M + L \rightleftarrows M_2L$$

and to define an overall binding constant

$$\beta_{21} = \frac{[M_2L]}{[M]^2[L]} \tag{7.4}$$

Generalization to a complex M_mL_l can easily be made:

$$mM + lL \rightleftarrows M_mL_l$$

$$\text{with } \beta_{ml} = \frac{[M_mL_l]}{[M]^m[L]^l} \tag{7.5}$$

In the following considerations, it will be assumed that the linear relationship between absorbance or fluorescence intensity and concentration is always fulfilled. Moreover, we will consider only the case where M does not absorb or emit light. In a titration experiment, the concentration of L is kept constant and M is gradually added. The absorption spectrum and/or the fluorescence spectrum are recorded as a function of the concentration of M. Changes in these spectra upon complexation allow one to determine the stability constant and the stoichiometry of the complexes.

In the case of simple stoichiometries, it may be sufficient to monitor the variations of absorbance or fluorescence intensity at an appropriate wavelength (chosen so that the changes are as large as possible), or the variations of the ratio of absorbances or fluorescence intensities at two wavelengths. The relevant formulae to be used can be found in Ref. 6.

Information on the stoichiometry of a complex can be obtained from the continuous variation method (Job's method) [6, 7], but this method is valid only when a single complex is formed.

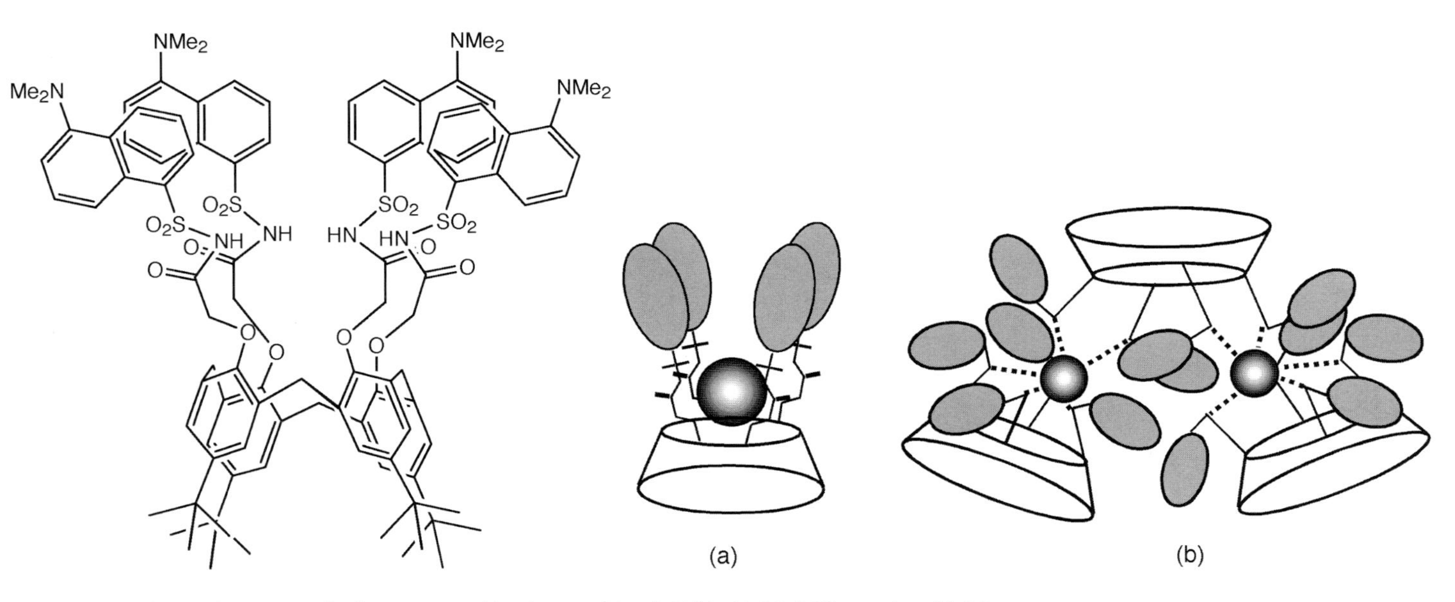

Scheme 7.1. Chemical structure of Calix-DANS4 and binding model with Pb^{2+}. (a) 1:1 (ML) complex; (b) 2:3 (M_2L_3) complex. (Reprinted with permission from ref. [9]).

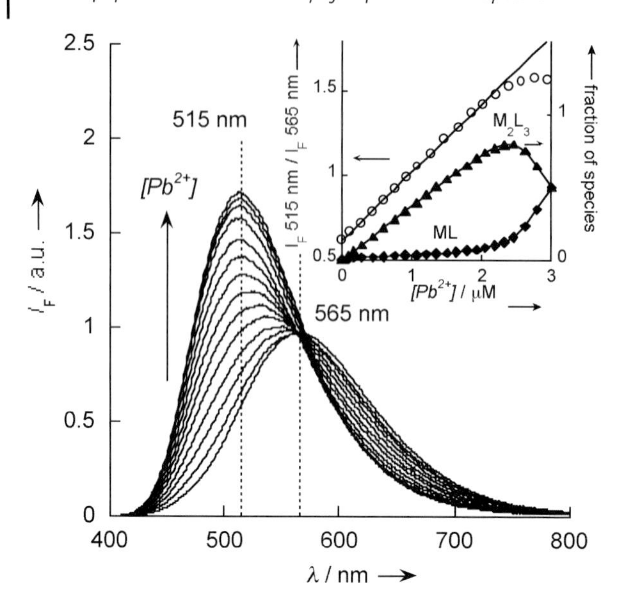

Fig. 7.1. Corrected emission spectra of Calix-DANS4 (3.6×10^{-6} M) in the presence of increasing concentration of Pb^{2+} in CH_3CN/H_2O (60:40 v/v) in lutidine buffer, pH 5.2. $\lambda_{exc} = 350$ nm. Inset: Ratiometric calibration curve (○), and fraction of the species M_2L_3 and ML as a function of lead concentration. (Reprinted with permission from ref. [10a]).

In the case of complex stoichiometries, and when several complexes can coexist in solution, data must be processed using several wavelengths simultaneously. This requires specific software. For instance, the commercially available SPECFIT Global Analysis System (V3.0 for 32-bit Window Systems) deserves attention. This software uses singular value decomposition and nonlinear regression modeling by the Levenberg–Marquardt method [8].

An example of the capabilities of this software is shown in Fig. 7.1. The fluorescent molecular sensor Calix-DANS4 consisting of a calix[4]arene bearing four dansyl groups (Scheme 7.1) offers an excellent selectivity for lead ion. Analysis by SPECFIT of the evolution of the whole fluorescence spectrum upon addition of lead ions revealed that a 2:3 complex and a 1:1 complex were successively formed with very high stability constants ($\log K_{23} = 33.5 \pm 1.5$ and $\log K_{11} = 10.0 \pm 0.5$) [9, 10].

7.2.2
Cooperativity and Anticooperativity

When L can complex more than one M, cooperative binding can occur. There are many definitions of cooperativity but they are all consistent with the following criterion [7]: a system is

Tab. 7.1. Relative values of the stability constants in the case of *n* identical and independent binding sites [7]

n	K_{11}	K_{21}	K_{31}	K_{41}	K_{51}	K_{61}
2	2	1/2				
3	3	1	1/3			
4	4	3/2	2/3	1/4		
5	5	2	1	1/2	1/5	
6	6	5/2	4/3	3/4	2/5	1/6

- *noncooperative* if the ratio $K_{(i+1)1}/K_{i1}$ is equal to the statistical value calculated when all binding sites are identical and independent. These statistical values are given in Table 7.1.
- *positively cooperative* if the ratio $K_{(i+1)1}/K_{i1}$ is larger than the statistical value.
- *negatively cooperative* (*or anticooperative*) if the ratio $K_{(i+1)1}/K_{i1}$ is smaller than the statistical value.

In particular, for a ditopic receptor that can bind successively two M (see above), the criterion for cooperativity is $K_{21}/K_{11} > 1/4$, i.e. complexation of a second M is made easier by the presence of a bound M.

For instance, a cooperative effect was observed with the bisanthraceno-crown ether **7.1** [11]. Complexation with a sodium ion brings closer together the two anthracene units which favors excimer formation, as revealed by the increase of the excimer band appearing in the fluorescence spectrum. This makes easier the binding of a second ion to form a 2:1 (metal:ligand) complex. The ratio of the stability constants K_{21}/K_{11} determined in acetonitrile from absorption spectra is equal to

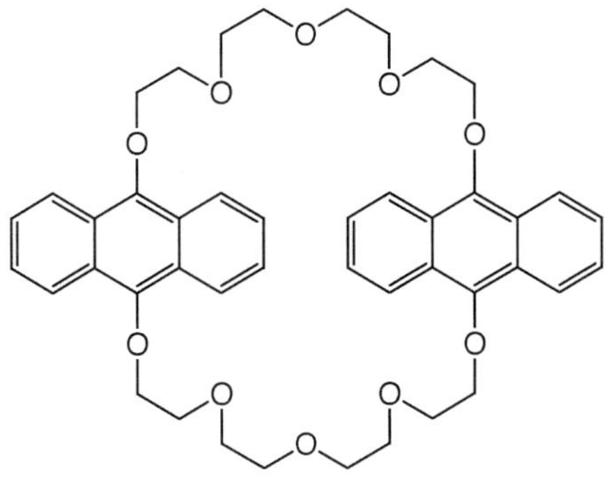

Compound 7.1

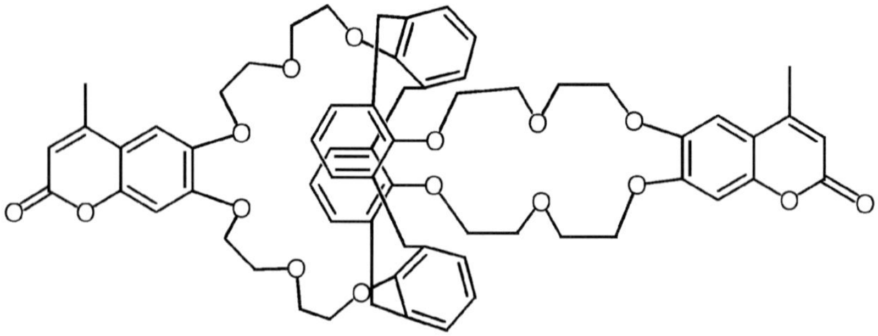

Compound 7.2

1.33, i.e. much greater than the statistical value of 1/4 that would be observed if the two binding sites were identical and independent.

In contrast, an anticooperative effect was observed in the bis-crown calixarene **7.2**, a fluorescent molecular sensor for cesium [12]. The complexation of a second cesium ion is made more difficult by the presence of a bound cesium ion, as shown by the value of K_{21} that is is in fact much smaller than K_{11} (log $K_{11} = 6.68$; log $K_{21} = 3.81$ in ethanol). The ratio K_{21}/K_{11} is 1.3×10^{-3}, i.e. much smaller than the statistical value of 1/4. Such an anticooperative effect is most likely due to electrostatic repulsion between the two cations; an unfavorable conformational change induced in the free crown by the bound cation in the other crown can also be invoked.

7.2.3 Possible Differences in Binding Constants in the Ground State and in the Excited State

In the case where a chromophore or a fluorophore of a supramolecular system is involved in the association with a species, the binding ability of this system may be different in the ground state and in the excited state. This occurs when a conformational change is induced by light, or when a redistribution of charges in the excited state leads to changes in electrostatic interactions in a host–guest complex.

If association and/or dissociation can occur during the lifetime of the excited state, the stability constants of the complex in the ground state and in the excited state may be different. The ground-state stability constant can be determined from titration using absorbance. The excited-state stability constant can be evaluated from titration using fluorescence, but only when the equilibrium is attained in the excited state, a condition that is rarely fulfilled because the lifetime of the excited state is often in the order of a few nanoseconds, which is too short for allowing diffusion-controlled association.

However, a few examples of differences between binding ability in the ground state and in the excited state have been described. For compound **7.1** in acetonitrile, the overall stability constant β_{21} of the complex with two sodium ions determined from absorption data (3.7×10^4) was found to be lower than that from fluorescence data (10^5), which means that the complexation ability is greater in the excited state [11]. This is likely to be caused by the formation of long-lived excimers that offer two complexing loops. In fact, the stability constant K_{21} for the binding of a second sodium ion (500) is larger than that found in the ground state (200).

7.2.4 Information on Photoinduced Processes from Fluorescence Spectra

Evidence for the existence of photoinduced processes occurring in supramolecular systems can be obtained in some cases:

- a decrease in fluorescence intensity (quenching) upon association may be related to electron transfer;
- a decrease in the fluorescence intensity of a donor and an increase in that of an acceptor can be used for the estimation of the energy transfer efficiency;
- an appearance of a new emission band may be due to excimer (or exciplex) formation.

Some examples will be now given.

7.2.4.1 Photoinduced Electron Transfer in a Calixarene-based Supermolecule Designed for Mercury Ion Sensing [10]

Calix-DANS2 (compound 7.3) bearing two dansyl fluorophores shows remarkable sensitivity and selectivity to Hg^{2+} with a detection limit of 0.3 μM.

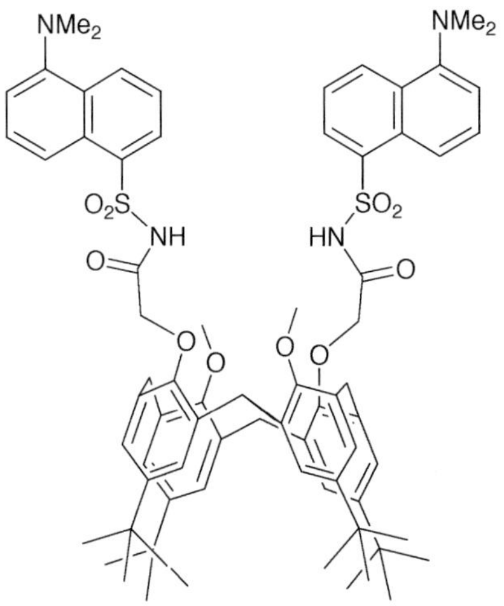

NMe2 NMe2 O2S SO2 NH HN O O O O O O

Compound 7.3

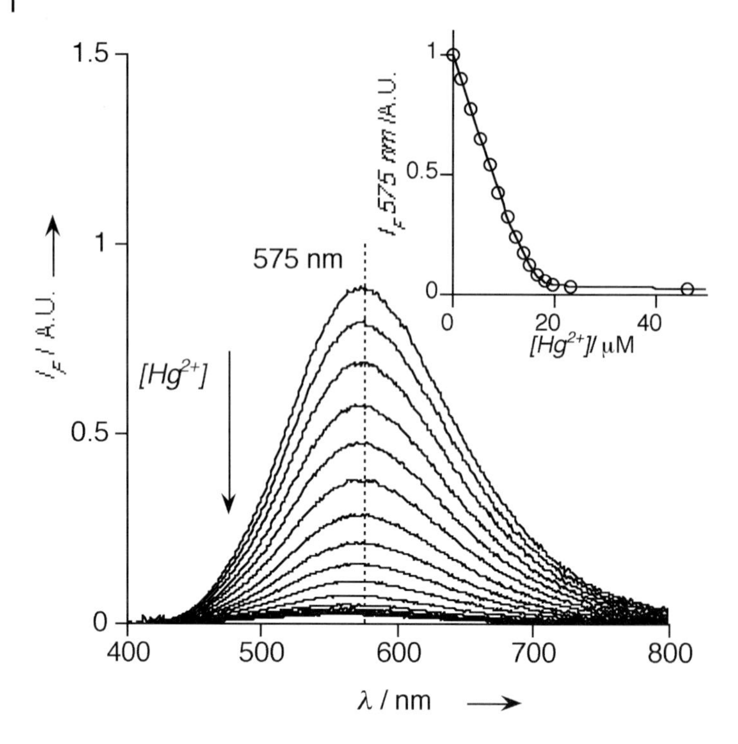

Fig. 7.2. Corrected emission spectra of Calix-DANS2 (1.6×10^{-5} M) in the presence of increasing concentration of Hg^{2+} in CH_3CN/H_2O (60:40 v/v) at pH 4.0. $\lambda_{exc} = 350$ nm. Inset: Calibration curve as a function of mercury concentration. (Reprinted with permission from ref. [9]).

Figure 7.2 shows the evolution of the emission spectra of Calix-DANS2 (1.6×10^{-5} mol L^{-1}) upon addition of Hg^{2+} in CH_3CN/H_2O (60:40 v/v) at pH 4.0 (in order to observe the strongest photophysical effects). It can be seen that the fluorescence of Calix-DANS2 is almost completely quenched upon Hg^{2+} addition. Analysis of the whole emission spectra upon mercury binding reveals the formation of a complex with a 1:1 stoichiometry and an apparent stability constant of 1.5×10^7. The quantum yield of the 1:1 complex is estimated to be only 2.5% of the fluorescence quantum yield of Calix-DANS2. The strong quenching upon mercury binding can be tentatively explained in terms of electron transfer from the excited dansyl fluorophore to the bound mercury ion.

The nature of the quenching mechanism can be easily confirmed by recording the emission spectrum in a frozen solution (EtOH/MeOH mixture (9:1) at liquid nitrogen temperature). Under these conditions, the relative decrease in fluorescence intensity expressed as I_F/I_0 is equal to 0.82, while at room temperature it is 0.23. Such a reduction in quenching efficiency in a frozen solution is characteristic of an electron transfer mechanism. In fact, immobilization of the solvent molecules in a frozen matrix prevents the reorganization of solvent molecules surrounding the charge separated pair, and raises its energy. Electron transfer is thus less efficient in a rigid matrix, which explains the higher fluorescence intensity in such a medium.

7.2.4.2 Excitation Energy Transfer in an Inclusion Complex of a Multichromophoric Cyclodextrin with a Fluorophore

Excitation energy transfer [13] is an important process occurring in multichromophoric supramolecular systems. A good example is provided by cyclodextrins containing several appended fluorophores with the aim at mimicking the photosynthetic light-harvesting antennae. In particular, the water-soluble β-cyclodextrin CD-St bearing seven steroidic naphthalene chromophores (Scheme 7.2) can form inclusion complexes with fluorescent dyes and in particular with oxazine 725 [14]. The formation of two complexes, CD-St:Dye and $(CD\text{-}St)_2$:Dye, was observed but at micromolar concentrations of CD-St and the dyes, the formation of the 1:1 complex can be ignored, and only the 2:1 complex will be considered here. This complex is an outstanding artificial antenna with 14 chromophores surrounding an energy acceptor [15].

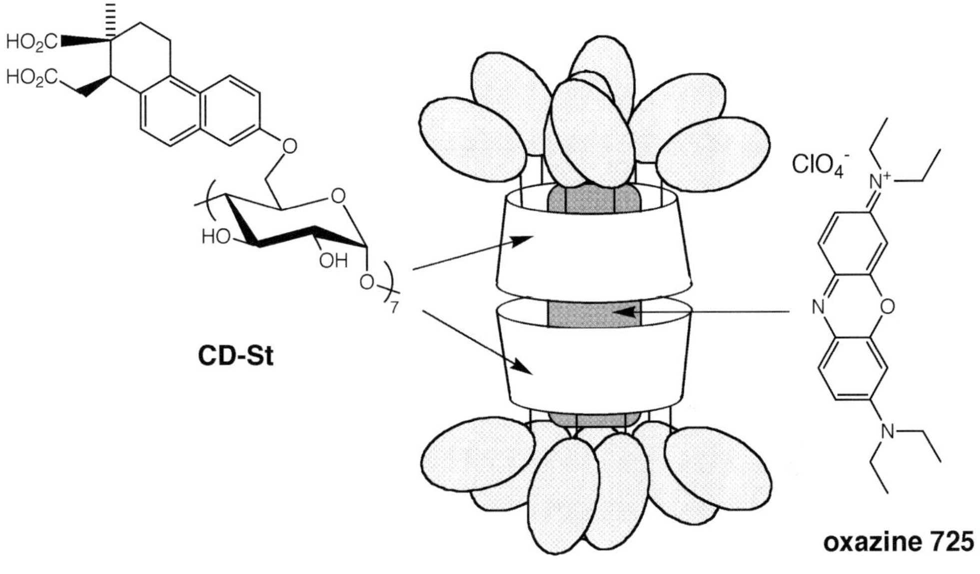

HO2C HO2C O O HO OH O 7 CD-St ClO_4^- N+ N O N oxazine 725

Scheme 7.2. Chemical structure of CD-St and oxazine 725, and schematic illustration of the 2:1 complex. (Adapted from ref. [15]).

Figure 7.3 displays the evolution of the fluorescence spectrum upon CD-St excitation (λ_{exc} = 320 nm) when increasing the oxazine concentration at constant CD-St concentration. A progressive quenching of CD-St fluorescence is shown to be accompanied by an increase of oxazine fluorescence. Moreover, the fluorescence

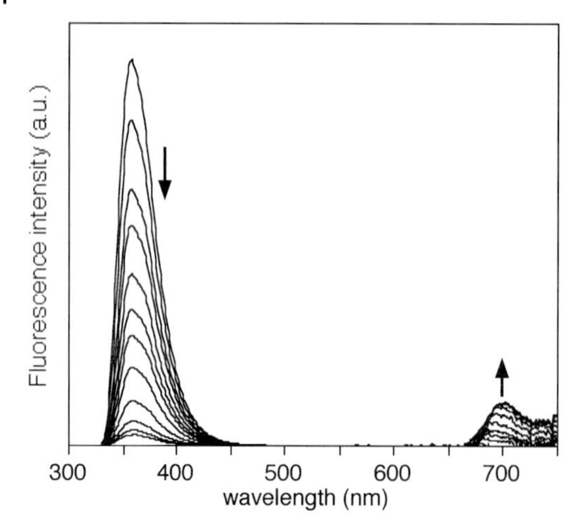

Fig. 7.3. Corrected fluorescence spectra of CD-St/oxazine 725 mixtures ($\lambda_{exc} = 320$ nm) as a function of oxazine concentration. The CD-St concentration is constant during these experiments (6.8 μM). The spectra were corrected for the inner filter effect. Solvent: mixture of buffer at pH 10 (I = 0.1 M) and ethanol 95:5 (v/v). (Reprinted with permission from ref. [15]. Copyright 2000 American Chemical Society).

emission arising from the dye exceeds the value expected from direct excitation of the dye. These observations show: (i) the formation of a complex between CD-St and oxazine; (ii) the process of energy transfer occurring from CD-St to oxazine.

The evolution of the fluorescence intensity of CD-St (at 357 nm) as a function of oxazine concentration contains information on the stability constant of the complex and on the energy transfer efficiency. In fact, the latter is directly related to the asymptotic value of the fluorescence intensity (corresponding to full complexation). After correction for the inner filter effect, data analysis provides the value of $(2.4 \pm 0.1) \times 10^{10}$ for the association constant $\beta = [(\text{CD-St})_2\text{:oxazine}]/[2\text{CD-St}]^2[\text{oxazine}]$. Moreover, the asymptotic value of the fluorescence intensity is very close to zero within experimental error which means that the energy transfer efficiency is close to 1. However, a full characterization of these photoinduced processes requires time-resolved techniques (see below).

7.3 Time-resolved Fluorescence Techniques

The determination of the rate constants for photoinduced processes in supramolecular systems is possible via time-resolved fluorescence techniques [16] provided that the characteristic times of these processes fall into the experimental time window that is defined by the lifetime of the involved excited states.

7.3.1 General Principles

The principles of time-resolved fluorometry are illustrated in Fig. 7.4. The *δ-pulse response* $I(t)$ of a fluorescent sample (i.e. the fluorescence intensity decay in response to an infinitely short light pulse mathematically represented by the Dirac function $\delta(t)$: delta excitation) is, in the simplest case, a single exponential whose time constant is the excited-state lifetime, but more frequently it is a sum of discrete exponentials, or a more complicated function; sometimes, the system is characterized by a distribution of decay times. For any excitation function $E(t)$, the response $R(t)$ of the sample is the convolution product of this function by the δ-pulse response:

$$R(t) = E(t) \otimes I(t) = \int_{-\infty}^{t} E(t')I(t - t')\,\mathrm{d}t' \tag{7.6}$$

In **pulse fluorometry**, the sample is excited by a short pulse of light and the fluorescence response is recorded as a function of time. If the duration of the pulse is not negligible with respect to the time constants of the fluorescence decay, the fluorescence response is the convolution product given by Eq. (7.6): the fluorescence intensity increases, goes through a maximum and becomes identical to the *true* δ-pulse response $I(t)$ as soon as the intensity of the light pulse is negligible. In this case, data analysis for the determination of the parameters characterizing the δ-pulse response requires a deconvolution of the fluorescence response.

In **phase-modulation fluorometry**, the sample is excited by a sinusoidally modulated light at high frequency. The fluorescence response, which is the convolution product (Eq. (7.6)) of the δ-pulse response by the sinusoidal excitation function, is sinusoidally modulated at the same frequency but delayed in phase and partially demodulated with respect to the excitation. The phase shift Φ and the modulation ratio M (equal to m/m_0), that is the ratio of the modulation depth m (AC/DC ratio) of the fluorescence and the modulation depth of the excitation m_0, characterize the *harmonic response* of the system. These parameters are measured as a function of the modulation frequency. No deconvolution is necessary because the data are directly analyzed in the frequency domain.

It should be noted that the harmonic response is the Fourier transform of the δ-pulse response, as expressed by the following relation

$$M \exp(-j\Phi) = \int_{0}^{\infty} i(t) \exp(-j\omega t)\,\mathrm{d}t \tag{7.7}$$

where ω is the angular frequency ($= 2\pi f$) and $i(t)$ is the normalized δ-pulse response according to

$$\int_{0}^{\infty} i(t)\,\mathrm{d}t = 1.$$

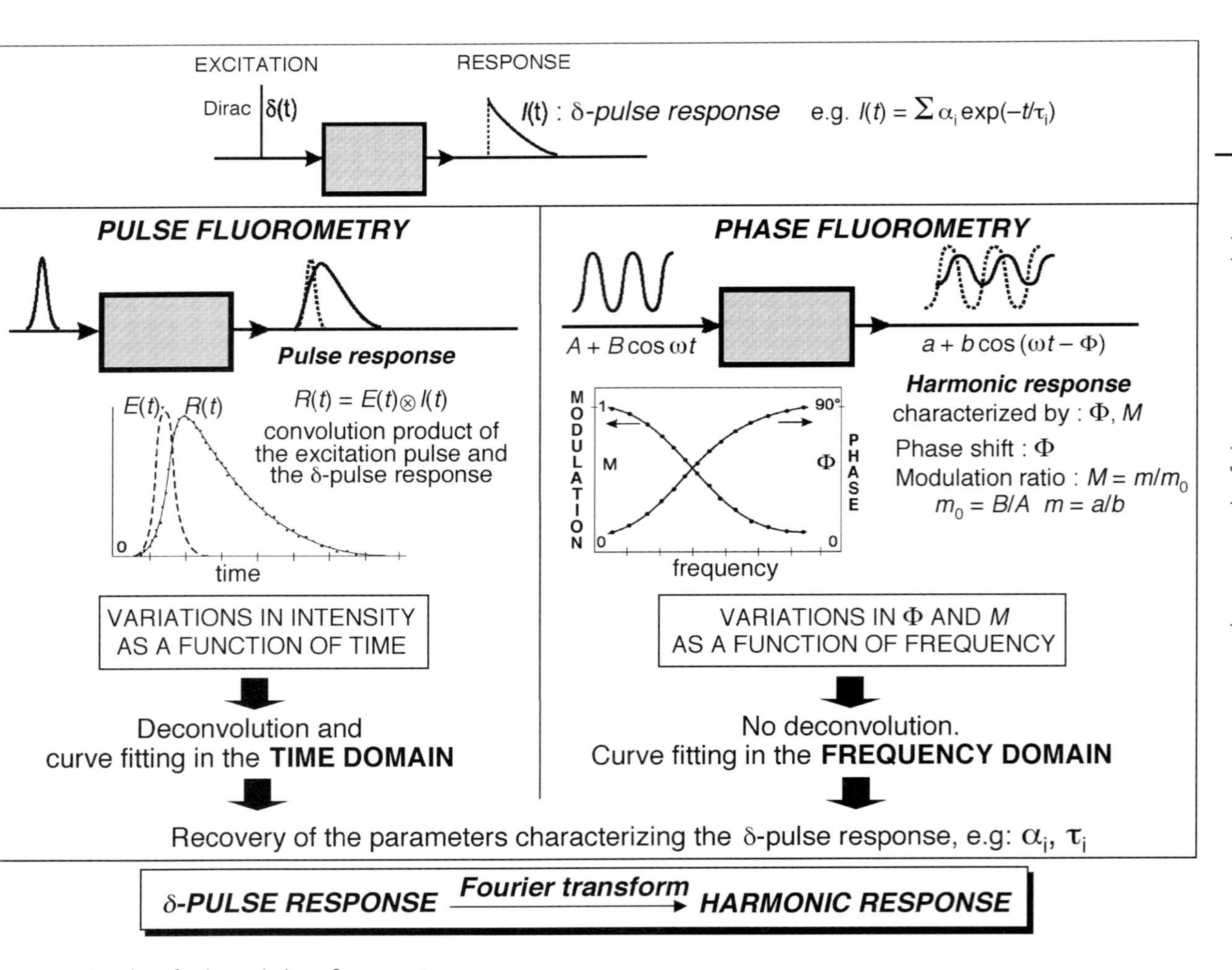

Fig. 7.4. Principles of pulse and phase fluorometries.

Consequently, the following useful relations are obtained:

$$\Phi = \tan^{-1}\left(\frac{P}{Q}\right) \quad \text{and} \quad M = [P^2 + Q^2]^{1/2} \tag{7.8}$$

where P and Q are the sine and cosine transforms of the δ-pulse response.

The preferred excitation-detection geometry for the study of macroscopic samples (typically a solution in a cuvette) is the right angle geometry, except when ultrashort laser pulses are used such as in fluorescence up-conversion experiments, where artifacts due to group velocity dispersion effects can spoil the measurements, as described in Section 7.5.2.

7.3.2
Pulse Fluorometry

The most widely used technique in time domain is the *time-correlated single-photon counting* technique, preferably called *single-photon timing* technique (*SPT*). *Time-gated* systems are less popular. *Streak cameras* offer a very good time resolution (a few picoseconds or less) but the dynamic range is smaller than that of the single photon-timing technique. The instruments that provide the best time resolution (a few tens of femtoseconds) are based on *fluorescence up-conversion*, but they are very expensive and not commercially available. Only the single photon timing technique will be presented below.

The basic principle of this technique relies on the fact that the probability of detecting a single photon at time t after an exciting pulse is proportional to the fluorescence intensity at that time. After timing and recording the single photons following a large number of exciting pulses, the decay of the fluorescence intensity is reconstructed.

Figure 7.5 shows a schematic diagram of a conventional single-photon timing instrument using a pulsed laser or a flash lamp and a monochromator. An electrical pulse associated to the optical pulse is generated (e.g. by a photodiode or the electronics associated to the excitation source) and routed – through a constant-fraction discriminator – to the start input of the time-to-amplitude converter (TAC). Meanwhile, the sample is excited by the optical pulse and emits fluorescence. The optics is tuned (e.g. by means of a neutral density filter) so that the photomultiplier detects no more than one fluorescence photon for each exciting pulse. The corresponding electrical pulse is routed – through a constant-fraction discriminator – to the stop input of the TAC. The latter generates an output pulse whose amplitude is directly proportional to the delay time between the start and the stop pulses. The height analysis of this pulse is achieved by an analogue-to-digital converter and a multichannel analyzer (MCA) which increases by one the contents of the memory channel corresponding to the digital value of the pulse. After a large number of excitation and detection events, the histogram of pulse heights represents the fluorescence decay curve.

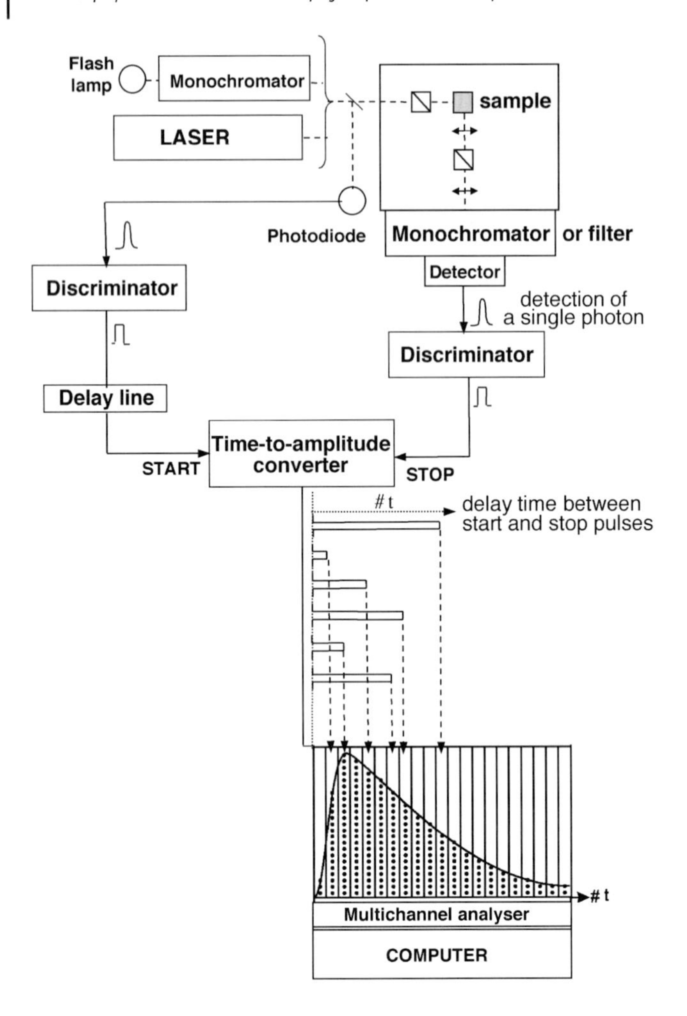

Fig. 7.5. Schematic diagram of a single photon timing instrument.

Obviously, the larger the number of events, the better the accuracy of the decay curve. The required accuracy depends on the complexity of the δ-pulse response of the system; for instance, a high accuracy is of course necessary for recovering a distribution of decay times.

When deconvolution is required, the time profile of the exciting pulse is recorded under the same conditions by replacing the sample by a scattering solution (e.g. a suspension of colloidal silica (Ludox)).

It is important to note that the number of detected fluorescence photons must be kept much smaller that the number of exciting pulses (<0.01–0.05 stops per pulse), so that the probability of detecting two fluorescence photons per exciting photon is negligible. Otherwise, the TAC will take into account only the first fluorescence pulse and the counting statistics will be distorted: the decay will appear shorter than it is in reality. This effect is called "pile-up effect".

Flash lamps running in air, or filled with gas, can be used, but the range of excitation wavelengths is limited, they deliver nanosecond pulses, and the repetition rate is not high (10^4–10^5 Hz). It is thus preferable to use mode-locked dye lasers and titanium-doped sapphire lasers: in fact, they can generate pulses over broad wavelength ranges, and the pulse widths are in the femtosecond and picosecond ranges with a high repetition rate. Dye lasers require handling of large volumes of dye solutions which must be replaced when the dye is bleached. In contrast, the Ti:sapphire laser is a solid-state laser that does not have this drawback and delivers shorter light pulses. Moreover, long operational lifetime, operational simplicity, reliability and broad tunability are distinct advantages. For these reasons, Ti:sapphire lasers are becoming more and more popular.

Picosecond diode laser heads offer an interesting alternative to mode-locked lasers. They are much less expensive. They can produce light pulses as short as 50–90 ps with repetition rates from single shot to 40–80 MHz. Peak powers up to 500 mW can be obtained. However, the main disadvantage is the absence of tunability, and the number of wavelengths is limited: 375, 400, 440, 635 to 1550 nm. No wavelength below 375 nm is presently available.

7.3.3
Phase-modulation Fluorometry

In the case of a single exponential decay, the decay time can be determined

- by phase measurements: $\tau_\Phi = \dfrac{1}{\omega}\tan\Phi$
- by modulation measurements: $\tau_M = \dfrac{1}{\omega}\left(\dfrac{1}{M^2} - 1\right)^{1/2}$

The values measured in these two ways should of course be *identical* and *independent of the modulation frequency*. This provides two criteria to check whether an instrument is correctly tuned by using a lifetime standard whose fluorescence decay is known to be a single exponential. Note that a significant difference between values obtained by means of the above equations is a compelling evidence of nonexponentiality of the fluorescence decay.

A single frequency suffices for measuring the decay time of a single exponential decay but the frequency should be chosen such that $\omega\tau$ is not too different from 1, i.e. $f \approx 1/(2\pi\tau)$. Therefore, for decay times of 10 ps, 1 ns, 100 ns, the optimum frequencies are about 16 GHz, 160 MHz, 1.6 MHz, respectively.

7.3.3.1 Phase Fluorometers using a Continuous Light Source and an Electro-optic Modulator

A schematic diagram of such instruments is given in Fig. 7.6. The light source can be a xenon lamp associated with a monochromator. The optical configuration

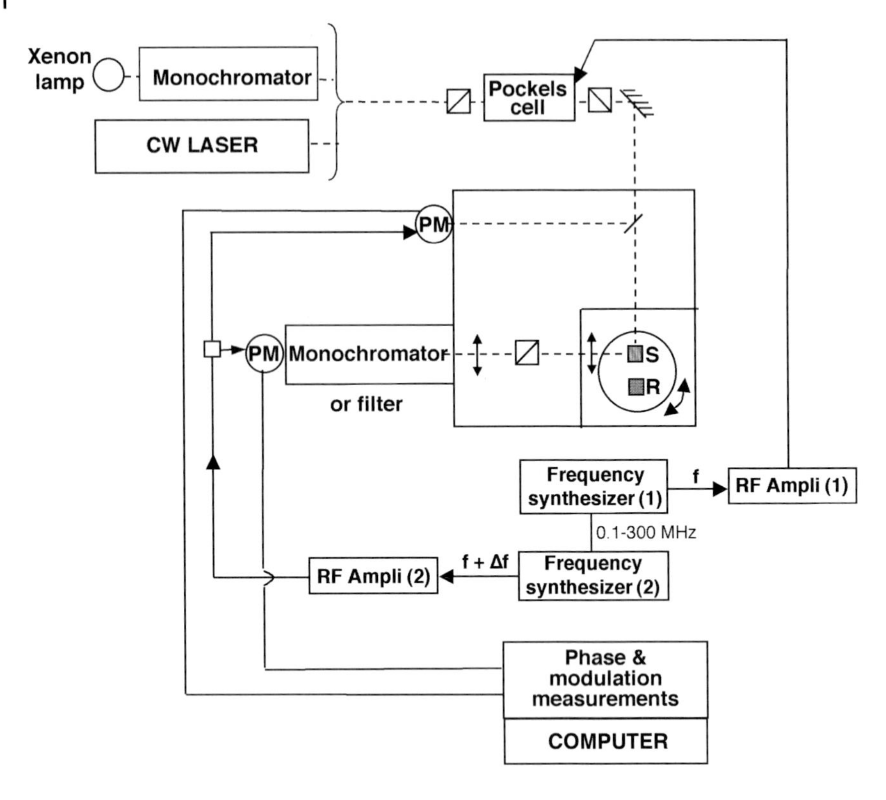

Fig. 7.6. Schematic diagram of a phase-modulation fluorometer using a CW light source (xenon lamp or laser) and a Pockels cell as a modulator.

should be carefully optimized because the electro-optic modulator (usually a Pockels cell) must work with a parallel light beam. The modulation frequency typically ranges from 0.1 to 300 MHz. The advantages are the low cost of the system and the broad range of excitation wavelengths.

In terms of light intensity and modulation, it is preferable to use a CW laser whose cost is not very high as compared to mode-locked pulsed lasers. The available wavelengths are the following: He-Cd: 325; 442.5 nm. He-Ne: 543.3; 594.1; 611.9; 632.8 nm. Ar^+: 457.9; 476.5; 488; 514.5 nm.

A beam splitter reflects a few percents of the incident light towards a reference photomultiplier (via or not a cuvette containing a reference scattering solution). The fluorescent sample and a reference solution (containing either a scatterer or a reference fluorescent compound) are placed in a rotating turret. The emitted fluorescence or scattered light is detected by a photomultiplier through a monochromator or an optical filter. The Pockels cell is driven by a frequency synthesizer and the photomultiplier response is modulated by varying the voltage at the second dynode by means of another frequency synthesizer locked in phase with the first one. The two synthesizers provide modulated signals that differ in frequency by a few tens of Hz in order to achieve cross-correlation. This procedure offers an excellent accuracy because the phase and modulation information contained in the signal is transposed to the low frequency domain where phase shifts and modulation depths can be measured with a much better accuracy than in the high frequency domain. For instruments using a CW laser, the standard deviations are currently 0.1–0.2° for the phase shift and 0.002–0.004 for the modulation ratio. When the light source is a xenon lamp, these standard deviations are ca. 0.5° and 0.005, respectively.

Practically, the phase delay φ_R and the modulation ratio m_R of the light scattered by a solution of glycogen or a suspension of colloidal silica are measured with respect to the signal detected by the reference photomultiplier. Then, after rotation of the turret, the phase delay φ_F and the modulation ratio m_F for the sample fluorescence are measured with respect to the signal detected by the reference photomultiplier. The absolute phase shift and modulation ratio of the sample are then $\Phi = \varphi_F - \varphi_R$ and $M = m_F/m_R$, respectively.

7.3.3.2 Phase Fluorometers using the Harmonic Content of a Pulsed Laser

A laser system that delivers pulses in the picosecond range with a repetition rate of a few MHz can be considered as an intrinsically modulated source. The harmonic content of the pulse train – which depends on the width of the pulses – extends to several gigahertz. The limitation is due to the detector. For high frequency measurements, it is absolutely necessary to use microchannel plate photomultipliers (that have a much faster response than usual photomultipliers). The highest available frequencies are then about 2 GHz. As for pulse fluorometry, Ti:sapphire lasers are most suitable for phase fluorometry, and decay times as short as 10–20 ps can be measured.

It should be noted that internal cross-correlation is not possible with microchannel plate photomultipliers, but an external mixing circuit can be used.

7.3.4 Data Analysis

In both pulse and phase fluorometries, the most widely used method of data analysis is based on a nonlinear least-squares method. The basic principle of this method is to minimize a quantity which expresses the mismatch between data and fitted function. This quantity is the reduced chisquare χ_r^2 defined as the weighted sum of the squares of the deviations of the experimental response $R(t_i)$ from the calculated ones $R_c(t_i)$:

$$\chi_r^2 = \frac{1}{v}\sum_{i=1}^{N}\left[\frac{R(t_i) - R_c(t_i)}{\sigma(i)}\right]^2 \tag{7.9}$$

where N is the total number of data points and $\sigma(i)$ is the standard deviation of the ith data point, i.e. the uncertainty expected from statistical considerations (noise). v is the *number of degrees of freedom* ($v = N - p$, where p is the number of fitted parameters). The value of χ_r^2 should be close to 1 for a good fit.

In the *single-photon timing* technique, the statistics obeys the Poisson distribution and the expected deviation $\sigma(i)$ is approximated to $[R(t_i)]^{1/2}$ so that Eq. (7.9) becomes

$$\chi_r^2 = \frac{1}{\nu}\sum_{i=1}^{N}\frac{[R(t_i) - R_c(t_i)]^2}{R(t_i)} \tag{7.10}$$

In *phase fluorometry*, no deconvolution is required: curve fitting is indeed performed in the frequency domain, i.e. directly using the variations of the phase shift Φ and the modulation ratio M as functions of the modulation frequency. Phase data and modulation data can be analyzed separately, or simultaneously. In the latter case the reduced chi-squared is given by

$$\chi_r^2 = \frac{1}{\nu}\left[\sum_{i=1}^{N}\left[\frac{\Phi(\omega_i) - \Phi_c(\omega_i)}{\sigma_\Phi(\omega_i)}\right]^2 + \sum_{i=1}^{N}\left[\frac{M(\omega_i) - M_c(\omega_i)}{\sigma_M(\omega_i)}\right]^2\right] \tag{7.11}$$

where N is the total number of frequencies. In this case, the number of data points is twice the number of frequencies, so that the number of degrees of freedom is $\nu = 2N-p$. In phase fluorometry data analysis requires the knowledge of the sine and cosine of the Fourier transforms of the δ-pulse response. This is of course not a problem for the most common case of multiexponential decays, but in some cases, the Fourier transforms may not have analytical expressions, and numerical calculations of the relevant integrals are then necessary.

In addition to the value of χ_r^2, it is useful to display graphical tests. The most important of them is the plot of the *weighted residuals* defined as

$$W(t_i) = \frac{R(t_i) - R_c(t_i)}{\sigma(i)} \tag{7.12}$$

where $\sigma(i) = [R(t_i)]^{1/2}$ for single-photon counting data. The fit is satisfactory when the weighted residuals are randomly distributed around zero.

In some cases, a distribution of decay times best describes the observed phenomena. The recovery of such distributions is very difficult. The data can be analyzed either by methods that do not require an *a priori* assumption of the distribution shape [17, 18], or by using a mathematical function describing the distribution [19].

7.3.5 Examples

7.3.5.1 Photoinduced Electron Transfer in a Self-assembled Zinc Naphthalocyanine–Fullerene Diad

The self-assembled diad formed by axial coordination of zinc naphthalocyanine (ZnNc) and fulleropyrrolidine bearing a coordinating imidazole ligand (C_{60}Im) is a good example of a supramolecular system undergoing photoinduced electron

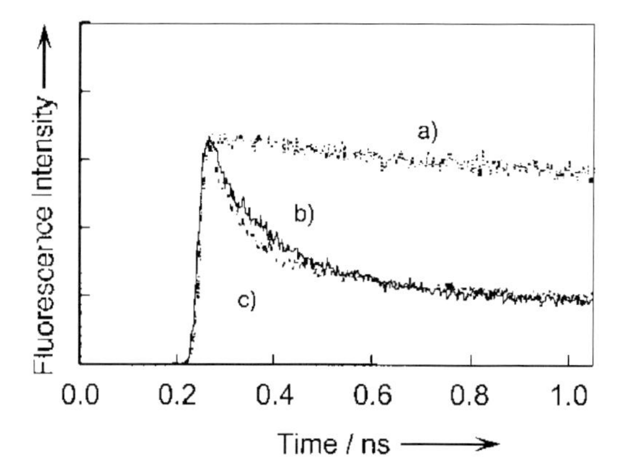

Fig. 7.7. Fluorescence decay profiles of a) ZnNc (0.03 mM) in toluene; b) and c) are the decay profiles of the diad ZnNc:C_{60}I (0.3 mM) in o-dichlorobenzene and toluene, respectively. $\lambda_{exc} = 410$ nm and $\lambda_{em} = 775$ nm. (Reprinted with permission from ref. [20]).

transfer (Scheme 7.3) [20]. This process was monitored by both time-resolved fluorescence and transient absorption techniques. Figure 7.7 shows the monoexponential decay of ZnNc ($\tau_0 = 2.42$ ns) and the biexponential decay of the diad ZnNc:C_{60}Im (71 ps (60%) and 2.14 ns (40%) in toluene).

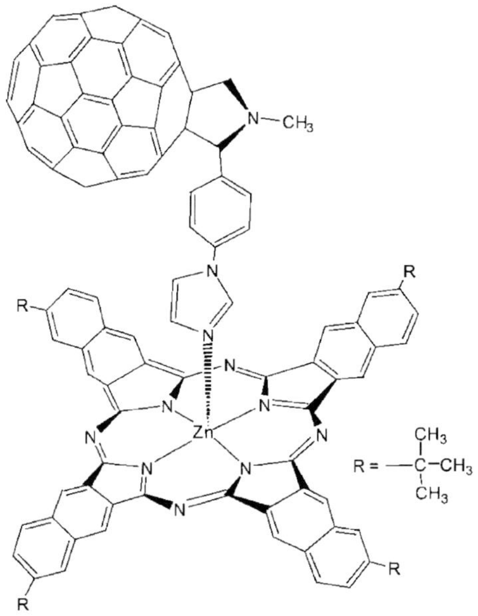

Scheme 7.3. Structure of the zinc naphthalocyanine-fullerene diad (ZnNc). (Reprinted with permission from ref. [20]).

The short component reveals a quenching due to photoinduced electron transfer from the singlet ZnNc to C_{60}Im, as proven by transient absorption spectroscopy. From the above values of decay times, the rate constant and quantum yield for electron transfer can be calculated according to the following relations:

$$k_{PET} = 1/\tau - 1/\tau_0$$

$$\Phi_{PET} = 1 - \tau/\tau_0$$

where τ_0 and τ are the decay times measured in the absence and the presence of the acceptor (C_{60}Im), respectively. In toluene, the values of k_{PET} and Φ_{PET} are $1.4 \times 10^{10}\ s^{-1}$ and 0.97, respectively, which suggest efficient charge separation in the diad.

7.3.5.2 **Excitation Energy Transfer in a Self-assembled Zinc Porphyrin–Free Base Porphyrin Diad**

The self-assembled diad ZnP-PH_2P consisting of a zinc porphyrin donor and a free base porphyrin acceptor (Scheme 7.4) was studied by time-resolved fluorescence [21]. The driving force of the assembly is the site selective binding of an imidazole connected to a free base porphyrin. Evidence for Förster back transfer was obtained from the analysis of the fluorescence decay (Fig. 7.8) and the relevant rate was quantitatively evaluated for the first time. The transfer efficiency [13] is 0.98, and the rate constants for direct and back transfer were found to be $24.4 \times 10^9\ s^{-1}$ and $0.6 \times 10^9\ s^{-1}$, respectively. These values are consistent with the Förster energy transfer mechanism.

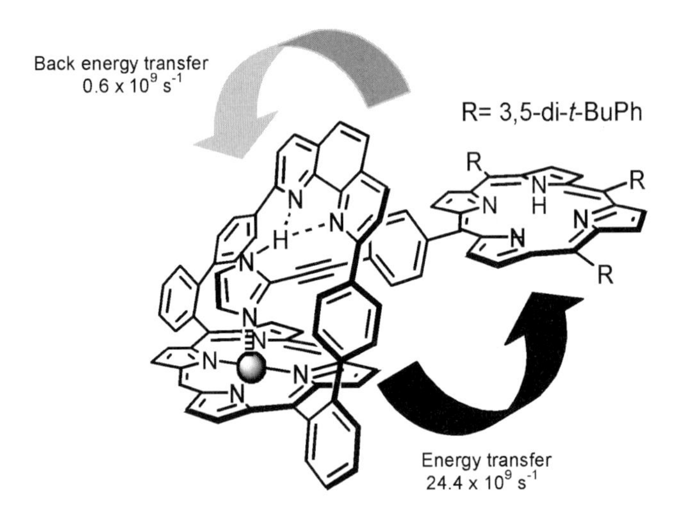

Scheme 7.4. Structure of ZnP-PH_2P diad.

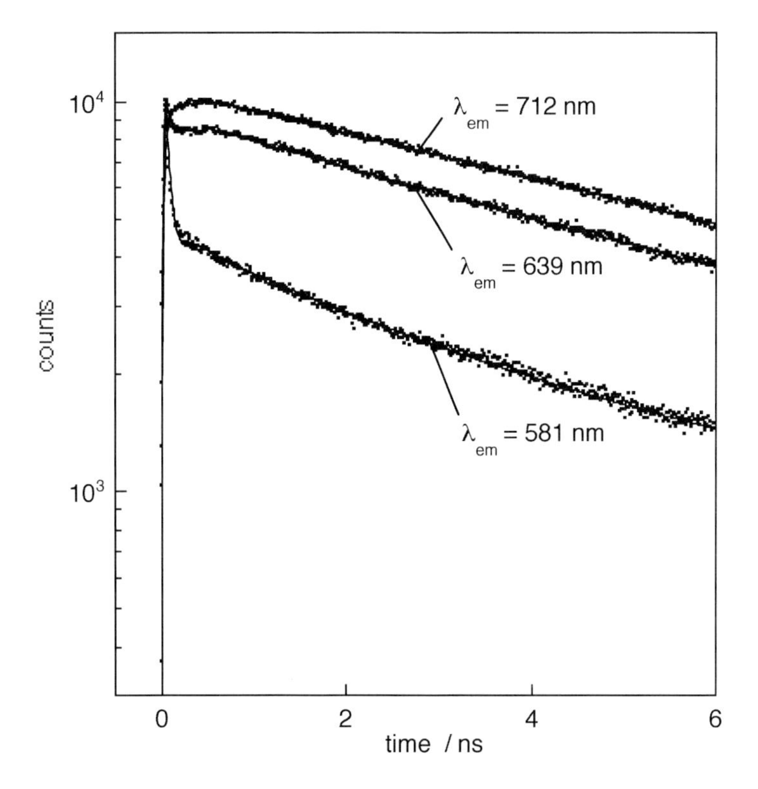

Fig. 7.8. Fluorescence decay curves of ZnP–PH_2P recorded at three wavelengths. The full lines represent the best fit resulting from a global analysis of the three curves with a sum of four exponentials (see text). λ_{exc}: 430 nm. The observation wavelengths are indicated in the figure. A risetime of the fluorescence intensity is clearly seen at 712 nm; it is due to the formation of excited acceptors resulting from energy transfer.

7.3.5.3 Excitation Energy Transfer in an Inclusion Complex of a Multichromophoric Cyclodextrin with a Fluorophore

As an example of excitation energy transfer studied by time-resolved fluorescence, let us take again the case of the inclusion complex of the multichromophoric cyclodextrin CD-St with oxazine 725 described in Section 7.2.4.2 [15]. Figure 7.9 shows the fluorescence decay of CD-St: the very first part of the decay is due to energy transfer [13] from the steroidic naphthalene fluorophores to oxazine 725. Data analysis led to an average decay time for transfer of about 25 ps, which is quite fast, as expected from the short average distance between donor and acceptor ($\approx$ 9–10 Å).

7.3.5.4 Excimer Formation of Cyanobiphenyls in a Calix[4]resorecinarene Derivative

In the supermolecule CRA-CB, four cyanobiphenyl groups (CB) are linked in a parallel direction to a calix[4]resorecinarene (CRA) [22] (Scheme 7.5) with the aim of

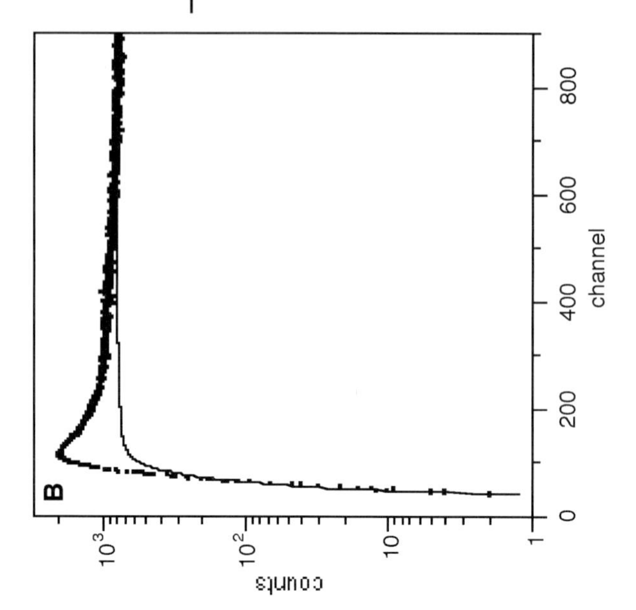

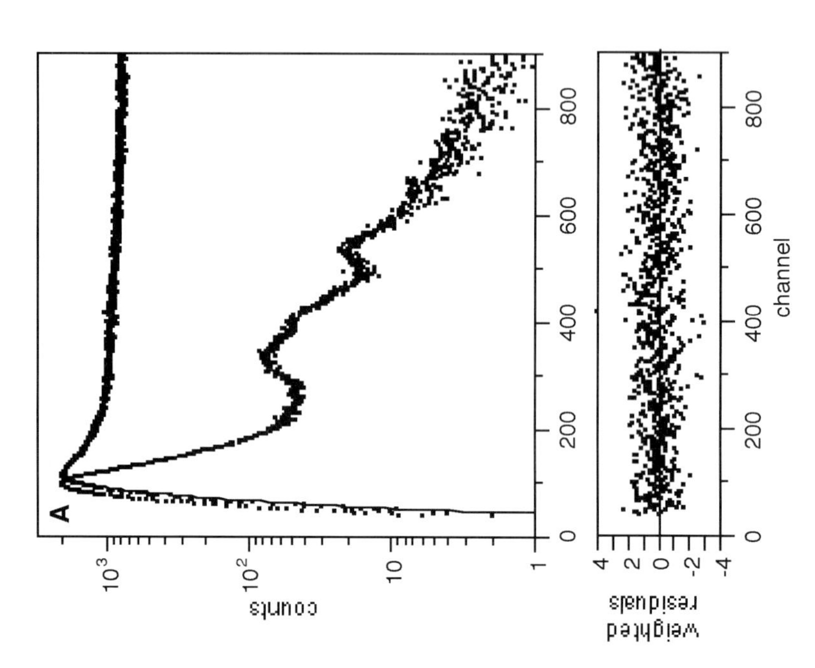

Fig. 7.9. Fluorescence decay of CD-St (16.6 μM) in the presence of Ox725 (80.5 μM). A. Experimental decay (points) and best fit (solid line) with a sum of four exponentials with two of them fixed at the values corresponding to the empty cyclodextrins (13.11 and 2.94 ns); the values of the two others are 12 and 90 ps. B. Reconvoluted curve (solid line) corresponding to only uncomplexed cyclodextrins (with normalization at long time) in order to visualize the effect of energy transfer. The difference between this curve and the experimental decay (points) represents the effect of energy transfer. $\lambda_{exc} = 326$ nm, $\lambda_{em} = 360$ nm. Channel width: 0.955 ps. (Reprinted with permission from ref. [15]. Copyright 2000 American Chemical Society).

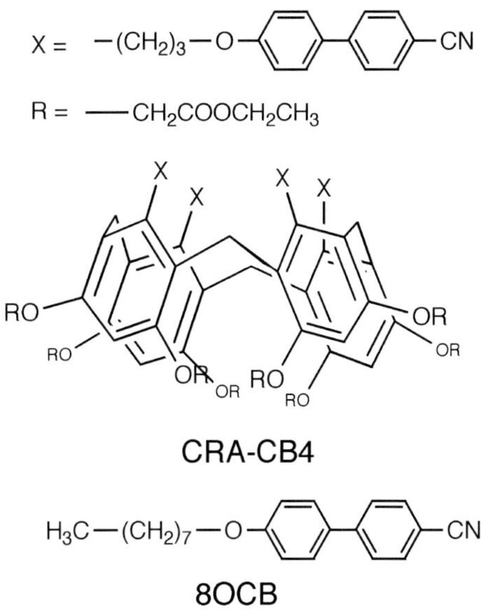

X = $-(CH_2)_3-O-$(biphenyl)$-CN$

R = $-CH_2COOCH_2CH_3$

CRA-CB4

$H_3C-(CH_2)_7-O-$(biphenyl)$-CN$

8OCB

Scheme 7.5. Structure of the calix[4]resorecinarene containing four cyanobiphenyl groups. (Redrawn from ref. [22]).

developing photoresponsive liquid crystals in which a monolayer of CRA on a solid substrate functions as the photo-driven command surface. Figure 7.10 shows the time-resolved fluorescence of the monomer and excimer bands. From data analysis, the rate constant for excimer formation and dissociation were estimated to be 1.9×10^9 s^{-1} and 4×10^7 s^{-1}, respectively. These values are significantly smaller than those for a liquid crystal of 4-cyano-4′-octyloxybiphenyl. The existence of preformed head-to-tail dimers in the latter case is likely to be responsible for the difference. On the other hand, in CRA-CB, the repulsive force between the fairly large dipole of the parallel CB groups hinders excimer formation that requires a twisting motion of these groups around a bond connecting to the CRA moiety.

7.4 Fluorescence Anisotropy

The size of a supramolecular object, as well as the orientational mobility of fluorophores can be determined by the measurement of fluorescence anisotropy. Energy transfer between identical fluorophores, also studied by fluorescence anisotropy, gives information on their relative distance. Fluorescence anisotropy can also be used to evaluate the state of association and the location of a fluorophore. Such measurements require the use of polarized light.

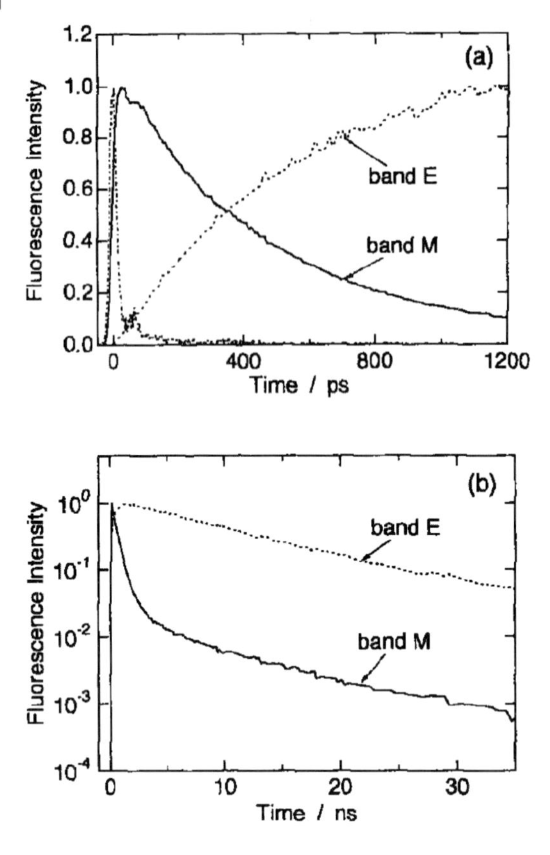

Fig. 7.10. Fluorescence rise and decay curves of band M (monomer) and band E (excimer) (at 365 and 420 nm, respectively) recorded at two different channel widths. λ_{exc}: 297 nm. The instrumental response function for the excitation laser pulse is shown in figure **a**. (Reprinted with permission from ref. [22]).

7.4.1
Principles

In classical terms, radiation is represented by an electromagnetic wave. The polarization of plane-wave radiation is defined by the way the oscillating electric field evolves in space, in a plane perpendicular to the direction of propagation. The most general polarization state is called elliptical polarization [23], but for luminescence applications the subset of *linear polarization* states usually suffices. In these cases the electric field vector oscillates along a well defined direction in a plane perpendicular to the direction of propagation. This direction is the *polarization direction*, and radiation with this characteristic is said to be *linearly polarized.*

If a beam of polarized radiation is passed through a linear polarizing filter, the intensity transmitted is a function of the angle between the filter *polarization axis* and the polarization direction of the beam. When the angle is zero, all radiation is transmitted (for an ideal filter; there are always losses by reflection, background absorption, etc); when the angle is 90°, all radiation is blocked (absorbed) by the filter. For intermediate angles, the fraction of intensity transmitted is given by the square of the co-sine of the angle (Malus law). This is easily understood as the electric field vector can be decomposed in two orthogonal components, one along the polarization axis that is transmitted, and one along the perpendicular direction that is absorbed. Since intensity is proportional to the square of the electric field amplitude, the Malus law follows.

Ordinary (*incoherent*) light is seldom fully polarized, and a beam of depolarized (or *natural*) light can be considered a mixture, in equal amounts, of two (incoherent) beams polarized in arbitrary but orthogonal directions.

In the general case, a beam of partially polarized incoherent radiation can be considered a mixture, in *unequal* amounts, of two (incoherent) beams polarized in orthogonal directions.

The *linear polarization* P of the beam is given by

$$P = \frac{I_{max} - I_{min}}{I_{max} + I_{min}} \tag{7.13}$$

where I_{max} is the maximum intensity recorded after passing the beam through a polarizing filter, and I_{min} is the minimum intensity recorded after passing the beam through a polarizing filter. The denominator is the total intensity of the beam. The filter orientations for maximum and minimum intensities, which are perpendicular, are *a priori* unknown, but correspond to perpendicular orientations. The polarization parameter P can take values between 0 and 1, $P = 0$ corresponding to unpolarized radiation, and $P = 1$ to fully linearly polarized radiation.

In fluorescence studies, spontaneous emission almost always dominates, hence radiation coming from the different fluorophores of a sample is incoherent.

The preferred excitation-detection geometry for the study of macroscopic samples is the right angle geometry [24]. In order to maximize the polarization of the emitted fluorescence, excitation is made with linearly polarized radiation. The direction of polarization is usually the vertical direction (see Fig. 7.11).

The polarization of the fluorescence emitted at 90° is determined by measuring the intensities of the vertically polarized fluorescence ($I_{||}$) and of the horizontally polarized fluorescence ($I_{\perp}$), i.e., the intensities after the fluorescence has passed through a polarizer with the polarizing axis set at the vertical and horizontal orientations, respectively. The resultant parameter is the *fluorescence polarization* p,

$$p = \frac{I_{||} - I_{\perp}}{I_{||} + I_{\perp}} \tag{7.14}$$

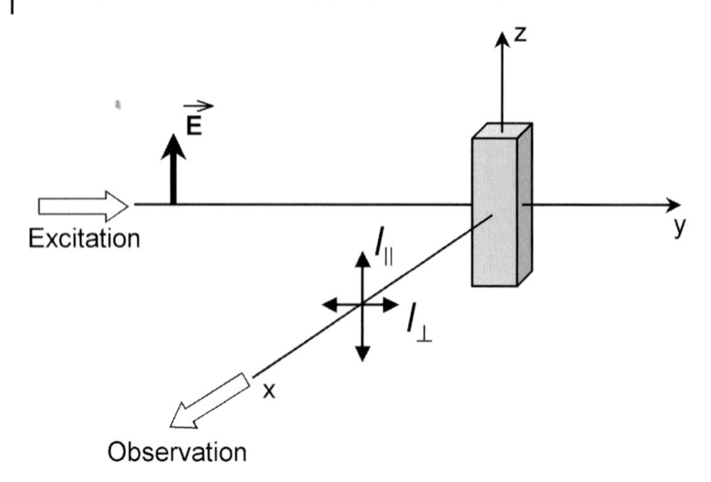

Fig. 7.11. Usual geometry for measuring the fluorescence anisotropy (right angle geometry).

In the common case of one-photon absorption, the fluorescence polarization can take values between $-1/3$ and $1/2$. Unlike the beam polarization P (one has $P = |p|$), the fluorescence polarization p can take negative values. This indicates that the direction of maximum intensity has rotated by $90°$, with respect to the excitation direction.

Fluorescence polarization cannot attain the ± 1 theoretical limits for maximum beam polarization owing to the nature of the absorption and emission processes, which usually correspond to electric dipole transitions. Although the excitation with linearly polarized radiation favours certain transition dipole orientations (hence certain fluorophore orientations, and the so-called *photoselection* process occurs), a fairly broad angular distribution is still obtained, the same happening afterwards with the angular distribution of the radiation of an electric dipole. The result being that, in the absence of fluorophore rotation and other depolarization processes, the polarization obeys the *Levshin–Perrin equation*,

$$p = \frac{3 \cos^2 \alpha - 1}{\cos^2 \alpha + 3} \tag{7.15}$$

where α is the angle made by the absorption and emission transition dipoles of the fluorophore. It is normally preferable to work with the quantity *anisotropy*, r, formally defined by Jablonski but already used by F. Perrin [25],

$$r = \frac{I_{||} - I_{\perp}}{I_{||} + 2I_{\perp}} = \frac{2p}{3 - p} \tag{7.16}$$

whose denominator is proportional to the fluorescence intensity, which is emitted in all directions and not only along the direction of observation. This quantity leads to simpler relations. For instance, Eq. (7.15) becomes

$$r = \frac{1}{5}(3\cos^2\alpha - 1) \tag{7.17}$$

The one-photon anisotropy takes values between $-1/5$ and $2/5$. The highest value is attained for collinear absorption and emission transition moments, i.e. namely when exciting in the S_0-S_1 transition.

In general, and following delta-pulse excitation, the fluorescence anisotropy can be written as a time-dependent ensemble average

$$r_\delta(t) = r_0\left\langle\frac{3\cos^2\theta(t) - 1}{2}\right\rangle \tag{7.18}$$

where r_0 is the so-called *fundamental anisotropy* that depends only on the angle α between the absorption and emission transition moments, Eq. (7.17), and θ is the angle between the emission transition moment at time $t = 0$ and the emission transition moment at a later time.

The two most important depolarization mechanisms that give rise to a time-dependent anisotropy are *fluorophore rotation* and *energy transfer*. Energy transfer leads to depolarization as the hopping of excitation from one fluorophore to another, when not parallel, is equivalent to an angular displacement.

For an excitation function $E(t)$, the anisotropy can still be defined by Eq. (7.16), and becomes [26]

$$r(t) = \frac{E(t) \otimes r_\delta(t) I(t)}{E(t) \otimes I(t)} \tag{7.19}$$

where $I(t)$ is the fluorescence intensity decay in response to delta excitation (see section 7.3.1), and $\otimes$ stands for the convolution between two functions, see Eq. (7.6). The *steady-state anisotropy*, $\bar{r}$, in particular, is obtained as

$$\bar{r} = \frac{\int_0^\infty r_\delta(t) I(t)\,dt}{\int_0^\infty I(t)\,dt} \tag{7.20}$$

For a molecule, macromolecule or supramolecular complex undergoing isotropic rotational diffusion, Eq. (7.18) reduces to [24]

$$r(t) = r_0 \exp(-t/\tau_r) \tag{7.21}$$

where τ_r is the *rotational correlation time*, approximately given by

$$\tau_r = \frac{\eta V}{kT} \tag{7.22}$$

where η is the viscosity of the medium, V is the hydrodynamic volume of the particle, k is the Boltzmann constant and T is the temperature. The corresponding steady-state anisotropy is given by the *Perrin equation*,

$$\bar{r} = \frac{r_0}{1 + \tau/\tau_r} \tag{7.23}$$

Measurement of the correlation time, and provided the viscosity of the medium is known, allows the determination of the hydrodynamic volume, hence the *size of the particle* where the fluorophore is embedded. This may in turn reflect an *association* process. For non-spherical particles, the anisotropy decay is given by more complex relations [24]. A time-dependent anisotropy may also indicate *intramolecular mobility.*

In the case of *Förster resonance energy transfer* (FRET) [13], i.e. energy transfer by the dipole-dipole mechanism, and for randomly oriented donor–acceptor pairs, the depolarization after one transfer step (ensemble average) is almost complete [27]. For this reason, fluorescence anisotropy is a good indicator of energy transfer between identical fluorophores, hence of *relative distances.* Existence of efficient FRET may therefore reflect an *association* process.

Fluorescence anisotropy is an *additive quantity* when weighted by the fraction of total intensity emitted, i.e. if several different fluorophores are present in a sample, or if a single fluorophore is present in several different environments, the total anisotropy measured at a given wavelength will be given by [26]

$$r(t) = \sum_i \alpha_i(t) r_i(t) \tag{7.24}$$

where $\alpha_i(t)$ is the fraction of the emitted radiation corresponding to the ith species, and $r_i(t)$ is the respective anisotropy. The same holds for the steady-state anisotropy, which can be written as

$$\bar{r} = \sum_i \alpha_i \bar{r}_i \tag{7.25}$$

In practice, the optics and the detection system respond differently to different polarizations, and a correction factor G must be introduced in Eq. (7.16) for the calculation of anisotropy [16a]:

$$r = \frac{I_{||} - GI_{\perp}}{I_{||} + 2GI_{\perp}} \tag{7.26}$$

In steady-state measurements, G is usually evaluated by additionally remeasuring the two polarized fluorescence intensities, but now using horizontally polarized excitation. These two components are identical, owing to the symmetry of the problem, and any difference in the recorded intensities must result from an instrumental polarization bias (assuming perfect time-stability of the instrument). In time-resolved measurements, use of a *depolarizer* (*scrambler*) after the emission polarizer but before the monochromator obviates the need for such a correction, provided exactly the same accumulation time is used for both polarized components. It is also possible to perform a determination of the G factor analogous to that described above for steady-state measurements, by using horizontally polarized excitation [16c]. The two recorded decays should now be integrated over time. Another, less general correction method in time-resolved decays is based on the principle of *tail matching* [16c]: For sufficiently long times, and if the anisotropy has already decayed to zero (this is not always the case, even when there is depolarization, see e.g. the cyclodextrin example below), the $I_{||}$ and $I_{\perp}$ decays must coincide. It is also common to perform a *global analysis* of the two polarized components [16a, 16c], without the need to previously compute the quantity anisotropy, hence by-passing the calculation of the G factor. This is however a risky procedure whenever the fluorescence intensity decay or the anisotropy decay is complex.

In time-resolved fluorescence, Eq. (7.19) must be used for directly fitting the anisotropy decay whenever the anisotropy decay is fast enough for an anisotropy decrease to be significant within the pulse duration. Equation (7.19) can also be used to compute the anisotropy corresponding to the phase-modulation technique [26].

7.4.2 Examples

7.4.2.1 Supramolecular Polymer Length

The coordination of ditopic fluorescent ligands **A** with a metal ion **M** (Zn^{2+}) leads to the formation of linear supramolecular polymers [28] of the $(A\text{-}M)_n$ type (Scheme 7.6).

The average length n is critically dependent on the [M]/[A] ratio, a maximum average length being expected for a 1:1 ratio. For $[M]/[A] \ll 1$, isolated **A** monomers predominate, while for $[M]/[A] \gg 1$ small complexes of the type **MAM** predominate. This can be shown by measuring the steady state anisotropy of solutions of **A** and **M** in a 3:2 chloroform-methanol mixture as a function of [M]/[A], see Fig. 7.12.

The perylene bisimide chromophore was excited in the S_0-S_1 transition, to ensure the highest possible anisotropy value, cf. Eq. (7.17). For the monomer **A**, the anisotropy has the value 0.045, that reflects a considerable depolarization owing to molecular rotation, cf. Eq. (7.23). However, the anisotropy steadily increases upon addition of the metal ion until a maximum of 0.133 is attained for a 1:1 ratio. This indicates the formation of progressively longer linear polymers whose rotation is increasingly slower, thus leading to higher anisotropy values, cf. Eq. (7.25). As expected, the anisotropy starts to decrease for $[M]/[A] > 1$, but stabilizes near 0.091, indicating that the dominant species present for $[M]/[A] \gg 1$ is larger than the monomer. Two important pieces of information had to be obtained before the

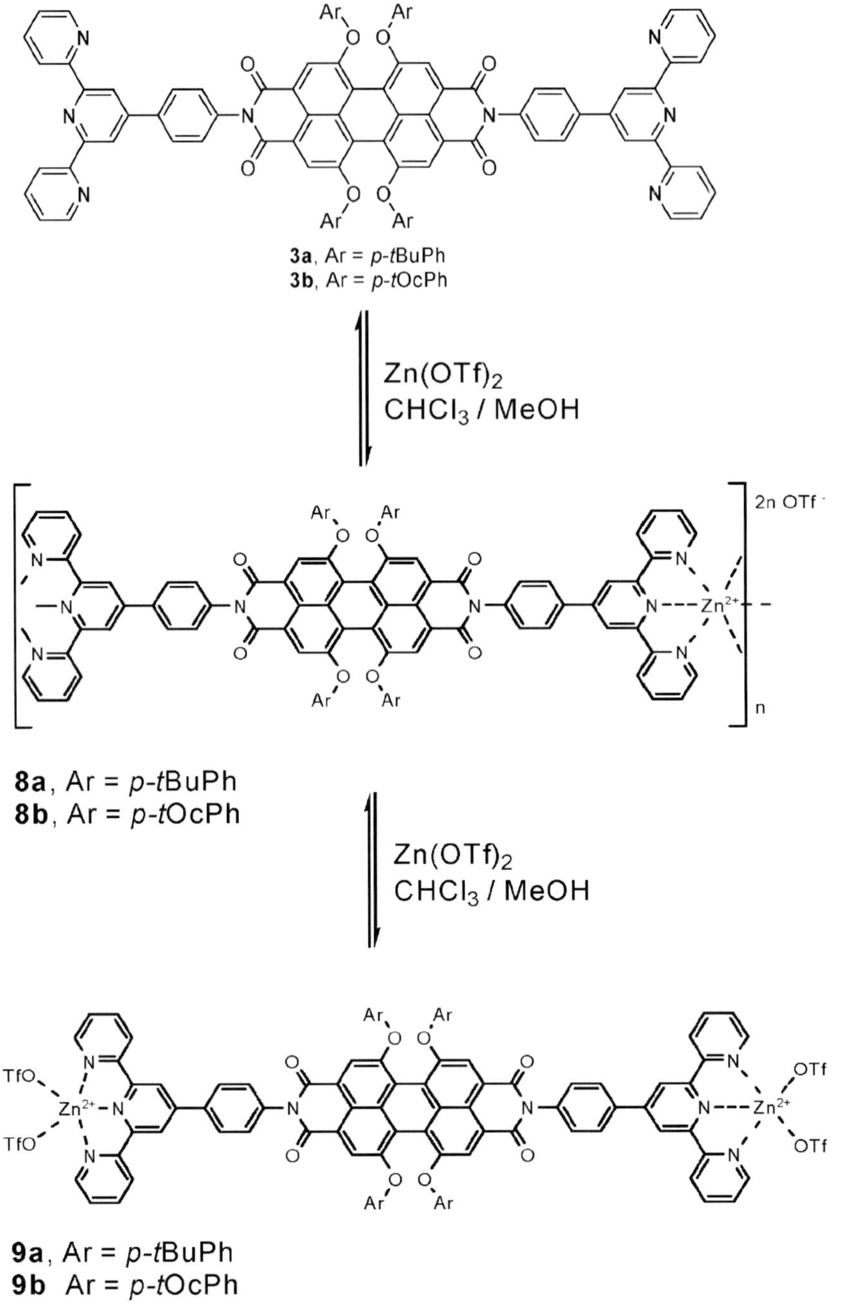

3a, Ar = p-tBuPh
3b, Ar = p-tOcPh
Zn(OTf)2
CHCl3 / MeOH
2n OTf⁻
n
8a, Ar = p-tBuPh
8b, Ar = p-tOcPh
Zn(OTf)2
CHCl3 / MeOH
TfO
TfO
OTf
OTf
9a, Ar = p-tBuPh
9b Ar = p-tOcPh

Scheme 7.6. Formation of supramolecular polymers upon addition of zinc triflate. When zinc(II) is in excess, the 2:1 complex dominates. (Reprinted with permission from ref. [28]. Copyright 2005 American Chemical Society).

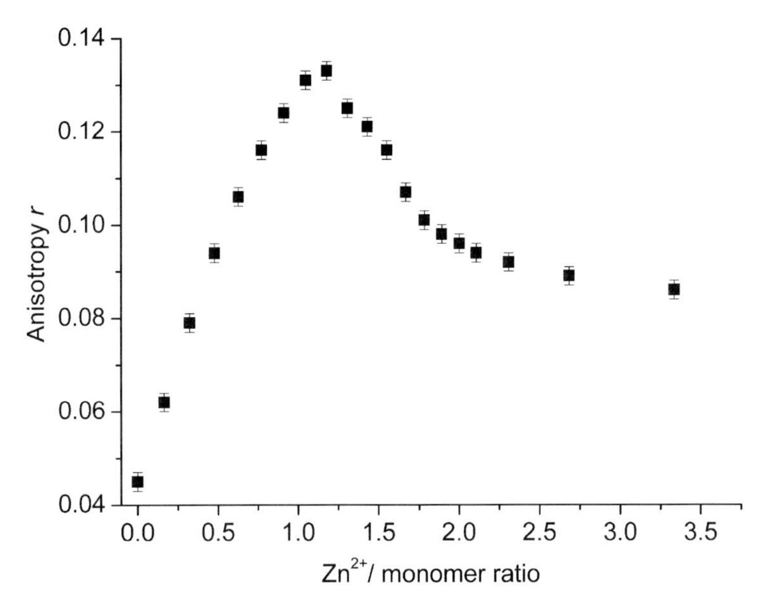

Fig. 7.12. Steady-state fluorescence anisotropy as a function of the Zn^{2+}/monomer ratio. (Reprinted with permission from ref. [28]. Copyright 2005 American Chemical Society).

above results could be interpreted. The first is that the monomer does not change its fluorescence properties (quantum yield, lifetime, spectrum) upon complexation with the metal ion. The second is that the depolarization by energy transfer between fluorophores in the polymer is not important, as the transition moments are aligned along the polymer longitudinal axis and are therefore all collinear.

7.4.2.2 **Excitation Energy Hopping in Multichromophoric Cyclodextrins**

To illustrate the use of fluorescence anisotropy in the elucidation of energy transfer between identical fluorophores, multichromophoric cyclodextrins [29] similar to the CD-St described in Section 7.2.4.2 are considered, but now in the absence of a dye.

In these multichromophoric cyclodextrins the fluorophores are randomly oriented. Excitation of one of the naphthoate fluorophores is followed by efficient dipole-dipole excitation energy transfer between the seven fluorophores, with a Förster radius of 14 Å. This process is not detectable by fluorescence intensity measurements, as neither the intensity nor the decay law are affected by energy transfer between identical fluorophores (also called *homotransfer*). The dynamics of energy hopping are on the other hand reflected in the fluorescence anisotropy. To avoid depolarization by rotational motion of the fluorophores, experiments were conducted in a low temperature and optically clear rigid glass (9:1 ethanol-methanol at 110 K).

The time-resolved anisotropy, Fig. 7.13, decays from an initial value of 0.291 to a constant value (0.042) that is reached in about 2 ns. The final value is exactly one-

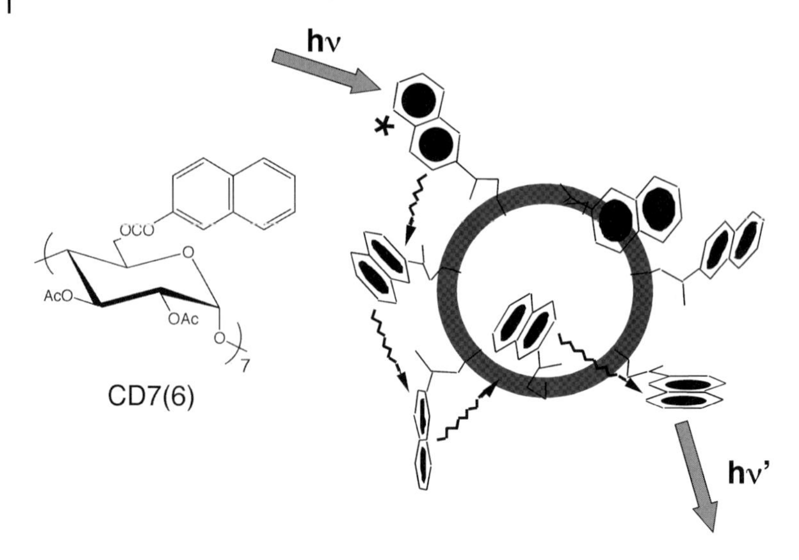

Scheme 7.7. Cyclodextrin CD7(6) and the excitation energy hopping process.

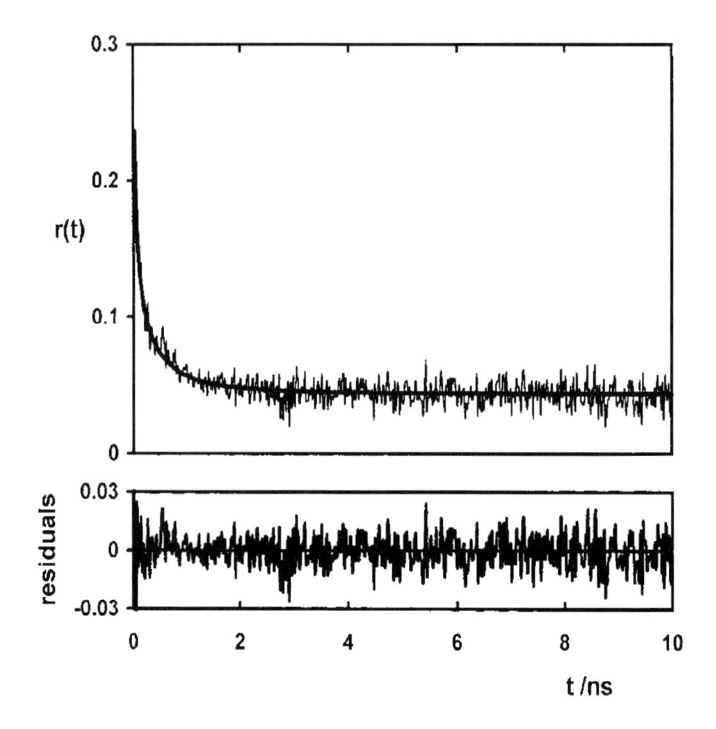

Fig. 7.13. Fluorescence anisotropy decay of the CD7(6) cyclodextrin in a low temperature rigid glass. (Reprinted with permission from ref. [29]. Copyright 1999 American Chemical Society).

seventh of the initial anisotropy, and can be understood on the basis of Eq. (7.24): for sufficiently long times, the excitation is equally shared by the seven fluorophores; however, the indirectly excited ones emit depolarized ($r = 0$) fluorescence, hence the overall anisotropy is $r_0/7$. From the time-dependent part of the anisotropy decay, an average reciprocal rate constant for transfer of 51 ps is obtained. A full analysis of the decay dynamics leads to an average distance between neighbouring fluorophores of 6 Å, in good agreement with the results of steady-state anisotropy and with molecular models.

Analogous results were obtained for the uncomplexed CD-St heptachromophoric cyclodextrin discussed in Section 7.2.4.2. In this case however, and owing to the larger Förster radius for homotransfer (22 Å), the leveling of the anisotropy occurs in ca. 100 ps, and the average reciprocal rate constant for transfer is only 2.4 ps [14]. Nevertheless, a similar nearest neighbor distance (7 Å) is recovered.

7.5 Transient Absorption Spectroscopy

In the previous sections, it has been shown how powerful the time-resolved fluorescence techniques are in real time probing of photoinduced processes and in allowing the determination of reaction rates from fluorescence lifetimes. The present section is devoted to the method of UV/vis transient absorption spectroscopy, which is a key method in probing non emissive species and is thus crucial to detect photoreaction products or intermediates following optical excitation of molecules in their electronic excited states. When carried out on short time scales, i.e. with femtosecond to subnanosecond excitation sources, fluorescent species can also be detected by their stimulated emission. Combining time-resolved fluorometry and transient absorption spectroscopy is ideal for the study of photochemical and photophysical molecular processes.

7.5.1 General Principles

Time-resolved UV/vis absorption spectroscopy has been initiated by Norrish and Porter who developed flash photolysis in the late 1940s, opening the way to the detection of transient chemical species with time resolution of a few microseconds [30, 31]. The present state of art transient absorption techniques allow detection of chemical intermediates with less than 10 fs resolution. The techniques used depend on the explored time scale but the principle, which is illustrated in Fig. 7.14, is the same.

The sample **S** is irradiated with a UV or visible light pulse (pump **P**) the wavelength of which is tuned to an electronic transition of the solute. The temporal behavior of the excited-state population and of the species produced is then probed with a second light pulse (probe **p**). The fluence of the probe pulse is much smaller than that of the pump pulse so that it does not perturb the existing populations but

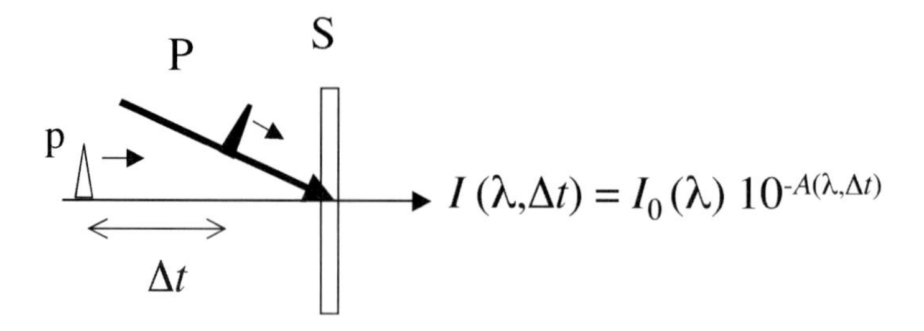

Fig. 7.14. Principle of pump-probe experiments with two short pulses. The photoinduced processes are probed by applying the Beer–Lambert law to the probe beam.

just allows their probing by using the Beer-Lambert law, at a single wavelength if the probe is monochromatic or on a broad spectral range if it is polychromatic. After having travelled through the sample, the probe beam is sent to the detection system which records the temporal change in the probe intensity upon excitation of the sample. The temporal behavior is obtained either by sending a short probe pulse on the sample at a variable delay **Δt** after the pump pulse (Fig. 7.14) and using a non time-resolved detection (picosecond and femtosecond laser photolysis) or by using a probe pulse of long duration and a time-resolved detection system (flash and nanosecond laser photolysis). In the next section we will describe in further detail typical pump-probe set-up for transient absorption spectroscopy with subpicosecond lasers. Further information on other techniques such as flash, nanosecond and picosecond photolysis can be found in Refs [30–33].

7.5.2 Pump-probe Spectroscopy with Subpicosecond Laser Excitation

High power, high repetition rate, ultrashort laser pulses, are now most often provided by amplified mode-locked solid-state lasers; harmonic generation, optical parametric amplification and frequency mixing allowing their tunability [34]. Different types of subpicosecond to femtosecond tunable dye lasers have also been developed and used for ultrafast spectroscopy during the last 25 years [33–35]. One of the main specificities of pump-probe spectroscopy with ultrashort light pulse lies in the possibility to easily generate an ultrashort white light continuum probe, which is an ideal tool for probing transient absorption spectra on a broad spectral range.

7.5.2.1 White Light Continuum Generation

If an intense ultrashort light pulse at a given wavelength (or frequency) is focused in a liquid or solid medium (water, glass, etc.), its spectrum broadens considerably while propagating in the medium and the output beam appears as a white light beam. This is mainly due to self phase modulation [36], a non linear process that results from the time-resolved change of the refractive index of the medium induced by the propagating short pulse $I(t)$:

$$n(t) = n_0 + 1/2 n_2 I(t) \tag{7.27}$$

where n_0 is the linear refractive index and $1/2\ n_2$, the effective nonlinear refractive index. The phase modulation that results from the index modulation leads to the creation of new frequencies in the spectrum of the propagating short pulse. At the output of a medium of length L, a pulse of initial frequency ω_0 will exhibit a time-resolved frequency given by:

$$\omega(t) = \omega_0 - (1/2 n_2 L \omega_0 / c) \frac{\partial I(t)}{\partial t} \tag{7.28}$$

In the leading edge of the pulse (positive derivative of the intensity), the new frequencies are lower than ω_0; they are higher in the trailing edge. This spectral broadening provides a polychromatic light source of short duration, called *continuum*, which is commonly used in pump-probe experiments on the femto-picosecond time scale.

When manipulating the continuum care must however be taken to avoid group velocity dispersion (GVD) through the optical components of the set-up [34]. Such a spectral dephasing (chirp) lengthens the probe pulse duration and the optimal time resolution of the set-up is lost. In addition, the spectral components of the probe do not reach simultaneously the sample, the longer wavelengths arriving first, which is a potential source of artifacts in measuring transient absorption spectra. Reflective optics, thin components, etc, are generally used to avoid GVD. For experiments requiring sub-100 fs time resolution, this is a crucial question since the ultrashort pump pulse itself is spectrally broad ($\Delta\nu.\Delta\tau = \text{constant}$) so that additional stages are needed in the pump-probe set-up for compensating GVD (rephasing the spectral components) in both pump and probe pulses (with a pair of prisms or gratings). In the present report we will limit ourselves to describe a subpicosecond set-up. Further information on femtosecond spectroscopy can be found in Refs [34, 35, 37].

7.5.2.2 **Subpicosecond Pump-continuum Probe Set-up**

Different types of set-up have been reported in the literature. A typical transient absorption set-up with a continuum probe and subpicosecond time resolution, which can be used with laser pulses of a few hundred femtoseconds of duration, is shown in Fig. 7.15. The sample is excited by the pump pulses **P**. The pump fluence (number of photons per cm^2) is set to be large enough to obtain an appreciable population of the excited state, which however can be small when an excitation source of high repetition rate is used (Ti-sapphire femtosecond lasers run at 1 kHz) because it allows fast accumulation of weak amplitude signals. With low repetition lasers (10 Hz) the excitation fluence should be close to the saturation fluence ($\approx 1/\sigma_a$, where σ_a is the absorption cross section of the solute at the pump wave-

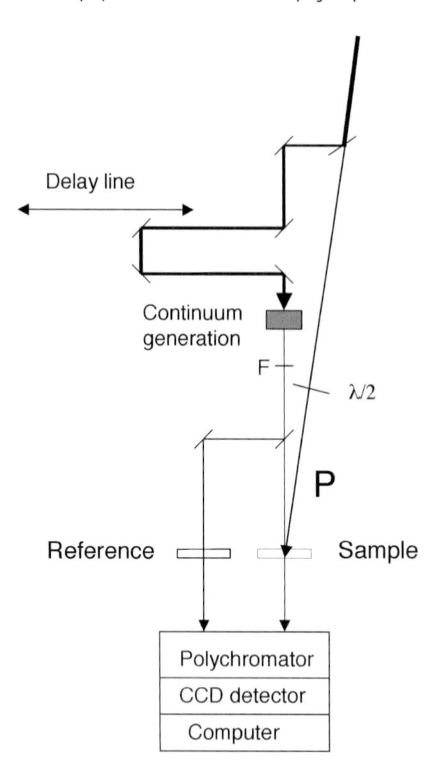

Fig. 7.15. Typical transient absorption set-up using a subpicosecond laser source as the pump and a continuum of white light as the probe. For ultrashort time resolution additional stages for compensating the group velocity dispersion in both light pulses are needed. P: pump, F: filter.

length). The sample is contained in a recirculating or moving cell, in order to avoid reexcitation of photoproducts. Short optical path must be used to avoid artifacts due to GVD of the probe and pump pulses within the sample cell, which implies the use of highly concentrated solutions.

The continuum probe can be generated either by using the same input laser beam as the pump or by using another beam generated by the laser system at a different wavelength. In the first case (see Fig. 7.15) a large amount of the input energy is extracted with a beam splitter, sent through a variable delay line then focused into a non linear medium, which, for example, can be a 0.5 or 1 cm water cell when 500 fs pulses are used or a 1 or 2 mm, sapphire or CaF_2, plate with 100 fs pulses or less. The continuum beam is spatially and spectrally filtered in order to eliminate the remaining input laser light. Its spectrum is flattened in the wavelength range used in the experiment. Pump and probe beams have a diameter of about 1 to 2 mm on the sample and cross at an angle of $\sim 10°$. The transmitted probe beams is sent to the entrance slit of a polychromator. It can be guided through optical fibers as in Ref. 38. The spectra of the two probe beams are simultaneously recorded by a computer-controlled CCD detector. The pump-probe delay time is adjusted by means of a stepper motor translation. The transient absorption signal of a reference solute having a long excited-state lifetime is used for zero delay time adjustment. Both pump and probe beams have linear polarization, the directions of which are set with a $\lambda/2$ plate either at the magic angle ($\sim 55°$) for spectrodynamics measurements, or parallel and orthogonal for anisotropy measurements. Data are accumulated over a fixed number of laser shots. Several methods exist to measure the chirp in the probe pulse when it reaches the sample. For example a chirp-plot can be obtained by two-photon absorption in a medium that is transparent at both the pulse and probe wavelengths, so that absorption is found only when the two pulses overlap in the sample [39]. One can deduce the chirp-plot from the delayed half-rise of the absorption signal across the spectrum. Once the chirp is known the spectra can be corrected numerically.

7.5.2.3 Time-resolved Differential Absorption Measurements

The excited state dynamics that are probed are expressed in terms of differential absorbance (ΔA), that is the difference between the absorbances of the excited (A_P) and unexcited (A) sample. The experiments provide the spectra of the probe beams which have travelled respectively through the sample (I_S) and reference (I_R) cell in the presence the pump pulse or without. The time resolved differential absorption across the explored spectral range is:

$$\Delta A(\lambda, t) = A_P(\lambda, t) - A(\lambda) = -\log[T_P(\lambda, t)/\log T(\lambda)] \tag{7.29}$$

where T is the transmission of the sample. In order to correct the measurements from background electrical noise, scattered light, and fluorescence noise, the signal of the detector must be recorded under four conditions [38]: (1) pump and probe off, (2) pump off and probe on, (3) pump on and probe off, (4) pump and probe on, so that ΔA is obtained from:

$$\Delta A = -\log(N/D) \tag{7.30}$$

with $N = ({I_S}^4 - {I_S}^3)/({I_R}^4 - {I_R}^3)$ and $D = ({I_S}^2 - {I_S}^1)/({I_R}^2 - {I_R}^1)$ where the numbers refer to one of the four conditions listed above.

7.5.2.4 Data Analysis

The absorbance is a direct signature of the existing populations of species (concentrations) in the probed sample. The unexcited and excited sample absorbances are respectively given by:

$$A(\lambda) = \varepsilon_a(\lambda)cL \tag{7.31}$$

$$A_P(\lambda, t) = \left[\varepsilon_a(\lambda)c_0(t) + \sum_i \{\varepsilon_{ui}(\lambda) - \varepsilon_{ei}(\lambda)\}c_i(t)\right]L \tag{7.32}$$

with $c_0(t) + \sum_i c_i(t) = c$, so that:

$$\Delta A(\lambda, t) = \sum_i \{\varepsilon_{ui}(\lambda) - \varepsilon_{ei}(\lambda) - \varepsilon_a(\lambda)\} c_i(t) L \tag{7.33}$$

where L is the sample optical path, c is the solute concentration, $c_0(t)$ and $c_i(t)$ are respectively the concentration of solute molecules in the ground state and of any other species, at time t after excitation. ε_a, ε_{ui} and ε_{ei} are respectively the absorption coefficient of the solute ground state, that of any other species and the stimulated emission coefficient of any excited species present at time t after excitation. In the spectral region where the fluorescence of the sample is expected, ΔA may be dominated by the stimulated emission signal (ε_e) and thus be negative. This is also true in the spectral region dominated by ε_a where the ground state absorbs, in such a case the negative signal would be due to a dominant bleaching signal.

Transient species are characterized by their spectra. If the spectra are not known, assignment can be made, for example, by comparison with analogues or from solvent effects. Reaction times are extracted from the analysis of $\Delta A(\lambda, t)$ as a single exponential function of time or as the sum of exponential functions. If some reaction times are very close to the time-resolution of the set-up the analysis is made by fitting the convolution product of the pump-probe cross correlation function $G^2(\tau)$ by a sum of exponential functions

$$\Delta A(t) = G^2(t) \otimes F(t) = \int G^2(t').F(t - t')\,\mathrm{d}t'$$

$$\text{with } G^2(\tau) = \int I_P(t).I_p(t - \tau)\,\mathrm{d}t \quad \text{and} \quad F(t) = \sum_i a_i \exp(-t/\tau_i) \tag{7.34}$$

Global analysis is also performed, like for example in reference [40]. Care must be taken however with hidden decay components resulting from spectral shifts due to solvation or vibrational cooling processes, which occur on this time scale.

7.5.3 Examples of Application

7.5.3.1 Charge Separation in Porphyrin–Fullerene Diads

Photoinduced electron transfer in fullerene based supramolecular systems has been described in Section 7.3.5 as an example of process that can be followed both by time-resolved fluorescence and transient absorption, and reaction time on the picosecond time scale was calculated from fluorescence decay measurements. Transient absorption with subpicosecond or picosecond resolution allows the characterization of the product formed while fluorescence is quenched. On a longer time-scale it provides information on the lifetime of the photoproduct. The porphyrin–fullerene (H_2P–C_{60}) diads in which the two chromophores are linked

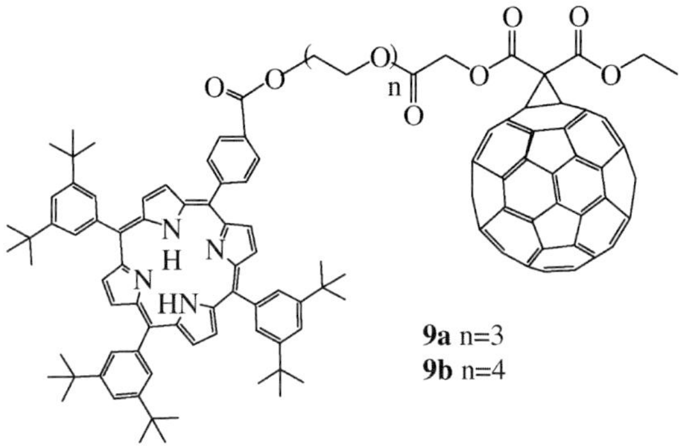

9a n=3
9b n=4

Scheme 7.8. Structure of the porphyrin-fullerene diads with a flexible polyether chain of variable length. (Reprinted with permission from ref. [41]).

by conformationally flexible polyether chains such, as those shown in Scheme 7.8, provide other examples of systems undergoing efficient charge separation [41]. Photoexcitation of the free base diads in polar solvents leads to charge-separated radical pair states, which are identified by their absorption. Figure 7.16 shows the transient absorption spectrum observed for the compound 9b in THF, 50 ns after

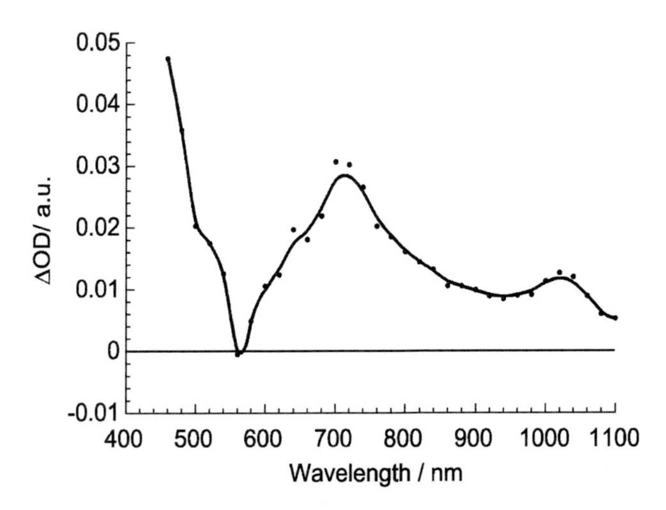

Fig. 7.16. Characterization of the photoinduced charge-separated state of a porphyrin-fullerene diads with flexible polyether linkers by transient absorption spectroscopy. The differential absorption spectrum is given for a 50 ns delay after excitation at 532 nm with a 10 ns laser. (Reprinted with permission from ref. [41]).

excitation at 532 nm with a 10 ns laser [41]. The spectrum corresponds to the radical pair $H_2P^{\bullet+}–C_{60}^{\bullet-}$, with characteristic bands of the porphyrin cation radical in the visible and of the fullerene anion radical in the near-infrared. The lifetime of the latter charge-separated radical pair is found to be 725 ns, one of the longest reported values for simple $H_2P–C_{60}$ diads in solution [41].

7.5.3.2 Cation Photorelease from a Crown-ether Complex

Finding a supramolecular system capable of fast and spatially controllable release of cations upon irradiation is of great interest for the study of intracellular process triggered by an ion concentration jump [42, 43, 45–47]. The complex of metal ions with a crown ether-linked merocyanine (DCM-crown) is an example of system (Scheme 7.9) which can temporarily release ions [46]. DCM-crown consists of DCM, a well-known laser dye, in which the dimethylamino group has been replaced by a macrocycle (monoaza-15-crown-5) that can bind metal ions [42].

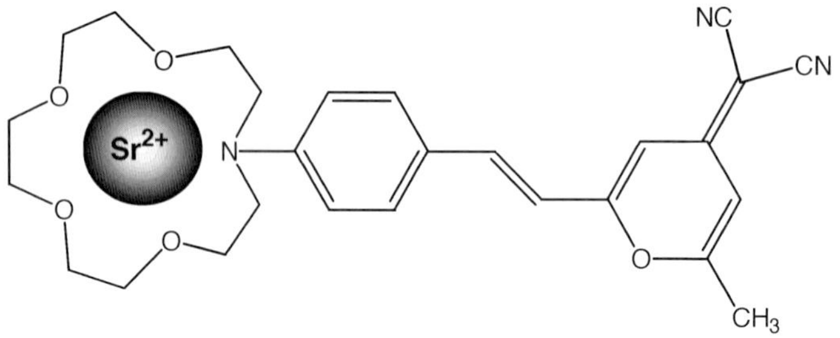

Scheme 7.9. Structure of the merocyanine DCM linked to an aza-crown.

Demonstration of the photorelease has been done in particular with Sr^{2+} [46]. This process was monitored on several time scales providing evidence for: (1) the delayed formation in 9 ps of the charge transfer state of the merocyanine chromophore following ultrafast photodisruption of the nitrogen – cation interaction, (2) the cation movement away from the excited chromophore into the bulk in 400 ps, (3) recombination of the complex in the ground in about 120 ns. These three steps are respectively illustrated in Fig. 7.17a, b, c (see caption for details). Similar transient absorption studies have been carried out on a PDS-crown-Ca^{2+} complex, where PDS is an aza-crown derivative of a substituted stilbene [47]. The spectrodynamics observed on the short time scale are very similar to those found in step (1) of the above description, with in particular a delayed rise of a stimulated emission band attributed to a solvent-separated cation-probe pair. Although the full scenario of the cation photoejection from the DCM-crown-Sr^{2+}, is complex [46], the spectra shown in Fig. 7.17 demonstrate that at least part of the photoexcited complexes does eject the ion into the bulk.

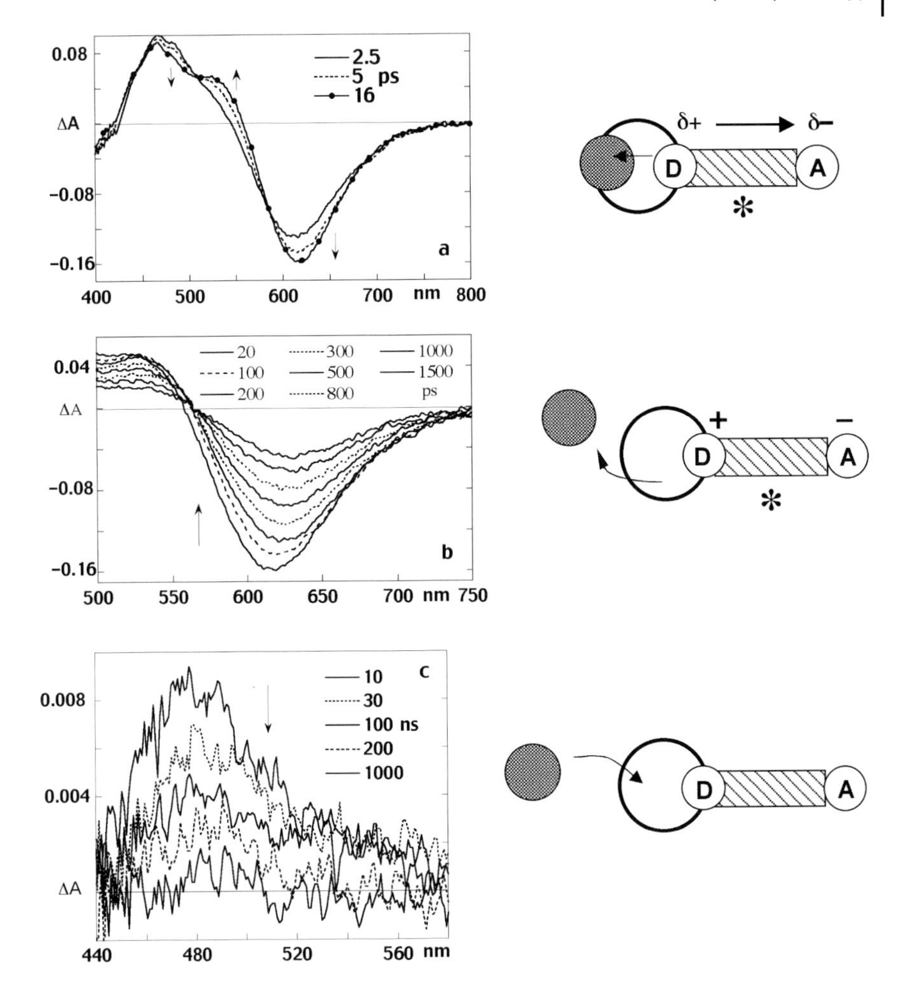

Fig. 7.17. Characterization of the photo-ejection of a Sr^{2+} ion from a complex with DCM-crown and of their recombination in the ground state. (Adapted from ref. [46]). A star indicates the excited state. (a) The picosecond change in the differential absorption spectrum, with the observation of isosbestic points, is attributed to the formation, delayed by the presence of the ion, of the intramolecular charge transfer state known for the free ligand [44]. (b) The time resolved subnanosecond red shift of the stimulated emission band, towards the wavelength range where that of the free ligand is expected, is interpreted by the motion of the cation away from the excited chromophore. (c) Once the excited state has decayed (2-ns lifetime), one observes the disappearance of the remaining free ligand in the ground state due to diffusion controlled recombination. This experiment has been done with two synchronized subpicosecond lasers, one as the pump, the other one for generating the continuum.

7.6 Concluding Remarks

The various examples of photoresponsive supramolecular systems that have been described in this chapter illustrate how these systems can be characterized by steady-state and time-resolved spectroscopic techniques based on either absorption or emission of light. Pertinent use of steady-state methods can provide important information in a simple way: stoichiometry and stability constant(s) of host–guest complexes, evidence for the existence of photoinduced processes such as electron transfer, energy transfer, excimer formation, etc. Investigation of the dynamics of these processes and characterization of reaction intermediates requires in most cases time-resolved techniques. Time-resolved fluorometry and transient absorption spectroscopy are frequently complementary, as illustrated by the study of photoinduced electron transfer processes. Time-resolved fluorometry is restricted to phenomena whose duration is of the same order of magnitude as the lifetime of the excited state of the fluorophores, whereas transient absorption spectroscopy allows one to monitor longer processes such as diffusion-controlled binding.

References and Notes

1 a) V. Balzani, F. Scandola, *Supramolecular Photochemistry*, Horwood, New York, **1990**. b), *Frontiers in Supramolecular Organic Chemistry* (H.-J. Schneider, H. Dürr, Eds.), VCH, Weinheim, **1991**.

2 a) A. P. de Silva, H. Q. N. Gunaratne, T. Gunnlaugsson, A. J. M. Huxley, C. P. McCoy, J. T. Rademacher, T. E. Rice, *Chem. Rev.* **1997**, *97*, 1515–1566; b) *Chemosensors of Ion and Molecule Recognition* (J. P. Desvergne, A. W. Czarnik, Eds.), Kluwer, Dordrecht, **1997**; c) B. Valeur, I. Leray, *Coord. Chem. Rev.* **2000**, *205*, 3–40; d) B. Valeur, *Molecular Fluorescence. Principles and Applications*, Wiley-VCH, Weinheim, **2002**, chap. 10.

3 a) A. P. de Silva, N. McClenaghan, in *Encyclopedia of Nanoscience and Nanotechnology* (J. A. Schwartz, C. Contescu and K. Putyera, Eds.), Dekker, New York, 2004, pp. 2749–2756. b) A. P. de Silva, N. D. McClenaghan, C. P. McCoy, in *Molecular Switches* (B. L. Feringa, Ed.) Wiley-VCH, New York, **2001**, pp. 339–361.

4 a) J. H. Alstrum-Acevedo, M. K. Brennaman, T. J. Meyer, *Inorganic Chemistry* **2005**, *44*, 6802–6827; b) T. A. Moore, A. L. Moore, D. Gust, *Advances in Photosynthesis*, **1999**, 8(Photochemistry of Carotenoids), 327–339; c) A. Harriman, *Photochemistry* **1998**, *29*, 425–452; d) L. Jullien, J. Canceill, B. Valeur, E. Bardez, J.-M. Lehn, *Angew. Chem. Int. Ed. Eng.* **1994**, *33*, 2438–2439; German version, **1994**, *106*, 2582.

5 In spite of these thermodynamic requirements, most of papers report equilibrium constants with dimensions for convenience (e.g. $dm^3\ mol^{-1}$ for the stability constant of a 1:1 complex or $mol\ dm^{-3}$ for its dissociation constant).

6 B. Valeur, *Molecular Fluorescence. Principles and Applications*, Wiley-VCH, Weinheim, **2002**, chap. 10, appendix A.

7 K. A. Connors, *Binding Constants. The Measurement of Molecular Complex Stability*, John Wiley & Sons, New York, **1987**.

8 H. Gampp, M. Maeder, C. J. Meyer, A. D. Zuberbühler, *Talanta* **1985**, *32*, 95–101.

9 R. Métivier, I. Leray, B. Valeur, *Chem. Commun.* **2003**, 996–997.

10 a) R. Métivier, I. Leray, B. Valeur, *Chem. Eur. J.* **2004**, *10*, 4480–4490. b) R. Métivier, I. Leray, B. Valeur, *Photochem. Photobiol. Sci.* **2004**, *3*, 374–380.

11 D. Marquis, J.-P. Desvergne, *Chem. Phys. Lett.* **1994**, *230*, 131–136.

12 I. Leray, Z. Asfari, J. Vicens, B. Valeur, *J. Chem. Soc., Perkin Trans. 2*, **2002**, 61429–61434.

13 Details on the mechanisms and theories of excitation energy transfer via dipole-dipole interaction (FRET: Förster resonance energy transfer) and via exchange interaction (Dexter's mechanism) can be found in B. Valeur, *Molecular Fluorescence. Principles and Applications*, Wiley-VCH, Weinheim, **2002**, chap. 4 and 9.

14 P. Choppinet, L. Jullien, B. Valeur, *Chem. Eur. J.* **1999**, *5*, 3666–3678.

15 M. N. Berberan-Santos, P. Choppinet, A. Fedorov, L. Jullien, B. Valeur, *J. Am. Chem. Soc.* **2000**, *122*, 11876–11886.

16 a) B. Valeur, *Molecular Fluorescence. Principles and Applications*, Wiley-VCH, Weinheim, **2002**, chap. 6. b) J. N. Demas, *Excited-state Lifetime Measurements*, Academic Press, New York, **1983**. c) D. V. O'Connor, D. Phillips, *Time-correlated Single Photon Counting*, Academic Press, London, **1984**. d) *Topics in Fluorescence Spectroscopy, Volume 1: Techniques* (J. R. Lakowicz, Ed.), Plenum Press, New York, **1991**. e) D. M. Jameson, E. Gratton, R. D. Hall, *Appl. Spectrosc. Rev.* **1984**, *20*, 55–106. e) J. R. Alcala, E. Gratton, D. M. Jameson, *Anal. Instrum.* **1985**, *14*, 225–250. f) J. R. Lakowicz, G. Laczko, I. Gryczynski, *Rev. Sci. Inst.* **1986**, *57*, 2499–2506.

17 A. K. Livesey, J. C. Brochon, *Biophys. J.* **1987**, *52*, 693–706.

18 A. Siemiarczuk, B. D. Wagner, W. R. Ware, *J. Phys. Chem.* **1990**, *94*, 1661–1666.

19 a) M. N. Berberan-Santos, E. N. Bodunov, B. Valeur, *Chem. Phys.* **2005**, *315*, 171–182; **2005**, *317*, 57–62. b) M. N. Berberan-Santos, B. Valeur, J. Lumin., **2006**, in press, doi:10.1016/j-jlumin.2006.07.04

20 M. E. El-Khouly, L. M. Rogers, M. E. Zandler, G. Suresh, M. Fujitsuka, O. Ito, F. D'Souza, *ChemPhysChem* **2003**, *4*, 474–481.

21 I. Leray, B. Valeur, D. Paul, E. Regnier, M. Koepf, J. A. Wytko, C. Boudon, J. Weiss, *Photochem. Photobiol. Sci.* **2005**, *4*, 280–286.

22 S. Akimoto, H. Nishizawa, T. Yamazaki, I. Yamazaki, Y. Hayashi, M. Fujimaki, K. Ichimura, *Chem. Phys. Lett.* **1997**, *276*, 405–410.

23 M. Born, E. Wolf, *Principles of Optics*, 7th ed., Cambridge University Press, **1999**.

24 B. Valeur, *Molecular Fluorescence. Principles and Applications*, Wiley-VCH, Weinheim, **2002**, chap. 5.

25 M. N. Berberan-Santos, in *New Trends in Fluorescence Spectroscopy. Applications to Chemical and Life Sciences* (B. Valeur and J.-C. Brochon, Eds.), Springer-Verlag, Berlin, **2001**.

26 M. N. Berberan-Santos, *J. Lumin.* **1991**, *50*, 83–87.

27 M. N. Berberan-Santos, B. Valeur, *J. Chem. Phys.* **1991**, *95*, 8048–8055.

28 R. Dobrawa, M. Lysetska, P. Ballester, M. Grüne, F. Würthner, *Macromolecules* **2005**, *38*, 1315–1325.

29 M. N. Berberan-Santos, P. Choppinet, A. Fedorov, L. Jullien, B. Valeur, *J. Am. Chem. Soc.* **1999**, *121*, 2526–2533.

30 R. G. W. Norrish, G. Porter, *Nature* **1949**, *164*, 658–658.

31 L. Lindqvist, *Arkiv Kemi* **1960**, *16*, 79–138.

32 G. Porter, M. R. Topp, *Nature* **1968**, *220*, 1228–1229.

33 G. R. Fleming, *Chemical Applications of Ultrafast Spectroscopy*, Oxford University Press, New York, **1986**.

34 *Femtosecond Laser Pulses. Principles and Experiments* (C. Rullière, Ed.), Springer-Verlag, Berlin, Heidelberg, New-York, **1998**.

35 *Femtosecond Lasers and Ultrafast Phenomena* (Y. H. Meyer, guest Ed.), *Rev. Phys. Appl.* **1987**, Special Issue *22*.

36 R. R. Alfano, S. L. Shapiro, *Phys. Rev. Lett.* **1970**, *24*, 592–594.
37 *Femtochemistry and Femtobiology. Ultrafast Events in Molecular Science* (M. M. Martin and J. T. Hynes, Eds.), Elsevier, Amsterdam, **2004**.
38 Y. H. Meyer and P. Plaza, *Chem. Phys.* **1995**, *200*, 235–243.
39 M. M. Martin, P. Plaza, N. Dai Hung, Y. H. Meyer, in *Ultrafast Phenomena VII* (C. B. Harris, E. P. Ippen, G. A. Mourou and A. H. Zewail, Eds.), *Springer Series in Chemical Physics*, Springer-Verlag, **1990**, *53*, 504–506.
40 N. P. Ernsting, S. A. Kovalenko, T. Senushkina, J. Saam, V. Farztdinov, *J. Phys. Chem.* **2001**, *105*, 3443–3453.
41 D. I. Schuster, S. MacMahon, D. M. Guldi, L. Echegoyen, S. E. Braslavsky, *Tetrahedron* **2006**, *62*, 1928–1936.
42 J. Bourson, B. Valeur, *J. Phys. Chem.* **1989**, *93*, 3871–3876.
43 M. M. Martin, P. Plaza, N. D. Hung, Y. H. Meyer, J. Bourson, B. Valeur, *Chem. Phys. Lett.* **1993**, *202*, 425–430.
44 M. M. Martin, P. Plaza, Y. H. Meyer, *Chem. Phys.* **1995**, *192*, 367–377.
45 M. M. Martin, P. Plaza, Y. H. Meyer, F. Badaoui, J. Bourson, J. P. Lefevre, B. Valeur, *J. Phys. Chem.* **1996**, *100*, 6879–6888.
46 P. Plaza, I. Leray, P. Changenet-Barret, M. Martin, B. Valeur, *ChemPhysChem* **2002**, *3*, 668–674.
47 P. Dumon, G. Jonusauskas, F. Dupuy, P. Pée, C. Rullière, J. F. Létard, R. Lapouyade, *J. Phys. Chem.* **1994**, *98*, 10391–1039.

8
Circular Dichroism Spectroscopy

Marie Urbanová and Petr Maloň

8.1
Basic Considerations

8.1.1
Circular Dichroism

Chiroptical methods, circular dichroism (CD) spectroscopy and optical rotatory dispersion (observed as optical rotation if the experiment is carried out at a single wavelength) are chiral variants of absorption spectroscopy and refractive index measurement. Contrary to parent non-chiral spectroscopies that use unpolarized electromagnetic radiation, they substantially expand information obtained by these conventional spectroscopic techniques at the cost of being applicable only to chiral samples. While parent spectroscopies have evolved over the years into general analytical tools exploring chemical, electronic and vibrational structure of molecular systems, chiroptical spectroscopies have specialized in the studies of the three dimensional state of matter, which they can probe with great sensitivity and given the current variety of chiroptical procedures, with reasonable selectivity as well.

Chiral objects absorb left and right circularly polarized light to slightly different extents. This phenomenon of *circular dichroism* [1] became the basis of the most widespread practical chiroptical method in the past few decades. There are some other chiroptical methods based on the interaction of chiral matter with circularly polarized light: *Optical rotatory dispersion* is based on the analogous difference in refraction. *Raman optical activity* measures differences in scattered light, and *circularly polarized luminescence* deals with the difference in emission.

In chiroptical spectroscopies, analogous quantities are used as in the parent spectroscopies. CD spectroscopy measures differences between absorptions for left and right circularly polarized light. The quantities used in absorption spectroscopy are therefore replaced by their differences in CD spectroscopy. Table 8.1 summarizes definitions, relations and units used in circular dichroism spectroscopy. In addition, linearly polarized light is converted into elliptically polarized light [2]. The corresponding quantity, ellipticity, $\theta = \arctan(b/a)$, where b and a are the lengths of the semiminor and semimajor ellipse axes, correlates with the differ-

Analytical Methods in Supramolecular Chemistry. Edited by Christoph Schalley

ISBN: 978-3-527-31505-5

Tab. 8.1. Quantities and units used in conventional and chiroptical spectroscopy. The spectral dependence of these quantities can be expressed as a function of wavelength λ, frequency ν, or wave-number $\tilde{\nu}$.

Name of quantity	Definition and relations	Units
Absorbance[a]	$A(\nu) = \log_{10}(I_0/I)$, I_0 is the intensity of light entering the cell, I is the intensity of light leaving the cell	dimensionless
Molar absorptivity, molar absortion coefficient	$\varepsilon(\nu) = \dfrac{A(\nu)}{cl}$, c is the molar (amount) concentration [mol L^{-1}], l is the pathlength [cm]	cm^{-1} L mol^{-1} $= 10^3$ cm^2 mol^{-1} $= 10^{-1}$ m^2 mol^{-1}
Circular dichroism[a]	$\Delta A(\nu) = A_L(\nu) - A_R(\nu)$, $A_L(\nu)$ and $A_R(\nu)$ are absorbances for left and right circularly polarized light	dimensionless
Molar circular dichroism	$\Delta\varepsilon(\nu) = \dfrac{\Delta A(\nu)}{cl}$, c is the molar (amount) concentration [mol L^{-1}], l is the pathlength [cm]	cm^{-1} L mol^{-1} $= 10^3$ cm^2 mol^{-1} $= 10^{-1}$ m^2 mol^{-1}
Ellipticity[a,b]	$\theta(\nu) = 32980\Delta A$	mdeg
Molar ellipticity[b]	$[\Theta](\nu) = \dfrac{\theta(\nu)}{10cl}$, $\theta(\nu)$ is elipticity in mdeg, l is the pathlength [cm], c is the molar (amount) concentration [mol L^{-1}], $[\Theta](\nu) = 3298\Delta\varepsilon$	deg cm^2 $dmol^{-1}$

[a] Used particularly when the concentration is not known.
[b] CD instruments are sometimes calibrated in θ, but all commercially available instruments record ΔA.

ence in absorption and represents another possibility how to express circular dichroism as a physical quantity. The relevant correlation is also given in Table 8.1. For merely traditional reasons, chiroptical data for small organic molecules are usually expressed as differences in absorption, while data describing the electronic circular dichroism of biopolymers are often given in ellipticities.

What was said about CD spectroscopy is in principle valid for all spectral regions used. However, the CD signal is 3 to 5 orders of magnitude weaker than the absorption signal in parent absorption. Consequently, CD spectra cannot be obtained directly by measuring two spectra at alternately left and right circular polarization of incident light and just taking their difference, but more specialized instruments are required. So far, CD instruments are commercially available for two spectral

Wavelength	200 nm	400 nm	800 nm	1 μm	10 μm
Wavenumber	5 x 10^4			1 x 10^4	1 x 10^3 /cm^{-1}
Spectral region	UV	VIS		IR	
Transitions	Electronic			Vibrational	
Spectroscopy	Electronic circular dichroism ECD			Vibrational circular dichroism VCD	
Signal observed for	UV-VIS chromophores e.g. aromatics			all compounds	

Fig. 8.1. Electronic and vibrational circular dichroism.

regions only, ultraviolet-visible (UV/vis) and infrared (IR). CD measured in the UV/vis region is called *electronic CD* (ECD or just CD), because electronic transitions are involved, CD measured in the IR region is called *vibrational* CD (VCD), because it refers to vibrational transitions (Fig. 8.1). Similarly, the pros and cons specific for parent spectroscopies are also valid for CD spectroscopies.

The result of a typical circular dichroism experiment is a circular dichroism spectrum. It consists of dichroic bands in a manner that is exactly analogous to absorption bands in the absorption spectrum. The individual absorption and CD bands are characterized by positions of their maxima ($\lambda_{\max}, \tilde{\nu}_{\max}$) and the maximum intensities ($A_{\max}, \Delta A_{\max}, \varepsilon_{\max}, \Delta\varepsilon_{\max}$). If we assume a suitable mathematical form for the band shape (usually Gaussian or Lorentzian), they are also characterized by a band width parameter ($\Delta\lambda, \Delta\tilde{\nu}$). However, there is a significant difference: the differential nature of the CD spectroscopy results in the possibility that some bands have negative signs, dichroic bands can be negative or positive depending on the absolute configuration of the studied sample. Bands of different signs can overlap within a single spectrum and consequently, the sum curve need not necessarily represent unequivocal information on all the individual bands present. For comparison with theoretical quantities, a suitable numerical parameter exists analogously in absorption and circular dichroism spectroscopies. It is called dipole strength D [C^2m^2] in absorption or optical rotational strength R [$CmJT^{-1} = C^2m^{-3}s^{-1}$] in circular dichroism and can be obtained experimentally by integrating the area under the corresponding band:

$$D = \frac{3 \ln 10 c \varepsilon_0 \hbar}{\pi N_A} \int \frac{\varepsilon(\nu)}{10\nu} d\nu = 1.022 \times 10^{-61} \int \frac{\varepsilon(\nu)}{\nu} d\nu \qquad (8.1)$$

$$R = \frac{3 \ln 10 c^2 \varepsilon_0 \hbar}{4\pi N_A} \int \frac{\varepsilon_L(\nu) - \varepsilon_R(\nu)}{10\nu} d\nu = 7.659 \times 10^{-54} \int \frac{\Delta\varepsilon(\nu)}{\nu} d\nu \qquad (8.2)$$

where $\varepsilon_0 = 8.854 \times 10^{-12}\ kg^{-1}m^{-3}s^2C^2$ is the permittivity of vacuum, N_A is Avogadro's number, c is the speed of light, $\hbar$ is the Planck constant divided by 2π, and ε is molar absorptivity given in the first unit listed in Table 8.1.

The ratio of the circular dichroism signal and the corresponding absorption is called *dissymmetry factor g*:

$$g = \frac{\Delta\varepsilon}{\varepsilon} = \frac{\Delta A}{A} \tag{8.3}$$

The dissymmetry factor g can be expressed using dipole and rotational strengths. When SI units are used:

$$g = \frac{4R}{cD} \tag{8.4}$$

8.1.2
Variants of Chiroptical Methods

A conceptually simple chiroptical experiment is to measure the optical rotation of the sample, that is, the angle of rotation of the plane of polarization of the linearly polarized light beam when passing through the optically active medium. This angle is a characteristic quantity and when expressed as a molar property at a suitable wavelength (usually the sodium D line), it is an important physical constant characterizing a particular chiral material. Attempts were made to interpret this quantity in structural terms and it has been also calculated by *ab initio* procedures [3–6]. Optical rotation, when measured in a range of wavelengths, becomes a curve of optical rotatory dispersion (ORD) and can also be calculated by *ab initio* procedures [4, 7–9]. For some time, it was primarily used for measuring chiroptical properties, but nowadays, it is largely abandoned and replaced by circular dichroism spectra. These spectra are easier to measure, their simpler shape is easier to analyze and they are mathematically equivalent (CD and ORD spectra can be transformed from one form into another using the Kramers–Kronig relationship; for a recent application, see [10]). Circular dichroism, either electronic or vibrational, is an absorption phenomenon and unless special precautions are taken, it provides information about three-dimensional structure of the system in its ground state. It can be modified (in a very simple manner, just by altering the detection system) to detect circular dichroism in fluorescence (FDCD) [11, 12]. Although the description seems to be different, these altered methods record exactly the same quantity: circular dichroism related to the sample in the ground state.

It is, however, possible to measure chiroptical properties related to the excited states of molecules – circularly polarized luminescence (CPL) [13]. In this case, the sample is excited by unpolarized radiation and the luminescence signal is analyzed by a circular analyzer. Several variants of this concept have been developed, but the applications are targeted more on electronic structure of the excited states of molecules than to their geometries.

Chiroptical properties related to molecular vibrations can be studied not only by vibrational circular dichroism, but also by using the chiral variant of the Raman spectroscopy – Raman optical activity (ROA) [14–16]. This method has been developed into practical use only recently, but it is very promising and similarly to the parent spectroscopies, seems to excel where VCD fails, as, for example, in the regions of low wavenumbers or aqueous solutions [17–23]. There are several arrangements of ROA experiment according to whether we induce ROA with circularly polarized incident light (incident circular polarization – ICP) or analyze the scattered light for circularly polarized components (scattered circular polarization – SCP) or both (double circular polarization – DCP) or according to scattering geometry (forward scattering, right angle scattering, backward scattering). At present, the method of choice seems to be ICP collected in the back scattering geometry.

8.1.3 Advantages and Limits of Circular Dichroism Spectroscopies

8.1.3.1 Chiral and Parent Non-chiral Spectroscopies

Chiroptical spectroscopies are based on the concept of chirality, the signals are exactly zero for non-chiral samples. In terms of molecular symmetry, this means that the studied system must not contain a rotation–reflection axis of symmetry. This lapidary definition implies that the more known symmetry elements (symmetry plane – equivalent to the one-fold rotation–reflection axis and the center of symmetry – equivalent to the two fold rotation–reflection axis) must also be absent and that the system must be able to exist at least formally in two mirror image-like forms. At first glance this limitation seems to be a disadvantage, however, this direct relation to molecular geometry gives chiroptical properties their enormous sensitivity to even minor and detailed changes in the three-dimensional structure. This property is absent in the parent non-chiral spectroscopies. Chiroptical methods sometimes provide enhanced resolution, because of the simple fact that dichroic bands can be positive and negative. Chiral spectroscopies give also a new dimension to the intensity parameter. The information about structure is also encoded in the sign, the absolute value and the width of spectral bands. Not only the positions of bands, but also the entire shape of the spectral pattern carries structural information on the sample. While parent spectroscopies are more oriented toward the positions of the spectral bands, chiroptical spectroscopies are primarily intensity oriented, although band positions are just as important as in the parent methods. Chiroptical spectroscopies can draw on substantial knowledge on electronic and vibrational molecular transitions that has been collected throughout the years of analytical use of the parent spectroscopies.

8.1.3.2 Electronic and Vibrational Circular Dichroism

If we mutually compare chiroptical methods, we find a remarkable complementarity that can be used as advantage when a particular system is investigated by the selection of chiroptical procedures. ECD [1, 24], which is a far reaching technique, tends to see a molecule as a whole and provides chiral information via chromophores and their properties [25–27]. This far-reaching nature of ECD can be utilized, for example, in supramolecular chemistry, where chirality is introduced into the system by a chiral, but spectroscopically neutral matrix. However, it is observed on an inherently nonchiral chromophore, like, for example, porphyrine [28–36] and others that have been reviewed recently [37]. This is possible because the entire system becomes chiral according to the matrix used. VCD is a more local phenomenon, although if there are strong coupled oscillators, then dipole coupling can also give rise to relatively far reaching interactions. But as a rule, each bond in the molecule is a vibrational chromophore and from VCD, we can get unbelievably detailed and local information (see section 8.5). The same becomes even more enhanced for Raman optical activity. These differences are nicely demonstrated by the dependence of chiroptical spectra on the chain length that was measured for regular polypeptide structures [38].

8.1.3.3 **Instrumentation**

Commercial instrumentation is available for both the ECD and VCD experiments and the commercial ROA spectrometer was introduced last year. To measure chiroptical spectra is relatively easy, provided that a good laboratory routine is worked out and consistently maintained. Electronic CD is easily measurable from ~175 nm to more than 1 μm (see Section 8.2.1), and custom-built instruments can go even further. At present, VCD can be measured from 4000 to about 900 cm^{-1}, routinely from 2000 cm^{-1} to about 900 cm^{-1} (see Section 8.2.2); measurements were reported below 600 cm^{-1} [39] and in the near IR up to 6150 cm^{-1} [40, 41]. The ROA spectra can be reasonably obtained from 2000 cm^{-1} up to a region close to the Rayleigh line [42].

8.1.3.4 **Calculations**

All these methods can be compared with the theoretical results obtained by direct *ab initio* calculations. Here, VCD has a distinct advantage because it usually refers to molecules in their well-defined electronic ground states. For small and medium-sized organic molecules, the precision of these calculations makes it possible to use them as an alternative method for the determination of absolute configurations [43–45]. For larger molecules, the method of transfer of molecular properties tensors from smaller molecular fragments can be used [46–48]. ROA can be calculated in a similar manner, but the computations are more demanding and therefore still limited to relatively small molecules [49, 50]. The *ab initio* calculations of the ECD spectra require wavefunction calculations of all electronic excited states participating in the ECD spectrum [51]. Although ECD is easier to measure compared to VCD, the *ab initio* calculations are a promising but more demanding concept. They represent recent contributions to our tools based on the procedures of the time dependent density functional theory (TDDFT) [4, 52–55].

8.2
Measurement Techniques (Methodology of CD Measurement)

CD spectra cannot be obtained from spectral responses to left and right circularly polarized light measured in series and subtraction of the two corresponding signals. However, this concept is implemented in the Raman optical activity measurement.

In the ECD and VCD measurements, where the difference in measured responses is 3 to 5 orders of magnitude smaller than individual signals, such serial procedure would require high stability of spectrometers that is by far not practically achievable. Therefore, the responses to left and right circularly polarized light are obtained synchronously and the difference is obtained directly using the most important part of a CD spectrometer – the photoelastic modulator (PEM), where the linearly polarized light achieved in the polarizer is converted alternately to the left and right circularly polarized light in a periodical manner with the frequency of tenths of kHz typically. In this apparatus, the piezoelectrically driven mechanical stress applied on the appropriate optical material causes periodic retardation of the beam from $-\lambda/4$ to $+\lambda/4$. At the output, the retardation produces alternately the left and right circularly polarized light at the working frequency of the PEM. The light with periodically varied state of polarization (but not the intensity) is introduced to sample and the difference in the two responses is detected with a lock-in amplifier at the PEM frequency. This principle of synchronous measurement is generally used in both spectral regions with the commercially available and laboratory made dispersive or Fourier transform CD spectrometers. In order to convert a raw chiroptical signal observed at the PEM frequency to a true CD signal, it is necessary to normalize it with a general transmission signal. This can be achieved by using several methods. It should be noted that the procedure is quite critical, because significant normalization errors can give rise to experimental artifacts that may be comparable in size to the true CD signal. In ECD, this is achieved by fine tuning the photomultiplier sensitivity, thus making the transmission signal constant throughout the wavelength range. In VCD, either the same principle is followed by analog electronic normalization of the PEM signal by the transmission signal (such a procedure is typical for dispersive instruments measuring the spectrum by a scanning procedure) or by digitizing signals separately and carrying out the normalization in a computer.

When performing CD experiments, one must consider that the range of acceptable conditions is more specific and narrower than in the conventional parent spectroscopies. The spectral region available in a particular arrangement of experiment is usually narrower than in the corresponding parent spectroscopies. This is caused by the material transparency, its mechanical properties and the narrow spectral sensitivities of cooled semiconductor detectors. The conventional absorption must be measurable very well to obtain acceptable and reliable CD spectra. This means that usual requirements for homogeneity, stability and transparency must be met. In addition, the absorption must neither be too high nor too small (the optimum sample absorbance is in a range of 0.6–0.8). Too high absorption implies the lack of incident light on the detector and sizable artifacts. On the contrary, too small absorption means that the noise level increases substantially. There are other adverse effects such as birefringence and unwanted polarization changes caused by reflections in the light path through all the optical elements in the spectrometer. These phenomena also arise within the cell windows and in the sample material. These effects are spectrally dependent and can cause false signals observed as artifacts.

Adhering to these rather strict requirements for optimal optical density, spectral range and minimizing spectral artifacts requires permanent verifications of the reality of the CD signal measured. Good CD results require formulation and keeping of the measurement protocol. It is recommended that the whole series of samples that are to be compared is measured under the same conditions. For example, a broken sample cell can distort the whole series.

Despite the above general features that are common to the both ECD and VCD, some attributes are characteristic of each CD technique.

8.2.1 Electronic Circular Dichroism Measurements

ECD spectra are measured with dispersive scanning instruments that are based on the general design of single beam absorption spectrometers. For CD measurements, the instrument is optimized for high light throughput and equipped with polarization optics and specialized detection systems. Polarization optics may be combined with the dispersive function, that is, the instrument can utilize a polarizing prism double monochromator with prisms made of a quartz single crystal. Alternating circular polarization of light is then achieved by the PEM.

Spectra are usually measured in solvents which are used in UV spectroscopy. Depending on the required polarity, we can use cyclohexane or another suitable highly purified hydrocarbon, acetonitrile, alcohols, certain ethers like 1,4-dioxane, water, various buffers and fluorinated alcohols (TFE, HFP). Special solvent mixtures are used for measurements at cryogenic temperatures. Due to the required high sensitivity, it is necessary to achieve a very stable and preferably flat baseline of the instrument. This is dependent even on particular sample cells. Usually, quartz cells with low birefringence windows are selected for CD measurements. Attention should be paid to avoid mechanical and temperature impacts on them as these can alter the baseline and invalidate sensitive experiments. Two kinds of cells are used: demountable (10–500 μm) and a fixed design. High stability can be achieved but this requires experience, consistency and attention to detail. Temperature variations should be minimized during the experiment.

8.2.2 Vibrational Circular Dichroism Measurements

Unlike ECD spectrometers that are built only as dispersive devices, a VCD spectrometer may be designed either as a scanning dispersive instrument or as a Fourier-transform (FT) instrument [56–59]. The latter instrument has well-known advantages, that is, multiplex (an interferometer does not separate light into individual frequencies before measurement), high throughput (more energy gets to the sample due to the simple optical path of the interferometer with no slits and fewer optical elements), and resolution (resolution is increased by lengthening the moving mirror stroke length with no decrease in energy throughput). The advantages result in much shorter scan times. All VCD spectrometers commercially available

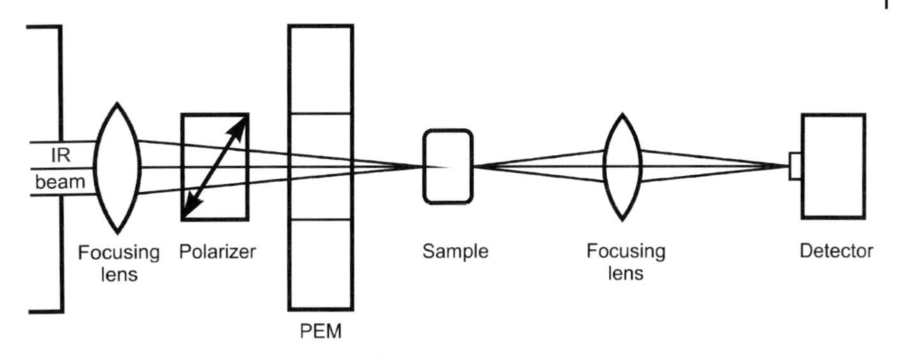

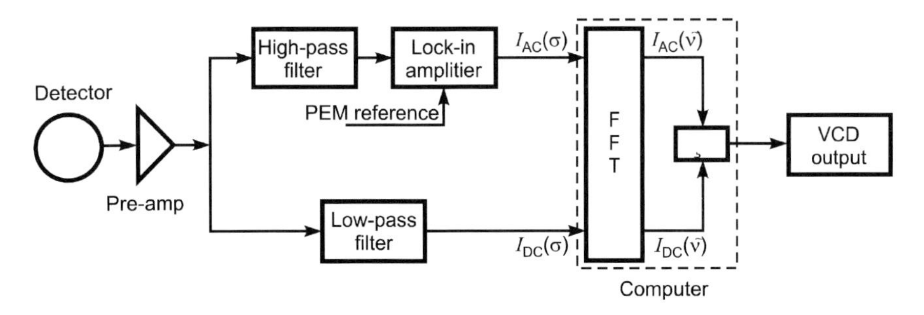

Fig. 8.2. Block diagram of optical and electronic layout of the VCD part of Fourier transform spectrometer.

since 1997 use the FT approach. Some of the commercially available FT–VCD instruments are VCD dedicated; others utilize a conventional FT IR spectrometer equipped with a VCD module. The scheme of the optical and electronic layout of both types of the FT–VCD instruments is in principle the same and is given in Fig. 8.2. Although optical and electronic signals are processed fully automatically in commercially available VCD instruments, a short explanation is given here to elucidate the advantages and limitations of the VCD.

The first part of the optical layout consists of a light source and a Michelson interferometer. The IR light beam is first modulated in the Michelson interferometer and then directed to the CD dedicated branch, where it passes through a linear polarizer and the PEM. The optical low-pass filter prevents the detector from being overloaded by a wide interval of FT frequencies. The combination of the linear polarizer with a PEM creates left and right circularly polarized light alternately at the PEM operating frequency that is much higher than the range of FT frequencies. At this moment, the beam is doubly modulated – with the interferometer frequencies (intensity modulation) and the PEM frequencies (polarization modulation). When passing through the sample, the polarization modulation of the beam is converted into the intensity modulation at the PEM frequency. The beam is then focused on the fast and sensitive, cooled semiconductor detector. In the electronic part, the signal from the detector is amplified and split into two paths. Signal in one path is Fourier transformed directly after passing through the low-pass filter. The resulting $I_{DC}(\tilde{\nu})$ gives the transmission spectrum. The signal in the other path after passing through a high-pass filter is directed into a lock-in amplifier that is tuned to the operating frequency of the PEM. The lock-in selectively amplifies the weak difference signal due to CD of the sample and that is then Fourier transformed into $I_{AC}(\tilde{\nu})$. The ratio $I_{AC}(\tilde{\nu})/I_{DC}(\tilde{\nu})$ is proportional to ΔA and is further processed to obtain the true VCD spectrum.

The spectral range accessible to FT–VCD using typical optical layout is somewhat limited due to the used materials (ZnSe PEM and ZnSe transmission optics and MCT detectors), together with the requirements for a high light throughput and nearly perfect optical alignment. FT–VCD is easiest to record in the mid-IR region (2000–800 cm^{-1}), which is coincidentally a very useful range where rich spectra of quite well resolved absorption and VCD bands are observed. Above 2000 cm^{-1}, the CaF_2 PEM and InSb detectors can be used, but it is more difficult to measure the FT–VCD spectra in this region.

FTIR spectrometers can record non-polarized absorption intensities over a wide dynamic range. For VCD, however, the available range of the total absorbance of measured solutions is only in the 0.1–1.0 range. The optimum absorbance value for the VCD measurements is about 0.4–0.6. This value has to be achieved by the optimal combination of the concentration, pathlength, and solvent. Practically, pathlengths from about 5 μm to several millimeters are used for the VCD measurements. If the sample gives absorption bands of very different intensities in the spectral region of interest, the experiment has to be repeated at different concentration/pathlength combinations to obtain complete VCD results. The solvents frequently used in the mid-IR region and their spectral windows are given in Table 8.2.

As a consequence, VCD measurements require rather high concentrations ~0.05–1 mol L^{-1}. However, due to a short pathlength, only small volumes of sam-

Tab. 8.2. IR solvents appropriate for VCD spectroscopy in the mid IR region and their spectral windows.

Solvent	Spectral window (cm^{-1})	Solvent	Spectral window (cm^{-1})
CCl_4	2000–850	DMSO-d_6	2000–1100; 970–700
CS_2	2000–1640; 1350–700	Methanol-d_6	2000–1200
$CHCl_3$	2000–1260; 1175–700	H_2O[a]	2000–1770; 1525–1000
$CDCl_3$	2000–975	D_2O	2000–1300; 1100–800[a]
Propanol	2000–1500	trifluorethanol	2000–1500

[a] only when a short pathlength < 6 μm and higher concentrations are used.

ple are needed. This leads to the requirement for ~1–15 mg of sample for a single VCD experiment, which is comparable with the requirements for NMR measurements. Depending on the solvent used, samples may be recoverable.

Alternatively, the dispersive VCD instrumentation can be used because in some cases the advantages of the FTIR technique, that is multiplex, throughput, and resolution, cannot be fully utilized. This is the case, for example, of very weak and broad signals or difficult solvents with strong absorption and only narrow spectral windows. In such cases, the resolution and multiplex advantages of FTIR are hardly utilized. Dispersive instruments are also used in the C–H, O–H region (3500–2000 cm^{-1}) [60]. The structural study of biomolecules and biopolymers such as sugars, peptides, proteins, oligonucleotides, DNA can serve as an example of the very successful use of dispersive VCD (for reviews see [38, 61–64]). In these studies, solutions in H_2O or D_2O are used. Because of the rather unfavorable absorption of water, only narrow spectral range is accessible for VCD measurements; moreover, the spectral bands are very weak and broad in this case. Some laboratories measure VCD with dispersive built in-house instruments and obtain remarkable results [65, 66]. The dispersive VCD instruments can be viewed as an extension of the ECD instrumentation described in Section 8.2.1 from the UV/vis to IR region. There are some differences between dispersive instruments used in the two spectral regions: Because of the transmission and dispersive problems, CD spectrometers in the IR region use mirrors instead of lenses. Unlike the ECD dispersive spectrometers that use prism monochromators, the dispersive VCD technique is based on the technology of gratings. Semiconductor cooled detectors are used instead of photomultipliers.

8.3 Processing of Circular Dichroism Spectra

Because CD is a single beam measurement, processing of CD spectra requires a baseline correction and intensity calibration, similar to other single beam techniques. When ECD is scanned with a commercially available spectrometer, the baseline is obtained as the CD spectrum of the corresponding (non-chiral) solvent measured in the same cell, using exactly the same experimental parameters as were used for the solution of the measured sample. In the majority of cases, when ECD does not have an extremely small intensity, the ECD baseline obtained as the solvent ECD is effectively very close to a zero line. Usually, the ECD intensity calibration should be checked at intervals of some months. The ECD instruments are calibrated by using standard samples (epiandrosterone or 10-camphorsulfonic acid). Current spectrometers are equipped with software that enables convenient baseline correction, unit conversion, averaging, and other simple data treatment. Multiple scanning and subsequent averaging is a common processing procedure applied to ECD and VCD measured with dispersive spectrometers. Averaging of N scans improves the signal/noise (S/N) ratio by a factor of $\sqrt{N}$. Weak signals require more accumulation than the strong ones.

8.3.1 Intensity Calibration in VCD Spectroscopy

VCD spectroscopy requires a few special procedures. The intensity calibration of a dispersive or FT–VCD spectrometer is carried out by a pseudosample that consists from a multiple quarter wave plate followed by a polarizer. Such a combination produces a large artificial pseudo CD signal of $\Delta A = \pm 1$ at some spectral points (Fig. 8.3). The theory of operation of this device is described in the literature [56, 57, 67]. Interpolation at all other wavenumbers gives a calibration curve that is used for the intensity calibration of spectra measured under the same conditions. With the software of the current FT–VCD instruments, the intensity calibration becomes an easy routine. The pseudosample measurement is carried out prior to the measurement of the sample and is then used by the software. This is an absolute calibration method based on first physical principles, which is its main advantage. In addition, from time to time, it is recommended that the calibration of the VCD instrument is verified by VCD measurements of samples whose VCD spectra have been carefully measured in several laboratories and whose proper shape and intensity have been published and are well known. The VCD spectra of these standards are known with a precision of a few percent and are generally accepted. For this purpose, (*S*)-(−)-α-pinene and (*R*)-(+)-α-pinene as neat liquids are used. Another appropriate standard also used in the region of the X–H stretching vibrations above 2000 cm^{-1} is the CCl_4 solution of camphor. The D_2O solution of hemoglobin serves as a VCD standard for VCD measurements of aqueous solutions of samples providing a very week signal in the mid-IR region and primarily in the carbonyl region.

Another purpose of the measurement of the pseudosample is that it provides a phase correction needed for Fourier transform of the AC interferogram. The self-correction conventionally used in Fourier transformation for one-sign absorption

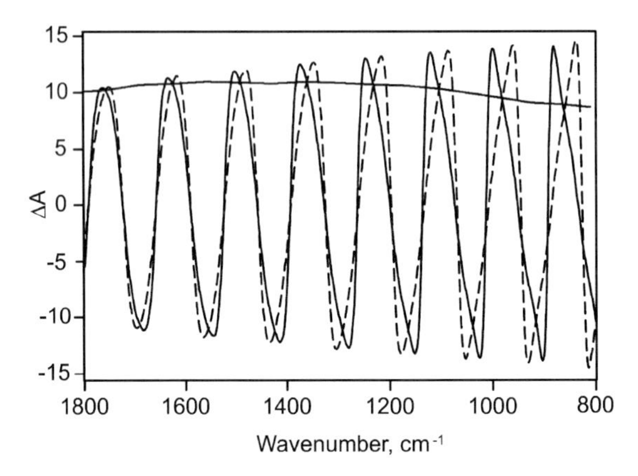

Fig. 8.3. The pseudo CD signal of calibration device. The cross-points have a CD signal of $\Delta A = \pm 1$.

signals causes difficulties if the VCD signal changes its sign. The processing procedure described above is usually included in the standard software supplied with the VCD spectrometer and can be easily used.

8.3.2 Baseline Corrections and Reliability in VCD

VCD is even more sensitive than ECD to any polarization distortion that can happen in a spectrometer during the light beam throughput. Birefringence in the optical material of optical lenses, filters, cells and other windows, and reflections may cause artifacts that may be comparable in size to the sample signals. As a consequence, the baseline differs from the zero line and a baseline correction should be applied for each VCD measurement. In the case of well optically tuned instruments, satisfactory baseline can be obtained as the VCD spectrum of the pure solvent measured under the same condition in the same sample cell [44, 45, 68].

Another source of artifacts is related to sample absorption, because in the region of the absorption band, anomalous dispersion takes place that may distort the equilibrium of the responses to right and left circularly polarized light, primarily in the case of a narrow sharp absorption band. The optimal baseline is ideally obtained by using a racemic material of the same absorption characteristics as the measured sample. Practically, racemic materials are rarely available, especially for biological samples like proteins, DNA and so forth. Should there be any doubt about the correctness of the VCD signal, it is necessary to check its reliability, for example, by recording a concentration-dependent series. Recently, a new method of reducing the artifacts has been announced [69] that uses a dual-PEM spectrometer.

To deal effectively with the weak nature of the VCD signals and the danger of possible artifacts, it is necessary to establish a reliable routine of permanent testing and documenting. The FT–VCD spectra are usually run in blocks that accumulate several thousands of interferograms. The number of blocks is determined by the signal intensity and the desired S/N level. Individual subsequent blocks can be tested for consistency and stability. In addition, in the case of a random or undesired event, some of the data blocks may be excluded but the rest can still be utilized. The measurement systems must be stable enough so that the sample and baseline are measured under strictly identical conditions. The subsequent subtraction is possible only if these requirements are met. In the VCD literature, noise curves usually accompany the published spectra. A noise curve is calculated as follows: The single blocks of raw VCD spectra are split into two groups. Then a half of their differences gives the noise spectrum and their averaging gives the mean VCD.

8.3.3 Advanced Processing of Circular Dichroism Spectra

In chiroptical spectroscopy, an important part of structural information is provided by the intensity parameters of the spectrum and also by its overall shape. Primarily

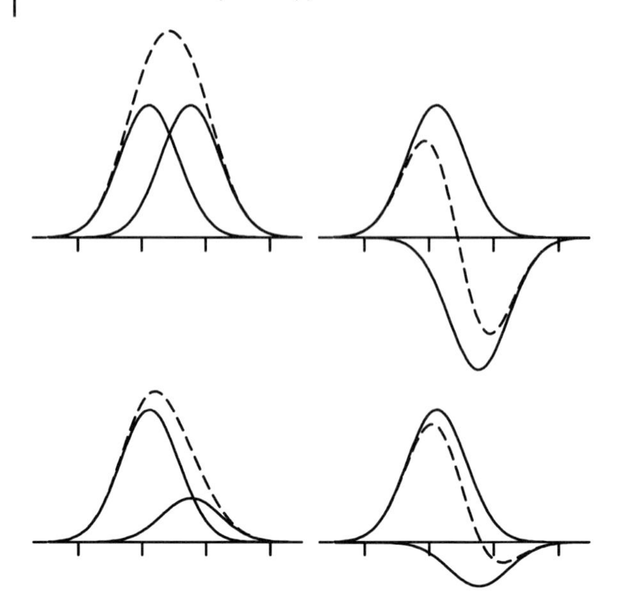

Fig. 8.4. Superposition of CD spectral bands of the same and mutually opposite signs.

in ECD, spectral bands are broadened by collisions in solutions, conformational equilibria and vibronic fine structures. The bands themselves may overlap quite strongly and we only observe their sum curves. This problem is typically more serious in chiroptical spectroscopy than in isotropic absorption. It is quite possible that neighboring bands in chiroptical spectra happen to be of mutually opposite signs and, in the extreme case, they can effectively cancel each other or at least seriously distort the resulting sum curve (Fig. 8.4). Thus, chiroptical spectra are more difficult to resolve into their components than similar absorption spectra, in which the component bands are always positive. Moreover, the result of a chiroptical experiment is seldom represented by a single curve. We usually measure series of spectra where particular components differ somewhat from one another as some structure related parameter is modified. Typical examples are changes in the chemical structure of the sample, solvent changes, pH changes, temperature changes, and so forth. For further analysis, it is therefore necessary to investigate the shape of individual spectra and extract systematic trends in order to distinguish them from random events, like experimental noise or to recognize spurious signals – artifacts.

In ECD or in limited spectral regions of VCD (for example amide I bands in peptides, C–H vibrations, etc.), it is possible to represent the sum curve by the summation of several Gaussian or Lorentzian bands and to adjust their parameters (central wavelength or wavenumber, intensity of the maximum and the band half-width) to experimental curve by a general curve fitting procedure as provided by several commercial computer programs like Spectracalc or Grams. However this procedure is quite tedious and highly uncertain, because of many statistical uncertainties and because various band parameters influence the sum curve with different sensitivity. In general, the curve is mostly influenced by the band width parameter. Unfortunately, the band width is the least informative of the three. The fitting procedure requires very careful weighing and additional information, like the number of component bands to be expected, their band shapes in the corresponding absorption spectra, and so forth. Fitting Gaussian or Lorentzian components to CD spectra provides experimental values of rotatory strengths that can be compared with theory and calculations [70, 71].

In vibrational optical activity, this comparison is usually carried out in a reversed manner. A common plausible value, for example, 10 cm^{-1}, is tentatively assigned to bandwidths, and the theoretical spectrum is simulated using calculated band positions and rotatory strengths or ROA intensities. The resulting theoretical VCD or ROA curve is then compared with the experiment and the difficult curve fitting is therefore avoided (Fig. 8.5).

The analysis of a series of chiroptical spectra and recovery of systematic trends in a given set can be carried out in several ways. In the past, the results strongly depended on the spectroscopist's personal experience; actually, this was the least objective part of the circular dichroism application. Nowadays, we can rely on general procedures of statistical data treatment like singular value decomposition, factor analysis (especially its first part, analysis of the correlation matrix and the projection of the experimental spectra onto the space of orthogonal components), cluster analysis and the use of neural networks. This field has been pioneered by Pancoska and Keiderling [72–76], and also by Johnson [77] when analyzing the chiroptical properties of biopolymers.

Experimental CD data, primarily when measured under difficult conditions, may appear quite noisy and it is tempting to smooth them for further analysis or presentation purposes. In general, we advise against such practices, although spectral software programs involve such procedures. CD analysis is heavily dependent on the shape of the spectra and any additional treatment tends to distort it. It is better to present data as they actually are.

8.4 Theory

8.4.1 Rotational Strength

Because circular dichroism is a difference in absorption for left and right circularly polarized light, its theoretical description includes subtraction of the transition probabilities induced by left and right circularly polarized radiation. The interaction Hamiltonian H^{INT} that determines transition probability includes electric, $\boldsymbol{E}$, and magnetic, $\boldsymbol{B}$, fields of electromagnetic circularly polarized radiation, and the electric, $\boldsymbol{\mu}$, and magnetic, $\boldsymbol{m}$, dipole moments of the molecule.

$$H^{INT} = -\boldsymbol{\mu}\boldsymbol{E} - \boldsymbol{m}\boldsymbol{B} + \text{higher orders} \tag{8.5}$$

The resulting relation for rotational strength R is called Rosenfeld equation and has been derived in 1928 [78]:

$$R = -\mathrm{Im}\{\boldsymbol{\mu}^{if} \cdot \boldsymbol{m}^{if}\} \tag{8.6}$$

where Im means the imaginary part of the scalar product in parentheses, vectors $\boldsymbol{\mu}^{if}$ and $\boldsymbol{m}^{if}$ are the respective electric and magnetic dipole transition moments that correspond to a transition from state i to f. We can imagine that the vector $\boldsymbol{\mu}^{if}$, the electric dipole transition moment, describes a shift of a charge cloud in the course of the transition (the direction and magnitude). Vector $\boldsymbol{m}^{if}$, the magnetic dipole transition moment, describes the direction around which the charge cloud rotates during a quantum transition. Two conclusions can be deduced from the Rosenfeld equation (8.6): (i) Because the charge density in a molecule is given by the structure of the molecule, rotational strength is directly determined by molecular structure. (ii) The scalar product nature of rotational strength implies that it possesses not only a magnitude but also sign. The value of R is not only determined by the size of both transition moment vectors, but also by the angle between them. This is where chiroptical methods get their enormous sensitivity to even minor changes of the geometrical arrangement of molecules. Rotational strength is non-zero when both vectors $\boldsymbol{\mu}^{if} \neq 0$, $\boldsymbol{m}^{if} \neq 0$ and when they are not perpendicular to each other.

The relatively simple Rosenfeld equation (8.6) determines the relation between the structure of a molecule and its interaction with circularly polarized radiation. Different methods are used for the computation of R, including the direct calculation by *ab initio* methods from first principles. However, at least within a limited range of applications, simplified approaches can be used that make a priori assumptions about a decisive mechanism by which optical activity of a molecule originates.

8.4.2
Mechanisms Generating Optical Activity

Electronic circular dichroism and absorption spectroscopy as well are largely based on a concept of chromophores, i.e. groups into which the active transitions in a molecule are localized. In electronic spectroscopy, this concept comes up rather naturally, because chromophores typically spread over several atoms often involving lone electron pairs or π-electron systems and such atom groupings may be usually identified with recognizable functional groups in a molecule. To a first approximation chromophores are usually not chiral (they possess local elements of symmetry, i.e. their own symmetry is higher than the symmetry of the whole chiral molecule), although there are notable exceptions, like helicenes, biphenyls, nonplanar amide groups, etc. The observed optical activity then arises from interaction between chromophores within the chiral geometrical framework of the molecule or from a perturbation imposed on a chromophore by means of potential field originating from the chiral skeleton. There are three such recognized mechanisms of optical activity origin and they can be used to obtain an at least semiquantitative interpretation of circular dichroism data in terms of chirality and geometrical arrangement of molecules provided that the concept is kept simple and the supposed mechanism is the decisive one: (i) dipole coupling mechanism; (ii) μ–m mechanism; and (iii) one-electron mechanism.

The dipole coupling requires two chromophoric groups having transitions with significant electric transition moments that couple via dipole–dipole interactions. If the mutual orientation of the two dipoles is chiral it results in a couplet of CD bands (a pair of CD bands equal in intensity but having opposite signs) belonging to in-phase and out-of-phase combinations of uncoupled components. Optical rotatory strengths can then be calculated or qualitatively estimated in a simple manner and is a function of electric transition moment size (can be determined from absorption) and of geometrical factors (distance and orientation of transition moment vectors). Within a molecule this mechanism is relatively far-reaching, because of the far-reaching nature of dipole–dipole interactions. The concept is rather well understood and has been applied to many structural situations with remarkable success and varying degree of sophistication. It can be applied to both ECD (for review see for example [26]) and VCD [79, 80]. π–π^* transitions in polypeptides can serve as a typical example.

The μ–m mechanism is similar in concept, but assumes interaction of a transition dipole within one group with a transition quadrupole (magnetic moment) within the other group. It is much more difficult to recognize and to apply. n–π^* transitions coupling with π–π^* transition in oligopeptide molecules depict such a situation [81, 82].

The one-electron mechanism is different from the above two situations, because it does not require two chirally oriented chromophores. Instead it describes mixing of (electronic) configurations within a single originally non-chiral chromophore under the influence of chirally oriented perturbation potential that breaks the original symmetry. The potential is usually expressed as a multipole series, where only terms having proper symmetry properties contribute to chiral mixing. In order to provide proper symmetry behavior, the potential must be a pseudoscalar representation of the symmetry point group of the chromophore in question [81, 82]. It also determines the multiplicity of a sector rule that describes the situation. The octant rule valid for the carbonyl chromophore is a typical situation where one electron theory applies. This mechanism has been successfully used in real calculations, however it must be used with caution. The rotatory strengths given by one electron mechanisms are usually quite small. If there is also another mechanism operative there is a danger that it might override the one-electron contribution.

The above mechanisms generating optical activity are not exactly quantitative and they are not additive. They can serve as a great way to explain simple chiral situations. They have many limitations and must be used carefully. They cannot be used for chromophores that are inherently chiral. However, complete methods of calculations for rather complex molecules based on this concept and a detailed analysis of properties of chromophores have been constructed [83, 84] and very successfully applied [85–87] (for a review see reference [88]). With the exception of simple dipole coupling, these mechanisms cannot be applied to VCD, although attempts towards similar concepts have been made [80, 89, 90].

8.4.3
Ab initio Calculations

Direct calculations of optical rotatory strengths based on the evaluation of matrix elements of both transition moments and the subsequent application of the Rosenfeld equation (8.6) were first attempted in the 1970s on the basis of semiempirical wave functions. The success of these early ECD calculations was limited, primarily because only a poor description of molecular excited states was available at that time. Nevertheless, in some relatively simple cases, they were successful (see, for example, the non-planar amide bond [91]). Such a calculation is in principle more demanding than an ordinary quantum chemical study of a molecule. It requires not only the calculation of the ground state energies and molecular geometries, but also energies and charge distributions pertaining to excited states. Systematically successful *ab initio* direct ECD calculations are a matter of the last five years and are closely related to recent advances in quantum descriptions of molecular excited states using advanced methods like TDDFT [4, 52–55].

Direct calculations of vibrational optical activity are younger, but once theoretical obstacles to them were solved, they turned out to be easier despite the fact that such a calculation is a many step procedure and its protocol is rather complicated. The use of this procedure for absolute configuration determination is described in Section 8.5.1. A fundamental advantage of VCD calculations consists in the fact that we are dealing with molecules usually in their well-defined electronic state – the ground state. For the calculation of vibrational optical activity, we must calculate atomic polar and axial tensors [16, 92].

The principal difficulty lies in the fact that atomic axial tensors are zero within Born Oppenheimer approximation. Therefore, special theories have to be developed; the most widespread is the magnetic field perturbation (MFP) method by Stephens et al. [93–95]. When combined with a reasonably sophisticated *ab initio* method (like density functional theory (DFT) with a functional like B3LYP or BWP91) and a reasonably large basis set (6-31G* and better), the calculation becomes quite reliable and can be used to assign absolute configurations to rigid molecules (see Section 8.5.1). Calculations at this level of approximation can now be performed with molecules up to 100 heavy atoms. For larger molecules containing repeat units (like biopolymers), it is possible to apply transfer of molecular properties tensors from smaller model molecules [46, 96, 97].

The calculation involves optimization of the molecular structure, computation of vibrational modes (by far the most demanding part computationally), computation of atomic polar and axial tensors and of all the sums leading to dipole and rotational strengths. As the last step, the theoretical VCD curve is simulated by using the empirical values for bandwidths. The quantum chemical part of the calculation is nowadays a standard part of commercial molecular computational programs such as Gaussian. There are even recent attempts to include solvent effects in the calculation [66, 96, 98].

8.5 Examples of Vibrational Circular Dichroism Applications

VCD has existed as a real spectroscopic technique since the 1970s, and its development led to significant improvement of structural knowledge. Almost from the very beginning, VCD was employed in structural studies of chiral molecules regardless of their size, from relatively simple organic molecules, through rigid alkaloids up to biomolecules, including large biopolymers such as polypeptides, proteins and nucleic acids. VCD applications include the simple use of VCD spectra to resolve opposite enantiomers, determination of enantiomeric excess [99], but also the determination of absolute configuration of rigid organic molecules, including pharmaceuticals and conformational analysis of flexible molecules (for a review see [45]; for examples see Section 8.5.1). Semiempirical structural studies of polypeptides, proteins and oligo- and polynucleotides represent a special and inspiring group of VCD applications (for reviews see [44, 64, 100–103]; for examples see Section 8.5.2). In recent years VCD was used as a tool to follow chiral supramolecular systems (for examples see Section 8.5.3).

8.5.1 Absolute Configuration and Detailed Structural Parameters

VCD has become a spectroscopic method capable of the direct determination of absolute configuration. The advantage of the method is particularly significant when only the solution phase of chiral matter is available and the preparation of a single crystal of sufficient quality appropriate for X-ray analysis is not feasible for some reason. In addition, the capability of solution structure determination is very desirable. The VCD calculation is derived from the basic physical principles and is independent of experiments. Since quantum mechanical methods are used for the calculations, the only empirical parameter is the bandwidth used in the spectra simulation as explained above. The methodology utilizing DFT [104, 105] is now well documented [106, 107] and extensively used for configurational and conformational analysis of chiral molecules. The survey of results obtained up to now is given in a review [45].

VCD configurational analysis of a rigid molecule typically consists of the following steps:

(i) Selection of the absolute configuration of the molecular structure that should be calculated. The number of chiral centers in a molecule determines the number of possible stereoisomers that must be considered.
(ii) Geometry optimization is followed by the determination of the normal vibrational modes, related to wavenumbers $\tilde{\nu}_{0,n}$ of IR and VCD bands, and the di-

pole and rotational strengths related to the integrated IR and VCD intensities. These calculations at the DFT level are incorporated into commercial software, as Gaussian 98 and its further versions. Detailed description and analysis of the methods used is given in a review article by Stephens et al. [108].

(iii) Conversion of the computed result into a theoretical spectrum can be done using commercial software. The Lorentzian band shape is used and the band-width is chosen depending on the experimental values.

(iv) Comparison of simulated – theoretical spectra with experiment can now be executed band-to-band. When the majority of the experimental and calculated IR bands correlate in terms of relative intensities and VCD bands in terms of both sign and relative intensity, the selected structure is with a very low uncertainty the real one. VCD spectra are very sensitive to configuration: Opposite enantiomers possess CD spectra with opposite signs. Even stereoisomers are distinguished almost unambiguously, because they differ significantly in sign and the relative intensities of a number of VCD bands.

For rigid molecules, the procedure described above determines whether the chosen configuration is the right one. Development of the calculation methods based on DFT and the current computer technology enable to carry out an *ab initio* VCD calculation for isolated gas-phase molecule up to ~100 heavy atoms. Solvent effects can also be included to a certain extent.

As an example of the VCD ability to discriminate absolute configurations, the experimental spectrum of 4-fluorophenyl-3-hydroxymethyl-1-methylpiperidine and the calculated spectra of the (3*R*,4*R*)- and (3*R*,4*S*)-isomer are presented in Fig. 8.5 [109]. An almost perfect band-to-band correspondence is evident between experimental and calculated spectra of the (3*R*,4*S*)-isomer, while the calculated spectrum of the low energy conformer of the (3*R*,4*R*)-isomer provides unrealistic VCD pattern with a poor or even no coincidence with the experiment. In the calculated spectrum of the (3*R*,4*S*)-isomer, four low-energy conformers were included in the spectral envelope. For comparison, the corresponding calculated and experimental IR absorption spectra are shown. It is evident that the variance with respect to configuration is much more pronounced in VCD spectra: While the calculated spectrum for the (3*R*,4*R*)-isomer significantly differs from experimental VCD, the unpolarized absorption spectrum of the (3*R*,4*R*)-isomer exhibits some common

→

Fig. 8.5. (a) VCD (b) and IR absorption spectra of 4-fluorophenyl-3-hydroxymethyl-1-methylpiperidine. From top to bottom: Calculated spectrum of low-energy conformation of (3R,4R)-isomer; calculated spectra of 7 conformations of (3R,4S)-isomer; an average spectrum of four lowest-energy conformations of (3R,4S)-isomer; and experimental spectrum. (Reprinted with permission from: P. Bouř, H. Navrátilová, V. Setnička, M. Urbanová, K. Volka, *J. Org. Chem.* **2002**, 67, 161–168. Copyright 2002 American Chemical Society.)

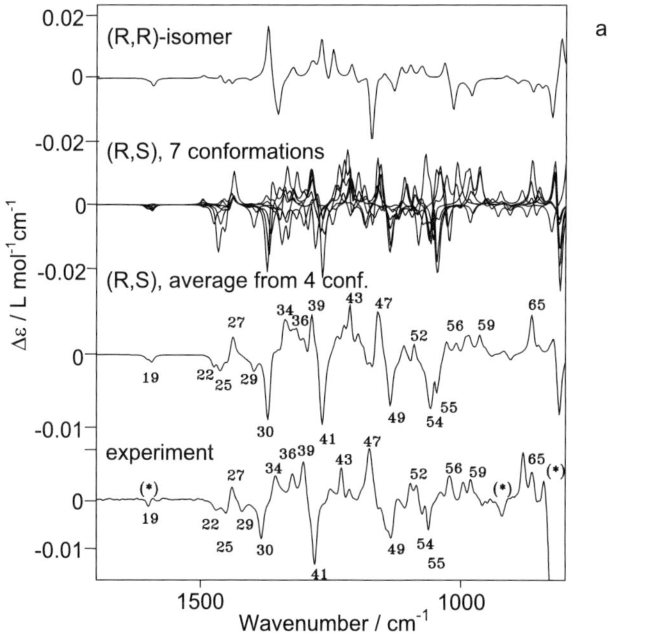

0.02
(R,R)-isomer
0
-0.02
(R,S), 7 conformations
0
-0.02
(R,S), average from 4 conf.
0
-0.01
experiment
0
-0.01
Δε / L mol⁻¹cm⁻¹
1500
1000
Wavenumber / cm⁻¹
a

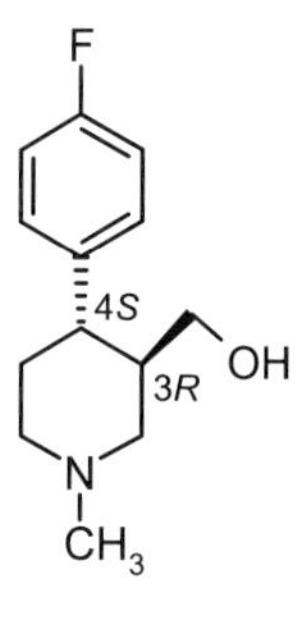

F
4S
3R
OH
N
CH3

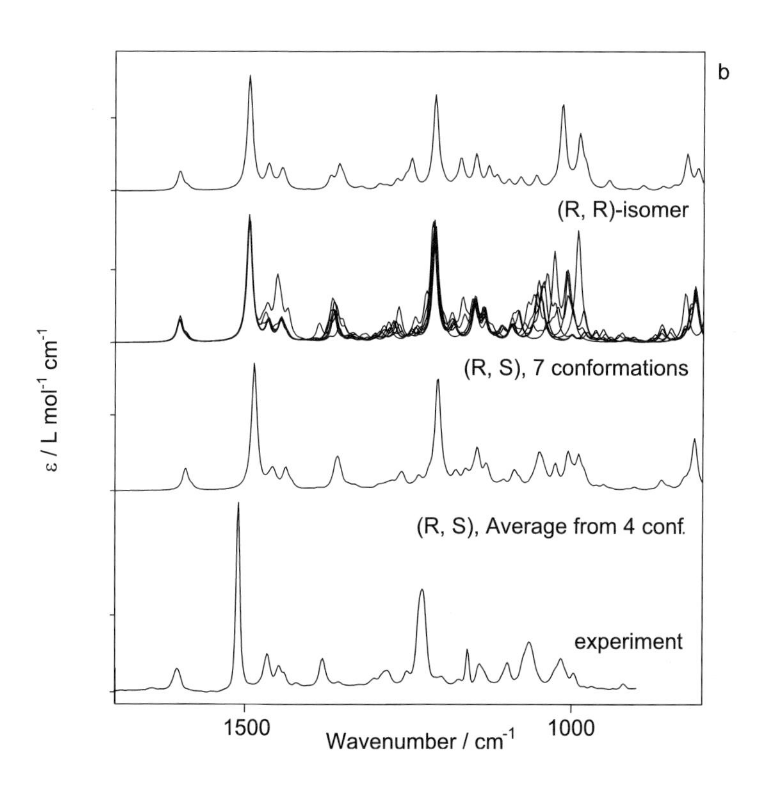

b
(R, R)-isomer
(R, S), 7 conformations
(R, S), Average from 4 conf.
experiment
ε / L mol⁻¹ cm⁻¹
1500
1000
Wavenumber / cm⁻¹

features with the experimental IR spectra, primarily in the higher-frequency region and it is not useful to distinguish between the isomers.

VCD is very helpful in the conformational analysis of chiral flexible molecules. If many conformers are identified, it is practical to search for low energy conformers at a semiempirical level. Figure 8.5 shows VCD and absorption spectra of low-energy conformers. VCD is again more sensitive to conformation than unpolarized absorption. Comparison of observed and experimental IR and VCD spectra can identify dominant conformations in solution. For the methylpiperidine derivative in Fig. 8.5, the average of the four lowest-energy conformations gives very good agreement with the absorption and VCD experiments. For obtaining a better result, the set of spectra for the low energy conformers are weighted according to their fractional populations at temperature T, which is given by Boltzmann relation $N_i/N_0 = \exp(-\Delta E_i/kT)$, where N_i/N_0 is the relative population of conformer i related to the population of the lowest energy conformer; ΔE_i are the relative energies.

The VCD study of the 1,1′-binaphthyl derivatives [110] serves as an example of other structural information that can be obtained by the comparison of experimental and computed VCD. This method allows monitoring not only of absolute chirality of the molecule, but also of the contributions of individual functional groups to the spectra, molecular conformations or some important structural parameter. As an example, we discuss chiral binaphthyls which represent popular building blocks, chiral recognition receptors and catalyst. Controlling the angle between naphthyl planes is important when supramolecular complexes based on these compounds are built.

The parent binaphthyl **1** exhibits rapid racemization (in minutes) and thus its VCD spectra cannot be obtained. However, other derivatives are stable and their VCD spectra were obtained in most cases for both enantiomers. Excellent agreement in the positions, signs and the relative intensity of the VCD bands for experimental and calculated spectra makes it possible to obtain detailed structural parameters, for example, the torsional angle between the naphthyl planes depending on the substituents given in Table 8.3. It is obvious that the chain-bridged deriva-

Tab. 8.3. Calculated torsion angles (deg) between two naphthyl planes. (Reprinted with the permission from V. Setnička, M. Urbanová, P. Bouř, V. Král, K. Volka, *J. Phys. Chem. A* **2001**, *105*, 8931–8938. Copyright 2001 American Chemical Society.)

Angle	Compound						
	1	**2**	**3**	**4**	**5**	**6**	**7**
9′–1′–1–9	74	87	95	104	90	85	57
2′–1′–1–2	71	86	93	108	89	82	52

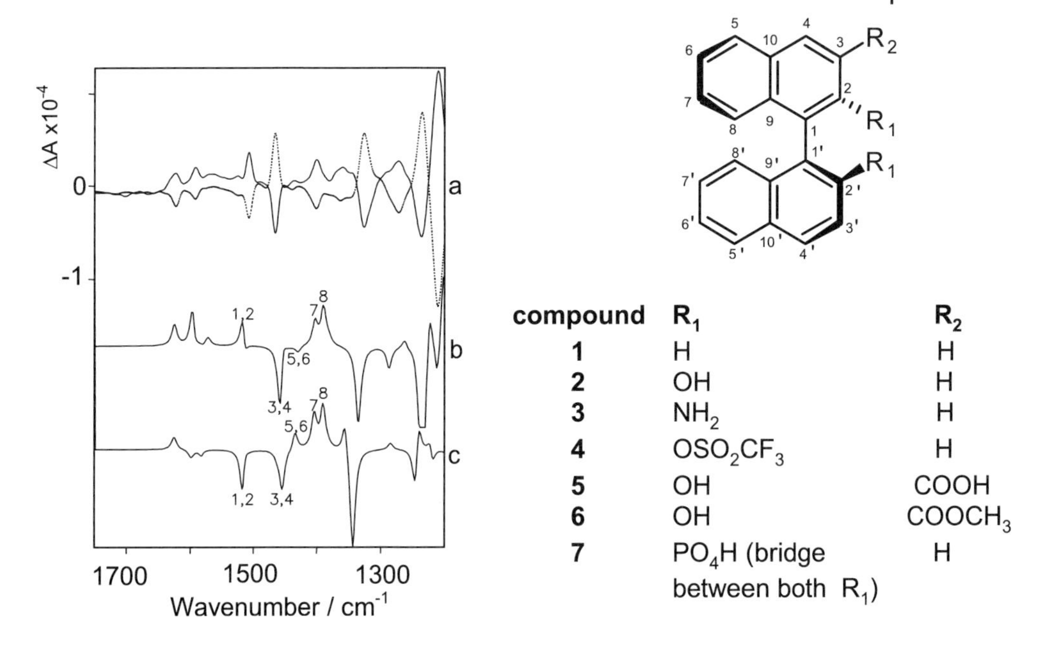

Fig. 8.6. (a) Experimental VCD spectra of 7: (S)-isomer, (R)-isomer, bold line; (b) calculated VCD spectra of 7: (S)-isomer; (c) calculated VCD spectrum of 1. (Reprinted with permission from: V. Setnička, M. Urbanová, P. Bouř, V. Král, K. Volka, *J. Phys. Chem. A* **2001**, 105, 8931–8938. Copyright 2001 American Chemical Society.)

tive **7** possesses a torsional angle $< 60^\circ$ that is quite different than for other derivatives. Figure 8.6 demonstrates how the structural difference is expressed in VCD. In addition, vibrational modes were assigned to individual groups of atoms by using the calculated distribution of vibrational kinetic energy and the nature of the corresponding normal modes was determined by a dynamic visualization of the normal mode displacements.

8.5.2
Solution Structure of Biomolecules

Biomolecules are mostly chiral, therefore chiroptical methods are the first choice among the spectroscopic methods applied to biomolecular structural studies. Characterization of peptides, nucleotides, proteins and nucleic acids by their ECD spectra became a generally used procedure and was repeatedly reviewed [1].

The n–π^* (210–230 nm) and π–π^* (185–200 nm) transitions of amide groups yield information about the secondary structure of oligopeptides, polypeptides and proteins. Three types of interactions between these two transitions are responsible for the observed ECD: exciton interactions between degenerate π–π^* transitions within different peptide groups, mixing of n–π^* and π–π^* transitions within different peptide groups, and mixing of n–π^* and π–π^* transitions within a single peptide group. The polypeptide structure implies that interactions of transitions result in typical shapes of UV ECD spectra for different elementary secondary structures such as α-helix, β-sheet, polyproline II-like. The methods analyzing protein spectra in terms of elementary secondary structures have been reviewed [111].

In DNA, the UV active chromophores are the bases of adenine (A), guanine (G), cytosine (C), and thymine (T). Their strong π–π^* and weak n–π^* transitions themselves situated in the 190–300 nm region are not inherently optically active. However, since bases are attached to optically active deoxyribose, the sugar component can induce the CD in the base chromophores. ECD of DNA brought very valuable information about different forms of single strands, double helixes, the dependence of the structure on physico-chemical conditions, as reviewed by Johnson [112], and also about non-classical conformations and modified oligonucleotides [113].

VCD does not suffer from the disadvantages characteristic for UV/vis region. Many reviews are available on the application of these techniques [16, 62, 63, 99, 101, 102], on detailed structural information about polypeptide (for a review up to 1995 see [61]) and on the potential of VCD for the study of DNA and sequential oligonucleotides [103, 114, 115]. For peptides and proteins [116–121], VCD particularly utilizes vibrational transitions related to the amide group: amide I, i.e. mostly the C=O stretching vibration, amide II, which is primarily the N–H deformation, and the C–N stretch. If the amide group is deuterated, the analogous characteristic vibrations are called amide I′ and amide II′. The VCD spectra for particular secondary structures, e.g. α-helix, β-sheet, or polyproline II-like structures, are characteristic. The two amide transitions are very sensitive to the structure of biopolymers alone but also enable to follow subtle changes in the local structure caused by physico-chemical conditions in solution and supramolecular interaction [34–36, 103, 122, 123]. Detailed structural studies of the local peptide structure were carried out using isotopic substitutions [124–127].

VCD was also applied to the study of DNA including sequential oligonucleotides and their interaction with metallic ions [79, 128–134]. Two mid-IR region are used for applications to the DNA structural studies: the nucleoside base region, 1800–1500 cm^{-1}, where bands are assigned to C=O and C=N vibrations of the DNA bases, and the phosphate region, 1300–1000 cm^{-1}, where vibrations typical to PO_2^- groups of the DNA backbone are observed. Aqueous solutions are very often the best environment for biopolymer studies, alternately H_2O or D_2O are used depending on the needed spectral window.

Recently, we have also used VCD and a combination of VCD and ECD to study the non-covalent interactions between two groups of important biomolecules, i.e. biopolymers – polypeptide and DNA – and cationic and anionic porphyrins [34–36].

The sensitivity of VCD to structural variations of the polypeptide–porphyrin complexes is demonstrated in Fig. 8.7. The secondary structure of the peptidic part of the poly-L-glutamic acid–cationic porphyrin TATP complex can be changed by pH variation. At acidic pH (~5), its α-helical structure is demonstrated by the

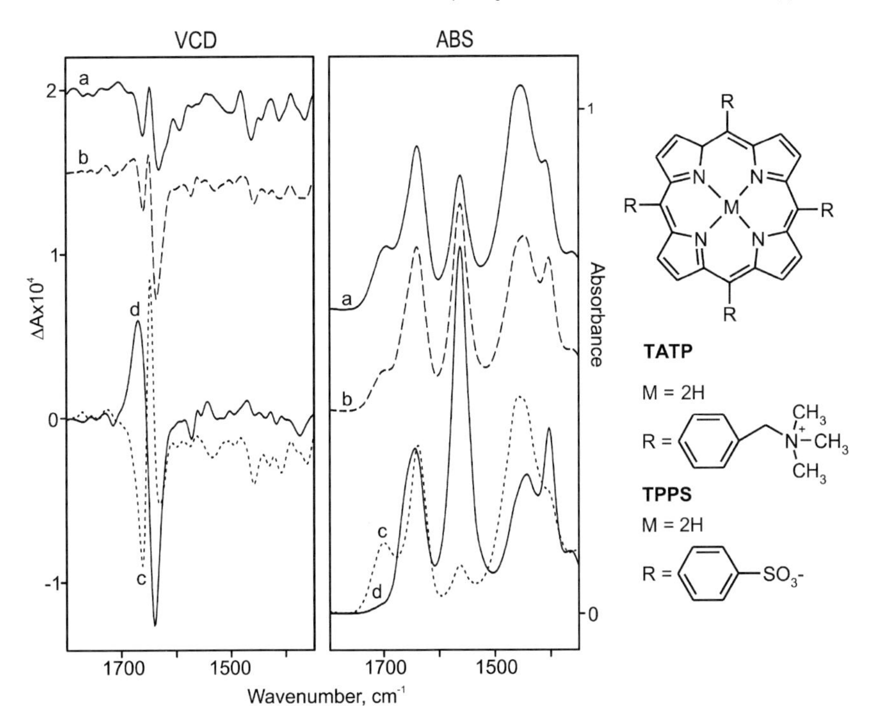

Fig. 8.7. VCD and absorption spectra of poly-L-glutamic acid-TATP-TPPS in D_2O at pH ~ 5 (a), simulated spectra (b). VCD and absorption spectra of poly-L-glutamic acid-TATP in 40% methanol-d_4/D_2O at pH ~ 5 (c) and pH ~ 8 (d). (Reprinted with permission from: L. Palivec, M. Urbanová, K. Volka, *J. Peptide Sci.* **2005**, 11, 536–545. Copyright 2001 John Wiley & Sons Ltd.)

typical shape of the amide I′ pattern (from small to high wavenumbers, −/+/−). At pH ~ 8, the same system adopts a polyproline II-like structure demonstrated by quite a different shape of the amide I′ pattern (−/+). If an anionic porphyrin is added to the poly-L-glutamic acid–cationic porphyrin complex, the structure of the peptidic part of the complex changes. Existence of the ternary complex is proved by UV/vis absorption. Its VCD spectrum in the amide I′ region can be simulated as a sum of both the α-helical and the polyproline II-like patterns (Fig. 8.7). Structural variation in this ternary complex is interpreted as partial destruction of the α-helical structure and the formation of left-handed polyproline II-like helical conformation.

As an example of complementarity between ECD and VCD, a study of interactions between calf thymus DNA and inherently achiral porphyrins is shown [34]. ECD (230–300 nm) reflects the DNA part of complexes; the optical activity of the porphyrin part of complexes induced by interaction with chiral DNA is observed in the Soret region (380–470 nm (Fig. 8.8). Although ECD spectra are treated generally as a reliable marker for different binding modes of porphyrins with DNA, VCD

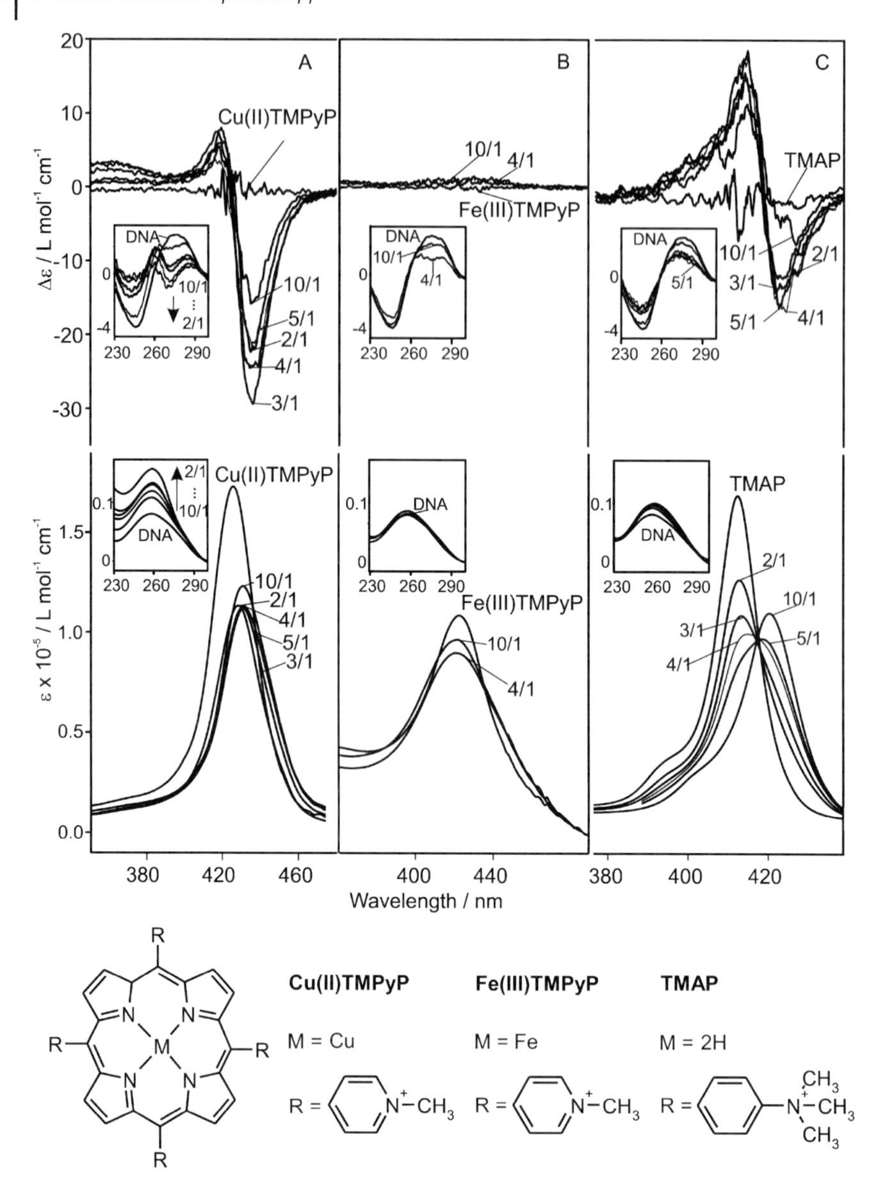

Fig. 8.8. ECD (top) and visible absorption (bottom) spectra of (A) DNA–Cu(II)TMPyP; (B) DNA–Fe(III)TMPyP; and (C) DNA–TMAP complexes in the Soret spectral region at different c(DNA)/ c(porphyrin) concentration ratios. UV spectra given as inset. (Reprinted with permission from: J. Nový, M. Urbanová, K. Volka, *Collect. Czech. Chem. Commun.* **2005**, 70, 1799–1810. Copyright 2005 Institute of Organic Chemistry and Biochemistry, Academy of Sciences of the Czech Republic.)

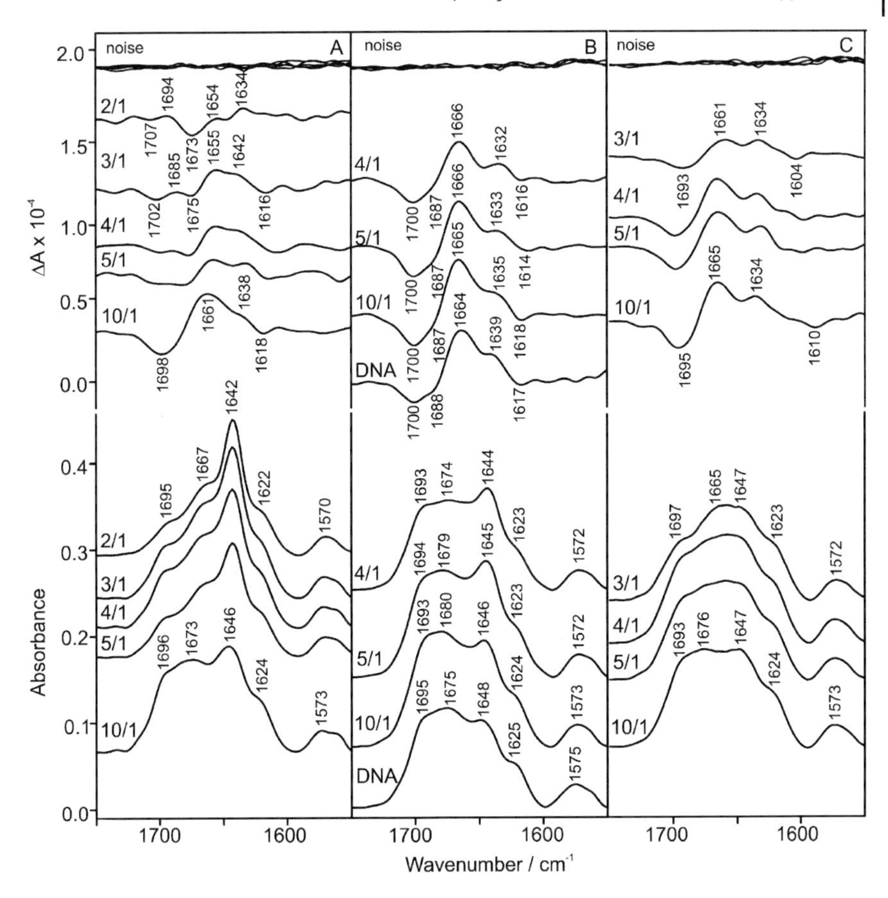

Fig. 8.9. VCD (top) and IR (bottom) spectra of (A) DNA–Cu(II)TMPyP; (B) DNA–Fe(III)TMPyP; and (C) DNA–TMAP complexes at different c(DNA)/c(porphyrin) concentration ratios. (Reprinted from: J. Nový, M. Urbanová, K. Volka, *Collect. Czech. Chem. Commun.* **2005**, 70, 1799–1810 with permission of the Institute of Organic Chemistry and Biochemistry, Academy of Sciences of the Czech Republic.)

spectra obtained for different porphyrin–DNA systems possess additional information about specific groups of atoms involved in the interactions (Fig. 8.9).

The interactions of Cu(II)TMPyP with DNA result in an intercalative binding mode that corresponds to a negative band at 426(−) nm in ECD spectra and a new positive band at 1690(+) cm^{-1} in VCD spectra. The binding of Cu(II)TMPyP affects substantially the DNA structure in the GC base-pair regions at higher Cu(II)TMPyP loadings. Intercalative sites are saturated at the 3/1 ratio, which is demonstrated by the fact that VCD and ECD resemble those of poly(dA-dT)$_2$ [112, 135].

Fe(III)TMPyP binds externally to a minor groove of DNA that is revealed by a positive band in ECD spectra and as a shift of the positive VCD band from 1639(+) cm^{-1} to 1632(+) cm^{-1}, accompanied by an intensity decrease.

ECD spectra show that TMAP binds to the minor groove binding at higher *c*(DNA)/*c*(TMAP) ratios and externally accompanied by self-stacking of TMAP along the phosphate backbone at smaller *c*(DNA)/*c*(TMAP) ratios, which are demonstrated in ECD as a positive band and as a couplet with both positive and negative bands of similar intensities. In VCD spectra, the minor groove binding is demonstrated by a shift of the band from ~1639(+) to 1634(+) cm^{-1} and the external major groove binding as a less intense positive band at 1661(+) cm^{-1}.

8.5.3 Supramolecular Systems

Most applications of ECD spectroscopy in supramolecular chemistry, which can be counted to hundreds, involve proofs of existence and determination of properties (chirality) of potentially chiral supramolecular structures. The convergence of the ECD amplitude is taken as a criterion of pure enantiomers. The loss of structure upon the change of physico-chemical properties can be followed by the decrease of CD signal. Another concept uses chromophores [24] for testing the three dimensional structure of supramolecular systems. If the isolated chromophore is not chiral and induced circular dichroism (ICD) is observed in the system, we can sometimes deduce the structure of the chromophoric environment (see section 8.4.2). *Ab initio* calculations open another possibility to employ ECD into supramolecular chemistry (Section 8.1.3).

Due to the advantages described previously, VCD is also suitable for applications in supramolecular chemistry in all situations where chiral functionalities are employed or a chiral arrangement of chiral or even non-chiral components takes place. Although not so widespread as ECD, VCD also starts to have impact in supramolecular chemistry. To illustrate the use of VCD in supramolecular chemistry, a few examples of recent research are provided.

VCD was used to study the sol–gel phase transition process of a brucine–porphyrin based gelator [136, 137]. In Fig. 8.10, VCD and IR absorption spectra of brucine, which represents a chiral moiety in the conjugate, are shown in three different solvents. VCD spectra reveal almost the same shape, position and intensity of the corresponding VCD bands independently of the solvent used. The band observed in absorption and VCD at 1680–1640 cm^{-1} was assigned to the C=O stretching vibration. The bands at 1604, 1501, 1467 and 1453 cm^{-1} were assigned to the C=C stretching and CH_3, CH_2 and CH deformation modes. Because VCD spectroscopy reflects the local arrangement of molecules in solution very sensitively, it is obvious that brucine alone is a rigid molecule with non-varying conformation in all solvents.

Tetrabrucine-appended porphyrin was proved to be an excellent gelator of CD_3OD; it entraps about 3000 molecules of solvent per one gelator molecule. Figure 8.11 shows its VCD and IR absorption spectra in the C=O region for the

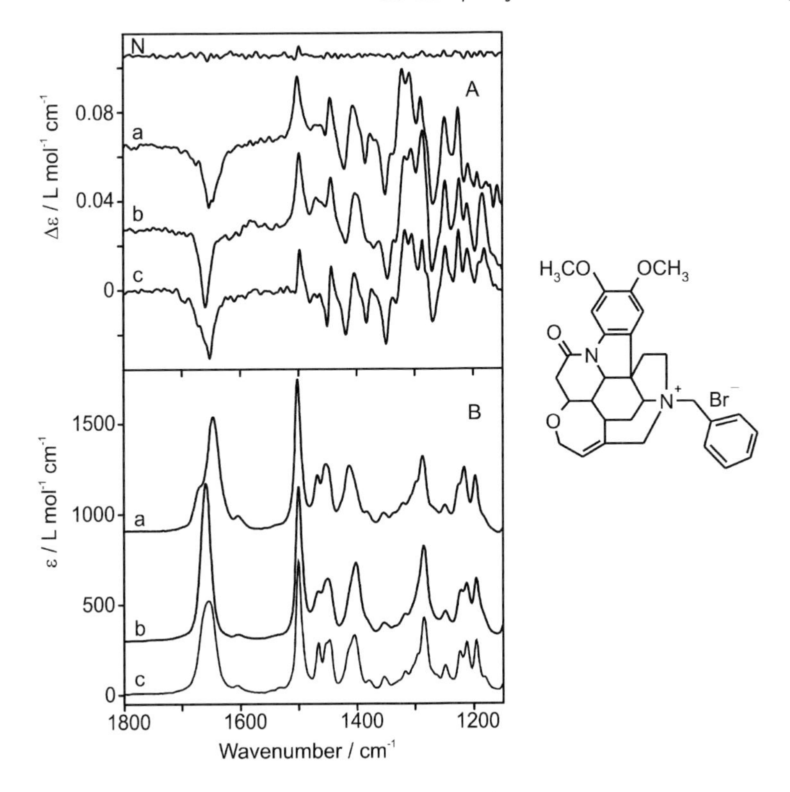

Fig. 8.10. VCD (A) and IR absorption spectra (B) of brucine in (a)CD_3OD; (b) DMSO-d_6; (c) $CDCl_3$, and (N) typical noise spectrum. (Reprinted with permission from: V. Setnička, M. Urbanová, S. Pataridis, V. Kral, K. Volka, *Tetrahedron: Asymmetry* **2002**, 13, 2661–2666. Copyright 2002 Elsevier.)

following states: when the gel phase is formed in CD_3OD, in the sol phase in DMSO-d_6, and in the CD_3OD/DMSO-d_6 mixture. It is evident that changes of the molecular chirality of tetrabrucine conjugate induced by the sol–gel phase transition are well observable in VCD spectra. The conjugate in DMSO-d_6 exists in the sol phase and provides the characteristic negative VCD signal in the C=O region very close in intensity and position to that observed for pure brucine shown in Fig. 8.10. This is in agreement with the fact that DMSO-d_6 prevents any interactions. Therefore, the rigid brucine moiety not participating in an interaction is manifest by its characteristic VCD signal. For CD_3OD/DMSO-d_6 = 4/1, the gel is formed and the intense positive VCD signal is observed. Its intensity is even higher in pure CD_3OD, when the more viscous gel is formed. The sign inversion and enhanced intensity indicate variation of rotational strength related to the C=O stretching modes as a consequence of the formation of highly ordered chiral assemblies. VCD sensitivity to sol–gel transition makes it possible to follow the temperature stability of the organogel and the sol–gel transition [136].

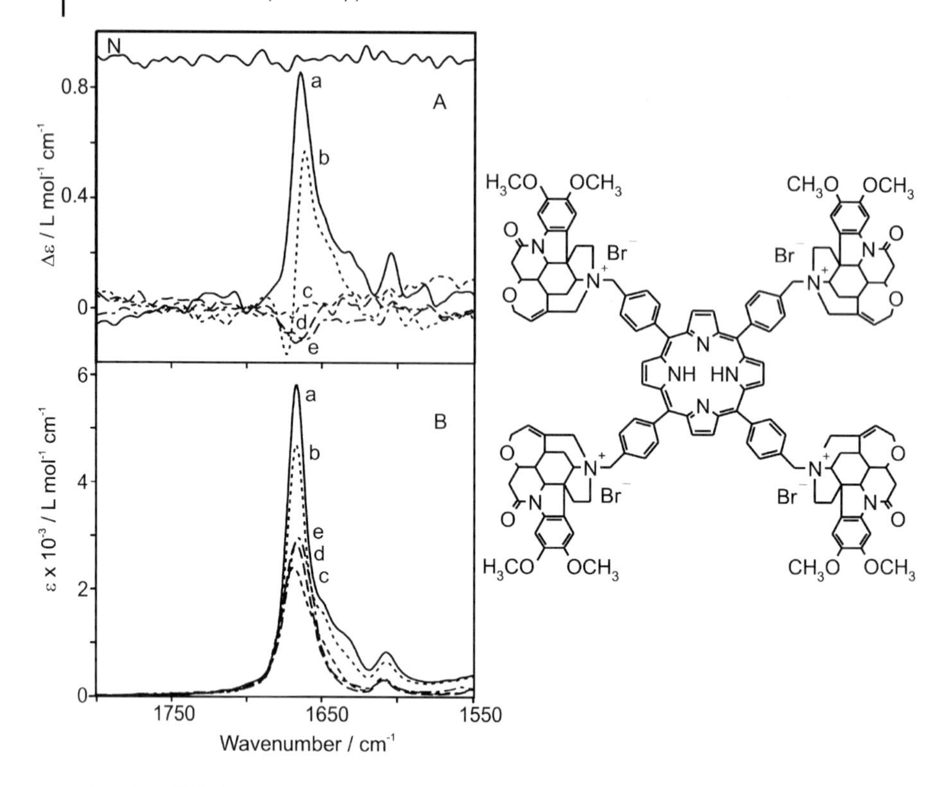

Fig. 8.11. VCD (A) IR absorption spectra (B) of tetrabrucine conjugate in (a) CD_3OD; (e) DMSO-d_6; and in the CD_3OD/DMSO-d_6 mixture solvent (v/v): (b) 4/1; (c) 1/1; (d) 1/4. (Reprinted with permission from: V. Setnička, M. Urbanová, S. Pataridis, V. Kral, K. Volka, *Tetrahedron: Asymmetry* **2002**, 13, 2661–2666. Copyright 2002 Elsevier.)

Another example of a semiempirical VCD study addresses an open question about the structural characterization of a challenging dendritic system, the substituted quinine QuiG0–G2 [138]. The chiral quinine moiety was substituted by dendritic polybenzyl ether units up to generation 2 (Fig. 8.12). Enhanced Cotton effects were observed in ECD spectra along with increasing size of inherently achiral dendritic polybenzyl ether branches attached to a chiral core. VCD is able to resolve individual characteristic vibrations and enables to follow "the dendritic effect" in detail. In IR absorption and VCD spectra, the signals marked as 1–6 are assigned to vibrations connected to the chiral quinine moiety (Table 8.4). Apart from these signals, additional IR bands 7–11 are assigned to vibrations localized in dendritic branches. Their intensities increase with the generation of the dendritic substituent attached to the quinine moiety. Only some vibrations localized in the formally achiral dendritic branches (7, 10, 11) acquire optical activity and the "dendritic effect" is observed. From a combination of IR absorption and VCD spectra, it is evident that the observed "dendritic effect" is due to the benzene and

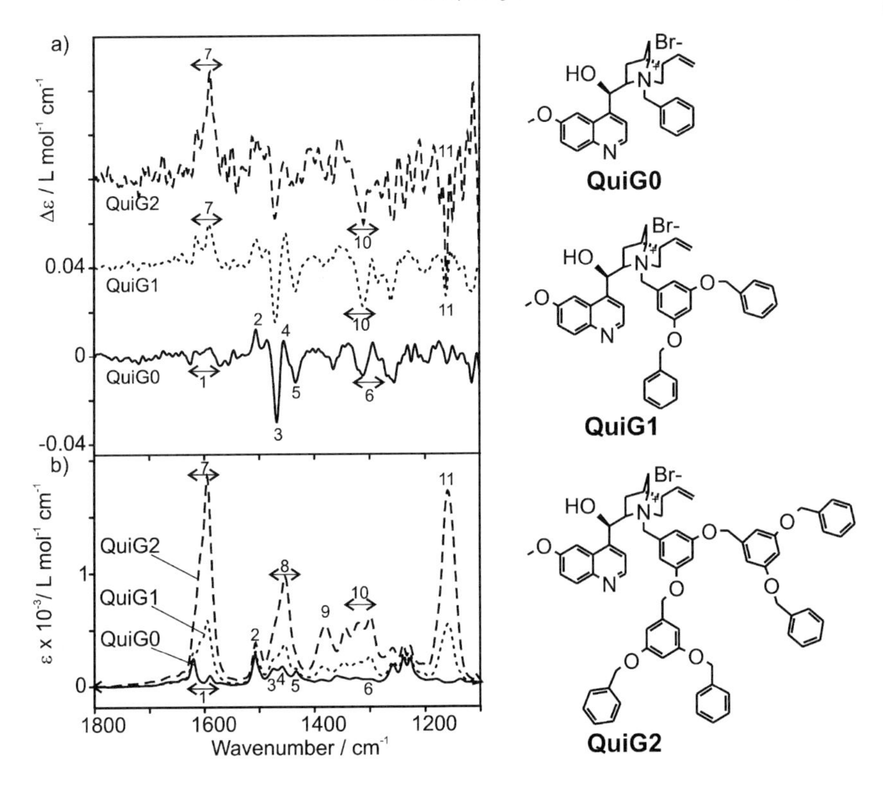

Fig. 8.12. VCD (A) and IR absorption (B) spectra of the quinine derivatives substituted with dendritic polybenzyl ether units for generations 0–2. (Reprinted with the permission from: U. Hahn, A. Kaufmann, M. Nieger, O. Julínek, M. Urbanová, F. Vögtle, *Eur. J. Org. Chem.* **2006**, 1237–1244. Copyright 2006 Wiley-VCH.)

C–O–C groups rather than connected with the CH_2 subunits of the dendritic branches. The VCD results support the explanation of the observed induced circular dichroism effect as a consequence of the chiral arrangement of the inherently achiral benzene rings within the dendritic branches that was induced by the chiral center.

In the following example, we describe the extension of the *ab initio* DFT based methodology to the field of supramolecular chemistry [71]. X-ray analysis [139] has shown that crystals of (*R*)-2,2′-dimethyl-biphenyl-6,6′-dicarboxylic acid contain intermolecular H-bonded supramolecular cyclotetramers and vapor phase osmometry [139] has indicated the degree of association ~4 in $CHCl_3$ solution. The methodology consisting in the combination of experimental and *ab initio* calculated VCD spectra was employed to confirm the tetrameric structure in the solution and elucidate it in more details.

Experimental absorption and VCD spectra in $CDCl_3$ and DMSO-d_6 are shown in Fig. 8.13, where the difference between the two solution spectra is especially

Tab. 8.4. Assignment of the VCD and IR absorption signals of quinine derivatives to characteristic vibrations. (Reprinted with the permission from: U. Hahn, A. Kaufmann, M. Nieger, O. Julínek, M. Urbanová, F. Vögtle, *Eur. J. Org. Chem.* **2006**, 1237–1244. Copyright 2006 Wiley-VCH.)

Pattern	ν (cm^{-1})	Peak assignment
1	1590–1620	C=C, C=N ring[a]
2	1507	quinoline[a]
3–5	1433–1474	quinuclidine ring and asymmetric CH_3[a]
6	1300	quinuclidine ring and asymmetric CH_3[a]
7	1595	dimethoxybenzene[b]
8	1453	scissoring CH_2[b]
9	1380	CH_2 overlapped by CH_3[b]
10	1300–1344	arom. O–R[b]
11	1158	C–O–C[b]

[a] Band assigned to the quinine moiety.
[b] Band assigned to the dendritic branch subunit.

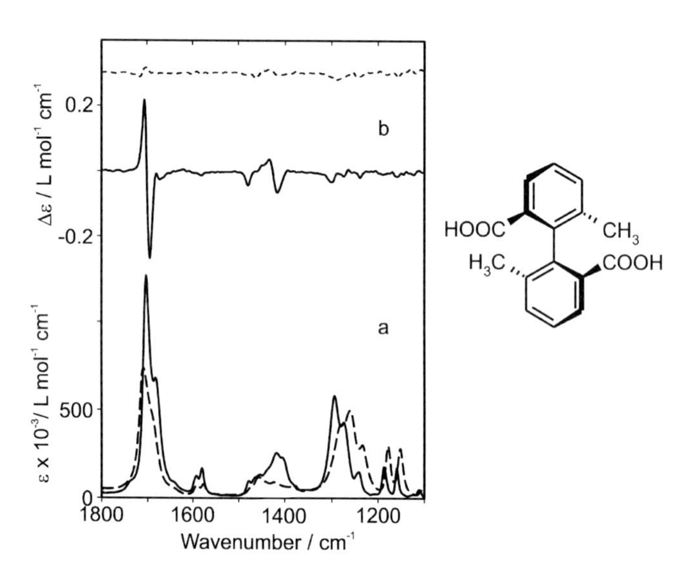

Fig. 8.13. (a) IR absorption and (b) VCD spectra of (S)-2,2′-dimethyl-biphenyl-6,6′-dicarboxylic acid in $CDCl_3$ (full line) and DMSO-d_6 (dashed line). (Reprinted with permission from: M. Urbanová, V. Setnička, F.J. Devlin, P.J. Stephens, *J. Am. Chem. Soc.* **2005**, 127, 6700–6711. Copyright 2005 American Chemical Society.)

pronounced in VCD in the carbonyl region: the VCD signal measured in $CDCl_3$ solution is more than 20 times higher than that obtained in DMSO-d_6.

The differences between the VCD spectra in DMSO-d_6, which prevents aggregation, and in $CDCl_3$, which supports aggregation, suggest that the observed effect is caused by the formation of supramolecular aggregates, most probably a cyclotetramer observed by X-ray crystallography. The formation of aggregates was also confirmed by the concentration dependence of IR absorption in the carbonyl and C–H, O–H stretching regions (not shown). Each tetramer contains four $(COOH)_2$ moieties which exist in two tautomeric forms labeled as a and b when the C=O groups are oriented "cis" and "trans", respectively, with respect to the central C–C bond. Each moiety can adopt each of the a or b forms, which leads to six distinct conformations of the cyclotetramer: aaaa, aaab, aabb, abab, abbb, bbbb. In reference [71], the structure of all conformers was optimized at the B3LYP/6-21B* level; vibrational frequencies, relative energies, enthalpies, entropies, and free energies (at 298 K) were calculated and the absorption and VCD spectra were computed. The relative energy is smallest for conformation aaaa with D_4 symmetry and increases quite regularly with the increasing number of $(COOH)_2$ moieties in conformation b. Surprisingly, the lowest relative free energy is calculated for the conformation aaab with the C_2 symmetry shown in Fig. 8.14.

The conformation responsible for the experimental spectra in $CDCl_3$ can be clearly identified by comparing the IR absorption and VCD spectra calculated for the six conformations mentioned above. A good coincidence is observed for confor-

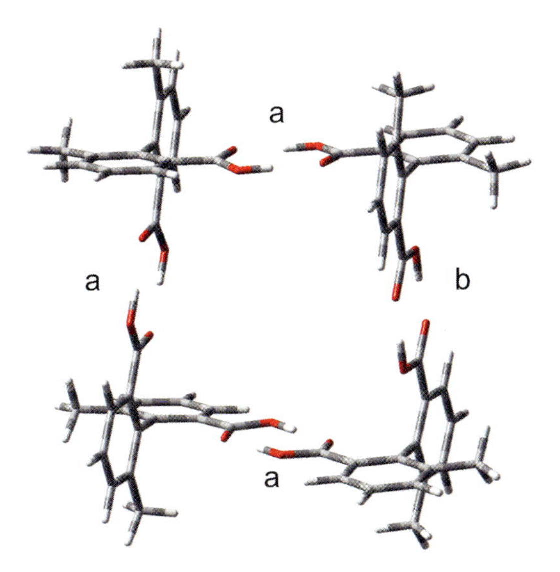

Fig. 8.14. B3LYP/6-21B* structure of the aaab conformers of cyclotetramer. (Reprinted with permission from: M. Urbanová, V. Setnička, F.J. Devlin, P.J. Stephens, *J. Am. Chem. Soc.* **2005**, 127, 6700–6711. Copyright 2005 American Chemical Society.)

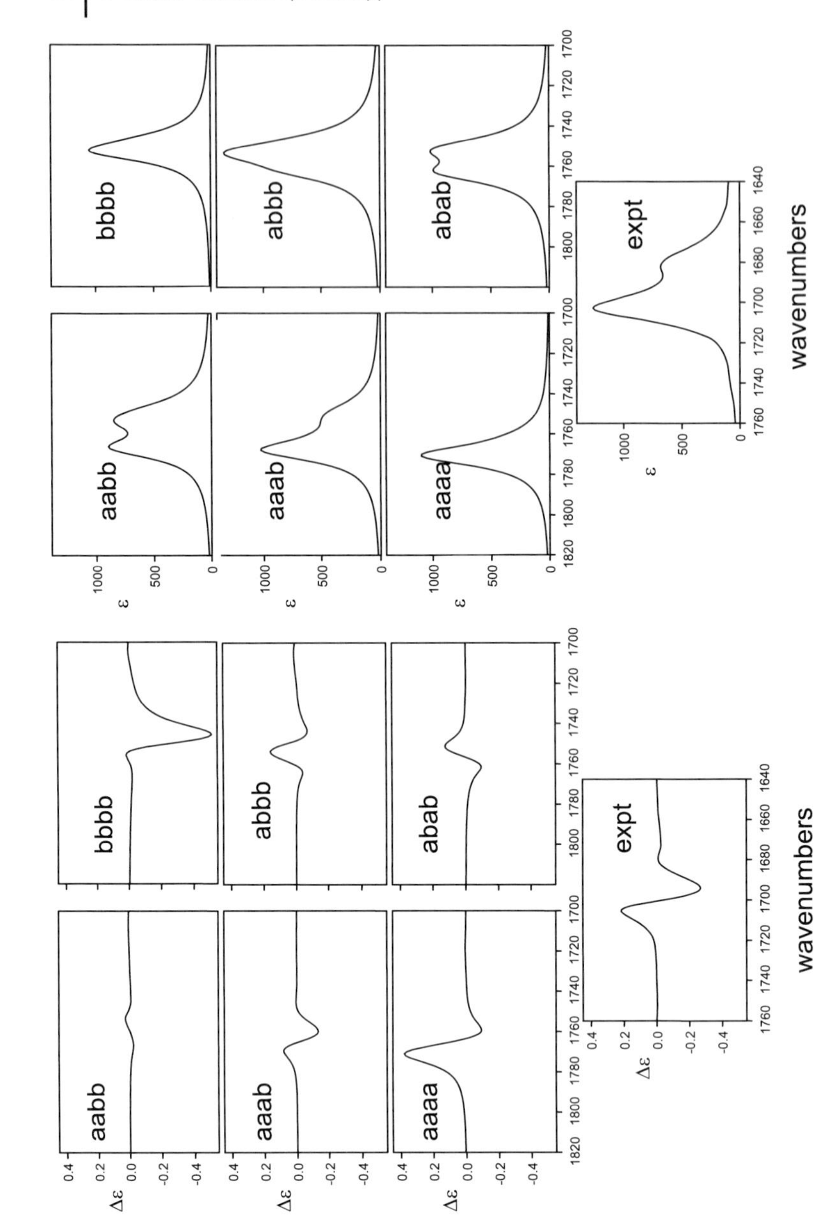

Fig. 8.15. B3LYP/6-21B* absorption and VCD spectra for the C=O stretching modes of the six conformations and the experimental spectra in $CDCl_3$. (Reprinted with permission from: M. Urbanová, V. Setnička, F.J. Devlin, P.J. Stephens, *J. Am. Chem. Soc.* **2005**, 127, 6700–6711. Copyright 2005 American Chemical Society.)

mation aaab. Any other conformation exhibits poor agreement between the calculated and experimental pattern. Part of the analysis concerning the C=O stretching modes is demonstrated in Fig. 8.15. A similar agreement was observed for the rest of the mid-IR region. This is the first analysis in which VCD spectroscopy coupled with DFT calculations was used to elucidate the structure of a supramolecular species; it clearly demonstrates the potential of VCD spectroscopy in the field of supramolecular chemistry.

8.6 Concluding Remarks

There are many methods of structural analysis which can be applied to problems related to supramolecular chemistry. They vary greatly in speed, resolution, structural capabilities and ease of application. In this respect, chiroptical methods belong somewhere in the middle. They are of relatively low resolution as far as structural detail is concerned. The level of available detail does not compare favorably with diffraction methods or NMR analysis. This is however fast improving along with the current progress made in the field of computational chemistry, particularly when molecular vibrations are examined. On the other hand chiroptical methods possess very distinct advantages: they are reasonably easy to apply; they do not require crystals; the possible range of solvents and solution concentrations is acceptably wide; the experiment can be carried out rather rapidly; and the time scale on which these methods operate is rather short, so they can be use to follow rather rapid structural changes, for example formation, disappearance or a change of a chiral supramolecular structure.

Particularly VCD and ROA provide not only static structural information, but also a picture of intermolecular forces as well. A comparison to molecular dynamics modeling is an evident application for the near future.

Abbreviations

CD	circular dichroism
CPL	circularly polarized luminescence
DCP	double circular polarization
DFT	density functional theory
ECD	electronic circular dichroism
FDCD	fluorescence detected circular dichroism
FFT	fast Fourier transform
FT	Fourier transform
ICD	induced Circular Dichroism
ICP	incident circular polarization
IR	infrared
MFP	magnetic field perturbation

ORD	optical rotatory dispersion
PEM	photoelastic modulator
ROA	Raman optical activity
SCP	scattered circular polarization
S/N	signal-to-noise ratio
TDDFT	time dependent density functional theory
UV	ultraviolet
VCD	vibrational circular dichroism
VIS	visible

References and Notes

1 N. Berova, K. Nakanishi, R. W. Woody, *Circular Dichroism. Principles and Applications*, 2nd edn John Wiley & Sons, Inc., 2000.

2 G. Snatzke, R. W. Woody, in *Circular Dichroism. Principles and Applications*, (Eds. N. Berova, K. Nakanishi, R. W. Woody), John Wiley & Sons, Inc., 2000, pp. 1–35.

3 P. J. Stephens, D. M. Mccann, J. R. Cheeseman, M. J. Frisch, *Chirality* **2005**, *17*, S52–S64.

4 P. J. Stephens, D. M. Mccann, E. Butkus, S. Stoncius, J. R. Cheeseman, M. J. Frisch, *J. Org. Chem.* **2004**, *69*, 1948–1958.

5 S. Grimme, A. Bahlmann, G. Haufe, *Chirality* **2002**, *14*, 793–797.

6 P. L. Polavarapu, *Chirality* **2002**, *14*, 768–781.

7 S. Grimme, F. Furche, R. Ahlrichs, *Chem. Phys. Lett.* **2002**, *361*, 321–328.

8 J. Autschbach, S. Patchkovskii, T. Ziegler, S. J. A. van Gisbergen, E. J. Baerends, *J. Chem. Phys.* **2002**, *117*, 581–592.

9 E. Giorgio, R. G. Viglione, R. Zanasi, C. Rosini, *J. Am. Chem. Soc.* **2004**, *126*, 12968–12976.

10 P. L. Polavarapu, *J. Phys. Chem. A* **2005**, *109*, 7013–7023.

11 K. Tanaka, Y. Itagaki, M. Satake, H. Naoki, T. Yasumoto, K. Nakanishi, N. Berova, *J. Am. Chem. Soc.* **2005**, *127*, 9561–9570.

12 T. Nehira, C. A. Parish, S. Jockusch, N. J. Turro, K. Nakanishi, N. Berova, *J. Am. Chem. Soc.* **1999**, *121*, 8681–8691.

13 P. J. M. Dekker, in *Circular Dichroism. Principles and Applications*, (Eds. N. Berova, K. Nakanishi, R. W. Woody), John Wiley & Sons, Inc., 2000, pp. 185–211.

14 L. D. Barron, L. Hecht, I. H. Mccoll, E. W. Blanch, *Mol. Phys.* **2004**, *102*, 731–744.

15 L. A. Nafie, G. S. Yu, X. Qu, T. B. Freedman, *Faraday Discuss.* **1994**, 13–34.

16 L. A. Nafie, T. B. Freedman, in *Circular Dichroism. Principles and Applications*, (Eds. N. Berova, K. Nakanishi, R. W. Woody), John Wiley & Sons, Inc., 2000, pp. 97–131.

17 E. W. Blanch, L. Hecht, L. D. Barron, *Methods* **2003**, *29*, 196–209.

18 E. W. Blanch, D. D. Kasarda, L. Hecht, K. Nielsen, L. D. Barron, *Biochemistry* **2003**, *42*, 5665–5673.

19 I. H. Mccoll, E. W. Blanch, L. Hecht, N. R. Kallenbach, L. D. Barron, *J. Am. Chem. Soc.* **2004**, *126*, 5076–5077.

20 I. H. Mccoll, E. W. Blanch, L. Hecht, L. D. Barron, *J. Am. Chem. Soc.* **2004**, *126*, 8181–8188.

21 F. J. Zhu, N. W. Isaacs, L. Hecht, L. D. Barron, *Structure* **2005**, *13*, 1409–1419.

22 F. J. Zhu, N. W. Isaacs, L. Hecht, L. D. Barron, *J. Am. Chem. Soc.* **2005**, *127*, 6142–6143.

23 J. Kapitan, V. Baumruk, V. Gut, J. Hlavacek, H. Dlouha, M. Urbanová,

E. Wunsch, P. Maloň, *Coll. Czech. Chem. Commun.* **2005**, *70*, 403–409.

24 A. Rodger, B. Norden, *Circular Dichroism and Linear Dichroism*, Oxford University Press, Oxford 1997.

25 J. K. Gawronski, in *Circular Dichroism. Principles and Applications*, (Eds. N. Berova, K. Nakanishi, R. W. Woody), John Wiley & Sons, Inc., 2000, pp. 305–336.

26 N. Berova, K. Nakanishi, in *Circular Dichroism. Principles and Applications*, (Eds. N. Berova, K. Nakanishi, R. W. Woody), John Wiley & Sons, Inc., 2000, pp. 337–382.

27 C. Andraud, C. Garcia, A. Collet, in *Circular Dichroism. Principles and Applications*, (Eds. N. Berova, K. Nakanishi, R. W. Woody), John Wiley & Sons, Inc., 2000, pp. 383–396.

28 R. F. Pasternack, J. I. Goldsmith, S. Szep, E. J. Gibbs, *Biophys. J.* **1998**, *75*, 1024–1031.

29 X. F. Huang, K. Nakanishi, N. Berova, *Chirality* **2000**, *12*, 237–255.

30 R. F. Pasternack, *Abstracts of Papers of the American Chemical Society* **2001**, *221*, U682.

31 R. F. Pasternack, *Chirality* **2003**, *15*, 329–332.

32 V. V. Borovkov, A. Muranaka, G. A. Hembury, Y. Origane, G. V. Ponomarev, N. Kobayashi, Y. Inoue, *Org. Lett.* **2005**, *7*, 1015–1018.

33 V. V. Borovkov, G. A. Hembury, Y. Inoue, *J. Org. Chem.* **2005**, *70*, 8743–8754.

34 J. Novy, M. Urbanová, K. Volka, *Coll. Czech. Chem. Commun.* **2005**, *70*, 1799–1810.

35 J. Novy, M. Urbanová, K. Volka, *J. Mol. Struct.* **2005**, *748*, 17–25.

36 L. Palivec, M. Urbanová, K. Volka, *J. Peptide Sci.* **2005**, *11*, 536–545.

37 S. Allenmark, *Chirality* **2003**, *15*, 409–422.

38 T. A. Keiderling, in *Biomolecular Spectroscopy, Part B*, (Eds. R. J. H. Clark, R. E. Hester), John Wiley, 1993, pp. 267–314.

39 P. L. Polavarapu, Z. Y. Deng, *Appl. Spectrosc.* **1996**, *50*, 686–692.

40 L. A. Nafie, R. K. Dukor, J. R. Roy, A. Rilling, X. Cao, H. Buijs, *Appl. Spectrosc.* **2003**, *57*, 1245–1249.

41 F. Long, T. Freedman, T. J. Tague, L. A. Nafie, *Appl. Spectrosc.* **1997**, *51*, 508–511.

42 W. Hug, G. Hangartner, *J. Raman Spectrosc.* **1999**, *30*, 841–852.

43 T. B. Freedman, F. J. Long, M. Citra, L. A. Nafie, *Enantiomer* **1999**, *4*, 103–119.

44 R. K. Dukor, L. A. Nafie, in *Encyclopedia of Analytical Chemistry*, (Ed. R. A. Meyers), John Wiley & Sons Ltd, 2000, pp. 662–676.

45 T. B. Freedman, X. L. Cao, R. K. Dukor, L. A. Nafie, *Chirality* **2003**, *15*, 743–758.

46 P. Bouř, J. Sopkova, L. Bednarova, P. Maloň, T. A. Keiderling, *J. Comput. Chem.* **1997**, *18*, 646–659.

47 P. Bouř, T. A. Keiderling, *J. Am. Chem. Soc.* **1993**, *115*, 9602–9607.

48 J. Kubelka, T. A. Keiderling, *J. Am. Chem. Soc.* **2001**, *123*, 12048–12058.

49 P. Bouř, J. Kapitan, V. Baumruk, *J. Phys. Chem. A* **2001**, *105*, 6362–6368.

50 L. A. Nafie, *Ann. Rev. Phys. Chem.* **1997**, *48*, 357–386.

51 S. Grimme, *Chem. Phys. Lett.* **1996**, *259*, 128–137.

52 F. Furche, R. Ahlrichs, C. Wachsmann, E. Weber, A. Sobanski, F. Vogtle, S. Grimme, *J. Am. Chem. Soc.* **2000**, *122*, 1717–1724.

53 J. Autschbach, T. Ziegler, S. J. A. van Gisbergen, E. J. Baerends, *J. Chem. Phys.* **2002**, *116*, 6930–6940.

54 E. Giorgio, K. Tanaka, W. D. Ding, G. Krishnamurthy, K. Pitts, G. A. Ellestad, C. Rosini, N. Berova, *Bioorg. Med. Chem.* **2005**, *13*, 5072–5079.

55 W. Schuhly, S. L. Crockett, W. M. F. Fabian, *Chirality* **2005**, *17*, 250–256.

56 L. A. Nafie, T. A. Keiderling, P. J. Stephens, *J. Am. Chem. Soc.* **1976**, *98*, 2715–2723.

57 L. A. Nafie, M. Diem, *Appl. Spectrosc.* **1979**, *33*, 130–135.

58 E. D. Lipp, C. G. Zimba, A. A. Nafie, *Chem. Phys. Lett.* **1982**, *90*, 1–4.

59 T. A. Keiderling, in *Practical Fourier Transform Infrared Spectroscopy*, (Eds. J. R. Ferraro, K. Krishnan), Academic Press, San Diego, CA., 1990, pp. 203–284.
60 T. B. Freedman, S. J. Cianciosi, N. Ragunathan, J. E. Baldwin, L. A. Nafie, *J. Am. Chem. Soc.* **1991**, *113*, 8298–8305.
61 T. B. Freedman, L. A. Nafie, T. A. Keiderling, *Biopolymers* **1995**, *37*, 265–279.
62 T. A. Keiderling, in *Circular Dichroism and the Conformational Analysis of Biomolecules*, (Ed. G. D. Fasman), Plenum Press, 1996, pp. 555–598.
63 T. A. Keiderling, in *Circular Dichroism. Principles and Applications*, (Eds. N. Berova, K. Nakanishi, R. W. Woody), John Wiley & Sons, Inc., 2000, pp. 621–656.
64 T. A. Keiderling, Q. Xu, *Adv. Protein Chem.* **2002**, *62*, 111–161.
65 T. A. Keiderling, Q. Xu, *Macromol. Symp.* **2005**, *220*, 17–31.
66 J. Kubelka, R. Huang, T. A. Keiderling, *J. Phys. Chem. B* **2005**, *109*, 8231–8243.
67 P. Maloň, T. A. Keiderling, *Appl. Optics* **1997**, *36*, 6141–6148.
68 M. Urbanová, V. Setnička, K. Volka, *Chirality* **2000**, *12*, 199–203.
69 L. A. Nafie, H. Buijs, A. Rilling, X. L. Cao, R. K. Dukor, *Appl. Spectrosc.* **2004**, *58*, 647–654.
70 X. H. Qu, E. A. Lee, G. S. Yu, T. B. Freedman, L. A. Nafie, *Appl. Spectrosc.* **1996**, *50*, 649–657.
71 M. Urbanová, V. Setnička, F. J. Devlin, P. J. Stephens, *J. Am. Chem. Soc.* **2005**, *127*, 6700–6711.
72 P. Pancoska, E. Bitto, V. Janota, M. Urbanová, V. P. Gupta, T. A. Keiderling, *Protein Sci.* **1995**, *4*, 1384–1401.
73 V. Baumruk, P. Pancoska, T. A. Keiderling, *J. Mol. Biol.* **1996**, *259*, 774–791.
74 J. Kubelka, P. Pancoska, T. A. Keiderling, *Appl. Spectrosc.* **1999**, *53*, 666–671.
75 P. Pancoska, V. Janota, T. A. Keiderling, *Anal. Biochem.* **1999**, *267*, 72–83.
76 P. Pancoska, J. Kubelka, T. A. Keiderling, *Appl. Spectrosc.* **1999**, *53*, 655–665.
77 W. C. Johnson, *Proteins* **1990**, *7*, 205–214.
78 L. Rosenfeld, *Z. Phys.* **1928**, *52*, 161–174.
79 L. J. Wang, P. Pancoska, T. A. Keiderling, *Biochemistry* **1994**, *33*, 8428–8435.
80 L. A. Nafie, T. B. Freedman, *Spectroscopy* **1987**, *2*, 24–29.
81 J. A. Schellman, *J. Chem. Phys.* **1996**, *44*, 55–63.
82 J. A. Schellman, *Acc. Chem. Res.* **1968**, *1*, 144–151.
83 P. M. Bayley, E. B. Nielsen, J. A. Schellman, *J. Phys. Chem.* **1969**, *73*, 228–243.
84 I. Jr. Tinoco, *Adv. Chem. Phys.* **1962**, *4*, 113–160.
85 R. Lyng, A. Rodger, B. Norden, *Biopolymers* **1992**, *32*, 1201–1214.
86 R. Lyng, A. Rodger, B. Norden, *Biopolymers* **1991**, *31*, 1709–1720.
87 V. Rizzo, J. A. Schellman, *Biopolymers* **1984**, *23*, 435–470.
88 A. Koslowski, N. Sreerama, R. W. Woody, in *Circular Dichroism. Principles and Applications*, (Eds. N. Berova, K. Nakanishi, R. W. Woody), John Wiley & Sons, Inc., 2000, pp. 55–95.
89 L. A. Nafie, T. B. Freedman, *J. Phys. Chem.* **1986**, *90*, 763–767.
90 M. G. Paterlini, T. B. Freedman, L. A. Nafie, *J. Am. Chem. Soc.* **1986**, *108*, 1389–1397.
91 P. Maloň, K. Bláha, *Coll. Czech. Chem. Commun.* **1977**, *42*, 687–898.
92 P. J. Stephens, F. J. Devlin, C. S. Ashvar, C. F. Chabalowski, M. J. Frisch, *Faraday Discuss.* **1994**, 103–119.
93 P. J. Stephens, *J. Phys. Chem.* **1985**, *89*, 748–752.
94 P. J. Stephens, *J. Phys. Chem.* **1987**, *91*, 1712–1715.
95 P. J. Stephens, F. J. Devlin, C. F. Chabalowski, M. J. Frisch, *J. Phys. Chem.* **1994**, *98*, 11623–11627.
96 P. Bouř, J. Kubelka, T. A. Keiderling, *Biopolymers* **2002**, *65*, 45–59.

97 J. Hilario, J. Kubelka, T. A. Keiderling, *J. Am. Chem. Soc.* **2003**, *125*, 7562–7574.
98 C. Cappelli, S. Monti, A. Rizzo, *Int. J. Quantum Chem.* **2005**, *104*, 744–757.
99 L. A. Nafie, T. B. Freedman, *Enantiomer* **1998**, *3*, 283–297.
100 T. A. Keiderling, P. Pancoska, L. J. Wang, R. Dukor, M. Urbanová, *Biophys. J.* **1993**, *64*, A245.
101 P. L. Polavarapu, C. X. Zhao, *Fresenius J. Anal. Chem.* **2000**, *366*, 727–734.
102 V. Andrushchenko, J. L. Mccann, J. H. van de Sande, H. Wieser, *Vib. Spectrosc.* **2000**, *22*, 101–109.
103 A. Polyanichko, H. Wieser, *Biopolymers* **2005**, *78*, 329–339.
104 J. R. Cheeseman, M. J. Frisch, F. J. Devlin, P. J. Stephens, *Chem. Phys. Lett.* **1996**, *252*, 211–220.
105 P. J. Stephens, C. S. Ashvar, F. J. Devlin, J. R. Cheeseman, M. J. Frisch, *Mol. Phys.* **1996**, *89*, 579–594.
106 C. S. Ashvar, F. J. Devlin, P. J. Stephens, K. L. Bak, T. Eggimann, H. Wieser, *J. Phys. Chem. A* **1998**, *102*, 6842–6857.
107 F. J. Devlin, P. J. Stephens, J. R. Cheeseman, M. J. Frisch, *J. Phys. Chem. A* **1997**, *101*, 6322–6333.
108 P. J. Stephens, F. J. Devlin, *Chirality* **2000**, *12*, 172–179.
109 P. Bouř, H. Navrátilová, V. Setnička, M. Urbanová, K. Volka, *J. Org. Chem.* **2002**, *67*, 161–168.
110 V. Setnička, M. Urbanová, P. Bouř, V. Král, K. Volka, *J. Phys. Chem. A* **2001**, *105*, 8931–8938.
111 S. Sreerama, R. W. Woody, in *Circular Dichroism. Principles and Applications*, (Eds. N. Berova, K. Nakanishi, R. W. Woody), John Wiley & Sons, Inc., 2000, pp. 601–620.
112 W. C. Johnson, in *Circular Dichroism. Principles and Applications*, (Eds. N. Berova, K. Nakanishi, R. W. Woody), John Wiley & Sons, Inc., 2000, pp. 703–718.
113 J. C. Maurizot, in *Circular Dichroism. Principles and Applications*, (Eds. N. Berova, K. Nakanishi, R. W. Woody), John Wiley & Sons, Inc., 2000, pp. 719–740.
114 V. Andrushchenko, Z. Leonenko, D. Cramb, H. van de Sande, H. Wieser, *Biopolymers* **2001**, *61*, 243–260.
115 A. M. Polyanichko, E. V. Chikhirzhina, V. V. Andrushchenko, H. Wieser, V. I. Vorob'ev, *Biofizika* **2005**, *50*, 810–817.
116 P. Pancoska, S. C. Yasui, T. A. Keiderling, *Biochemistry* **1989**, *28*, 5917–5923.
117 M. Urbanová, R. K. Dukor, P. Pancoska, V. P. Gupta, T. A. Keiderling, *Biochemistry* **1991**, *30*, 10479–10485.
118 V. P. Gupta, T. A. Keiderling, *Biopolymers* **1992**, *32*, 239–248.
119 V. Baumruk, T. A. Keiderling, *J. Am. Chem. Soc.* **1993**, *115*, 6939–6942.
120 M. Urbanová, P. Pancoska, T. A. Keiderling, *Biochim. Biophys. Acta* **1993**, *1203*, 290–294.
121 M. Urbanová, T. A. Keiderling, P. Pancoska, *Bioelectrochem. Bioenerg.* **1996**, *41*, 77–80.
122 M. Urbanová, V. Setnička, V. Král, K. Volka, *Biopolymers* **2001**, *60*, 307–316.
123 A. M. Polyanichko, V. V. Andrushchenko, E. V. Chikhirzhina, V. I. Vorob'ev, H. Wieser, *Nucl. Acids Res.* **2004**, *32*, 989–996.
124 J. Kubelka, T. A. Keiderling, *J. Am. Chem. Soc.* **2001**, *123*, 6142–6150.
125 R. Huang, J. Kubelka, W. Barber-Armstrong, R. A. G. D. Silva, S. M. Decatur, T. A. Keiderling, *J. Am. Chem. Soc.* **2004**, *126*, 2346–2354.
126 R. Huang, V. Setnička, C. L. Thomas, M. A. Etienne, R. P. Hammer, T. A. Keiderling, *Biophys. J.* **2005**, *88*, 159A.
127 V. Setnička, R. Huang, C. L. Thomas, M. A. Etienne, J. Kubelka, R. P. Hammer, T. A. Keiderling, *J. Am. Chem. Soc.* **2005**, *127*, 4992–4993.
128 L. J. Wang, T. A. Keiderling, *Biochemistry* **1992**, *31*, 10265–10271.
129 L. J. Wang, T. A. Keiderling, *Nucl. Acids Res.* **1993**, *21*, 4127–4132.

130 L. J. Wang, T. A. Keiderling, *Biophys. J.* **1993**, *64*, A279.

131 L. J. Wang, L. G. Yang, T. A. Keiderling, *Biophys. J.* **1994**, *67*, 2460–2467.

132 V. Andrushchenko, H. Wieser, P. Bouř, *J. Phys. Chem. B* **2002**, *106*, 12623–12634.

133 V. Andrushchenko, H. V. de Sande, H. Wieser, *Biopolymers* **2003**, *69*, 529–545.

134 V. Andrushchenko, D. Tsankov, H. Wieser, *J. Mol. Struct.* **2003**, *661*, 541–560.

135 D. R. Yazbeck, K. L. Min, M. J. Damha, *Nucl. Acids Res.* **2002**, *30*, 3015–3025.

136 V. Setnička, M. Urbanová, S. Pataridis, V. Král, K. Volka, *Tetrahedron: Asymmetry* **2002**, *13*, 2661–2666.

137 V. Král, S. Pataridis, V. Setnička, K. Zaruba, M. Urbanová, K. Volka, *Tetrahedron* **2005**, *61*, 5499–5506.

138 U. Hahn, A. Kaufmann, M. Nieger, O. Julínek, M. Urbanová, F. Vögtle, *Eur. J. Org. Chem.* **2006**, 1237–1244.

139 M. Tichý, T. Kraus, J. Závada, I. Cisařová, J. Podlaha, *Tetrahedron: Asymmetry* **1999**, *10*, 3277–3280.

9
Crystallography and Crystal Engineering

Kari Rissanen

9.1
Introduction

X-ray crystallography is at present the most powerful tool for the detailed structural analysis of crystalline supramolecular compounds, complexes and intermolecular interactions. However, only a few decades ago a single crystal X-ray diffraction study of a crystalline solid was considered to be a very tedious and time-consuming task performed by well-educated and experienced crystallographers. Certainly, X-ray crystallography was not generally accepted as a routine analytical tool for structural analysis, except for well-diffracting small molecule crystals. Meanwhile, the extremely fast development of computers has had a direct impact on the speed and ease of X-ray diffraction analysis. Modern area detector based diffractometers allow very fast and accurate data collection and processing, so that several, even very large, data sets can be collected and processed in one day. Fast computers also speed up the structure solution and refinement, so that even very complicated problems can be tackled within a reasonable time. The simultaneous development of Supramolecular Chemistry and crystallographic techniques has merged the supramolecular ideology with classical crystallography giving birth to a new research area called Crystal Engineering. Crystal Engineering is a very young research field and still in its infancy but can already be considered as a scientific genre.

Even though the first use of the term "Crystal Engineering" can be found from proceedings (as abstract) of the American Physical Society Meeting in 1955 [1], it was as late as 1971 when the term crystal engineering was used by G. M. J. Schmidt [2] in a full scientific paper. In this article, crystal engineering was used for the first time as an explicit term, also the article postulated that, under suitable conditions, molecular recognition events, *viz.* self-assembly, could be the major factor leading to crystal formation. The rapid development of supramolecular chemistry – elegantly defined by J.-M. Lehn [3, 4] as the "chemistry beyond the molecule, bearing on the organized entities of higher complexity that result from the association of two or more chemical species held together by intermolecular forces" – had a major impact on the birth of crystal engineering. The weak inter-

Analytical Methods in Supramolecular Chemistry. Edited by Christoph Schalley

ISBN: 978-3-527-31505-5

molecular interactions which form supramolecules or supramolecular assemblies are the same as those that act in the formation of crystals, thus a link between supramolecular chemistry and crystal engineering became apparent. This concept was taken to the crystallographic extreme by J. Dunitz [5, 6] when he referred to organic crystals as "supermolecule(s) *par excellence*".

Since the publication "*Crystal Engineering: the Design of Organic Solids*", a single author book [7] on crystal engineering principles, Professor Gautam Desiraju (Hyderabad, India) has had a major role in the definition and development of "Crystal Engineering". More recently, Professor Dario Braga (Bologna, Italy) has contributed significantly to the future trends in crystal engineering [8]. Very recently, two journals, *Crystal Engineering Communications* (CEC, Royal Society of Chemistry journal, started late 1999) and *Crystal Growth and Design* (CGD, American Chemical Society journal, started late 2000) have been established, focusing on all aspects of crystal engineering. Also, a large number of purely crystal engineering research papers appear frequently in the top chemical journals like *Journal of American Chemical Society, Chemistry European Journal, Chemical Communications, Inorganic Chemistry, Journal of Organic Chemistry, New Journal of Chemistry, European Journal of Inorganic Chemistry, European Journal of Organic Chemistry*, just to mention a few. The Institute of Scientific Information (ISI), Web of Science general search under topic "crystal engineering" between years 1986–2006 references 1136 articles containing "crystal engineering" in the article titles, keywords or abstracts [9]. The quantity of publications is not yet very big, but is very rapidly increasing and will soon reach a level of several hundred publications per year.

Especially, large supramolecular systems impose a very demanding task on X-ray crystallography. The size of the structures to be solved are comparable to small proteins (up to *FW* 10 000), yet protein crystallographic techniques cannot be applied to supramolecular systems. The methods developed for small molecule crystallography have to be adapted to large supramolecular systems thus making the structural analysis of such systems both demanding and, when successful, also very rewarding.

9.2 Crystallography

9.2.1 Introduction

This chapter is written for graduate and post graduate students and researchers with no previous knowledge of structural analysis based on single crystal X-ray diffraction. The aim is not to give a comprehensive treatise of the crystallography *per se*, but to give a non-mathematical description of the method, practical hints and advice to those supramolecular chemists designing and synthesizing systems, whose structures they want to study and analyze using X-ray crystallography with or without the help of a service crystallographer or a crystallographic collaborator.

This concise treatise is not intended to make readers experts on hardcore crystallography, but to help them to get a qualitative picture of the method and understand the underlying principles and drawbacks of single crystal X-ray diffraction analysis. To study the theoretical basis of X-ray crystallography in detail, the reader is encouraged to read some excellent X-ray crystallography text books to become fully acquainted with the theory involved at introductory [10], intermediate [11] or advanced [12] level.

Modern single crystal X-ray crystallography has two major application areas. Small molecule crystallography focuses on target structures having molecular weights between 100 and 700 (in crystallographic terms more appropriate definition is "a weight of the asymmetric unit", W_{au} [13], more about this later in the text), resulting in very precise and accurate bond distances and angles with low crystallographic agreement factor, so called *R*-factor (normally <0.07) due to good quality crystals with no disorder and a high (usually >10) ratio of the number of observed reflections to the number of refined parameters. The *R*-factor is a measure of the level of agreement between the properly scaled observed structure factors, F_{obs}, (from the intensity data obtained during data collection phase) and calculated structure factors (F_{calc}) (see Section 9.2.2.5 for explanation of a structure factor). It can also be expressed as a percentage, i.e. an *R*-factor of 0.07 can be reported as 7%. The *R*-factor, also known as residual factor, is mathematically defined (Eq. 9.1) as the sum of all discrepancies between the corresponding observed and calculated structure factors (ignoring the signs of the differences) and normalizing the sum by dividing by the sum of observed structure factors. The *R*-factor is thus structure independent and can be compared with different structures.

$$R = \frac{\sum_{hkl} \left| |F_{obs}| - |F_{calc}| \right|}{\sum_{hkl} |F_{obs}|} \tag{9.1}$$

The other, technically quite different application area, is protein crystallography which deals with very large structures, even virus structures, where W_{au} can be as high as 100 000. However these structures cannot be considered to be crystallographically very accurate, often the resolution is not high enough to determine individual atom positions, and instead amino acid residues are located. Protein *R*-values are seldom less than 0.15 and the ratio of the number of observed reflections and number of refined parameters is very low (<5). However protein X-ray structures are considered to be reliable structural information about the conformation and overall structure of the protein in the crystalline state.

Accurate determination of the structures of molecular inclusion complexes and molecular assemblies has gained in importance in supramolecular chemistry and recently in crystal engineering. Detailed information about the nature of these weak intermolecular interactions is crucial in order to understand and further develop supramolecular systems and understand crystal growth and the properties of crystalline materials. Particularly accurate structural information can be gained in the solid state by single crystal X-ray crystallography.

9.2.2
A Walk through a Single Crystal Structural Determination

Single crystal structural analysis is a multistep procedure and has many key points. The aim of this section is to give a non-mathematical "walkthrough" the structural analysis which goes as follows (after the synthesis and purification of the target compound): *crystal* → *selecting and mounting of the crystal* → *unit cell determination and preliminary space group selection* → *data collection (measurement of the intensity data), data processing and final space group determination* → *data reduction, structure solution and refinement* → *analysis of the structure* (tables, figures, etc.). Some basic crystallographic definitions are given in the appropriate sections, but as mentioned in the introduction, for a mathematical explanation the reader is encouraged to check Refs. [10]–[12]. Modern area detector based diffractometers (Fig. 9.1) are fully automated robots that will do the data collection automatically with relatively little effort from the structural analyst.

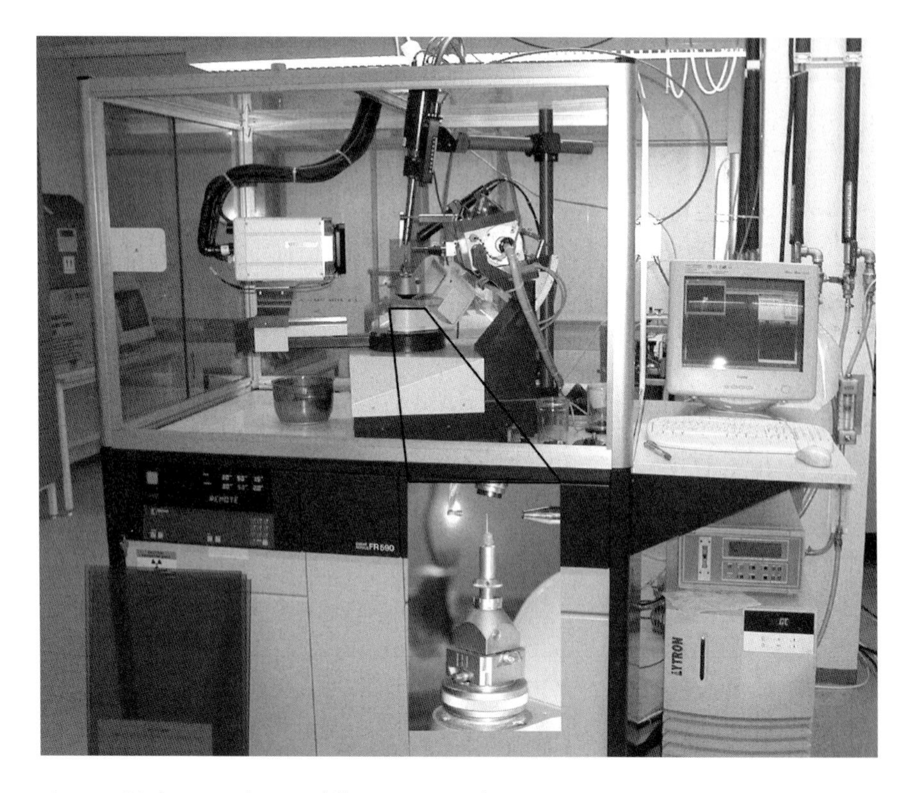

Fig. 9.1. Modern area detector diffractometer (Bruker-Nonius Kappa APEX II) with sealed-tube Mo-radiation source and low-temperature cooling device in Jyväskylä University X-ray diffraction laboratory. The blow-up picture shows the goniometer head into which the crystal under study has been mounted.

9.2.2.1 The (Single) Crystal

By far the most important item in structural analysis of a particular supramolecular system by X-ray crystallography is, of course, the crystal. The need of a single crystal, good enough for the structural analysis of a given system is crucial; this is at the same time the major drawback of X-ray crystallographic structural analysis. The lack of suitable crystals will make the structural analysis completely impossible, unlike in almost all other analytical techniques, where "running" the measurement is possible, *e.g.* one can run an NMR or MS spectrum from virtually any soluble sample. This, however, does not mean that reasonable results are obtained from the NMR or MS measurement even if the spectrum is of fairly good quality. In X-ray crystallography the measurement (called data collection) cannot be done without the crystal. In other words without a good enough *single* crystal, structural analysis by X-ray crystallography *cannot* be performed. It is worth noticing that not all material that looks like a crystal is a good single crystal. Examples of good diffraction quality crystals can be seen when either table salt (NaCl) or granulated sugar is inspected under a microscope. It is quite often impossible to obtain such good looking (and diffracting) crystals as salt or sugar, but the reader is reminded that the "looks" of the crystal is not a guarantee that it will be of good diffraction quality. The final check should always to be done using the diffractometer.

The starting step in X-ray diffraction structural analysis is the production of the crystals from the target molecule/system. It might be that, during the purification step of the synthesis, precipitation has already produced material which contains at least one good enough single crystal or that the recrystallization of the compound can easily be done, *i.e.* growing a crystal from the target compound by specific procedure(s) (see below). The most important property of the crystal is a good enough periodicity of the crystal lattice, *i.e.* packing of the molecules and solvent molecules (if present), to gain enough diffraction power. The periodicity is induced by the crystal growth, a very subtle and more or less spontaneous ordering/desolvation process of molecules, ions and/or solvent molecules. No matter what technique (evaporation, diffusion, sublimation, etc.) is used to obtain the crystal, the only goal is to produce the best possible periodicity of the entities to be studied. The success of this on the other hand solely depends on the nature of the system (molecules and solvents) from which the crystals are intended to be formed. Supramolecules normally behave like large organic molecules, being soluble in certain solvents, and the same techniques that are used for organic molecules can normally be applied to supramolecules, too. Before starting, the reader is encouraged to read appropriate chapters in the crystallographic text books [10–12], or from articles/websites on crystal growing [14, 15] or search the World Wide Web for information using the search term "crystal growth". With modern area detector diffractometers the size of the crystal is not as important as it was earlier, typically the size is from $0.1 \times 0.1 \times 0.1$ mm to $0.5 \times 0.5 \times 0.5$ mm (well-diffracting crystals as small as $0.05 \times 0.05 \times 0.05$ mm can give good data sets).

In addition to the periodicity of the crystal lattice, there are other factors affecting the diffraction quality of the crystal, *i.e.* the type of the ordered atoms have a significant influence on the diffraction power of the crystal. The heavier (more electrons)

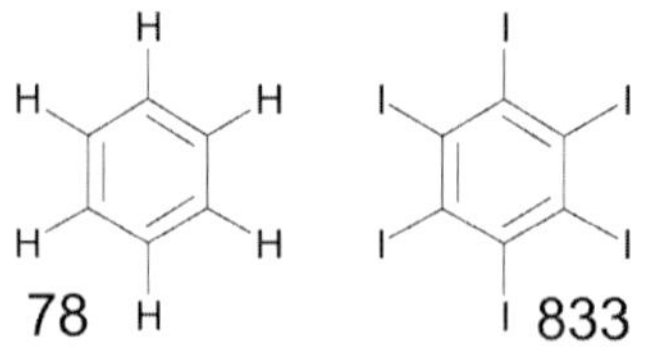

Fig. 9.2. The chemical structures and molecular weights of benzene (left) and hexaiodobenzene (right).

and more well-ordered atoms the crystal lattice contains, the better will be the diffraction/scattering power. In supramolecular systems, the crystal lattice consists mainly of light elements like carbon, hydrogen, oxygen and nitrogen, and only when metallo-supramolecular systems are studied are heavier elements, like transition metals, bromide and iodine, usually present. A very rough but practical estimate for the intrinsic diffraction power (dp) [16] (if ideal periodicity is assumed) of the compound can be estimated simply by dividing its formula weight by the number of atoms in the molecule/assembly (N), $dp = FW/N$, e.g. benzene $dp = 6.5$ ($78/12 = 6.5$); hexaiodobenzene $dp = 69.4$ ($833/12 = 69.4$).

Therefore the quality of the crystal, i.e. perfect periodicity of the crystal lattice, is extremely crucial for light-atom structures with high W_{au} (>1000) values, as is frequently the case in supramolecular systems.

The third important point about the crystal is its stability. Even if the two first requirements related to diffraction quality of the crystal are met, crystal structure analysis may be hampered or prevented by crystals that are highly unstable under normal laboratory conditions. Large supramolecular assemblies are held together by weak intermolecular interactions, and crystals consisting of such entities are often very sensitive to the environment and fragile to handle. Solvent molecules usually play a crucial role in the crystal growth of supramolecular systems and are needed in the construction of the periodical crystal lattice. Thus, many supramolecular crystals are only stable in solvent(s) from which they have been (re)crystallized. It is impractical to change laboratory conditions to enhance crystal stability since crystals can easily be handled inside the mother solvent and/or protected against decomposition. When no information about the stability of the crystals exists, all new crystals should be treated as unstable. This already has to be taken into account at the recrystallization phase. When the crystals are formed and existing in solution, the crystallization vessel should be closed to ensure that the mother solvent does not evaporate and thus the crystals stay in solution. Mailing or transportation of crystals, if needed, must accordingly be done in well packed sealed tubes, flasks or ampoules.

9.2.2.2 Mounting of the Crystal

For unit cell determination and data collection, the crystal must be mounted into the diffractometer on the goniometer head (see Fig. 9.1, blow-up part), this is done by using a specially constructed brass pin crystal holder (Figs. 9.3. and 9.4).

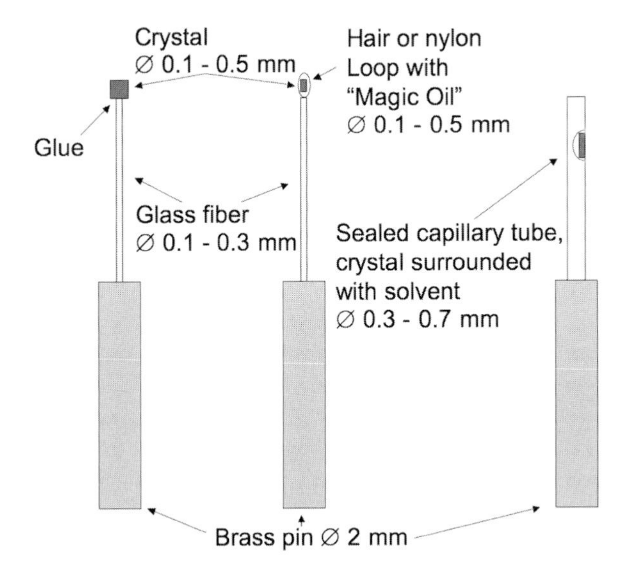

Fig. 9.3. Schematic presentation of different crystal holders (to be placed into the goniometer head: Fig. 9.1).

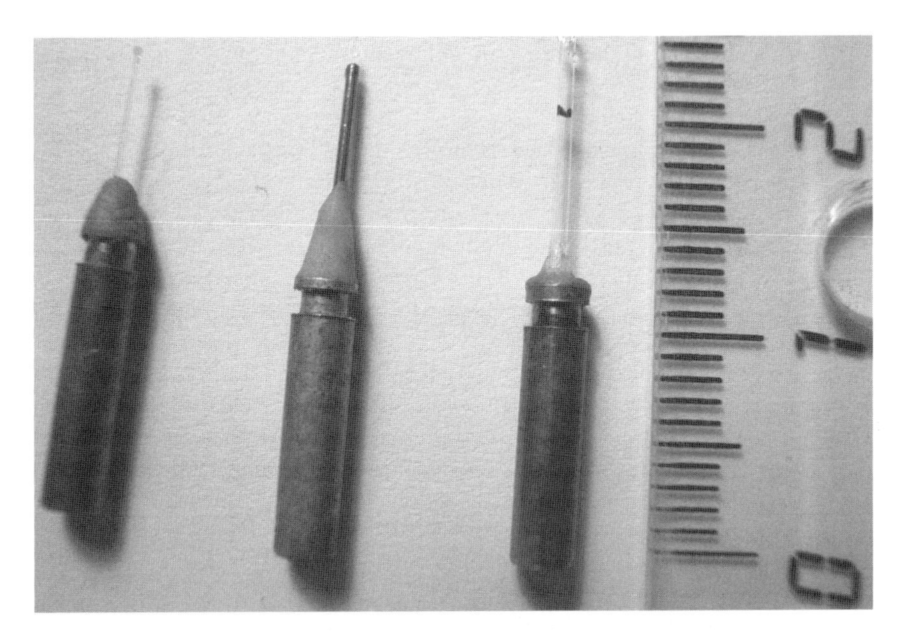

Fig. 9.4. A photo of the actual crystal holders with a cm scale for comparison.

Stable crystals for room temperature measurements can be simply glued with non-crystalline glue onto the tip of a glass fiber fixed on a brass pin (see Fig. 9.3, left). At present, all measurements are or should be routinely done at low temperatures (80–170 K) and then a loop (see below and Fig. 9.3, middle) is used. For low temperature measurements or mounting sensitive and unstable crystals the selected crystal is "fished" under a microscope from the mother solvent or from the microscope plate by a "loop" crystal holder dipped into commercially available perfluorinated polyether [17] (so called "Magic Oil"). This "oil" is viscous, oxygen free and unreactive and forms a film around the crystal, protecting it from the hostile environment. During unit cell determination and data collection the crystal must not move inside the loop. This is best achieved when the crystal is cooled (with cooling devices available from X-ray diffractometer manufacturers) below 193 K (−80 °C). This procedure has an additional advantage since the low temperature during unit cell determination and data collection reduces the thermal movement of the atoms in the crystal thus improving the quality of the data set.

If the crystal does not survive cooling (if cracks appear and/or the intensity of the reflections is lost) or it cannot be transferred intact into the loop (or seems to react or decompose inside the oil as happens occasionally), the crystal can be sealed into a capillary crystal holder (see Figs. 9.3 and 9.4, right).

The advantages of this procedure is that the crystal is in contact only with the mother solvent (actually surrounded by it all the time) and that the unit cell determination and data collection can also be performed at room temperature. Disadvantages are the extreme care needed to seal and manipulate the fragile capillary crystal holder, which is unique for each crystal, and the possible movement of the crystal inside the capillary tube during measurement. After successfully placing the crystal into the diffractometer, unit cell determination can commence as the next step.

9.2.2.3 Unit Cell Determination and Preliminary Space Group Selection

The essence, after a suitable crystal is obtained and properly mounted, is to check using the diffractometer, if the crystal diffracts X-rays. The crystal in its proper holder is first placed into the goniometer head (Fig. 9.1, blow-up) and then the goniometer head with the crystal is mounted into the goniometer (the part that rotates the crystal during measurements) in the center of the diffractometer (Fig. 9.1). The goniometer head with the crystal is carefully transferred into the goniometer and then, using the screws in the goniometer head (Fig. 9.1, blow-up), adjusted to the exact center of the X-ray beam so that when rotated the center of the crystal always stays in the same place (in the center of the X-ray beam).

The X-ray beam will diffract (scatter) from the crystal and the reflected X-ray beams (reflections) are collected with a detector. Figure 9.5 schematically shows the principle; the reader is encouraged to read Refs. [10–12] for a more profound discussion of the diffraction phenomenon.

Before the unit cell determination and the actual data collection, the diffraction pattern image from the crystal should be very carefully evaluated for three factors: (i) Does the crystal diffract at all (even very nice looking crystals might not dif-

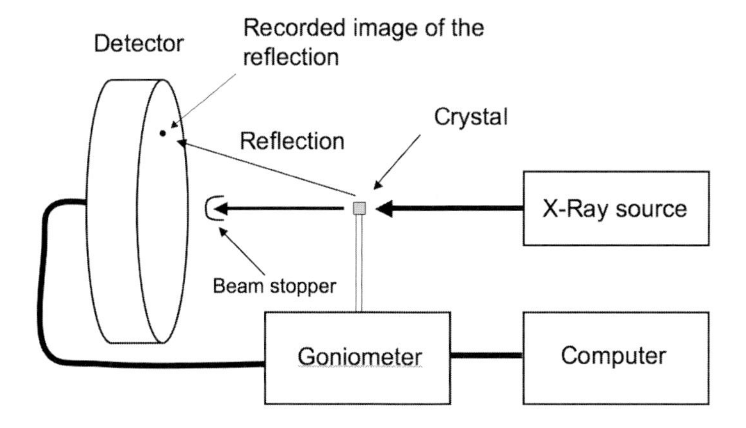

Fig. 9.5. Schematic presentation of diffraction of an X-ray beam from the crystal.

fract)? (ii) If so, does the diffraction pattern indicate twinning of the crystal ("twin peaks" or two reflections systematically too close to each other)? (iii) How high is the resolution with which the crystal diffracts (visible reflections near the edges of the image, Fig. 9.6). The diffracting image is obtained by rotating the crystal around the brass pin axis during the exposure with the X-rays and should result many reflections ("spots") on the detector. The detector image should look like the sky full of stars in a dark cloudless night (Fig. 9.6). If such an image is obtained, the crystal diffracts. The quality of the image will show if it is a well-diffracting single crystal. If very few or no "spots" appear in the image, the crystal does not diffract well enough and has to be replaced.

If in the diffraction image either one of the two factors is not acceptable, *i.e.* the crystal is clearly twinned or does not diffract well enough or not at all, a new crystal has to be selected. Only a very persistent selection procedure for the best possible crystal will bring success. In his/her own work, the reader is encouraged not to give up with the first five crystals if they fail, but to test at least 15–20 crystals until a well-diffracting single crystal is found. Remember that even crystals which look bad when viewed under the microscope, can diffract very well and give excellent result and *vice versa*, samples that look very good might not diffract at all. If the batch of crystals available does not have any good single crystals, then of the whole batch has to be recrystallized or other solvents used. Working with a bad or faulty data set obtained from poor quality crystals will result in frustration and loss of time. It is much more advisable to start from the beginning and focus more on getting good crystals than to force unit cell determination and data collection on a bad crystal. The authors' laboratory has many examples where only a very persistent crystal selection procedure has produced acceptable results. With a poorly-diffracting crystal giving a bad data set, the overall skeleton of the structure might be resolved, but will not produce a structure which can be published.

Even though modern area detector diffractometers are able to collect a data set without determining the unit cell, it is always advisable to determine the unit

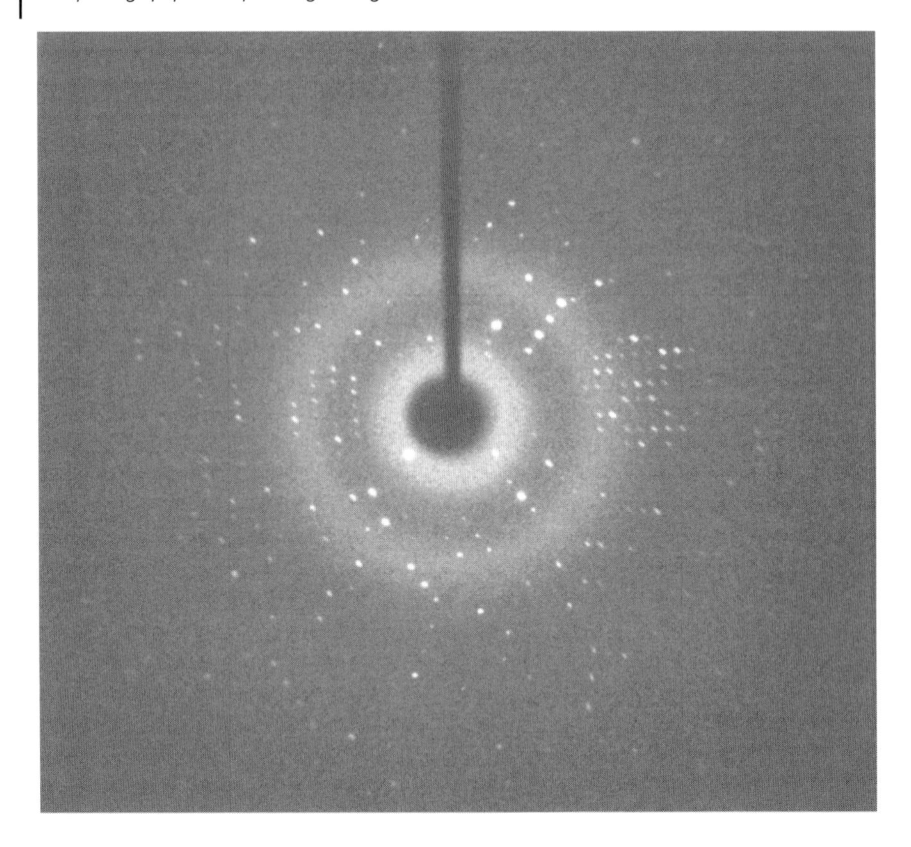

Fig. 9.6. Well-resolved reflections on an area detector from a moderately diffracting crystal.

cell first. The unit cell can be non-mathematically defined as the smallest virtual/imaginary 3-D unit into which the atoms in the crystal can be placed so that, by multiplying, the whole macroscopic crystal will be defined. The unit cell can only have a certain shape (Fig. 9.7) defined by six variables: a, b, c, α, β, γ, which define a 3-D entity called unit cell. The variables a, b and c express the length of the unit cell axes, normally in Ångstrom units ($1\,\text{Å} = 10^{-10}$ m), or sometimes in picometers ($1\ \text{pm} = 10^{-12}$ m). Typical values for the unit cell axes for a medium sized structure ($FW \approx 750$) can vary between 5–35 Å (longer axes sometimes occur). The three other variables, denoted by Greek letters α, β and γ, (alpha, beta and gamma) define the angles (in degrees) between the unit cell axes: α is the angle between the b and c axes; β is the angle between the a and c axes and γ is the angle between the a and b axes. All six variables combined define the volume (normally denoted as V) of the unit cell. Due to the crystal symmetry (see Refs. [10]–[12]) only seven types of primitive unit cells (lattice points only at the edges of the unit cell) can exist (actually there are seven 3-D crystal lattices which are represented by the primitive unit cells). The unit cell can be visualized as a (nano-sized) box, i.e. a match box, whose dimensions are defined by the six above mentioned variables.

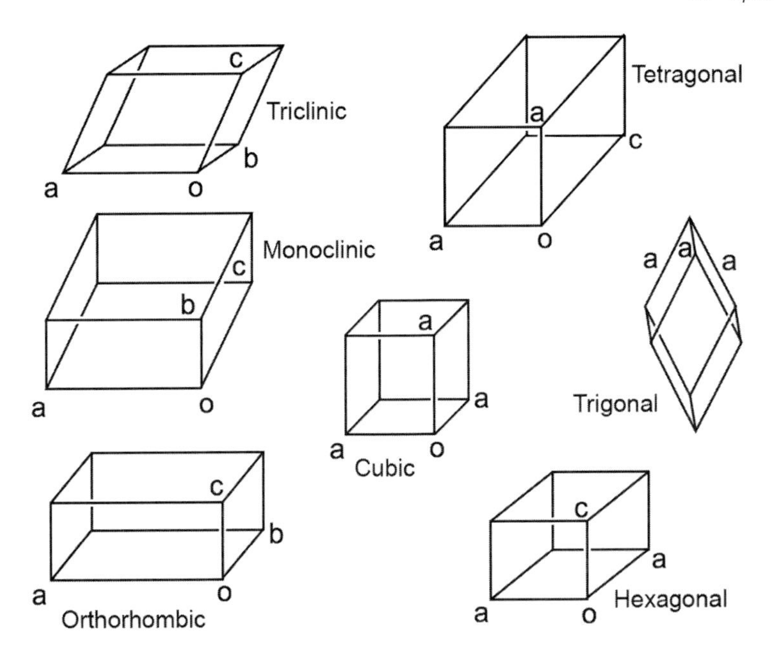

Fig. 9.7. Seven primitive unit cells.

Figure 9.7 shows the seven primitive space lattices (unit cells). The variables *a*, *b*, *c*, α, β and γ, are free, *viz.* they can have whatever value which define a 3-D object which can be multiplied to produce the macroscopic crystal. In all but the triclinic unit cell some variables are correlated (the axis) or restricted (the angles). Those unit cells which have no axis correlations are of lower symmetry (Fig. 9.7, left), triclinic being the least symmetrical. In higher symmetry space groups (Fig. 9.7, right) one or more correlations between the variable exist (see Table 9.1). The unit cells will differentiate from each other by the correlations of the six variables. Table 9.1 gives the definitions for each variable of the seven unit cells.

Only the seven primitive unit cell are described above. If the lattice points on the faces of the unit cell (named as *A*-, *B*- or *C*-centered unit cell) or in the center of the unit cell (body-centered, *viz.* *I*-centered unit cell) are taken into account, a total of 14 so called Bravais lattices are obtained. The treatment of the Bravais lattices is out of the scope of this Chapter and the reader is encouraged to check Refs. [10]–[12] for further details.

The actual unit cell determination will be made by fully automatic or semi-automatic protocols from the diffractometer manufacturers and these protocols will include measurement of a set of reflections (see Fig. 9.6), which are then processed with a computer program and a set of possible unit cells is shown on the computer screen. In the case of good single crystal, the computer program can often output the correct unit cell. However, it has to be stressed that the data collection, *viz.* the intensity measurement from the crystal, can be started without

Tab. 9.1. Unit cell names and correlation of the variables.

Name	a	b	c (Å)	α	β	γ (°)
Triclinic	$a \neq$	$b \neq$	c	$\alpha \neq$	$\beta \neq$	γ
Monoclinic	$a \neq$	$b \neq$	c	90	β	90
Orthorhombic	$a \neq$	$b \neq$	c	90	90	90
Tetragonal	$a =$	$b \neq$	c	90	90	90
Hexagonal	$a =$	$b \neq$	c	90	90	120
Trigonal	$a =$	$b =$	c	α	α	α ($\neq$ 90)
Cubic	$a =$	$b =$	c	90	90	90

prior knowledge about the unit cell, but it is always advisable to do the unit cell determination since this will give some information about the unit cell contents (W_{au}) and will help to foresee possible difficulties in the later stage of the X-ray structural analysis.

The size of the unit cell, *viz.* the volume, will tell something about the constitution of the crystal. Before starting the data collection, exact knowledge about the W_{au} [13] of crystal to be measured is missing (although is can be estimated if the FW of the studied compound is know). However, knowing W_{au} provides an estimate about the possible difficulties during the structure solution. The unit cell determination will give the unit cell volume (V), which is needed while evaluating W_{au}. A simple equation (Eq. 9.2) holds [10–12]:

$$d_{calc} = Z \times W_{au}/0.6023 \times V \tag{9.2}$$

with d_{calc} = the calculated density of the crystal (Mg m^{-3}); V = unit cell volume (Å^3) and Z = number of equivalent positions in a given space group.

Reorganizing (Eq. 9.2) we get:

$$W_{au} = (0.6023 \times d_{calc} \times V)/Z \tag{9.3}$$

For lattice symmetry reasons Z must be an integer (symmetry fixed, see Refs. [10–12]) and can only have values 1, 2, 3, 4, 6, 8, 9, 12, 16, 24, 32, 48 and 96. The calculated density of the crystal, d_{calc}, can only be obtained if Z, W_{au} and V are known, but the density of the crystal can be measured (d_m) by flotation technique, but just a plain intelligent guess (d_g) ("light" organic $d_g = 1.0$–1.25, metal complex $d_g = 1.3$–1.6, inorganic complex $d_g = 2$–3) is enough. The following example will illustrate the estimation of W_{au}. Crystallizing C-ethylresorcinarene from pyridine (Fig. 9.8) produces very nicely diffracting crystals and the unit cell determination yields a monoclinic unit cell (see above and Refs. [10]–[12]) with $V = 4952$ Å^3. The FW of C-ethylresorcinarene is 601 g mol^{-1} and that of pyridine 79 g mol^{-1}.

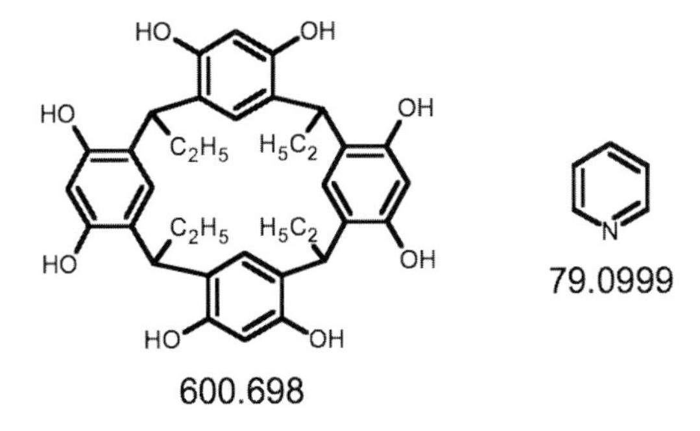

Fig. 9.8. Molecular structure and formula weight (*FW*) of C-ethylresorcinarene (left) and pyridine (right).

For a monoclinic unit cell the integer Z can have values of 1, 2, 4 or 8 (see Refs. [10]–[12]). So let's take $d_g = 1.2$ (intelligent guess, light organic structure) and $Z = 1$, we get the estimate for W_{au} (EW_{au} in Eq. 9.4).

$$EW_{au} = (0.6023 \times d_g \times V)/1 = (0.6023 \times 1.2 \times 4952)/1 = 3579 \tag{9.4}$$

In the monoclinic unit cells $Z = 1$ is very rare, so let's recalculate the same with more common $Z = 2$ and $Z = 4$. We get $Z = 2$ which gives $EW_{au} = 1790$ or $Z = 4$ which gives $EW_{au} = 895$. The EW_{au} of 1790 fits reasonably well with asymmetric unit contents of three resorcinarenes or two resorcinarenes and 7 pyridines (sum of $FWs = 1802$ and 1755, respectively) and EW_{au} of 895 fits equally well with the sum of one resorcinarene and 4 pyridines (sum of $FWs = 917$). So obtaining here an estimate for the W_{au} tells us that either we have three resorcinarenes (without solvent pyridines), two resorcinarenes and 7 pyridines or alternatively one resorcinarene and 4 pyridines in the asymmetric unit depending on the space group. This is not so crucial at this point, since after data collection the space group (see next section and Refs. [1]–[3]) will be determined and Z will be fixed. In this example the space group is monoclinic $P2_1/c$, $Z = 4$ and $d_{calc} = 1.23$ and in the crystal (*viz.* unit cell) the resorcinarene is solvated with 4 pyridines with $W_{au} = 917$ (EW_{au} was 895).

To get a more practical view on the whole procedure, a real X-ray structural analysis of compound **1** (Fig. 9.9, an imine obtained from the reaction of salicylaldehyde and *p*-chloroaniline) is presented here. Compound **1** is a small, purely organic molecule with an elemental composition of $C_{13}H_{10}ClNO$ and *FW* of 231.67 and forms very well-diffracting single crystals.

The unit cell determination gives a unit cell with parameters $a = 13.3674$ Å, $b = 5.7367$ Å, $c = 14.4758$ Å, $\alpha = 90°$, $\beta = 105.485°$ $\gamma = 90°$ and $V = 1069.78$ Å^3. Thus, compound **1** crystallizes in a monoclinic crystal system (unit cell), we can now estimate W_{au} with various Z values and the diffraction power by calculating dp. Since the compound **1** contains one heavier element (chlorine), let's make an

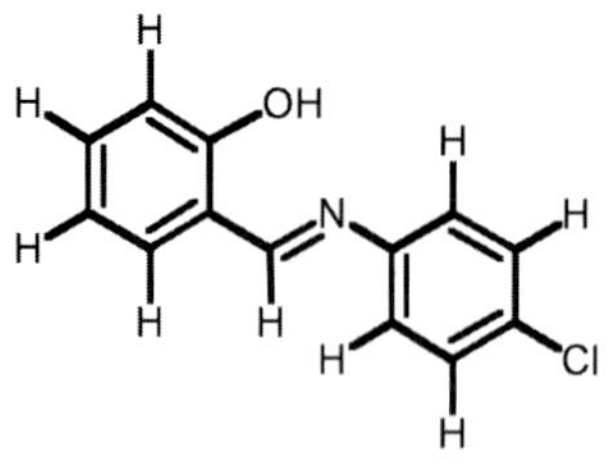

Fig. 9.9. The proposed molecular structure of compound **1**.

intelligent guess that the density $d = 1.25$, now we can calculate EW_{au} $(Z = 1)$ from (Eq. 9.4):

$$EW_{au} = (0.6023 \times d_g \times V)/1 = (0.6023 \times 1.25 \times 1069.78)/1 = 805.41 \qquad (9.5)$$

In the monoclinic unit cells $Z = 1$ is very rare, so let's recalculate the same with more common $Z = 2$ and $Z = 4$. We get $Z = 2$ which gives $EW_{au} = 402.71$ or $Z = 4$ which gives $EW_{au} = 201.35$. Since the FW of compound **1** is 231.67 we can, within acceptable error in the density, assume that the correct $Z = 4$ (this information can used during determination of space group, next section). The estimated diffraction power is $\text{dp} = 231.67/25 = 9.27$, so we have a typical organic structure to be solved and analyzed. If the true diffraction quality of the crystal is good, we might get a relatively good data set (reflections) and would not have difficulties in the structure solution phase.

9.2.2.4 **Data Collection, Data Processing and Final Space Group Determination**

The data collection, as mentioned in the previous section, is a fully automatic process after mounting and centering of the crystal and unit cell determination. The X-rays will reflect from the crystal producing a reflection image with tens or hundreds of reflections in it (as in Fig. 9.6) when recorded with the area detector (older detectors could measure only one reflection at the time and the data collection times were much longer). The complex mathematics of the diffraction phenomenon and data processing will not be discussed in detail here; an exact treatise will be found in the text books [10–12]. However, some basic facts about the diffraction from the crystal on a very descriptive level will be discussed before we can proceed to data collection, processing and space group determination.

As defined in the previous section, the unit cell is the smallest unit containing the atoms of the structure, unit cell has a definite shape and the orientation and lengths of the axes and defined relative to the coordinate origin, marked O in Fig. 9.7. The whole unit cell can be divided into imaginary planes with which the atoms in the unit cell coincide. The X-ray beam diffracts from the electrons of the atoms in the unit cell, thus the atoms on a particular imaginary plane diffract the X-ray beam from that particular plane. The intensity of the reflection from such a plane is directly proportional to the amount and type (how many electrons) of the atoms in that plane. If no atoms coincide with such a plane, there is no reflection from that plane. These reflection planes are simply obtained by dividing the unit axes with an integer, in 3-D such a plane is thus defined by three integers written as h k l and are called Miller indices. Miller indices can have all values from $-h$ to h, $-k$ to k and $-l$ to l, but due to lattice symmetry reasons the negative Miller indices will coincide with their corresponding positive indices in higher symmetry lattices (e.g. in orthorhombic lattices $h\ k\ l = -h\ -k\ -l$, more details in Refs. [10–12]). The integers h, k and l indicate how many equally spaced planes are formed after the intersection of each of the unit cell axes, *viz.*

$$hkl; \quad \text{where } h = (a'/x)^{-1},\ k = (b'/y)^{-1} \text{ and } l = (c'/z)^{-1}$$

where a', b' and c' are the unit cell axis normalized to unity (viz. a', b' and $c' = 1$) and x, y and $z = 0, \pm 1, \pm 2, \pm 3, \ldots \pm n$. The integer 0 in index h, k, or l means that no intersection with the respected unit cell axis occurs, extreme case being reflection plane 000 which is a point, *viz.* the origin. The integer 1 means that the plane runs parallel to the corresponding unit cell axis (-1 that the plane run to the opposite direction). These planes are truly imaginary and can be derived for any unit cell. An example will clarify the principle, imagine that the a-axis is intersected three (3) times, the b-axis is intersected two (2) times and the c-axis is intersected three (3) times, this reflection plane would then be called *3 2 3* (three, two, three). A reflection plane *−1 0 0* (minus one, zero, zero) refers to a plane which runs along the a-axis and do not intersect with either b- or c-axis. Consider an orthorhombic unit cell (Fig. 9.7) with $a = 10$, $b = 15$ and $c = 20$ Å (α, β and $\gamma = 90°$). We can now count how many theoretical reflections we should have from such a unit cell, *viz.* count all possible Miller indices (reflection planes, remember here $h\ k\ l = -h\ -k\ -l$ holds, so we do not have to collect reflections with negative Miller indices), so that the spacing of the planes is 1 Å or more. Resolution can be expressed as:

$$r = a/h, b/k, c/l, \text{ here } r = 1 \quad \text{which gives } h = 10/1,\ k = 15/1 \text{ and } l = 20/1 \quad (9.6)$$

thus the maximum Miller index with 1 Å spacing ($r = 1$) is *10 15 20*, so the first Miller indices are *0 0 1* and the last *10 15 20*. Multiplying we get $10 \times 15 \times 20 = 3000$ possible reflection planes. Be aware that this is the maximum number of reflections to be measured for this orthorhombic unit cell, in practice there is always less than the maximum number of observed reflections. During data collection, all possible reflections must be measured, the reflection might be very weak or non-existing, but those also have to be measured. This imposes a great technological demand for the data collection procedure. All the reflection planes have to be placed into a reflecting position by rotating the crystal along all unit cell axes and simultaneously placing the detector into a correct position. This is done with a help of fully automated mechanical robot (the goniometer, Figs. 9.1 and 9.5), which is able to rotate the crystal in all positions coupled with an area detector capable of recording tens or hundreds of reflections on a single exposure of an X-ray beam. The data collection can last, depending on the total amount of the reflections and on the exposure time needed to record the reflection image, from few hours to some tens of hours.

The data collection produces a set of reflection images with several hundreds or thousands of reflection spots. These reflection spots (Fig. 9.6) have to be converted into a format suitable for computer programs for space group determination and structure solution. This stage is called the data processing and is done with a help of sophisticated computer programs, which convert the reflection spots into intensity values with specific Miller indices, also the error or estimated standard deviation (*e.s.d.*) of the intensity value (in practical work called the sigma of the intensity) is provided in the computer output. The Miller indices will contain the spatial information about the atoms in that particular reflection plane, the intensities of the reflections tell how much these atoms diffract, sigma is the statistical error of the intensity value. The data collection and subsequent processing will therefore give all information that can be obtained from the experiment. Unfortunately, one very important variable in the calculated structure factor (F_{calc}), the so-called phase of the reflection, needed for the calculation of the electron density map of the unit cell, cannot be experimentally obtained and has to be derived by other means (more about this in the next section). A typical reflection listing from a unit cell is given in Table 9.2. These intensity data with their *e.s.d* values together with Miller indices and the unit cell parameters, a, b, c, α, β and γ, constitute so-called raw intensity data which will be used for space group determination.

After getting good raw intensity data, the most important step will be space group determination. The space group is a mathematically derived set of symmetry operations within given crystal lattice (unit cell). There are altogether 230 space groups; fortunately most of them do not appear at all or are very rare in experimental crystallography. The symmetry operations in a given space group define the asymmetric unit of the unit cell, *i.e.* asymmetric unit is the part of the unit cell which, when treated with all symmetry operations in the given space group, will produce the whole content of the unit cell. In more simple words, the unit cell can be divided into symmetry related sections and when the symmetry operations between the sections are known (mathematically fixed in the given space group) we only have to define the contents of that section, *viz.* the asymmetric unit of the unit cell. This is a major advantage, since even for a very large unit cells we have to determine only a small fraction of it (in higher symmetry space groups). Earlier, we have defined the weight of the asymmetric unit, W_{au} [13], this actually means the atoms we have to locate and define within the asymmetric unit of the unit cell.

The crystal lattice (unit cell) itself or – more importantly – in combination with the symmetry operations of a given space group can cause a disappearance of some reflections. This is especially pronounced in the high symmetry unit cells due to symmetry reasons. These missing reflections are called *systematic extinctions* or sometimes systematic absences. This is manifested in the intensity data by a systematic absence of certain reflections. When looking at the reflection listing example (Table 9.2), it can be seen that every second reflection is very strong and every second is practically zero (intensity almost equal to its σ). These zero reflections,

Tab. 9.2. Miller indices, intensity and its sigma for compound **1**.

Miller indices			Intensity	Sigma
(*h*)	(*k*)	(*l*)		
0	0	2	75.20	3.20
0	0	−3	2.20	0.90
0	0	3	2.50	0.80
0	0	−4	1666.00	58.50
0	0	4	1655.00	60.30
0	0	−5	−0.10	0.80
0	0	5	0.90	0.80
0	0	−6	1896.80	65.60
0	0	6	1897.40	71.70
0	0	−7	2.80	1.50
0	0	7	−0.60	1.20
0	0	−8	122.30	5.20
0	0	8	112.30	5.10
0	0	−9	−0.60	1.20
0	0	9	−2.10	1.60
0	0	−10	174.00	7.30
0	0	10	183.70	8.60
etc.				
15	2	−4	14.00	3.30
15	2	−3	−1.50	2.00
15	2	−2	55.80	5.10
0	0	0	0.00	0.00

if systematically observed with certain Miller indices, are those systematic extinctions. Thus, by carefully analyzing the intensity data for systematic extinctions, the space group of that unit cell can be determined. Unfortunately, in some cases the systematic extinctions of different space groups are the same or there are no systematic extinctions. In these cases the space group cannot be unambiguously determined. Fortunately, modern computer programs provided by the diffractometer manufacturers, or obtained from other sources, will do the intensity data analysis and list out the systematic extinctions suggesting a suitable space group or a few that will fit best to the systematic extinctions.

In lower symmetry space groups, the choice of the space group is easier than in higher symmetry cases, due to a smaller number of possible symmetry operations. The triclinic unit cell has no systematic extinctions, but there are only two possible space groups. The first (denoted as $P1$, no. 1) does not have any symmetry operations, *viz.* the unit cell is at the same time the asymmetric unit (symmetry operation x, y, z), the second unit cell has a center of symmetry as the symmetry operation, leading to two symmetry related parts of the unit cell, (x, y, z) and $(-x, -y, -z)$. This triclinic space group is denoted as $P{-}1$ (minus sign marks the presence of center of symmetry) and its asymmetric unit is half of the unit cell. Space group $P{-}1$ does not have symmetry operation which would produce the mirror image of the original object, thus $P{-}1$ is an acentric, chiral space group. If an optically pure compound crystallizes in a triclinic space group, it can do it only in $P{-}1$. The same applies also in the higher symmetry space groups, but more choices are available. The most common space group for monoclinic unit cells (Fig. 9.7) is denoted as $P2_1/c$ with two systematic extinctions. The first being $0k0$, $k = 2n + 1$ (this means that all reflections with Miller indices where k is odd are systematically extinct, viz. *0 1 0, 0 3 0, 0 5 0, 0 7 0*, etc.). The second is with Miller indices $h0l$, $l = 2n + 1$, viz. $0\ 0\ 1,\ 0\ 0\ 3, \ldots,\ 1\ 0\ 1,\ 1\ 0\ 3, \ldots,\ 2\ 0\ 1,\ 2\ 0\ 3, \ldots.$ etc. are extinct. In monoclinic unit cells, these two simultaneous extinctions uniquely define the space group as $P2_1/c$. All space groups, their systematic extinctions and symmetry operations are tabulated in the Tables for X-ray Crystallography, Vol C [18].

The data collection at 173 K for compound **1** produces a reflection list (so called "data set", Table 9.2) with 6040 reflections, however only 1882 of these are *independent reflections* (routinely, many reflections are measured more than once during the data collection due to crystal lattice symmetry, *viz.* reflection *0 0 2* is the same as *0 0* −2, see Table 9.2). From the systematic extinctions of this monoclinic data set for compound **1**, the computer program finds out two systematic extinctions, namely $h0l$, $l = 2n + 1$ and $0k0$, $k = 2n + 1$. As the crystal system is monoclinic, these two systematic extinctions define unambiguously that the space group for compound **1** is $P2_1/c$ and that $Z = 4$. This information is then used in the structure solution phase.

9.2.2.5 Data Reduction, Structure Solution and Refinement

After the space group has been determined and selected from the raw intensity data obtained in the previous section, the next step before structure solution and refinement is the so-called data reduction. In this stage, the measured intensities are converted, again by automatic computer programs, to so-called observed structure factors. A structure factor is a very complex mathematical entity which includes data from all atoms, their scattering (diffraction) power, 3-D coordinates, thermal movement, phase of the reflection etc. The detailed description of structure factors is outside of the scope of this text: the reader is strongly encouraged to consult the text books [10–12]. The integrated intensities, I_m, (from the reflection list, Table 9.2) of a particular reflection (e.g. *1 1 2*) from the data collection are proportional to the square of the amplitude of the structure factor (F) of that reflection, when all corrections for the diffraction phenomenon have been taken into account. These corrections are Lorenz, polarization (these two are normally merged as *Lp* correction), absorption and radiation damage being the most important ones. In reality, only the *Lp* correction can be calculated, both absorption correction and radiation decay have to be empirically measured. For this reason, the following relation (Eq. 9.7) is normally presented.

$$I_m \propto F^2 \quad (9.7)$$

After *Lp*, absorption and radiation damage decay corrections to the integrated intensities the observed structure factors (F_{obs}) are obtained. Each measured reflection thus will have the corresponding F_{obs}. In other words, F_{obs} is the square root of the measured intensity of a given reflection corrected for effects that come from the properties of the crystal itself.

In a case where the absorption of X-rays is very small and the crystal does not decompose (the decomposition can be monitored by measuring the intensity of so called "standard" reflection many times during the data collection, however modern area detector equipped diffractometers are so fast that virtually no X-ray radiation decay is normally observed), the data reduction is just a fast computer calculation of the *Lp* correction to the measured intensities and F_{obs} are obtained. Improperly done data reduction or exclusion of the effects of absorption or decay can render the structure solution impossible, so care must be taken in order to do proper data reduction.

The unit cell determination, data collection and data reduction contain all the information which can be obtained from experiments and mathematical corrections. The only, and thus the most important, data present in the structure factor that cannot be obtained experimentally is the phase of the reflection. This imposes the so-called "Phase Problem", since without phase, the other components of the structure factor cannot be calculated. On the contrary, if we know the phase of the reflection, we can calculate the 3-D coordinates of the atoms contributing to the corresponding structure factor.

In the structure solution step, (see details in Refs. [10]–[12]) some method to obtain or estimate the phases has to be used. Nowadays, this is done using very sophisticated (semi)automatic computer programs which use very complex mathematics and statistics to "guess" and evaluate the phases by computing a very large number of possible sets of phases. Since the data collection can produce thousands of reflections, the computer program has to calculate and evaluate tens of thousands of phases and test statistically their suitability as a correct phase set (again see Refs. [10]–[12]). If the computer program is able to calculate such a proper *ab initio* phase set, then this set of phases together with the atom types present can be used to calculate an approximate 3-D electron density map of the asymmetric unit (by summing up all structure factors).

Modern structure solution programs (the most used being SHELXS/SHELXD [19], SIR97 [20] and SIR2000 [21], others exist, too) need as input only the W_{au}, in a form of reasonable chemical composition, and the scaled structure factors ($= F_{obs}$, actually programs use $F^2{}_{obs}$). The success of the structure solution, *viz.* is the program able to find proper phases, depends mainly on two factors, the quality of the data collected (which in turn depends on the quality of the crystal), *viz.* how much of the data is observed over a certain threshold value. The routinely used criterion for an observed reflection is $I > 2\sigma I$, *i.e.* the measured intensity value is more than two times bigger than its sigma). Generally speaking, it can be said that the higher the percentage of the observed reflections, the better the data is (severe

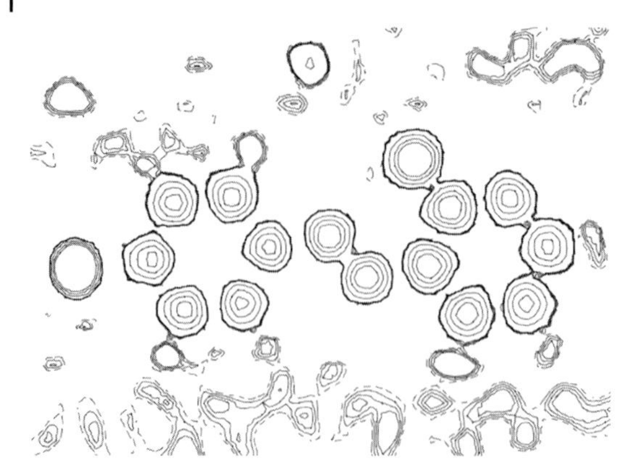

Fig. 9.10. The calculated electron density map for compound **1**.

systematic errors or improper data reduction of the data can overrule this). For compound **1** 1686 reflections out of 1882 (89.6%) are larger than 2σ, so the data set is excellent. The more reflections are observed above the 2σ intensity level, it is more likely that the structure solution programs can produce a feasible model for the experimental electron density. Figure 9.10 shows the calculated electron density map for compound **1**.

From the well-resolved electron density map, the structural analyst has then to assign the atom types and the 3-D coordinates for the best structure solution (= the phase set, which with given atom types fits best with observed structural factors, *viz.* the measured reflection data). The assignment is based on the relative height of the peaks in the electron density map coupled with the proposed chemical structure diagram of the studied structure. For compound **1**, the structural scheme presented in Fig. 9.9 is used. It turns out that all heavier atoms can be located from the electron density map and assigned unambiguously (Fig. 9.11).

Frequently, with smaller and well-diffracting structures ($W_{au} < 700$ and $dp > 10$), all atoms of the structure can be written out as the initial model by the program and they just have to be named correctly (as in Fig. 9.11) and refined. The refinement process (see Refs. [10]–[12]) uses incremental movement of the atom coordinates and atomic displacement parameters (commonly called as "thermal parameters") of the structure solution model using the so-called least-squares method. The model (the calculated structure factors) is fitted against the measured data (the observed structure factors) and the *R*-factor (see above, Section 9.2.1) is calculated. With larger structures or if the unit cell contains light atom solvent molecules (C, H, O, N atoms only), some atoms, sometimes even 50% of all atoms, cannot be located from the first very crude electron density map (calculated from the *ab initio* phase set). However, those atoms which are chemically feasible (based on the proposed molecular structure) can be fed into the calculation of the calculated structure factors F_{calc} (F_{calc} will approach F_{obs} when a more accurate model is

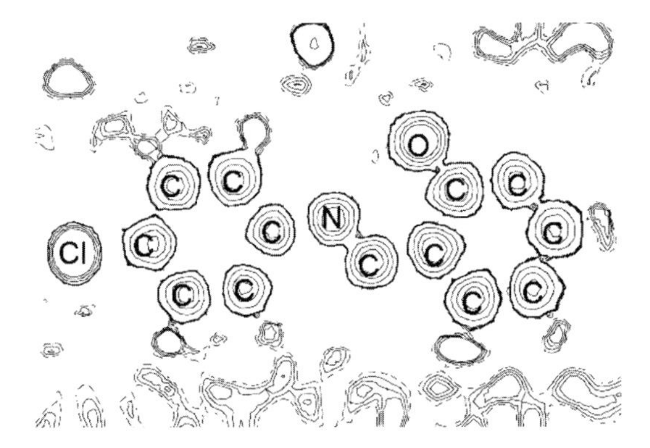

Fig. 9.11. Assigned of atoms from the electron density map for compound **1**.

used). The F_{calc} are then used for the computation of the "calculated electron density map". This calculated electron density map (from F_{calc}) is then extracted from the "observed electron density map" (from F_{obs}). The map obtained is called difference electron density map and the missing atoms can be located from it. This process is called difference Fourier analysis (DFA) and after finding some missing atoms, all the parameters of the old and the new atoms are subjected to new refinement cycle. This cycle, refinement – DFA – refinement – DFA . . . , is repeated until all atoms of the structure are located and until the difference electron density does not show any more peaks (non-H atom cut-off level is *ca.* 1.0 $e\text{Å}^{-3}$) and the atomic parameters no longer shift during the refinement (convergence < 0.05). When all atoms of the structure, *i.e.* the atoms in the molecule(s)/ions and the atoms of the solvents molecules are located and refined to the end, we can proceed to the analysis of the structure. This stage however, might not be so easy to achieve as problems are often encountered in large supramolecular systems.

All the above procedures and calculations apply to structures with good quality and well-behaved data. Unfortunately, this is not true in most cases for supramolecular structures because crystal quality is often poor. Large supramolecules do not form well-ordered crystal lattices (periodicity!), the dp values are low (<10), the W_{au} are large (>1000) and the crystal lattice normally contains a substantial amount of (different) solvent molecules, which are quite often severely disordered, further decreasing the crystal quality.

When the reader has ensured that the best possible data set from the supramolecular system under study has been obtained, it is feasible to proceed to the structure solution phase. The reader is reminded that with higher W_{au} structures the structure solution programs will have increasing difficulties in producing a good model for the structure. Here, the same principle applies as with crystals: do not give up if the first try with the default parameters of the structure solution program do not produce a reasonable result. The programs have many other sets of parameters which will affect the structure solution procedure. A rule of thumb as to what set of parameters will actually produce success with particular supramolecular system cannot be given, as each crystal structure determination is unique and it might be that various settings and even different structure solution programs have to be tried, in order to get a proper structure solution. After unsuccessfully trying the best with varying sets of parameters, the reader should consider that either the space group or crystal system is incorrect (check carefully the data for systematic extinctions and redetermine the crystal system/unit cell, and the space group). Sometimes, a change to lower symmetry might help to solve the structure (*e.g.* going from monoclinic to triclinic). This however will double, triple or quadruple the W_{au}, thus making structure solution more difficult (if W_{au} grows very large, this "simplification" will make structure solution impossible).

If the space group and the crystal system are correct and no alternative can be found, the reader must accept the fact that the quality of the data is not good enough for structure solution and return to the recrystallization phase again.

Even if structure solution giving most of atoms was successful, the task of completing the structure is not yet over. Supramolecular structures almost by definition contain included guest or solvent molecules either inside the cavity of a host compound or in the crystal lattice. This is, as such, not a problem, but quite often these solvent molecules are crystallographically disordered. This means that the electron density resulting from the guest or the solvent molecule cannot be clearly determined and in severe cases the guest or solvent cannot be located at all (although *e.g.* NMR or MS shows it to be there!). This unwanted disorder is most prominent, when the guest or solvent molecule has very weak or non-existing interactions with the host, *e.g.* inside molecular capsules and carcerands. Also if the host compound contains long aliphatic chains, these can quite often be seriously disordered and it can be particularly difficult to solve and refine the structure to an acceptable level. However, very careful control of the crystal growing process and/or change of the recrystallization solvent(s) can reduce the disorder and thus improve the quality of the crystal. One example of such is a work from authors' laboratory where a very large ($FW = 7106.50$) cavitand cage compound, $[(C_{268}H_{320}N_8O_{16}P_8Pt_4)@CF_3SO_3]^{7+}\cdot 7\ CF_3SO_3^{-}\cdot 12.5\ C_6H_6$ was studied [22]. The first data collections produced a result where the core of the cage was clearly determined but all together 88 carbon atoms from the eight $C_{11}H_{23}$ alkyl chains could not be located at all. After 6 months work of repeated recrystallizations and after 20 data collections from 20 different crystals, an acceptable data set was obtained which finally produced a structure solution and location of all non-hydrogen atoms; however the $C_{11}H_{23}$ alkyl chains showed very large thermal movement.

When all the atoms of the target structure have been found and refined so that no movement/variation of any structural parameter occurs, the refinement programs will output a file which can be used for the analysis of the structure and deposition to databases, like Cambridge Crystallographic Data Centre. This file is called Crystallographic Information File, i.e. CIF file, and it has a specific and regulated format.

The crystallographic, chemical graphics or some molecular modeling programs can read CIF files. However, many crystallographic programs can also output Brokehaven Protein Data Bank file, *viz.* PDB file, which is a routine input file to the majority of chemical visualization and graphics programs. An example of the CIF file format for compound **1** is presented below (Table 9.3).

9.2.2.6 Analysis of Structure

After successful structure solution and refinement, the final and very important phase completes the structural analysis. This is the detailed inspection of the structure looking for the intramolecular and intermolecular structural features of the system studied. Traditionally a plot of the molecule or a representative part of the structure (in a case continuous structures, like metallo-organic or organic frameworks, MOF or OF, respectively) with thermal ellipsoids (*viz.* anisotropic displacement parameters) is presented. Such a plot drawn using program ORTEP [23] for compound **1** is presented in Fig. 9.12 (top).

To show the total size of the molecule and to illustrate intramolecular interactions such as hydrogen bonds, a plot is best used where atoms are drawn with their van der Waals radii (CPK model, Fig. 9.12, bottom).

Compound **1** has two intramolecular features which are revealed by X-ray structural analysis. Firstly the compound **1** can exist in two different tautomeric forms, namely enol form and keto form. The equilibrium between these different forms is influenced both by the chemical composition of the compound and the environment (polarity of the solution or crystal lattice). Figure 9.13 shows the structures of the enol and keto tautomers of compound **1**. The X-ray structure of compound **1** (Fig. 9.12) unambiguously shows that the hydrogen atom is at the phenolic OH and the intramolecular hydrogen bond is O(1)–H(1) $\cdots$ N(1), the O $\cdots$ N distance being 2.618(2) Å and the O–H $\cdots$ N angle 149(2)° *viz.* compound **1** exists in solid state as an enol tautomer (Fig. 9.13). Generally the location of the hydrogen atom by X-rays can be ambiguous because hydrogen has only one electron. However small light atom organic compound with very good diffraction quality (like **1**) will make it possible to locate (and even refine) hydrogen atoms using difference Fourier analysis, *i.e.* the electron density map will reveal the position of the hydrogen(s). In addition careful inspection of the bond distances on the structural moiety will give further evidence about the position of the hydrogen atom(s). (It is easy to differentiate the phenolic O-atom from the carbonyl O-atom, C–O bond distance is 1.35 Å when the C=O is ca. 1.25 Å). The other intramolecular aspect is the planarity of the compound, since **1** is fully conjugated, it should be planar. Based on the 3-D coordinates, the planarity of **1** can easily be calculated and it turns out that the compound is perfectly flat with deviation from planarity being very small, 0.0586(13) Å below [C(7)] and 0.0723(17) Å [C(10)] above the least-squares plane, calculated for all non-H atoms.

Compound **1** shows many intermolecular interactions. Firstly, the packing plot (but also the 3-D atom coordinate list) reveals that no solvent molecules are included in the crystal lattice. The crystallization was done from ethanol, but ethanol molecules are not included into the crystal lattice. The packing plot (Fig. 9.14) shows that an intermolecular interaction between the moleclues, *i.e.* π–π stacking, creates an antiparallel pair of **1** with short non-bonded contact distance of 3.30 Å

Tab. 9.3. First lines of the CIF file for compound **1**.

```
data_Compound 1
_audit_creation_method       SHELXL-97
_chemical_name_systematic
;
?
;
_chemical_name_common          ?
_chemical_melting_point        ?
_chemical_formula_moiety       ?
_chemical_formula_sum
'C13 H10 Cl N O'
_chemical_formula_weight       231.67
loop_
_atom_type_symbol
_atom_type_description
_atom_type_scat_dispersion_real
_atom_type_scat_dispersion_imag
_atom_type_scat_source
'C'  'C'  0.0033  0.0016
'International Tables Vol C Tables 4.2.6.8 and 6.1.1.4'
'H'  'H'  0.0000  0.0000
'International Tables Vol C Tables 4.2.6.8 and 6.1.1.4'
'Cl'  'Cl'  0.1484  0.1585
'International Tables Vol C Tables 4.2.6.8 and 6.1.1.4'
'N'  'N'  0.0061  0.0033
'International Tables Vol C Tables 4.2.6.8 and 6.1.1.4'
'O'  'O'  0.0106  0.0060
'International Tables Vol C Tables 4.2.6.8 and 6.1.1.4'
_symmetry_cell_setting         Monoclinic
_symmetry_space_group_name_H-M       P 21/c
loop_
_symmetry_equiv_pos_as_xyz
'x, y, z'
'-x, y+1/2, -z+1/2'
'-x, -y, -z'
'x, -y-1/2, z-1/2'
_cell_length_a                      13.3674(2)
_cell_length_b                      5.73670(10)
_cell_length_c                      14.4758(3)
_cell_angle_alpha                   90.00
_cell_angle_beta                    105.4850(10)
_cell_angle_gamma                   90.00
_cell_volume                        1069.78(3)
_cell_formula_units_Z               4
_cell_measurement_temperature       173(2)
```

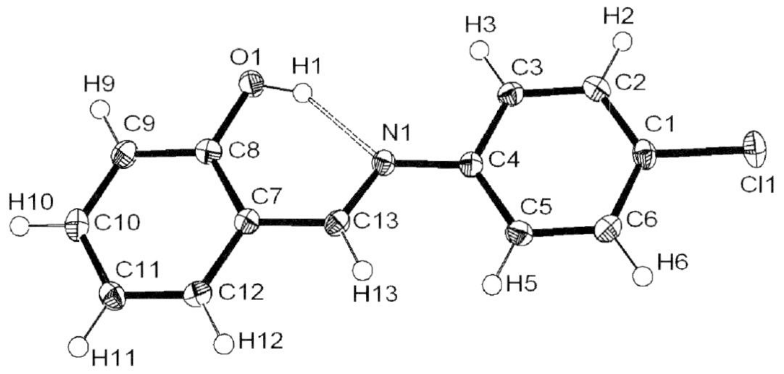

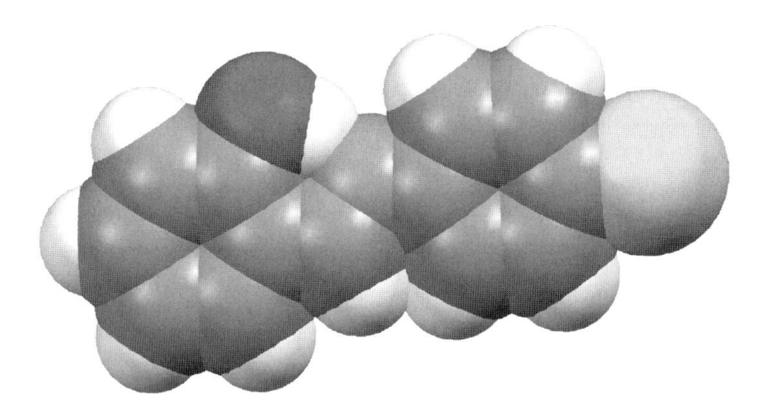

Fig. 9.12. A plot of the X-ray structure of compound **1** with thermal ellipsoids at 50% propability level (top) and with VDW radii (below).

[C(4) to C(8)], sum of VDW radii of two benzene rings is 3.6 Å. These pairs are then sideways connected to the adjacent pairs by C–H $\cdots\pi$ (edge-to-face) interactions to the adjacent molecules [from H(6) to C(11) and C(12), 2.836 and 2.888 Å, respectively and *vice versa*] and C–H $\cdots$ O hydrogen bonds [H(3) $\cdots$ O(1), 2.705 Å].

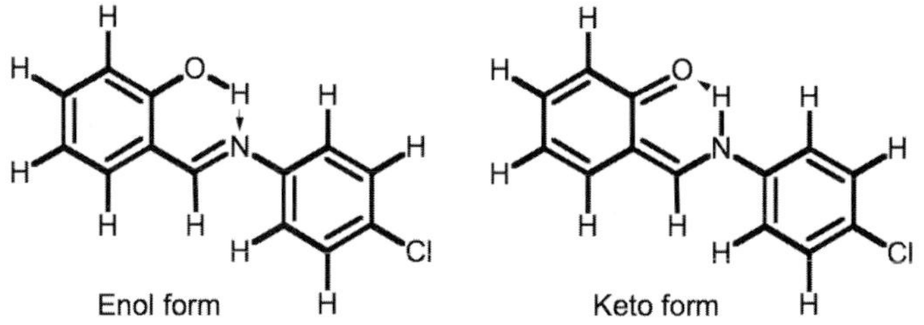

Fig. 9.13. The structures of the enol and the keto tautomers of compound **1**.

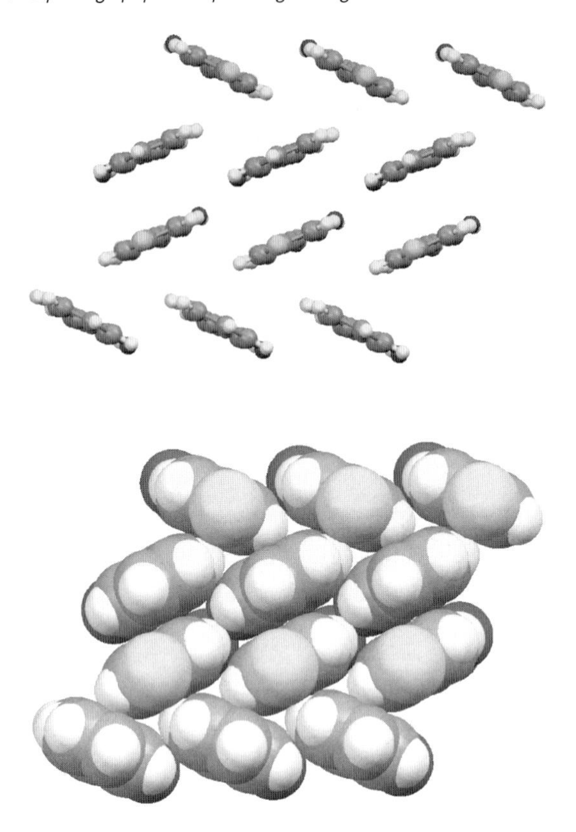

Fig. 9.14. The packing of compound **1**, ball-and-stick (top) and VDW radii (below).

The third dimension of the crystal is build up from a C–H···Cl interaction of 2.942 Å [Cl(1) to H(2) and *vice versa*].

Generally in X-ray crystallographic work, the table of crystal data (= unit cell, space group, experimental parameters), fractional coordinates [fractional coordinates refer to the atomic position as fractions of the corresponding unity defined unit cell axis, e.g. $x = 0.25$, $y = 0.25$, $z = 0.25$ means that the atom is located $\frac{1}{4}$ of the length of each unit cell axis starting from the origin ($x = 0, y = 0, z = 0$), in order to calculate absolute coordinates each fractional coordinate has to be multiplied by the corresponding unit cell length], bond lengths and distances, the numeric values of anisotropic displacement parameters and intermolecular distances shorter than the sum of van der Waals radii are typically prepared, but not nowadays anymore included into a publication. The above mentioned data are commonly submitted to the Cambridge Crystallographic Data Centre during the submission to journal as the CIF file. From these data and by inspecting the structure by means of crystallographic graphics programs, many important features about the structure can be extracted (see above for compound **1**), typically in supramolecular structures these are H-bond distances (because it is difficult to determine the positions of hydrogen atoms, this is usually given as the heteroatom–heteroatom distances), $\pi \cdots \pi$–, C–H $\cdots \pi$–, and cation–π interactions. Sometimes, a packing plot of the structure is presented, but not so frequently encountered in modern chemical literature, more frequently they appear in crystal engineering journals.

The easiest way to inspect the structure (in addition to the tables mentioned above) utilizes graphical representations of the structure on a computer screen. Some very nice and easy to use crystallographic graphics programs are available in the Internet. One such a program is Mercury, provided by Cambridge Crystallographic Data Centre (http://www.ccdc.cam.ac.uk/products/csd_system/mercury/downloads/).

The solid state X-ray structure will also give information about the configurational, stereochemical or tautomeric information about the studied structure. In the case of optically pure compounds, the absolute configuration of an unknown compound can be obtained, if the molecule studied contains at least one heavier element (like Cl, S, or metal ion) per 20 carbon atoms. If compound **1** would be optically active, the chlorine atom would enable the determination of the absolute configuration.

9.3 Crystal Engineering

9.3.1 Introduction

The aim of this section, like the previous one, is not to give a complete treatise or even overview of crystal engineering. The very versatile nature and variety of chemical systems and interactions, which can be considered to form the basis of crystal engineering, mean that it is impossible to cover all these aspects with this section. The current text offers a very rough outline of the basic principles of crystal engineering, *viz.* what do we mean when we speak about crystal engineering? Section 9.2.2.1 described the crystal and properties required for the X-ray diffraction phenomenon to occur. As underlined in Section 9.2.2.1, X-rays will not diffract from a crystal unless it is an entity with specific features, *i.e.* periodicity of the components (molecules or ions) which constitute the crystal lattice and thus the complete macroscopic crystal, so X-ray diffraction analysis cannot be performed on an amorphous material (neither by single crystal X-ray diffraction analysis, as described earlier in this text, nor by a powder diffraction method, which is not covered in this text).

9.3.2 Definition

The spontaneously or "engineering"-induced organization of the components into a crystal is crucial for the structural features and physical/chemical properties of the crystal. Therefore, it is of utmost importance that the components that are intended to form the crystal must show some, even very weak, attractive interactions under the condition where the organization happens. Due to the subtle and often very weak interactions of the components, the formation of a crystal lattice is a very complex self-assembly phenomenon including many, some even mystical, simultaneously acting processes (thermal motion, kinetics, diffusion, adsorption, absorption, viscosity, etc.) or their combinations. Therefore, crystallization or the ability to grow single crystals has been and still is considered as an "art", impossible to describe using accurate scientific terms. This is mainly due to the fact that even though most of the conditions leading to crystal growth can be controlled and monitored, the formation of a good single crystal(s) still remains unpredictable and unexplainable. Quite often the surprised person, after a successful recrystallization experiment, cannot really explain what was different in the successful event when compared with previous unsuccessful events. As the process is so subtle, chances for a failure are much more evident than those for success. After tens of frustrating crystallization experiments the success might be interpreted as an "Act of God", especially in those cases where exactly same conditions (*i.e.* those which have or can be controlled) on first trial result in microcrystalline or amorphous powder, but the successful trial does give beautiful single crystals.

As all intermolecular interactions, except covalent bond formation (a very special case of crystal engineering), are by nature supramolecular and the crystal itself can be considered as "*supermolecule par excellence*" [5, 6], Crystal Engineering can be defined as "*a set of designed processes which aim to a formation of a crystalline material utilizing the supramolecular interactions and self-assembly of the components*", in a very short form "*making crystal by design*" [24]. The above definition does not include nor require any specific interactions, physical or chemical conditions, crystallization techniques, additives or serendipity [25].

In crystal engineering research, this denotes that all means and methods possible can and should be used in order to produce a suitable crystalline material. Of course, it must be emphasized that the means and methods to be used should not be too drastic, *e.g.* that the components and the supramolecular interactions between them have to survive during the crystallization experiments and must be able to self-assemble. If the system studied in the solid state (crystal engineering is by virtue only possible in the solid state) is formed by hydrogen bonding, great care has to be taken NOT to break the hydrogen bonding system by the use of heavily competing solvents (like H_2O, DMSO, DMF) or not to try to grow crystals from highly acidic conditions for an acid labile compound/system. Figure 9.15 illustrates the general principle of crystal engineering: that is the components are designed to self-assemble into crystal lattice by intermolecular supramolecular interactions. Chemically, the component(s) can vary over the whole range of chemical functionalities, from metal ions and anions via organic molecules and oligomers to biomolecules.

The properties of the components will define the methods and conditions to be used, but in all cases, the only target is to guide the components into a self-assembly process which leads to periodical 3-D lattice of the components. Many

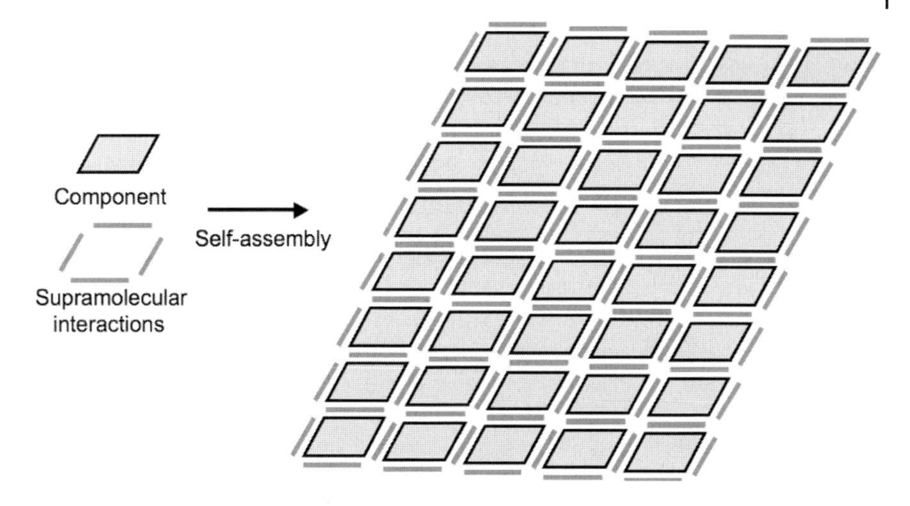

Fig. 9.15. The formation of macroscopic crystal from components and intermolecular interactions.

aspects how the design of crystallization processes will be adopted depends of the system studied. The reader is urged to check some very good web sites [15] for basic principles of crystal growing methods.

To define and to aid the utilization of the supramolecular interactions and to design them in crystal engineering, Desiraju presented [26] a concept of *supramolecular synthons*, mainly developed for hydrogen bonds which maybe are the most widely exploited of the supramolecluar interactions in the crystal engineering genre. More recently, the palette of intermolecular interactions in crystal engineering has been extended to virtually all attractive intermolecular forces such as covalent bonds between atoms (not supramolecular interaction), coordination bonds between metal centers and ligands, Coulombic attractions and repulsions between ions, strong hydrogen bonds, weak hydrogen bonds, CH–π-interactions, cation–π-interactions, π–π-interactions, halogen–halogen interactions, halogen–oxygen interaction, halogen–nitrogen interaction, C–H–fluorine interactions and van der Waals and all possible combinations of these interactions [26].

The above mentioned interactions extend over a very wide energy range: from the very high energy covalent bonds via the intermediate coordinative bond through hydrogen bonds to very weak van der Waals interactions (dispersion forces). The difference in the interaction types and strengths allows the differentiation of the target materials for their designed self-assembly (*viz.* synthesis) from the components to macroscopic superstructure (*i.e.* crystal).

The design of the processes needed in crystal engineering can be divided into two phases. In the first phase, the component (Fig. 9.15) has to be either selected from the pool of available compounds, prepared by somebody else or from commercial sources, or it has to be synthesized via normal synthesis chemistry methods. Prior to the synthesis, a very careful analysis of the desired (and possible) supramolecular features of the component has to be conducted. When the component is ready then the second, nearly always the more difficult, phase can be started. This is the crystal growing/recrystallization phase where either with or without solvents (green chemistry), via supramolecular synthesis, self-assembly of the components is initiated. In the simplest case, this might be the crystallization of one component from suitable solvent (like compound **1** from EtOH). This process might involve slow evaporation of the solvent, slow cooling of the heated solvent, or slow cooling of the saturated room temperature solution in a refrigerator. The success of this simple process depends on the intermolecular interactions between the components (crystallizes without solvent) or between the component and the solvent molecules (crystallizes as a solvate, component with solvent molecules).

The fact that each component (molecule or ion) will behave differently, will not allow a presentation of "a general procedure" or "a rule of thumb". The reader is encouraged to study the solubility of the component(s) and then just try to grow crystals from suitable solvent (for volatile components sublimation might produce good quality crystals). When binary or ternary systems with solvent mixtures are being processed, the only way is just to use enough time to learn how the component system behaves with various solvent systems and physical conditions.

This latter phase requires the characterization of a solid, hopefully, single crystalline product(s) with X-ray diffraction analysis, but in many cases not good enough quality single crystals can be obtained, instead microcrystalline or amorphous powders emerge. Then, routine analytical and spectroscopic laboratory tools are much less useful than in the case of solution chemistry. This means the crystal engineer has to master or have good collaborators on methods that are not routine in chemistry laboratories (CPMAS NMR, DSC, TGA, AFM, STM, XPD, XFS, etc).

To get a glimpse of what kind of chemical system crystal engineers have studied over the recent years, the reader is encouraged to study some selected prospects [8, 27], books [7, 28] and reviews [29–31]. More special topics of crystal engineering, such as polymorphism, green chemistry applications, and structural networks are very nicely reviewed by Bernstein [32], Braga *et al.* [33] and Öhrström and Larsson [34].

9.4
Conclusions

The X-ray crystallography of supramolecular systems is a difficult but rewarding task. The success of an X-ray study of a supramolecular system requires a well-diffracting single crystal. Obtaining it is often the most difficult and time-consuming part of the work. Being very persistent with the diffraction quality of the crystals, to obtain the best possible periodicity of the crystal lattice with no twinning, is the key to success. Do not collect data on a bad crystal, test several different crystals and do data collection only for the best possible crystal. Repeat the crystal growing process and/or change the solvent system if needed. Do not worry which solvent you use, aim for a well-ordered and nicely diffracting single crystal.

Only good data can bring you a structure solution. Even though the modern structure solution programs are extremely powerful, do not underestimate the role of the crystal. Even good data does not necessarily produce instant structure solution, be as persistent with the structure solution programs as you should be with the crystal quality. The larger the W_{au} of your system under study, the more difficulties you will face, but successful X-ray analysis of an interesting supramolecular system is always worth the time you put into it.

The field of crystal engineering offers nearly limitless possibilities for using single crystal X-ray diffraction analysis, provided that the crystal growth process can be "engineered" to a successful production of a good quality single crystals. Other techniques can be used to study the non-single crystalline materials, but the structural information and the details of supramolecular interactions cannot be accurately determined. The core of crystal engineering, the ability to control and direct crystal growth, is still mostly unresolved phenomenon, and only very persistent laboratory work will bring "craftmanship" to those who are willing to spend a lot of time with crystallization experiments. However frustrating and time-consuming the crystal growing process might be, the successful result will always be worth the time put into it.

Acknowledgements

Professor Jussi Valkonen, Laboratory of Inorganic Chemistry, Department of Chemistry, University of Jyväskylä, Finland and Professor Joao Rodrigues, Department of Chemistry, University of Madeira, Portugal, are kindly thanked for their valuable suggestions and hints. Dr. Luca Russo is thanked for the technical assistance in some figures and the refinement of compound **1**.

References and Notes

1 Pepinsky, P. *Phys. Rev.* **1955**, *100*, 952.
2 Schmidt, G. M. J. *Pure Appl. Chem.* **1971**, *27*, 647–678.
3 Lehn, J. M. *Pure Appl. Chem.* **1978**, *50*, 871–892.
4 Lehn, J. M. *Supramolecular Chemistry: Concepts and Perspectives*; VCH: Weinheim, **1995**.
5 Dunitz, J. D. *Pure Appl. Chem.* **1991**, *63*, 177–185.
6 Dunitz, J. D. *Perspectives in Supramolecular Chemistry*; Ed. Desiraju, G. R., Wiley: New York, **1996**; Vol. 2.
7 Desiraju, G. R. *Crystal Engineering: the Design of Organic Solids*; Elsevier: Amsterdam, **1989**.
8 Braga, D. *Chem. Comm.* **2003**, 2751–2754.
9 Institute of Scientific Information, Web of Knowledge, Web of Science, www.isiknowledge.com, 04.04.2006.
10 Clegg, W. *Crystal Structure Determination*; Oxford University Press, Oxford, 1998.
11 Massa, W. *Crystal Structure Determination*; Springer, Berlin Heidelberg, 2000.
12 Giacovazzo, C. (Ed.), *Fundamentals of Crystallography*; IUCR, Oxford University Press, Oxford, 1992.
13 W_{au} is defined as the sum of the molecular weight of all atoms in the

asymmetric unit expressed as a plain number. W_{au} of the studied system can be much higher than the *FW* of the (supra)molecule itself. For example three molecules with *FW* of 250+ four solvent molecules with *FW* of 78 in the asymmetric unit will give W_{au} as high as 1062 ($3 \times 250 + 478 = 1062$).

14 a) Jones, P. G. *Crystal Growing*. Chemistry in Britain **1981**, *17*, 222–225. b) Hulliger, J., Chemistry and Crystal Growth, *Angew. Chem. Int. Ed. Engl.* **1994**, *33*, 143–162.

15 a) http://www.xray.ncsu.edu/GrowXtal.html, 04.04.2006; b) http://www.cryst.chem.uu.nl/growing.html, 04.04.2006.

16 More precisely dp should be calculated as $dp = W_{au}/N_a$, where N_a = number of atoms in the asymmetric unit. This however can be done only after the completed X-ray analysis when the total contents of the asymmetric unit is known.

17 For example: Sigma-Aldrich, Fomblin Y HVAC 140/13, CAS: 69991-67-9.

18 INTERNATIONAL TABLES, VOL C – Wilson, A. J. C., Ed., International Tables for Crystallography, Volume C, Kluwer Academic Publishers, Dordrecht, The Netherlands, **1995**.

19 SHELX97, Programs for Crystal Structure Analysis (Release 97-2). Sheldrick, G. M., Institüt für Anorganische Chemie der Universität, Tammanstrasse 4, D-3400 Göttingen, Germany, **1998**, http://shelx.uni-ac.gwdg.de/SHELX/.

20 SIR97 – Altomare, A., Burla, M. C., Camalli, M., Cascarano, G. L., Giacovazzo, C., Guagliardi, A., Moliterni, A. G. G., Polidori, G., Spagna, R. *J. Appl. Cryst.* **1999**, *32*, 115–119, http://www.irmec.ba.cnr.it/.

21 SIR2000 – Burla, M. C., Camalli, M., Carrozzini, B., Cascarano, G. L., Giacovazzo, C., Polidori G., and Spagna, R. *Acta Cryst.* **2000**, *A56*, 451–457.

22 Fochi, F., Jacopozzi, P., Wegelius, E., Rissanen, K., Cozzini, P., Marastoni, E., Fisicaro, E., Manini, P., Fokkens, R., and Dalcanale, E. Self-Assembly and Anion Encapsulation Properties of Cavitand-Based Coordination Cages, *J. Am. Chem. Soc.* **2001**, *123*, 7539–7552.

23 ORTEP-III – Burnett, M. N., Johnson, C. K., Report ORNL-6895. Oak Ridge National Laboratory, Oak Ridge, Tennessee, **1996**.

24 Braga, D. *Chem. Comm.* **2003**, 2751–2754.

25 Merton, R. K. and Barber, E. *The travels and adventures of serendipity*, Princeton University Press: Princeton, **2004**.

26 Desiraju, G. R. *Angew. Chem., Int. Ed. Engl.* **1995**, *34*, 2311–2327.

27 Desiraju, G., *Current Science* **2001**, *81*, 1038–1042.

28 Tiekink, E. R. T. and Vittal, J. (Eds.), *Frontier in Crystal Engineering*, John Wiley & Sons, **2005**.

29 Moulton, B. and Zaworotko, M. J. *Chem. Rev.* **2001**, *101*, 1629–1658.

30 Braga, D., Grepioni, F. and Desiraju, G. *Chem. Rev.* **1998**, *98*, 1375–1406.

31 Aakeröy, C. B. and Beatty, A. M. *Aus. J. Chem.* **2001**, *54*, 409–421.

32 Bernstein, J. *Chem. Comm.* **2005**, 5007–5012.

33 Braga, D., D'Addario, D., Maini, L., Polito, M., Giaffreda, S., Rubini, K. and Grepioni, F. *Applications of Crystal Engineering Strategies in Solvent-free Reactions: Towards a Supramolecular Green Chemistry*, in *Frontiers in Crystal Engineering* Ed. E. Tielink and J. Vittal, John Wiley & Sons, **2006**.

34 Öhrström, L. and Larsson, K. *Molecule Based Materials – the Structural Network Approach*; Elsevier: Amsterdam, **2005**.

10
Scanning Probe Microscopy

B. A. Hermann

10.1
Introduction: What is the Strength of Scanning Probe Techniques?

With the invention of scanning tunneling microscopy in 1981, it became possible to image surfaces with atomic resolution. Since then, an entire family of scanning probe microscopy (SPM) techniques has developed. In recent years, the field of chemistry has benefited from these developments more than any other discipline. SPM techniques characterized, manipulated and tore apart molecules to learn about the molecular structure. SPM probes induced chemical reactions, observed surface chirality, changed molecular conformation and charge states of molecules – to name some of its implications on the field of chemistry. This chapter highlights various applications of scanning probe microscopy in the field of *supramolecular* chemistry. The majority of results giving high resolution insight into the world of supramolecular chemistry were obtained with the scanning tunneling microscope. This article therefore focuses primarily on scanning tunneling microscopy (STM). An introduction to STM is given in order to familiarize the reader with this method listing references to reviews and books on the subject. Some additional remarks on STM tips, instrument designs, vibrational damping and piezoelectric transducers are delivered in tutorial form. Atomic force microscopy (AFM) and single molecule force distance measurements are briefly covered by giving a short introduction with some exemplary experimental data. In the main part of the article measurements from various research groups in the field highlight the multifaceted and diverse applications of scanning probe methods in supramolecular chemistry. Sample preparation methods are discussed along with the experimental results.

The increasing number of interdisciplinary collaborations between experts in scanning probe microscopy (SPM) and supramolecular chemistry has already given much useful information about real space arrangements of supramolecular assemblies and supermolecules. While large and flexible molecules (the typical constituent of supramolecular chemistry) were widely considered too problematic to be analyzed with scanning probe techniques during the previous decade, data with almost atomic resolution has since then been obtained (even at room temperature and in air), see Fig. 10.1. This opens great possibilities to gain even further insight into the world of supramolecular chemistry.

Analytical Methods in Supramolecular Chemistry. Edited by Christoph Schalley

ISBN: 978-3-527-31505-5

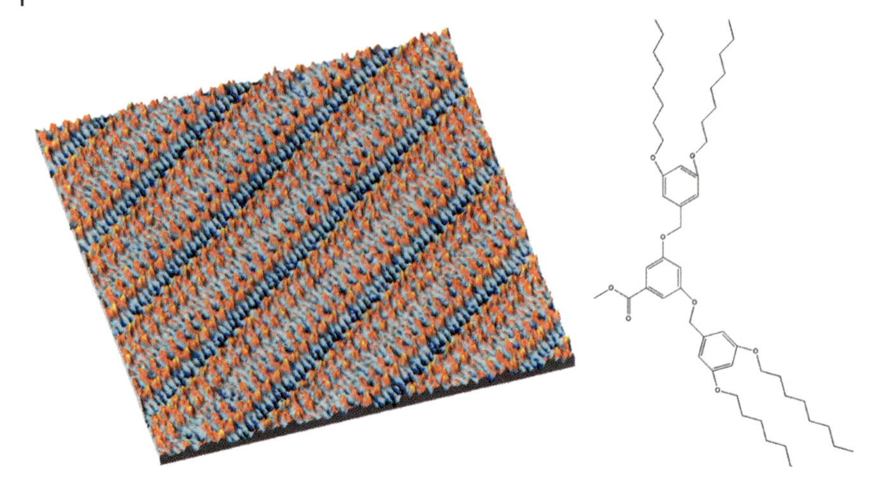

Fig. 10.1. Self-organized monolayer prepared by solution casting of octyl-decorated Fréchet dendritic wedges functionalized by a methyl ester group (30 nm × 30 nm, bias voltage: $U_{bias} = 500$ mV, tunnelling current: $|I_t| = 70$ pA; 0.2 mM in hexane), C. Rohr, B.A.H., LMU Munich, Germany and M. Malarek, L. Scherer, C.E. Housecroft, E.C. Constable, Uni. Basel, Switzerland.

Scanning probe microscopy (SPM) methods, especially scanning tunneling microscopy (STM), are unbeaten in determining the spatial resolution of molecules. The term microscopy may be misleading, since the technique does not rely on optics or any other elements typical for conventional microscopy. As the name suggests, at the heart of an SPM, there is a local probe (a tip), which scans over the surface of a sample at a distance of less than 10 nm. The data obtained during scanning (by measuring a particular physical quantity, e.g. a tunneling current or a force) are used to generate an image on a computer. These images, depending on the type of microscopy and its operation mode, can display many different properties of the sample, e.g. information on the height of molecules on a sample (also called molecular topography), but also electronic, mechanical, electrostatic, optical or magnetic properties and frictional forces can be studied [1]. As a result, SPM is applied in many different fields of science [2–12].

SPMs allow one to obtain local information in real space in some cases in atomic resolution. In contrast, diffraction methods achieve high resolution in reciprocal space by averaged information on large ensembles. While at first sight, the diffraction methods appear advantageous when investigating regular arrangements (crystalline over μm distance), even in this area scanning tunneling microscopy (STM) is gaining more and more ground for the study of 2-D crystallization [13–15]. The particular advantage of SPM lies in the study of local effects, such as structural defects, local domain formation or chirality and dynamic rearrangements in a very high (sometimes submolecular) resolution. Its additional ability to manipulate molecular or atomic entities and to switch charge states and conformations in a controlled way makes the STM an attractive tool for supramolecular chemistry.

Interactions between the probe and the sample are responsible for the information gathered as the tip is scanned over the surface. Based on the nature of these interactions (tunneling of electrons, electric and magnetic forces, chemical forces, van der Waals forces, ...) only parts of the sample in the ultimate vicinity contribute to what is probed by the tip. If the interaction has a strong distance dependence, such as the electron tunneling current used in scanning tunneling microscopy, and an atomically sharp tip is used, images in sub molecular or even atomic resolution may be obtained.

Beyond imaging many more applications of SPM exist [1, 16, 17]. Each SPM realization has a spectroscopy mode where, for example, one parameter is determined as a function of another (*e.g.* current as a function of voltage) [18, 19]. Objects of nanometer size, can be moved (manipulated) with the scanning probe tips through the forces acting between the tip (or the electric field of the tip) and the objects [20–22]. Additionally "on-site on-time" [23], and even "real-time" [24] observation of the processes on surfaces, e.g. adsorbate diffusion, is possible. On top of this, nano-lithography can be performed both with STM and AFM [25, 26].

SPM techniques are applicable under atmospheric conditions (in air), with the probe submerged in liquid [27], inside special electrochemistry set-ups [28], in ultra-high vacuum [29, 30] and under physiological conditions [31]. The measurements can be performed at room temperature, at elevated temperatures [32], and at low temperature [33–36]. The measurement set-up can be combined with optics [37, 38] and/or magnetic fields. Light emission induced from a STM tip can be measured [39]. SPM is readily applied to study catalysis [40–42], surface coatings [4], photonic materials [43], sensor surfaces and biological specimens [31, 44]. Besides many physical and engineering applications, SPM has proven useful in studying large molecules [45] and molecules from supramolecular chemistry [15, 46, 47].

10.2 How do Scanning Probe Microscopes Work?

Tunneling of electrons has been studied for more than 40 years by investigating two conducting electrodes separated by a thin insulating barrier[1] [9]. If the barrier is thin enough, electrons can travel through it by a quantum-mechanical mechanism called electron tunneling. Scanning tunneling microscopy (STM) makes use of electron tunneling by replacing one of the electrodes by a sharp metal tip. Starting at the tip, the tunneling electrons proceed through vacuum (air or liquid), through the layer of atoms or molecules of interest and end in the substrate or vice versa[2], when a negative or positive voltage is applied between the tip and the

1) Molecules have been embedded in the insulating layer to obtain (spatially averaged) information about electronic states of these molecules.

2) Contrary to meaning of the term "substrate" in catalysis, here substrate denotes the support of the atoms or molecules which are studied. Later in the chapter, it is explained, that the electronic states of the substrate have a large contribution to the tunnelling current. Hence, the substrate should be homogenous on the atomic scale. Great attention goes into best possible preparation of single crystalline surfaces to be used as substrate when employing a STM.

substrate, respectively. The substrate corresponds to the second electrode in this arrangement. Scanning tunneling microscopy with the tip acting as a local probe was the first in a family of new scanning probe methods [2–12] (e.g. atomic force microscopy, AFM, scanning near field optical microscopy, SNOM). All scanning probe methods have in common that a probe is scanned over the sample surface in very close vicinity, and the microscopes must therefore be carefully isolated from external vibrations in order to prevent the local probe from accidentally touching the surface (and thus losing its sharpness).

Tutorial 1. Vibrational Damping

The distance between tip and sample must typically be held within 0.01 Å to atomically resolve a metal surface with an STM. Hence it is mandatory to work in a compact and rigid design and to isolate the system from external vibrations, which is technically realized by a damping stage [48, 49]. Vibrations can be caused by the building (15–20 Hz), walking of people (1–4 Hz), vacuum pumps and other apparatus as well as sounds [50]. Needless to say, a silent STM lab (no one walks or speaks, no air-conditioning) is best for the data acquisition process. Each mechanical set-up, for example an STM and its surrounding apparatus, has its own resonance frequency and can be looked upon as a harmonic oscillator. External vibrations constitute a sinusoidal driving force. The resonant frequency of the microscope set-up should therefore not match the frequencies of external driving forces, and preferably lie below 1 Hz [50].

The technical isolation of the measurement set-up from building vibration is typically realized in multiple stages: the entire apparatus is positioned on a stone plate/table, which is either damped pneumatically, or using a layered wood-silicon structure or it is suspended on elastic cordage. An SPM may additionally be suspended on springs, with the springs being actively or passively damped. Preferably, an additional eddy-current damping[3] is installed. Large-amplitude low-frequency perturbations may still be relatively unaffected by conventional damping systems, i.e. an SPM design needs to have a sufficient internal stiffness. Any unavoidable connection between the damping stage and the outside world (e.g. via cables running from the table to the electronics) should be as flexible as possible in order to minimize mechanical coupling. External sounds are minimized by reflecting the sound waves from a hard surface and internal sound waves are damped by soft materials; this is often realized by layered hoods. Alternatively, the microscope can be placed in a silent room.

3) For eddy current damping typically conductive metal sheets are placed in close vicinity to magnets. When the sheets move or vibrate eddy currents are induced. The Lenz rule tells us that the system tries to oppose the cause of the eddy currents, thus damps the movement or vibration.

10.2.1
Scanning Tunneling Microscopy (STM)

An STM contains a tunnel junction in which the barrier width can be tuned continuously over a range, where quantum mechanical tunneling is possible (the nanometer range). Materials used for the electrodes, i.e. for the tip and for the substrate are metals, semimetals or doped semiconductors. Almost any kind of materials can be adsorbed at the tip or the substrate with one restriction: it should at least be weakly conducting[4] [51].

10.2.1.1 Working Principle of STM

When a sharp tip is positioned as a local probe above the surface and a voltage is applied, the tunneling of electrons results in a current, which exponentially decays with increasing distance between tip and surface. An increase in the distance of one Ångstrom leads to a decrease in tunneling current by approximately one order of magnitude. It is this sensitivity, which ultimately causes the unique resolution of STM.

A tunneling current occurs when electrons move through a barrier which classically cannot be overcome. In classical terms, if a particle does not have enough energy to move over a barrier, it will not. However, since electrons are quantum mechanical objects they have wavelike properties besides their particle like properties. These waves do not end abruptly at a wall or barrier, but decay exponentially (Fig. 10.2). If the barrier is thin enough, the probability function may extend through the barrier into the region beyond it and electrons are thus able to tunnel through the barrier. Electron tunneling can be used in conjunction with feedback to control the tip height above the surface. A so controlled STM tip is scanned over the surface line by line. From the scan lines a computer can generate a representative image.

The concept of a scanning tunneling microscope along with the first STM data were first published in the year 1982 by G. Binnig, H. Rohrer, C. Gerber and E. Weibel [52, 53]. Only five years later [54], its invention was awarded the Nobel prize in physics.[5]

How can the tunneling barrier width and the lateral position of the tip be controlled so accurately? The tip is mounted on an actuator consisting of piezoelectric ceramics for reliable and exact positioning in all three dimensions (particular for the scanning). Only the nanometer positioning ability of piezo ceramics (see Tutorial 2 on Piezoelectric Tube Scanners and Translational Stages) made SPM techniques initially possible.

4) STM on adsorbates and molecules is also performed on thin layers of insulators for application in catalysis and molecular electronics.

5) More correctly: half the Nobel prize was awarded for the invention of STM, the other half of the Nobel prize for the invention of the electron microscope to Ernst Ruska.

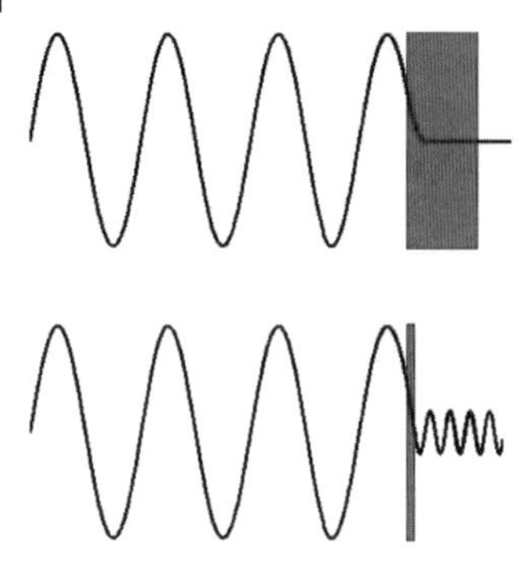

Fig. 10.2. STM is based on quantum mechanical tunneling, which can be best explained by taking into account the wave character of electrons. Here the sinusoidal wave on the left side of both pictures denotes the quantum mechanical state of a freely propagating electron. At a thick barrier the quantum mechanical state associated with the electron decays exponentially within the barrier (top picture). An electron propagating through a thin barrier has a quantum mechanical probability of tunneling through the barrier (bottom picture).

Tutorial 2. Piezoelectric Tube Scanners and Translational Stages

Highly oriented ferroelectric ceramics[6] polarized in an electric field (>2000 V mm^{-1}) are best for piezoelectric actuators. In such a piezoelectric material, even in a *non-polarized* state, neighboring electric dipoles can obey a parallel orientation. Such a group is referred to as Weiss domain. However, the overall dipole moments of several Weiss domains in point random directions (Fig. 10.3a) – hence, the material is not polarized. During the polarization process (when a strong electric field is applied) the electric dipole moments of all the Weiss domains align (Fig. 10.3b) and the material slightly expands along the axis of the external electric field and contracts perpendicular to it. Even upon canceling the electric field the dipoles stay roughly in alignment (Fig. 10.3c).[7] When afterwards an electric voltage is applied to a polarized piezoelectric material, the Weiss domains increase their alignment proportional to the electric field generated by the applied voltage (this is referred to as the converse piezoelectric effect[8]). The result is a very small change of the dimensions (stretching, contraction) of the piezoelectric material in the order of nm and micrometer. That is the dimensional change a scanning probe microscope makes use of.

For SPM applications, a piezo stage must be designed to move in x-, y- and z-directions. Several types of scanners have been technically realized (Fig. 10.4). Most common is a tube scanner[9] in configuration B. Overall, piezo materials show a reliable performance for thousands of operation hours[10], however, one has to be aware of possible hysteretic behavior[11] and piezo creep[12].

Tutorial 2 *(continued)*

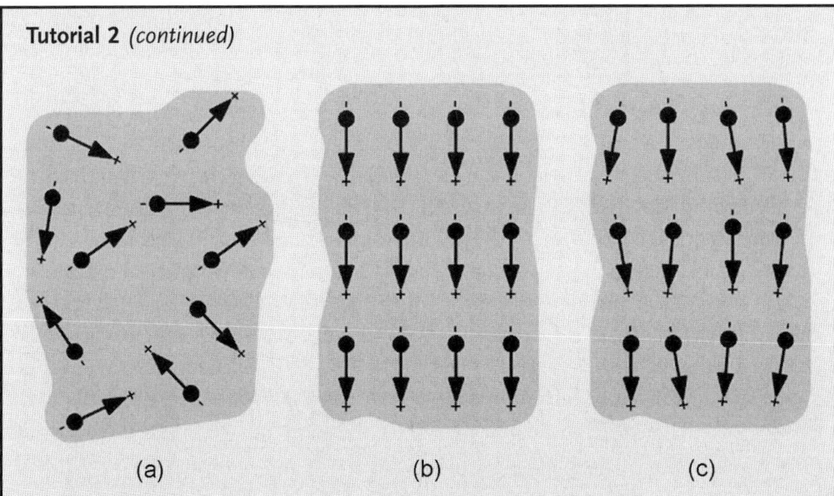

Fig. 10.3. Piezoelectric ceramic with groups of oriented dipoles (a group is indicated with one arrow): (a) Prior to orientation in an electric field, the dipoles point in random directions; (b) while a strong electric field is applied the dipoles align; (c) a polarized piezoelectric ceramic ready to use.

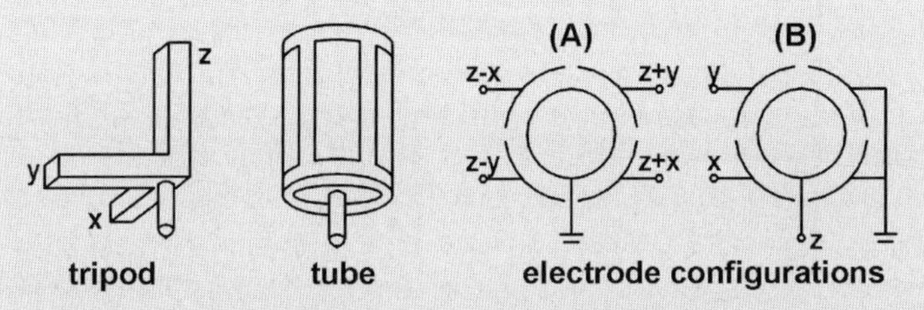

Fig. 10.4. Two scanner realizations: tripod and tube, (A) contact scheme with the z-signal added to x- and y-signal, (B) contact scheme with z-electrode in the middle.

The maximum scan size that can be achieved with a particular piezoelectric scanner depends upon the length of the scanner tube, the diameter of the tube, its wall thickness, the strain coefficients of the piezoelectric ceramic and the applied voltage. The sensitivity of the piezo depends on temperature: its maximum scan range is approximately reduced by a factor 5–6 by cooling the piezo material from room temperature to liquid helium temperature (4.2 K). The process of calibration is described in Tutorial 3.

For approaching a STM tip to a surface (or vice versa) [55, 56] a mechanical set-up with micrometer adjustment [57] or a piezo-based linear translation stage is necessary. In principle, three classes of piezo-based motors exist for linear movement: caterpillars, impact drives (driven by inertia) and slip-stick motors [58]. Modern slip-stick inertia translation stages combine the latter two principles.[13)]

Tutorial 2 *(continued)*

Piezos can be operated in the kHz regime and can travel up to several mm in distance, if needed. Piezos work under applied voltages between 60 and 500 V depending on the type and technical realization.

6) Lead zirconium titanate (PZT) is the most common material used for piezoelectric actuators.
7) If a piezoelectric material is heated above the so-called Curie temperature, the overall alignment of the Weiss domains is lost and the piezoelectric properties disappear.
8) Contrary to the (non-linear) *converse* piezo-electric effect: when pressure is exerted upon a piezo material a voltage is generated – this is called the (linear) piezoelectric effect.
9) A tube scanner moves the tip along the section of a circle or a parabola. This non-linearity can lead to distortions towards the edges of an image, especially when dimensions of the scan areas come close to the maximum elongation specified for that piezoelectric ceramic at the typical operation voltage (approximately μm × μm even for small scanners). Most commercial SPM set-ups correct for that by scanning a larger window than used for displaying as the image.
10) The operation can be ended by depolarization (see footnote above) or by a hairline crack in the piezo materials.
11) A piezo material can show a hysteretic behavior (due to internal heating) when alignment of the electric dipoles leads to differences between forward and backward scans.
12) When a new scan size is actuated with a piezo scanner (e.g. 60 μm), the piezo will further stretch (to 61.8 μm) over time. This effect causes an apparent bending of rows at the beginning of an SPM image. Hence, to avoid piezo creep in an image, the measurement should be restarted after the first 20% of the imaged area, when a new scan range has been selected.
13) A mounting block slips along a guided (movable) shaft, to which it is otherwise sticking because of friction. In the following procedure, a net step is obtained: the sliding shaft is first accelerated very rapidly over a short period of time (typically microseconds) so that the inertia of the sliding block overcomes the friction. The sliding block unlocks from the accelerated shaft and remains in its place. Subsequently the guiding shaft moves back to its initial position slowly enough for the mounting block to stick to the shaft and to move a step. Periodic repetition of this sequence leads to a step-by-step motion of the mounting block in one direction.

10.2.1.2 Operation Modes of STM

When a bias voltage is applied between tip and substrate to induce electron tunneling, STM images can be obtained in several operation modes:

Constant Current Mode Setting the instrument to hold a specified tunneling current is one mode of operating an STM (Fig. 10.5a). In this mode, a feedback loop adjusts the distance between tip and sample, the so-called z-height, in order to keep a particular current (the set-point current I_t). Ideally, if the surface were electronically homogeneous, the z-height of the tip would directly yield the topography of the surface. In fact, the STM actually senses the number of filled or unfilled electron states near the Fermi level (the highest energy of occupied states at zero temperature) within an energy range defined by the applied bias voltage. Rather

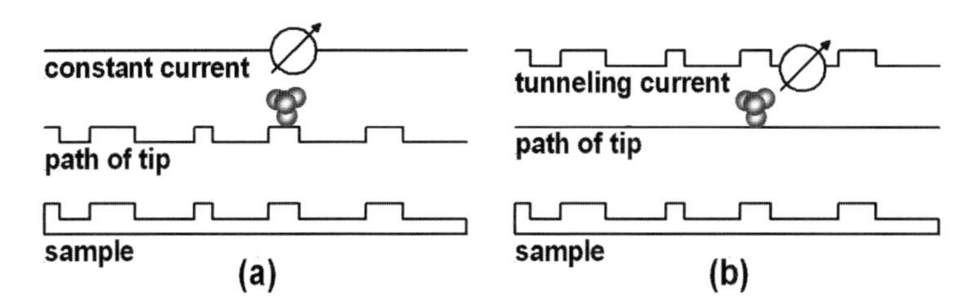

Fig. 10.5. (a) constant current modus for obtaining a "topographic" image, (b) constant height modus for measuring current images; only applicable on (atomically) flat surfaces.

than measuring physical topography, the tip follows the surface of constant tunneling probability (related to the electron density). This can be exploited to study molecules, because the chemistry of molecules is mostly determined by the electronic properties of the bonds.

Constant Height Mode Alternatively, on (atomically) flat surfaces, the tip–sample distance can be kept constant by disabling the feedback (Fig. 10.5b). The variations of the tunneling current are then recorded. Constant height mode is often used to atomically resolve surfaces and does not give any direct topographic information. However, it is particularly useful, when high scan speeds are desired (e.g. for video STM).

STM in Liquids Under suitable conditions (see below), STM can be applied with the tip submerged in liquid. For imaging molecules on surfaces at liquid–solid interfaces, STM is performed in nearly saturated solutions [59]. Typically the solvent is non-conductive, has a low vapor pressure and shows a lower affinity to the substrate than the molecules to be studied. The scanning parameters, i.e. tunneling voltage and current that determine the distance of the tip to the sample, are optimized such that an ordered monolayer of the molecules fits between tip and substrate and can be imaged with high resolution.[14] A voltage pulse (applied outside the imaging area) can help to initiate ordering of molecules at the liquid–solid interface. The dynamic exchange of molecules adsorbed on the surface with molecules in the liquid face eases repair in self-organized structures. Thus, it is challenging to distinguish spontaneous ordering of the molecules under the scanning tip from a substantially ordered architecture, which could also in rare cases include defects.

STM at the liquid–solid interface bears many advantages compared to operation under ultrahigh vacuum conditions: First of all, it combines ease in operation and low infrastructure costs, but also it is better suited for investigations of large and

14) When the parameters are optimized for scanning the surface, the tip is low enough to penetrate the molecular layer. Atomically resolved images of the surface then become possible although the molecules are still present.

fragile molecules[15]. However, the samples can only be cooled down to temperatures close to the solidification temperature and heated to the evaporation temperature of the solvent.

When an STM is operated in air, in general a water film is present on all surfaces. STM at the liquid–solid interface on the other hand offers control and tuning of the solvent properties as a function of the particular substrate and/or molecules. When measuring at the liquid–solid interface, STM is performed in a sealed fluid cell, otherwise the operation time is limited by the evaporation of the solvent.

Due to its many advantages, STM at the liquid–solid interface [27] has provided detailed insight into the molecule–substrate (epitaxy [60]) and molecule–molecule interactions (e.g. hydrogen bonding, metal complexation) responsible for the ordering of molecules on the atomically flat surface. It is possible to induce chemical reactions at the liquid–solid interface, via external stimuli (e.g. light), or by manipulation with the STM tip, where the location and orientation of functional groups could be controlled [27].

10.2.1.3 **Imaging with STM**

The following paragraphs mainly concentrate on the "constant current mode", and only briefly mention peculiarities of the "constant height mode". As already mentioned, in electronically homogeneous samples constant tunneling current images allow to map the topography. More precisely speaking, the tunneling tip follows a surface of constant tunneling probability (or electronic density). STM actually senses the number of filled or unfilled electron states near the Fermi level (the highest energy of occupied states at zero temperature) within the energy range determined by the bias voltage.

During the scanning process, a bias voltage is applied between tip and substrate to establish the tunneling current. Usually, this small tunneling current of the order of nA is transformed into a voltage and amplified by a factor of 10^9 to 10^{10} to facilitate the measurement. When the set-up is operated with an analog electronic in "constant current mode", the so obtained voltage[16] is compared (Fig. 10.6) to the reference voltage (corresponding to the set-point current) and the difference is taken. An electronic feedback circuit minimizes this difference by adjusting the vertical position of the tip over the surface: for this the difference is multiplied with a predetermined gain (typically containing a proportional and an integral part) and high voltage amplified, and finally added or subtracted to the piezo voltage to regulate the tip. The speed and quality of the feedback circuit is optimized by tuning the interplay between these proportional[17] and integral[18] pa-

15) Thermal evaporation or organic molecular beam epitaxy (OMBE) are the most common methods to introduce molecules into an ultrahigh vacuum environment. Most molecules in the field of supramolecular chemistry are too fragile for these methods, because they decompose.

16) In general applications, the logarithm of the tunneling current signal is calculated. For imaging in atomic or sub-molecular resolution, comparing with a linear function of the tunneling current signal may be advantageous.

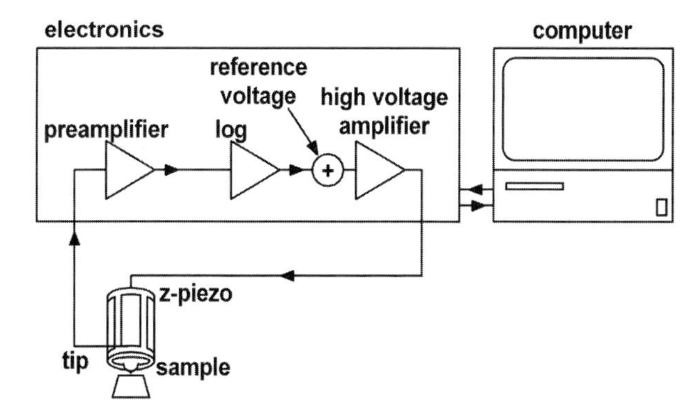

Fig. 10.6. STM tip and piezo tube in the feedback circuit use in "constant current mode".

rameters[19] [24]. Typically, sub-molecular resolution[20] is only obtained for ideal parameters (bias voltage, set-point current, proportional- and integral parameter).

By applying appropriate voltage ramps to the x- and y-piezo-drives (Fig. 10.7), the tip is scanning along the surface line by line, controlled by the feedback and thus follows the electronic contour of the studied object.

Perpendicular to the scan direction, the scanner steps from one line to the next line.[21] The scan speed is typically defined as the number of lines acquired per unit of time and is usually given in Hz.

The step size is given by the size of the scan area and the pixel resolution of the image[22]. The acquired signal – either the voltage needed to move the tip up and down in z direction (constant current mode) or the tunneling current itself (con-

17) also called proportional gain. Optimal values for the proportional gain may easily span two orders of magnitude (e.g. 0.2 to 20). When the proportional gain is large, a large difference between signal and set-point will lead to large adjustment in the tip height.

18) also called integral gain. Concerning the optimal values, see proportional gain. For the integral gain, the difference between signal and set-point is integrated over time. This gain so to speak "remembers what happened in the past". Only when a significant integral gain is used, can the difference reach zero, otherwise a permanent difference between set-point and tunneling current remains.

19) In real-time applications, optimizing the feedback parameters is particularly crucial. With respect to this a lot can be learned from the papers of the SPM community.

20) An additional enhancement in resolution can be achieved by operating the microscope in the low current regime (0.2 to 10 pA, in contrast to typical values of some ten pA to some ten nA). Applying a bias voltage to the tip leads to a strong inhomogeneous electric field, which in turn exerts an electrostatic force on the molecules on top of the substrate and disturbs the measurement. When investigating soft and flexible molecules, this electrostatic force can be minimized by lowering the tunneling currents, because for smaller tunneling currents the tip is further away from the object of study (see discussion later in the chapter).

21) The direction of a line is often referred to as the "fast scan direction", the perpendicular direction consequently as the "slow scan direction".

22) Typical sizes are 256 points by 256 points (256×256) and 512 points by 512 points (512×512).

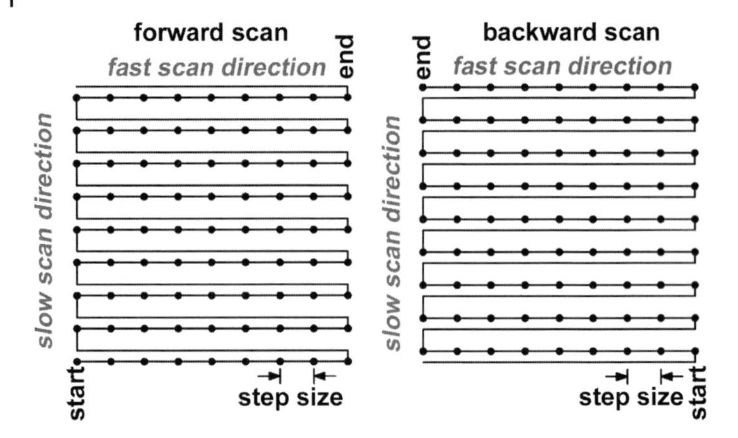

Fig. 10.7. Scanning process: The surface is imaged line by line. While the tip continuously follows the surface, the measurement is taken at the positions denoted by black circles.

stant height mode) – is converted into an image by assigning different colors or gray scales to the data values.

In the first case, the z-height has to be calibrated on reference samples (see Tutorial 3 on calibration) in order to receive meaningful results.

Tutorial 3. Calibration

Scanning probe instruments are used to study surfaces quantitatively. They have to be calibrated in order to give accurate length dimensions. Because of peculiarities of piezo scanners, i.e. non-linearities, temperature and scan speed dependences and hysteresis, this process is usually not straightforward and has to be done thoroughly to obtain accurate results. Thermal drift[23] can lead to much greater distortions on an SPM image than piezo non-linearity. Hence, one is advised to establish thermal equilibrium[24], when calibrating a system.

Lateral calibration (x-y) The lateral calibration in the x-y-scan direction of scanning probe microscopes is usually performed by a set of measurements on appropriate physical standards e.g. a micro/nano grid and/or a standard sample that can easily be imaged with atomic resolution, in most cases HOPG (highly oriented pyrolytic graphite) for STMs. Such standards consist of regular periodic structures of well-known dimensions in one or two lateral directions and are used to determine the voltage necessary to extend the piezo tube to a certain distance. From the images of these standards, the conversion factors of voltage to distance for the x and the y axes as well as correction fac-

Tutorial 3 *(continued)*

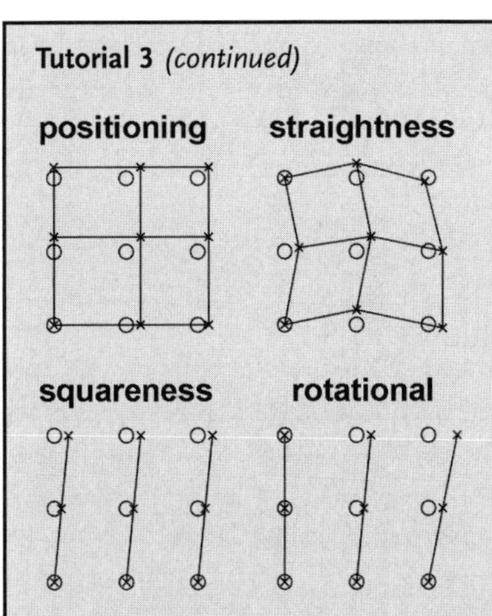

Fig. 10.8. Different types of image distortions. The small circles mark the reference positions in an ideal 2-D grid, the crosses indicate the measured positions. Depending on their origin, these errors are either temporary, permanent or of dynamic nature. Hence, the dynamic behavior of the scan system has to be taken into account when performing a lateral calibration. The scan speed may even have a greater influence on the actual correction factor than the scan range. Further to be considered are the digital resolution (in z-direction) and the number of pixels (in x-y-direction).

tors for non-linearity are obtained (see below). For calibration purposes, these standards are systematically scanned on various length scales and at various scanning speeds (as well as different temperature ranges in variable temperature set-ups). In the case of an STM, the smallest scale is often controlled with graphite (HOPG) in atomic resolution or with a clean silicon (111) surface showing the well-known (7×7) reconstruction (for ultrahigh vacuum compatible STM).

Non-linear distortions in a measured image can be of following type: positioning, straightness, squareness, and rotational errors (Fig. 10.8).

Height calibration (z) Monoatomic steps on the Au(111) surface and a nanogrid with defined z-height have proven to be good standards for the z-calibration of an STM. In many respects the calibration procedure for the vertical axis (z-axis)[25] is comparable to that for the lateral axes, however, the vertical movements are often much faster (in order to follow the object's topography or more accurately its electronic contour), than in the lateral directions. In order to obtain accurate results, the parameters of the feedback control system (e.g. proportional factor P and integration time constant I), that influence the dynamic behavior of the SPM in z-direction, need to be adjusted

Tutorial 3 *(continued)*

carefully for the particular scan speed used e.g. by monitoring the error signal and subsequently minimizing it.

The z-calibration is performed by measuring the z-height profile across a monatomic step on a well-known sample, in most cases a Au(111) surface,[26] carefully taking into account the sample tilt.[27] When measuring the step height in a line scan, the area directly adjacent to the step edge should be neglected due to dynamic effects and convolution[28] of the scanning tip geometry with the actual topography. Additionally, a z-calibration with a defined nanogrid is usually performed in order to obtain data on a larger scale.

Due to piezo wear, the z-height needs to be inspected in regular intervals and also the lateral calibration should be checked at least every three months to ensure correct quantitative data.

23) The working temperature may not be equal to the surrounding temperature, when the STM is heated up during the measurement. A STM, like any other instrument, is composed of various materials, each of which adopt the working temperature with a different speed. Variations in the temperature lead to elongation or compression of all these materials. Though thermal expansion (compression) is small in most materials for temperature changes of the order of 1 °C, the length changes are nevertheless large enough to create distortions on the length scale of nanometers.

24) Thermal drift can be controlled in comparing top to bottom with subsequent bottom to top scans: no differences in angles (e.g. observed between the same row) and no compression or elongation should be observed, when the set-up is in thermal equilibrium.

25) If possible, it is best to adjust the average z position (or z offset) in a way that the z actuator operates in the middle of its range, i.e. symmetrically around 50% of its maximal elongation during each of the calibration measurements.

26) Gold surfaces tend to deteriorate rather quickly; hence, freshly prepared samples should be used for the calibration.

27) Choose a flat area near a step as scan area to determine the tilt of the sample. For correcting the tilt in the step height image obtained nearby, subtract the plane obtained in this way.

28) A common problem in all scanning probe microscopy methods is that tips have some width (in the best case they are still at least one atom wide) and that the width of the tip leads to broadening of the structures observed on the surface by convolution of the tip with the surface. This is explained in more details in the Tutorial on STM tips.

10.2.1.4 **Tunneling Spectroscopy**

The electronic structure of a metal is shown in a simplified diagram (Fig. 10.9). The electrons of the metal occupy all available energy levels up to the Fermi energy E_F, at which they precisely compensate the positive charge of the ionic cores of metal atoms. For an electron to leave the metal and thus to obtain the energy of the vacuum level, it needs to acquire an extra amount of energy ϕ above the Fermi energy (Fig. 10.9). The energy ϕ is known as the work function of the metal.

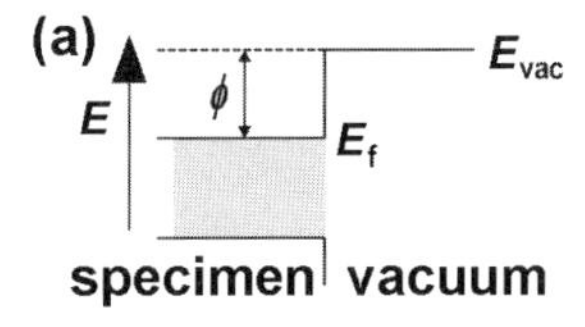

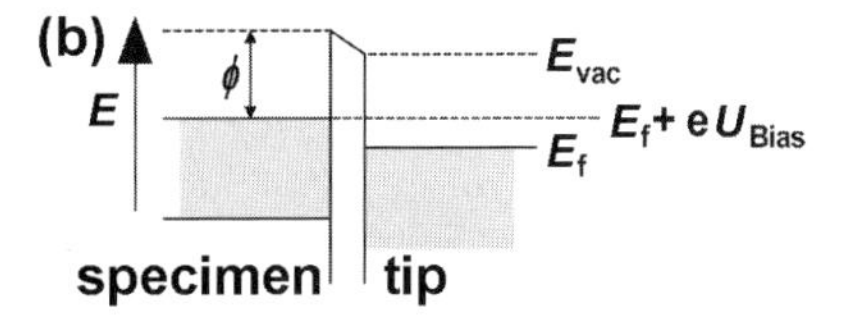

Fig. 10.9. (a) A metallic surface (denoted specimen) with a simplified density of states with occupied states up to the Fermi energy level E_F showing that the work function ϕ has to be overcome to free an electron from the metal into the vacuum space. (b) The same metallic surface with a scanning tunnelling microscope tip (the density of states of the tip is similarly simplified) in close vicinity and a voltage U_{Bias} is applied. Now electrons can directly tunnel from the sample into unoccupied states of the tip without overcoming the barrier. Only electrons within the energy window between E_F and $E_F + eU_{Bias}$ can tunnel.

Classically, electrons still need to have an extra energy ϕ above the Fermi energy to move from the specimen to the tip or vice versa, even if the tip and the specimen come in close proximity (provided there is no conductive connection).

In quantum mechanics: if the distance d between specimen and tip is small enough, electrons can "tunnel" through the vacuum barrier (see above) without having an extra energy ϕ. When a bias voltage $+U_{Bias}$ is applied between specimen and tip, the tunneling effect results in a net electron current from the occupied states within eU_{Bias} of the Fermi energy into the unoccupied states to the tip[29]. This so-called tunneling current can also flow from the tip into the specimen when the polarity of the bias voltage is reversed. When the electrons tunnel from the specimen into the tip, the highest occupied states (or the HOMO – highest occupied molecular orbital) of the specimen are probed. When the electrons tunnel from the tip into the specimen, the lowest unoccupied states (or the LUMO – lowest unoccupied molecular orbital) are filled first. Although being strongly simplified, the HOMO/LUMO treatment holds for many experimental observations [61]. However, a quantitative analysis should take into account shifts and broadening of energy levels due to the electronic coupling between adsorbate and substrate as well as the finite temperature[30] and the electric field in the vicinity of the STM

29) Here and in the following discussion it is assumed that the positive pole of the voltage is applied to the tip!

30) Strictly speaking, the theory presented in the following is valid only at $T = 0$ K. A finite temperature always leads to a level broadening.

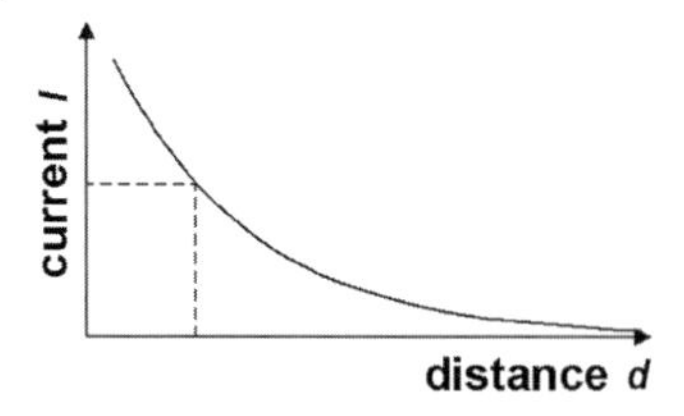

Fig. 10.10. Tunneling current versus tip-sample distance in an STM.

tip ($\sim 10^7$ V cm^{-1}). Furthermore, the presence of polar solvents can lower the energy due to local uncompensated charges.

In a simplified treatment (see Eq. 10.1), the tunneling current depends on the tip-surface distance d, on the voltage U, and on the height of the barrier ϕ (Fig. 10.10):

$$I_t(d) \propto eU_{Bias} \exp[-(2d/\hbar)(2m_e\phi)^{1/2}] \tag{10.1}$$

This approximation[31] shows that the tunneling current depends linearly on the bias voltage U_{Bias} and exponentially on the distance d. The other quantities in the equation, namely the work function ϕ, the electron charge e and mass m_e, and Planck's constant $\hbar$, are constant for a given material. For a typical value of the work function ϕ of 4 electron Volts (eV), the tunneling current reduces by a factor 7.7 for every 0.1 nm increase in d. This means that over a typical atomic diameter of 0.3 nm, the tunneling current changes by a factor 456! The tunneling current depends on the distance so strongly that it is dominated by the contribution flowing between the last atom of the tip and the nearest atom in the specimen. As mentioned previously this is the primary reason for the high resolution of the STM.

In order to observe images in atomic or sub-molecular resolution with STM, a high sensitivity is essential and, in particular, the tip must be held in a stable position in close vicinity to the surface. To remain as sharp as possible, the tip must approach to the surface without uncontrolled contact. This can be realized in various experimental STM set-ups (see Tutorial 4 on STM designs).

31) The formula was derived for metals, where one can assume a s-orbital shaped apex atom and a one dimensional symmetric tunneling barrier in the limit of a small bias voltage.

Tutorial 4. STM Designs

The components of an STM set-up: piezoelectric tubes and tripods as well as linear translational drives were explained in Tutorial 2 on piezoelectric tubes and translational drives. This tutorial focuses on how the components can be combined and operated to perform STM.

To better understand advantages and disadvantages of possible designs, first the operation is discussed, which consists of two phases: (a) the approach of the tip – discussed in this tutorial and (b) the scanning (eventually combined with spectroscopy, manipulation, or lithography) – discussed in the following paragraphs. Approaching the tip towards the sample without touching, is possible due to a controlled interplay between coarse approaches, e.g. piezo translational stage[32], and fine approaches realized by over-extending the scanner tube or tripod. The procedure of an automatic approach is a repetition of these coarse and fine approaches: With the coarse approach, the tip is brought a step closer to the sample (i.e. the step size is smaller than the range of the z piezo). Then in the fine approach, the tube/tripod scanner is slowly extended in z-direction, with the feedback turned on. Then, as long as no tunneling current is established, the tip is fully pulled back. With the coarse approach the tip is moved another step towards the sample. This is repeated, until the tip reaches the surface, i.e. a tunneling current is detected.[33]

In order to allow the above described procedure, the components can be combined in various designs. Among the fully piezoelectrically driven scanning tunneling microscopes, two main categories of designs exist: based on linear piezo-translational stages and a beetle or Besocke set-up (Fig. 10.11).

A Besocke STM (beetle) consists of a scanner tube plate attached to three piezoelectric walkers, which move on three heliclines (denoted ramp in Fig. 10.11) along the circumference of the round sample holder plate. The sample

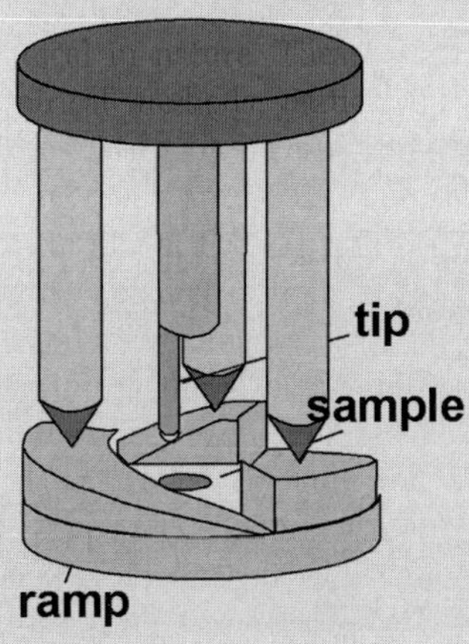

Fig. 10.11. Besocke set-up.

Tutorial 4 *(continued)*

is mounted in the middle of the sample holder, while the scanner tube is attached in the middle of the scanner tube plate. By applying an appropriate voltage to each of the three piezoelectric walkers, they stretch in a counterclockwise direction around the circumference of the sample holder and the scanner tube plate with the three walkers has made a step down the ramp. When the voltage is then quickly switched off, the scanner tube plate stays in place due to inertia. Repeated many times, the scanner tube plate (and thus the tip) approaches towards the surface over the walkers moving down the helices in this slip-stick motion. A disadvantage of the beetle design is that positioning in x–y-direction with the walkers is hardly reproducible because the tip circles[34].

The other type of design relies on linear piezoelectric translational stages, which either move the sample or the scanner tube, while the other remains fixed. These designs do not suffer from the drawback mentioned above and can reproducibly position the scanner tube and thus the tip in x-y-direction. However, concerning stability/drift upon cooling, the beetle design is superior, because it is a very well balanced system.

32) In some designs, the translational stage is replaced by a mechanical set-up e.g. with a screw with fine pitch thread.

33) For automatically approaching a surface, it is crucial that the noise on the measured tunneling current is as small as possible.

34) Due to technical limitations, it is not possible to mount a tip in the middle and straight with sufficient precision to avoid circling, when the plate (where it is attached to) is rotated.

The scanning tunneling microscope is a tool not only capable of imaging individual atoms and molecules, but also of manipulating and spectroscopically characterizing them. Ho [62] looks upon the tunnel junction as a nanocavity allowing the versatile STM to interact with atoms and molecules. Since the electrons coupled to the nuclei of molecules are the driving force for chemical transformations and reactions, the STM with its tunneling electrons can be used to characterize and induce atomic and molecular motions [20–22] and to form and dissociate chemical bonds and thus introduce chemical reactions [63–65].

The spectroscopic (I-U) information obtained with STM [18] is best valued by comparing it with conventional methods to obtain I-U data: information on electronic properties of the sample is gathered by regular electric transport or magneto-transport measurement set-ups over a comparatively large sample area (of dimensions of some 10 nm to a few millimeters) with the microscopic structure of the sample largely unknown. STM can be used as a spectroscopic tool, probing

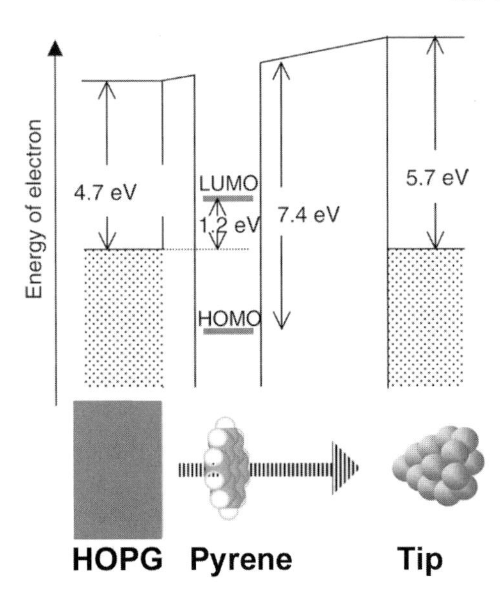

Fig. 10.12. Schematic representation of the energy diagrams of pyrene between a STM tip and HOPG. Based on [66], Copyright (2005), with permission from Elsevier.

the electronic properties of a material at a specific position with atomic precision, while the surrounding area is known[35].

The electronic structure of an atom depends on its chemical nature and on its environment, i.e. how many nearest neighbors surround the atom, their chemical nature and the symmetry of their arrangement. Scanning tunneling spectroscopy in general shows the same spatial resolution as is obtained in imaging, i.e. atomic resolution or molecular or sub-molecular resolution.

The different scanning tunneling spectroscopy (STS) modes [18] are discussed below.

As an example: STS and STM have been used to characterize and interpret the π-stacking of 1-pyrenehexadecanoic acid on HOPG at the liquid–solid interface (Fig. 10.12). The tunneling electrons were found to couple to the π-stacked pyrene moiety at lower voltage than for isolated molecules [66]. The conductance of an α-helical peptide trapped between a gold-tip and gold-surface via S-Au linkage was determined with STS and found to be comparable to the reference molecule dodecanedisulfide [67].

35) The position is controlled by imaging the area prior to the spectroscopic measurement and re-checked after finishing the spectroscopic measurement.

d*I*/d*U*: Scanning Tunneling Spectroscopy (STS) – Electronic Density of States The local electronic and chemical environment can be probed with scanning tunneling spectroscopy (STS) using various methods: (a) In the first mode discussed here, the feedback is turned off and the tip is positioned above the object/position of interest while ramping the bias voltage and simultaneously recording the tunneling current.[36] In this way, the electronic structure at a specific location on the sample surface is obtained, because the first derivative $\mathrm{d}I/\mathrm{d}U$ is proportional to the electronic density of states (see below). STS measurements can be performed at every point in an image, providing a two-dimensional map of the electronic structure (Fig. 10.13a) [68]. (b) Such information is also obtained in the second mode discussed here by taking "topographic" (constant-current) images employing different bias voltages and by subsequently comparing them (Fig. 10.13b).

A lock-in amplifier allows to directly measure $\mathrm{d}I/\mathrm{d}U$, yielding a direct access to the density of states at the Fermi-level, when plotting $(\mathrm{d}I/\mathrm{d}U)/(I/U)$ versus U (Eq. 10.2):[37]

$$(\mathrm{d}I/\mathrm{d}U)/(I/U) \approx [\rho_\mathrm{S}(e\ U)\rho_\mathrm{T}(0) + A(U)]/[B(U)] \tag{10.2}$$

with ρ_S being the density of states of the sample and ρ_T being the density of states of the tip. $A(U)$ and $B(U)$ are functions only weakly depending on U in the case of metals [69].

d²*I*/d*U*²: Inelastic Electron Tunneling Spectroscopy (IETS) – Phonon States Local chemical analysis with the STM is possible by identifying molecules through their vibrational eigenmodes using inelastic electron tunneling spectroscopy (IETS). The ability to measure a local vibrational signal with sub-Angstrom resolution in single molecules makes it possible to directly characterize the chemical bonds and their interactions with the surrounding medium. In addition, the molecules are usually unaffected by IETS since the tunneling electrons involved have only small energies. To determine the IETS signal, the tip is placed accurately over the molecule

Fig. 10.13. (a) In steps A to G the formation of a metal-porphyrin-metal bridge is observed . The $\mathrm{d}I/\mathrm{d}U$ versus bias voltage curves are obtained at the denoted points (a–d) in the corresponding images; (b) images obtained at different bias voltages during the bridge forming process (A–G). (Reprinted with permission from [68]. Copyright 2003 AAAS). (c) Manipulation of a cyclodextrin necklace with an STM tip. Displayed is a pair shuttling process and a reversible bending of the necklace. (Reprinted in parts with permission from [75]. Copyright 2000 American Chemical Society)

36) Each I-U curve is obtained within fractions of a second, therefore thermal drift can be neglected.

37) The formula is valid for low bias voltages in WKB (Wentzel-Kramers-Brillouin) approximation.

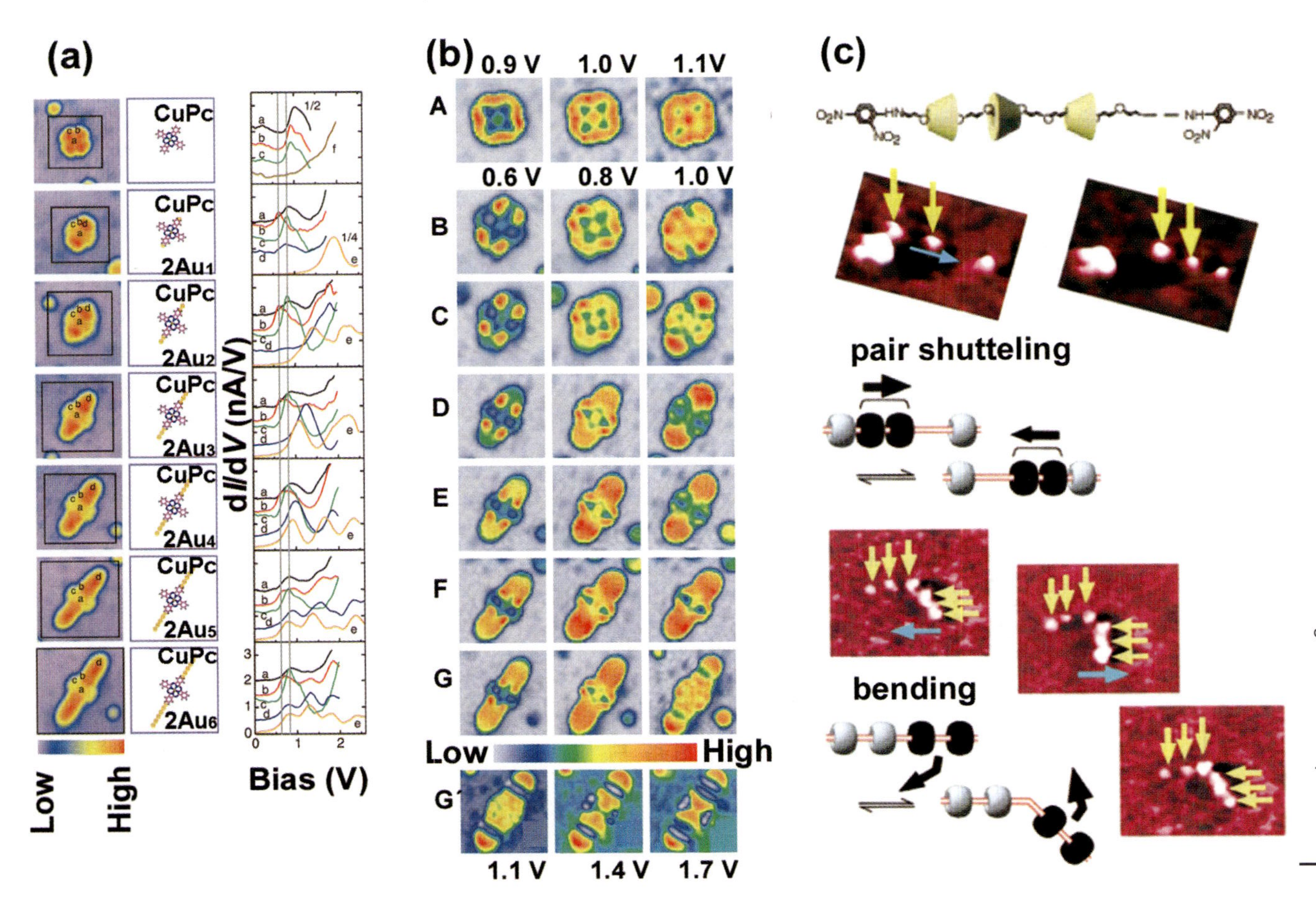
(a)
CuPc
CuPc 2Au1
CuPc 2Au2
CuPc 2Au3
CuPc 2Au4
CuPc 2Au5
CuPc 2Au6
Low
High
dI/dV (nA/V)
Bias (V)
(b)
0.9 V 1.0 V 1.1V
0.6 V 0.8 V 1.0 V
A
B
C
D
E
F
G
G'
Low
High
1.1 V 1.4 V 1.7 V
(c)
pair shutteling
bending

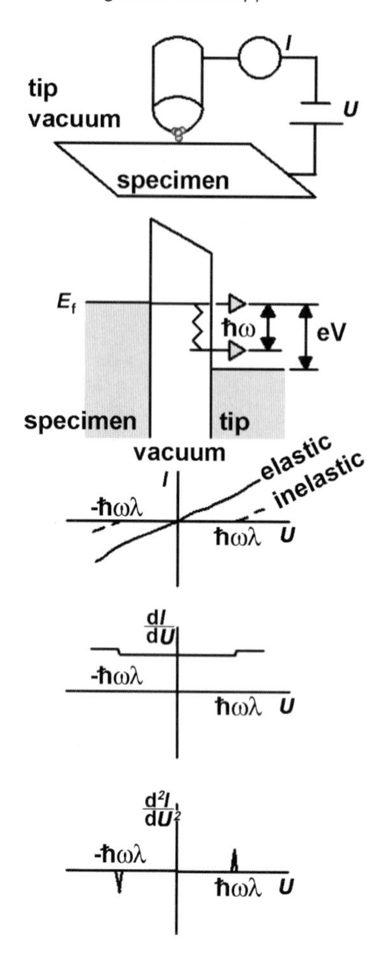

Fig. 10.14. Contribution of a vibrational mode to I/U, dI/dU, d^2I/dU^2 by opening an additional tunneling channel.

of interest and dI/dU curves are obtained e.g. with lock-in at each point of interest. When the threshold energy of $\hbar\omega$ needed to excite a vibrational mode is reached, an additional tunneling channel opens, leading to a peak in the d^2I/dU^2 signal for positive bias at $+\hbar\omega$ and a depression in the d^2I/dU^2 signal for negative bias at the symmetric value of $-\hbar\omega$ (Fig. 10.14).

A drawback of IETS are the small signals, which require high statistics to improve the signal-to-noise ratio and observe the peaks in the second derivative of the signal. The requirements imposed on the measurement setup are extremely high, because a long data-taking time without external disturbances are mandatory for reliable results. Additionally, the interpretation of the results appears to be difficult, therefore, single-molecule vibrational spectroscopy in solids has not been extended far beyond a demonstration of feasibility [62].

dI/dz: Current versus Distance – Work Function dI/dz measurements provide information about the z-dependence of the tunnelling current, which is correlated with the work function at a specific location x, y of the surface (equation 10.1). Note that ϕ in eq. 10.1 is the barrier height, which corresponds to the average work function $\phi = (\phi_s + \phi_t)/2$, where ϕ_t and ϕ_s are the tip and sample work functions, respectively). dI/dz can as well be probed in two modes of operation: (a) in the first mode, the tip is retracted over the object/position of interest while recording the tunneling current as function of the tip-sample distance. (b) In the second mode, the tunneling current is measured at different heights in constant height mode. With dI/dz measurements, the characteristics of the tunnelling gap and the STM tip can be tested, since surface/tip contaminations and water films (in general present when measuring under ambient conditions) can influence the results.[38)]

10.2.1.5 Manipulating Atoms and Molecules with STM

With an STM tip as "engineering tool" for surface modification, artificial atomic-scale architectures can be fabricated [70–72], chemical reactions on surfaces can be induced [63–65] and properties of single molecules can be studied at an atomic level [20, 21]. Manipulations with an STM tip can be performed by precisely controlling tip-sample interactions using tunneling electrons or the inhomogeneous electric field between the tip and sample.

The following discussion introduces surface/sample modifications through the STM tip by focusing on manipulation of atoms. Manipulation of molecules is based on the same principles; examples are listed at the end of this section.

The STM tip has to be particularly well-defined for manipulation experiments. A controlled contact with the surface is often performed prior to manipulation experiments (see Tutorial 4 on STM tips). An STM manipulation process used to relocate single atoms or molecules across a surface is called a lateral manipulation. Such a lateral manipulation process involves: (a) vertically approaching the tip towards the manipulated atom to increase the tip–atom interaction, (b) moving the tip parallel to the surface while the atom stays under the influence of the tip and moves laterally, (c) retracting the tip back to the normal scan height and thereby leaving the atom in its desired final location on the surface.

The atom can be moved under the influence of the STM by either pulling, pushing or sliding the atom with the tip.

Which of the movements is initiated can be controlled by choosing appropriate components of the force of the STM tip in the direction perpendicular ($F\perp$) and parallel (F_{II}) to the surface.

Pulling: the tip starts from directly above (no F_{II} component) and moves over the atom in direction of the planned lateral movement. Under the increasing F_{II} component of the force, the atom hops into the next energetically favored position on the substrate. At this point, the STM feedback system retracts the tip in order to

38) As a rule of thumb for measurements in air, if the tunnelling current I_T drops to one half within a distance of $z < 0.3$ nm, the tip can be considered to be very good, if it drops within 1 nm $< z <$ 2 nm, the tip may still lead to atomic resolution on HOPG, for any values $z > 2$ nm, the tip should be replaced.

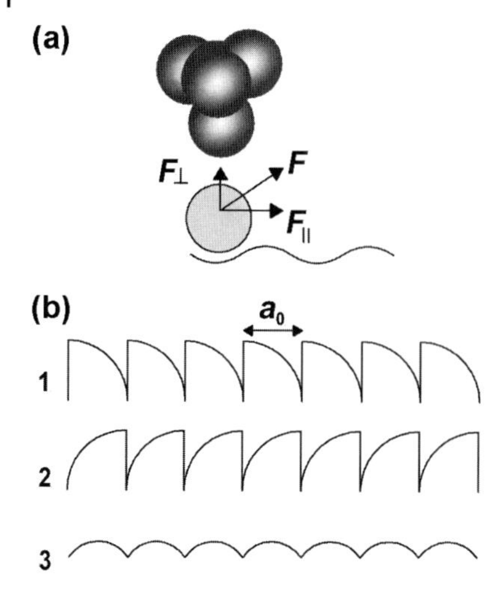

Fig. 10.15. (a) schematic drawing explaining the force component parallel and perpendicular to the surface. (b) height signal (in constant current mode) during the lateral movement of the atom with the tip (1) pulling, (2) pushing, (3) sliding. Adapted from [20].

maintain the constant current causing an abrupt increase in the tip height. Repeating the sequence provides a saw-tooth-like tip height curve (Fig. 10.15b1). The atoms follow the lateral movement of the STM tip.

Pushing: the tip starts behind the atom (large F_{II} component) and moves over the atom in direction of the planned lateral movement. At a certain distance the $F\perp$ and the F_{II} component of the force push the atom to the next energetically favored position on the substrate. The atom moves while staying in front of the STM tip. When the atom has moved away, the STM feedback system initiates a sudden approach of the tip to the surface in order to maintain the constant current causing again an abrupt decrease in the tip height. Repeating the sequence also provides a saw-tooth-like tip height curve, but of opposite form (Fig. 10.15b2).

Sliding: the atom stays under the tip during the lateral movement (Fig. 10.15b3).

As an example, Moresco and co-workers [73] have manipulated a six-leg single hexa(4-*t*-butylphenyl)benzene (HB-HPB) molecule[39] to adsorb five copper adatoms on a Cu(111) one at a time at low temperatures. When the molecule was picked up after the manipulation with the STM tip, the HB-HPB molecule released a Cu_5-cluster.

Crommie and co-workers [74] have reversibly built K_xC_{60} ($x = 1$–4) complexes by moving C_{60} molecules over potassium adatoms on a Ag(001) surface with an

39) The HB-HPB molecule consists of a central benzene ring connected to six phenyl groups each connected to a *t*-butyl end group. The phenyls can rotate whereas steric hindrance between the phenyl groups forbids a full planar conformation of the molecule. Moresco and co-workers believe that the *t*-butyl end groups elevate the central HPB, creating a cage in which the Cu adatoms are trapped between the central HPB group and the metal surface underneath.

STM tip at low temperatures. Further manipulation of the so obtained K_xC_{60} complexes allowed moving them without detaching the potassium atom, thus proving that indeed a complex had been formed. When accidentally picking up the C_{60} molecule, the potassium atom was liberated.

Manipulation has also been applied to supermolecules: Shigekawa *et al.* have manipulated a cyclodextrin necklace with the STM tip [75]. They were able to move one and two cyclodextrin molecules at a time. Additionally, they were able to reversibly bend the necklace (Fig. 10.13c).

Tutorial 5. STM tips

STM tips can be made of e.g. platinum iridium (PtIr), tungsten (W), titanium carbide, gold, silver, etc. *Tungsten* has a step-like density of states with no features and therefore expected to have the least influence on the tunneling process ($\rho_T(0)$ in eq. 10.2). However, tungsten oxidizes quickly in air (leading to much reduced conductance). For reproducible results, the oxide layer needs to be removed by special treatments in vacuum, i.e. heating, Ar ion sputtering, field emission, or electron bombardment (some of these methods are discussed in greater detail near to the end of this tutorial). In order to avoid reoxidation after such a tip cleaning treatment, tungsten tips are only used in vacuum. *PtIr* tips with a very steep geometry (see theoretical simulations of tips at the end of the tutorial) can be obtained by cutting and pulling a PtIr wire at the same time, by electrochemical etching or by ion-milling.

STM tips operated in electrochemistry environments have to be coated with wax or with a solvent resistant polymer (e.g. nail varnish, epoxy) leaving the tip apex free.

However small the structures under study are, the STM tip always consists of atoms, which have a finite dimension. Hence, the structure of the tip has an influence of what is observed with STM. If one assumes in a Gedanken-experiment that an infinitesimally sharp object is standing on the surface (mathematically a delta-function), then the tip will be convoluted at the object (Fig. 10.16, far right) and visible in the STM image will be the front most tip-end (up-side down) and not the sharp object itself. Hence, sharp tips are crucial for high-resolution in STM and, more general, in all Scanning Probe Methods (SPM).

In ultrahigh vacuum, a number of methods can be used in order to improve the sharpness of the STM tips prior to and during the investigations:

When performed outside the scan area of interest, a *sudden increase of the bias voltage* to about −7 to −10 V (at the tip) for 2–4 scan lines (on metal surfaces) leads to a controlled field emission of electrons[40]. By this treatment some atoms may move down to the tip apex due to the non-uniform electric field and form a nanotip (Fig. 10.17) [76].

Tutorial 5 *(continued)*

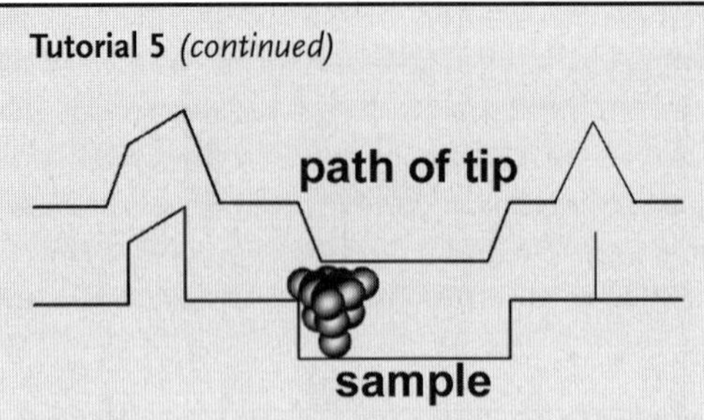

Fig. 10.16. Convolution of the tip shape into the measured profile.

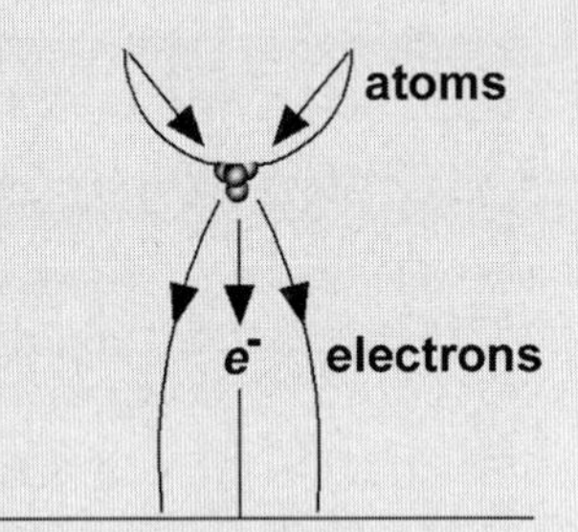

Fig. 10.17. Field emission of electrons from a tip by applying a large negative bias to the tip.

During measurements on silicon, a *controlled collision* of the tip with the Si surface can be used to improve tip sharpness. The tip may pick up a Si-cluster, which forms a monoatomic apex with a p_z-like dangling bond.

Another possibility sharpening an STM tip is to place the tip in a heating filament and *bombard* it with electrons.

Recently, the method of a two step *controlled tip sample contact* described to extract atoms and form a sharp tip on Ag(111)[41)] [77].

For modeling purposes, the electronic structure of a tip is approximated by a linear combination of atomic orbital (LCAO) method. Cluster models of 10–20 atoms are utilized [78]. It has been found that the tunneling current is concentrated on a single apex atom, if the other front most atoms of a tip are not located on the same level. Hence, the apex atom of the tip matters (examples are in Fig. 10.18) [79].

Tutorial 5 *(continued)*

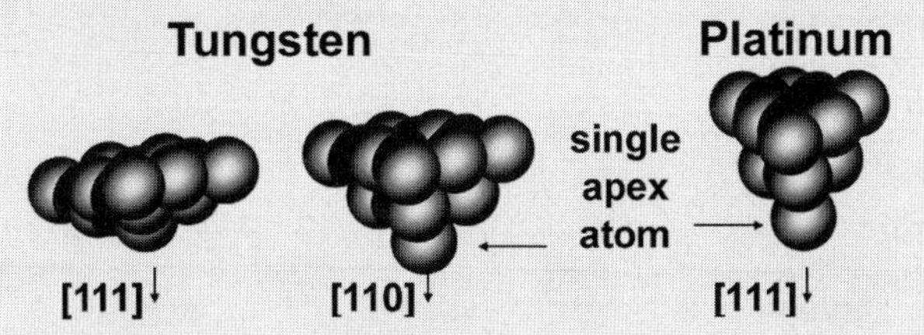

Fig. 10.18. Cluster models of "theoretical" Tungsten and Platinum tips. The apex atom points in specific atomic directions. The apex can be replaced by an adsorbate (not shown) [79]. Adapted from reference [78].

40) For this procedure usually the tunneling current preamplifier has to be either protected by a special circuit design or disconnected.

41) A suitable 'z' piezo voltage is applied to dip the tip into the substrate by 3.5 nm while the tip is over an atomically flat surface area. The dipping will coat the tip with Ag atoms. Dipping speeds between 0.2 and 2.5 nm/s and tunneling biases between +3 and +8 V have been reported by Hla and co-workers. For the second tip-sample contact (on a flat area at the side), the tip penetration depth is precisely controlled to be again 3.5 nm. The tunneling voltage applied for this second tip-crash is 1.5 V. The newly obtained crash profile shows a smaller indentation and the authors concluded they had obtained a sharper tip.

10.2.2 Atomic Force Microscopy (AFM)

The AFM makes use of a sharp tip attached to a cantilever, acting as a spring. Unlike STM, where the tunneling current is a measure of the interaction, the force between tip and sample is detected via its mechanical influence on the cantilever deflection or resonance. The AFM can be used to study insulating, semiconducting and conducting samples.

10.2.2.1 Function Principle of AFM

The AFM tip is a couple of microns long and often less than 10 nm in diameter at the point of the tip. It is attached to the free end of a cantilever, which is typically 100 to 200 μm long [80]. While the tip is scanned over the surface by piezo transducers, the tip-sample interactions change the deflection or the resonance of the cantilever and can be measured. e.g. by laser beam deflection with a photo-detector. In static mode the bending of the cantilever is a measure for the interaction with the surface. In dynamic mode the amplitude, frequency and phase of the oscillating cantilever respond to the tip-sample interaction. A shift of resonance frequency, the damping and the bending can be measured and used to control the z-position of the cantilever tip. Finally, a computer generates a representative image (Fig.

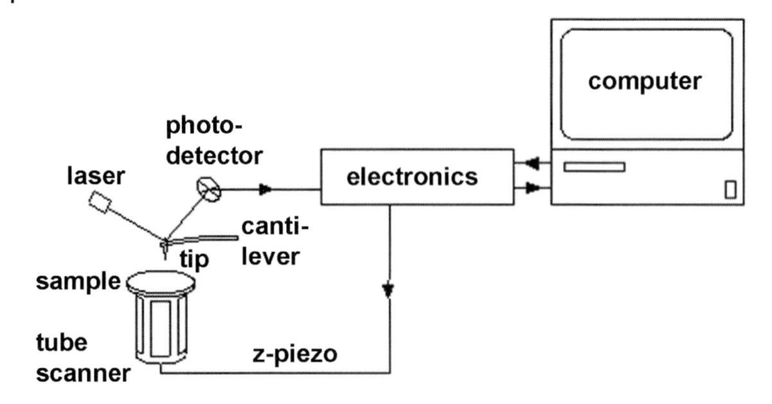

Fig. 10.19. Schematic set-up of an AFM using the beam deflection method. The piezo tube scans and approaches the sample to the tip attached to a cantilever. The force acting on the highly flexible cantilever is transduced into an electronic signal via a beam deflected onto a four quadrant photo-detector.

10.19). A large number of methods are commonly used to detect the cantilever resonance or bending – either incorporated into the cantilever, e.g. piezo-resistive, or external, e.g. capacitive, interferometric, beam deflection (Fig. 10.19) [81]. The cantilever should be soft and have a high eigenfrequency (typical cantilever properties vary from 10 mN m^{-1} and a few kHz resonance frequency in contact mode up to 100 N m^{-1} and 400 kHz in dynamic mode) [82] – both requirements being realized by its small dimensions. Cantilevers engineered for very specific applications (e.g. to be operated in liquid, ultrasharp tips, nanotube tips, magnetic tips) can be bought off-the-shelf.

10.2.2.2 **Various Operation Modes of AFM**

In vacuum, the long-range attractive van der Waals interactions as well as the short-range chemical forces are the predominant interaction between sample and tip (provided that electrostatic Coulomb interactions are carefully compensated or negligible and the tip is non-magnetic). The van der Waals forces are caused by the interaction of fluctuations in the electromagnetic field and are attractive when the tip approaches the surface, the chemical forces originate from Pauli-exclusion and nuclear-repulsion. The attraction increases until tip and sample have approximately the distance of a chemical bond. When distance between tip and surface is further decreased the interaction becomes repulsive[42] (for more details see [83]).

The different force ranges define the operation modes of the AFM. In contact mode, the repulsive chemical force is predominant. In the dynamic mode, the oscillation amplitude can be wider than the force potential displayed in Fig. 10.20. The interactions thus must be integrated over the oscillation cycle [84, 85].

42) Even with a very stiff cantilever, the distance between tip and sample cannot be decreased below a few tenths of a nanometer due to the strong repulsive forces – the surface would rather deform.

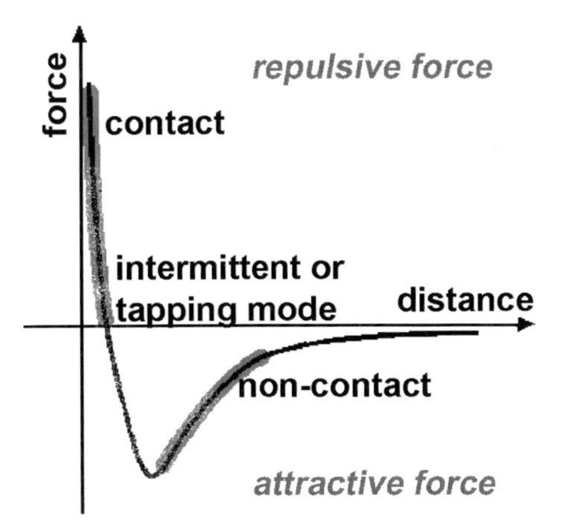

Fig. 10.20. The different force regimes to operate an AFM.

When the dynamic mode is applied in the non-contact regime (typically in vacuum), the attractive forces dominate.[43] When the dynamic mode is applied in the intermittent regime (typically in air or liquids) repulsive forces are the predominant interaction.

Static Mode in Contact Regime In contact mode, a cantilever with a low spring constant[44] is employed allowing a so-called soft contact between tip and surface with a contact area of typically several nanometers in diameter. This contact forces the cantilever to trace the surface topography when scanned over the surface. Similar to STM (which measures a surface of constant tunneling probability), the cantilever actually measures a surface of constant repulsive force in the contact mode. When operated in air, a thin water film on the surface exerts an attractive force on the tip. Assuming a homogenous water layer, this capillary force can be looked upon as staying constant in a measurement. Both the attractive capillary force and the cantilever spring balance the repulsive chemical forces. Constant force and constant height images are possible. The elasticity of the surface can be measured by distance modulation.

Constant force: the deflection of the cantilever is kept constant by the feedback circuit, which moves the scanner up and down in z-direction. In this case, an image of constant force is generated from the movement of the scanner. The advantage of this method is, that with a cantilever deflection held constant, the total force exerted on the sample remains constant during the measurement. This is the preferred mode for most applications.

43) For non-contact dynamic AFM, the cantilever tip stays in the attractive force regime of the surface. To be more specific integral over the cantilever tip attractive forces dominate, however, considering only the front most tip end (the apex atom), repulsive forces may dominate there.

44) The effective spring constant holding together the atoms of the material investigated has to be greater.

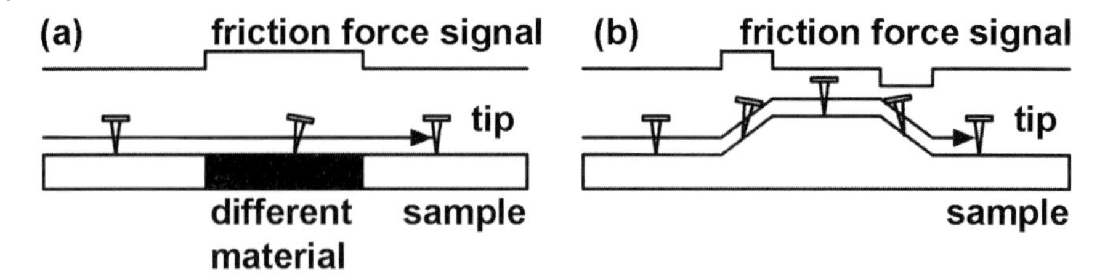

Fig. 10.21. (a) Friction force microscopy on an inhomogeneous material and (b) on steep slopes.

Constant height: similar to the STM constant height mode the feedback is turned off, and the height of the scanner is fixed. This mode is preferably used on atomically flat surfaces and in particular when high scan speeds are desirable. In constant height mode, the spatial variation of the force generates a change in the bending of the cantilever, which is recorded and translated into a representative image by the computer.

Static Mode in Liquids In liquids, the capillary force is not present allowing a more straightforward interpretation of the results. Furthermore, the contact area is as small as a single-atom contact. Special cantilevers have to be used for the operation in liquid. When reused, those cantilevers must be carefully cleaned and are best kept wet until employed again.

Lateral Force or Friction Force Microscopy (FFM) By looking at the torsion of the cantilever, not only vertical but also lateral forces can be detected[45)] [86]. Lateral deflections of a cantilever either arise from the surface friction (Fig. 10.21a) or from a steep change in slope (Fig. 10.21b). The friction signal is determined by subtracting the mean signal in the reverse direction from the mean signal in the forward direction and multiplying the result by one half. By comparing both a regular image and a friction image, much insight can be gained on liquid crystal molecules [87], self-assembled monolayers [88], material contamination and oxidation, etc. [89].

Dynamic Mode in Non-Contact Regime In the non-contact mode, the cantilever is held within a 0.5 to a few nm distance of the surface and the cantilever is oscillated. Here, the tip only exerts a very low force on the sample (10^{-12} N up to 10^{-9} N). The large distance between tip and sample is advantageous for studying soft and elastic materials. Furthermore, the probability for contamination originating from the tip is reduced.

The cantilever is operated at its resonance frequency (several hundred kHz) with amplitudes ranging from a few nm to 100 nm. The resonance frequency of the

45) An AFM equipped with a beam deflection monitoring of the cantilever bending, typically uses a four quadrant sensitive photo detector, which is able to discriminate between up and down and torsional movement the cantilever.

cantilever varies with the square root of its spring constant, which in turn depends on the force gradient (the derivative of the force/distance curve) and therefore strongly depends on the tip-sample distance.

The scanner is moved up and down in z-direction during the scan, in order to keep the resonance frequency constant. Similarly to the constant force mode in contact mode a surface of constant force gradient[46] is measured. The computer generates image as a function of the motion of the scanner.

Non-contact AFM is performed in liquid, under vacuum and in air [90]. In the non-contact regime, the water film present on a sample in air will typically not be penetrated. If for a particular sample this is a problem, and contact mode also happens to be a problem (for example: a soft sample could be damaged by a dragging tip) then a tapping mode operation of the cantilever in the intermittent regime may be the best choice.

Dynamic (or Tapping) Mode in Intermittent Regime

In the intermittent regime, an oscillating cantilever is operated closer to the sample surface than in non-contact mode. Typically, the cantilever exerts strong repulsive forces at its closest position to the surface – so this mode received the name tapping mode (it seems that the cantilever taps on the sample). Herein, too, the cantilever oscillation is affected by the sample surface, and topography images can be generated as described above. Especially when large areas are scanned and for applications, where both contact and non-contact AFM have limitations, tapping mode operation is the method of choice.

Reinhoudt and co-workers [91] have unraveled the nanostructure of a supramolecular assembly of hydrogen-bonded tetrarosettes on graphite[47] in a tapping mode AFM study. They obtained the high resolution by thermally equilibrating their instrument over a period of 1 to 2 days in a contact mode operation before turning to a tapping mode operation.

10.2.2.3 Single Molecule Force Spectroscopy – Force-Distance Measurements

Measuring the force as a function of tip-sample distance can yield information on the elasticity of individual supermolecules, on conformational transitions (e.g. of proteins), on the mechanical stability of chemical bonds and secondary structures, as well as of the desorption of the molecules from the solid substrate [82, 92–94]. Moreover, information on the chemical bond formation of the tip cluster *with* a particular bonding site on the sample surface can be obtained [95].

The forces acting between the molecule and the tip are caused by interactions with the tip apex atom[48] and the mesoscopic part of the tip. The latter contribu-

46) Strictly speaking this is only true in the limit of low amplitudes, otherwise the forces need to be integrated over the oscillation period.

47) (tetramelamine)$_{3'}$(5,5-diethyl barbituric acid)$_{12}$.

48) The main contributions on the tip apex atom are the short range repulsive chemical interactions: Pauli exclusion, nuclear repulsion, steric hindrance, and at somewhat larger distances: attractive van der Waals interactions as well as chemical bonding interactions.

49) Of course, magnetic interactions are only detectable, when tip and sample consist of magnetic materials or are coated with magnetic layers.

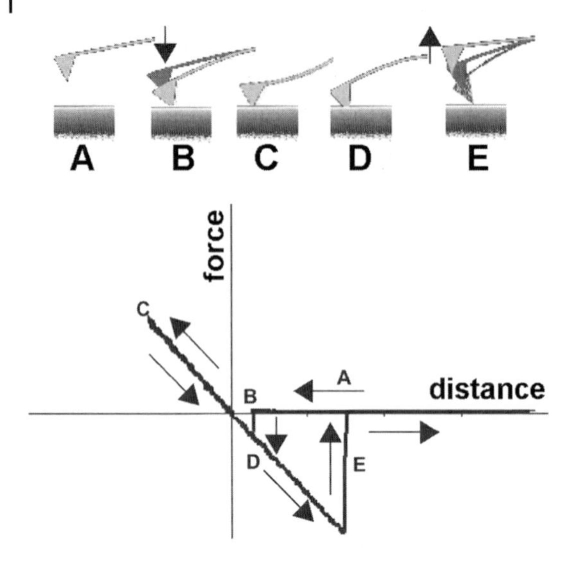

Fig. 10.22. Top: Cantilever movement while obtaining a force distance curve (top). Bottom: Schematic of force versus cantilever displacement (distance) for a tip attached to a soft cantilever approaching (and being retracted from) a hard surface.

tions arise from the long range van der Waals forces and electrostatic, magnetic,[49] capillary forces (in air, when a water film is present). Since the forces on the mesoscopic part of the tip are in many cases considered an undesired background, they are reduced by using a tip that is as sharp as possible. For avoiding the influence of capillary forces, performing the force versus tip-sample distance measurements in ultrahigh vacuum can help. A well-known way to reduce long range forces is to perform the measurements with the entire tip submerged in a liquid, because then the capillary force is absent and the liquid may compensate some of the long-range van der Waals forces acting on the mesoscopic part of the tip.

Figure 10.22 serves as instructive example to understand how a tip attached to a soft cantilever interacts with a hard surface. At the top of Fig. 10.22 the cantilever tip and surface are drawn schematically (cases A to E), the force between the tip and the surface, measured by the cantilever deflection, is displayed in the graph below.

When approaching the surface with the tip (A), the tip experiences an attractive force gradient[50] larger than the cantilever stiffness, it snaps to the surface by bending the cantilever (B). The AFM tip moves closer to the surface, the attractive forces between tip and sample become smaller. As the tip is approached further, the forces can even become repulsive (C).[51] When the cantilever is then retrieved (D),

50) Note that the interaction between tip and surface can also be repulsive, for example if tip and/or surface are electrically charged.

51) Deviations from a slope of -1 can be attributed to the elastic properties of the surface.

the deflection recovers and even changes sign because the cantilever sticks to the surface, until the restoring force resulting from the bending exceeds this binding force and the cantilever snaps out of contact in a characteristic hysteretic motion (E).

When a large molecule, for example a protein, is bound to the tip with one end, and to the sample with another end, this process can be used for study of the elasticity and chemical bonding configuration of the protein. Typically, force versus piezo displacement (distance) is then recorded during retrieving the tip. Depending on the elasticity and involved processes (secondary folding) the curve part between the marked cases A and B as well as D and E in figure D has many more features. Particularly, when single molecules are studied "snap-in" and "snap-out" processes are observed seldom,[52)] rather the elastic properties and conformational unfolding dominate the obtained curve, which ends with a rupture event.

Since molecular unbinding/unfolding processes are of stochastic nature, rupture forces from many rupture events (typically several hundreds) are compiled in a histogram and multiple curves are plotted. The results can be interpreted e.g. by thermal activation of the rupture [93]. The use of AFM in force distance measurements has emerged from the study of biopolymers and was applied to the field of chemistry several years ago [82]. However, its specific application to the field of supramolecular chemistry is relatively new.

Anselmetti and co-workers have investigated a resorc-[4]arene host-guest system with force-distance spectroscopy (Fig. 10.23) [96]. While the calixarene hosts were attached to a gold surface, ammonium, trimethyl ammonium and triethyl ammonium acted as guests. The guests were attached with a PEG polymer to the AFM tip. The stretching of the PEG polymer before the rupture event can be explained treating the polymer as a wormlike elastic chain. The polymer stretching served to discriminate real single binding events from unspecific adhesion. The single-molecule kinetic reaction rates obtained in this way are consistent with the expected nature host-guest interaction of a moderate-affinity between the calixarene and the ammonium hosts.

10.3 Which Molecules can be Studied?

STM research initially looked at adsorbates (atoms, small gas molecules, small clusters) on terraces of single crystalline metal substrates [97, 98]. So it is not surprising, that it had been thought that molecules should be compact, flat and small in order to be suitable for being studied with scanning tunneling microscopy.

52) The snap-in and snap-out processes occur only because the stiffness (i.e. the second z-derivative of the chemical bonding interaction) of the chemical bond is superior to that of the cantilever. If such processes occur, the chemical bonding interaction cannot be mapped continuously, because the energy landscape between the two stable states of the tip-sample contact remains unknown. This can in some cases be avoided by using stiffer cantilevers and/or (when imaging is performed) by oscillating the tip, i.e. in the non-contact mode.

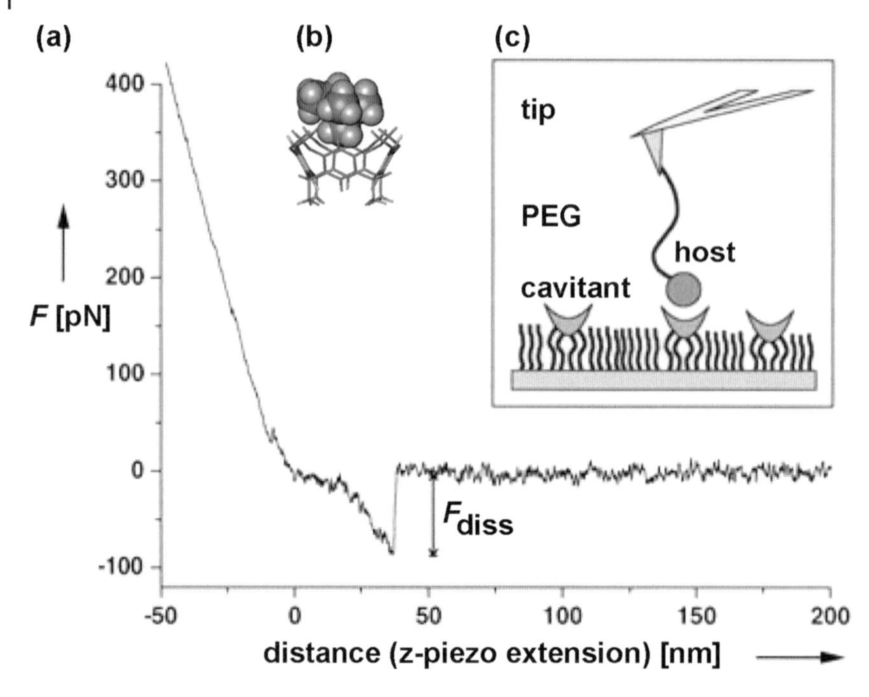

Fig. 10.23. Force versus distance when pulling a host molecule attached via a PEG linker to the AFM-tip from a cavitant. Around a piezo displacement of 38 nm a rupture event occurs, before that the PEG polymer stretches. (Reprinted in parts from [96] with permission of Wiley, VCH, copyright 2005.)

However, large and floppy molecules have since studied with high resolution, provided the interaction with the substrate or neighboring molecules (in a self-assembled or self-organized monolayer) is strong enough to prevent that the molecules are moved under the tip during the investigation (examples see below and section 10.4). The STM has the restriction that tunneling electrons should be able to travel through molecule and substrate, hence both should be at least weakly conducting. The resolution of molecules can be enhanced by studying molecules with STM on top of a few layers of oxide [51], on an otherwise electrically conducting substrate [61]. Of course, with AFM, molecular self-organization even on thick insulators can be studied [99, 100]. Large bio-molecules have predominantly been studied with AFM in the past [10, 101, 102]. However, recent results with STM on DNA show excellent resolution as well [103].

10.3.1
Differences between STM and AFM

Usually, *AFM* is the method of choice to gain a fast large scale overview on a sample. The operation mode should be carefully chosen according to the criteria mentioned above. It may be desirable to avoid the capillary force originating from the water film in air (see discussion above) by measuring in liquid or in ultrahigh vacuum. On smaller molecules high resolution images have been obtained with non-contact AFM under ultrahigh vacuum conditions [99, 100, 104–107]. AFM in liquid and UHV-AFM usually require operation by an expert because the methods are either technically challenging or demand extensive experience. It is important to recall, that the AFM measures the force between (ideally the apex atom of) the AFM tip and the molecules on the surfaces (van der Waals, chemical, electrostatic and magnetic forces)[53)].

Typically, *STM* is the method of choice to receive high resolution images and to gain insight into the electronic structure of molecules on conductive[54)] substrates. STM data depend on the electronic interaction of the molecule with the substrate and therefore require a careful choice and preparation of the substrate. For STM, it is relevant how well different parts of the molecule allow current passage to the substrate. For example, the middle part of a molecule standing on legs may show no contrast in an image, despite offering many electronic states, because the current will solely pass through the legs of the molecule e.g. a porphyrin molecule [108]. When all parameters are finally optimized, the sharpness of the STM tip is responsible for the obtained resolution. Particularly good resolution can be obtained with low tunneling currents (see footnote 20) and/or at low temperatures, again both technically challenging and demanding experienced users.

10.3.2
Exemplary Results on Smaller Molecules

Manifold self-organized and self-assembled mono- and multilayers of smaller molecules have been reported in literature, to name some examples: multiple structured monolayers of iodobenzene[55)] [109], multilayers of perylene molecules[56)] [110], mulitilayer and monolayers of dialkylamino hydroxylated squaraines[57)] [111], monolayers employing marker groups [112], and self-assembled monolayers of thiolate bound molecules [113, 114].

Within organized monolayers, segregation in domains of different conformers [115], segregation in rows of different structure [116], dislocations and defects [117] as well as chirality as emergent property of self-organized layers [118–122]

53) The capillary force resulting from the water film always present in air belongs also in this listing of forces. However, it is generally an unwanted interaction, because the interaction is not reflecting the molecular or surface properties.

54) As discussed above: the underlying substrate has to be at least weakly conducting; suitable are metals, semi metals, doped semiconductors, and oxides on metal surfaces (maximum 3 layers).

55) organic molecular beam epitaxy (OMBE) onto Cu(110) at 55 K in ultrahigh vacuum; studied with low temperature STM.

56) organic molecular beam expitaxy (OMBE) on Cu(110) in ultrahigh vacuum.

57) in phenyloctane on highly oriented pyrolytic graphite (HOPG).

have been observed with SPM methods. STM has also been used to study dynamic effects e.g. rotations [123], hindered rotations [124], conductance [125] and conformational switching of molecules [126, 115] or dynamic reconfiguration of self-organized monolayers [14, 122].

Tutorial 6. Preparation of Molecular Adlayers

For STM investigations, the substrates[58] need to have atomically flat terraces of the size of some hundred molecules. Several ways to deposit molecules on these substrates are briefly discussed here. These methods not only strongly differ in their technical realization, but also in the properties of the resulting molecular adlayers. As will be discussed below, some methods are only suitable for particular classes of molecules.

STM investigations historically started with small molecules deposited under controlled conditions (ultrahigh vacuum) and Langmuir-Blodgett films studied in air. AFM investigations of bio-molecules were first performed with the tip submerged in liquid.

Thermal evaporation[59] (Fig. 10.24a) and organic molecular beam epitaxy (OMBE)[60] [127] allow monomers (dimers, trimers, ...) of typically uncharged molecules on surfaces, having a sublimation temperature in a suitable range

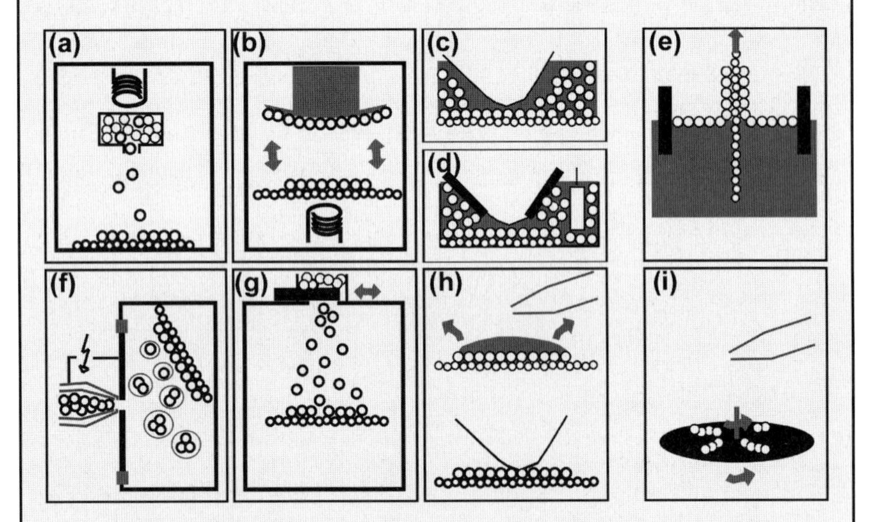

Fig. 10.24. Simplified schematics illustrating some deposition techniques: (a) thermal evaporation; (b) stamping; (c) liquid-solid interface; (d) electro-chemistry set-up; (e) Langmuir-Blodgett technique; (f) electro-spray deposition; (g) pulse injection method; (h) solution casting; (i) spin-coating.

Tutorial 6 *(continued)*

i.e. between room temperature and 700 °C. Deposition is possible on heated or cooled substrates. Often an ordering of the molecules depends on careful heat curing after the deposition. This method is particularly suited for ultrahigh vacuum application. However, this method can lead to decomposition of larger molecules, hence it is rarely applied to study supramolecular adsorbates.

Stamping (Fig. 10.24b) or dry contact transfer (DCT) is an alternative to deposit large fragile molecules directly under ultrahigh vacuum conditions (related methods performed in air: microcontact printing (μCP) [128] and dip-pen nanolithography (DPN) [129, 130]). The molecules are physisorbed on a flexible stamp, e.g. a fiberglass sheath, which is degassed in ultrahigh vacuum and then manipulated into direct contact with the substrate in order to deposit some molecules onto the sample surface [131]. The molecules should have a greater adhesion to the substrate than to the stamp. In a thermal annealing step, a molecular order can be initiated (typically in sub monolayer coverage).

The *liquid–solid interface* (Fig. 10.24c) can directly be investigated operating the tip submerged in liquid [59]. This important method was already introduced in the above section "operation modes of STM". Additionally, an electric potential can be applied to the surface: bio-molecules successfully 2-D crystallized and imaged with AFM on mica in a pH-controlled surrounding buffer [31].

Even more sophisticated:

Electro-chemical environment (Fig. 10.24d) [132, 28], combine an electrochemistry set-up (electrode, counter-electrode – not shown in the schematic drawing – and reference electrode) with the STM tip as a forth electrode. Similar to electro-voltammetry, the potential between the electrode (in this case the substrate) and the counter-electrode can be tuned to induce molecular order. Electrochemistry tips have to be specially prepared i.e. coated (see Tutorial 5 on tips).

In *Langmuir-Blodgett* deposition (Fig. 10.24e), a film of molecules (with amphiphilic properties) is forced in a densely packed layer at a liquid-air interface or at a liquid-liquid (lipophilic/hydrophilic) interface. The substrate (submerged into the liquid beforehand) is pulled from the liquid to air to adsorb a monolayer. With this technique, the monolayer is forced into an organization by narrowing the organization area with movable sidewalls. The technique has been particular successful on molecules with liquid crystal properties [133, 87].

Electro-spray deposition [134, 135] (Fig. 10.24f) is commonly used in mass spectrometry investigations of super-molecules.[61)] With a nozzle formed to a micropipette, molecules are sprayed through a hole into the vacuum with the help of a carrying gas. Between pipette and the hole a high voltage is applied. The molecules are ionized and the droplets contain one to several molecules. A geometry with a substrate slightly above the spraying axis is preferable to catch mostly small droplets carrying only a single molecule each.

Tutorial 6 *(continued)*

The *pulse injection* method (Fig. 10.24g) uses a fast valve (open only for milliseconds) to inject the solvent with the molecules directly into vacuum onto a sample mounted approximately 100 mm under the opening [136]. The molecules remain uncharged and the pressure in the chamber substantially increases during the deposition. This method is successful even for very large and fragile supramolecular assemblies (see discussion below).

Via *solution casting* (Fig. 10.24h) a droplet of solvent containing the molecules is deposited onto the substrate and the solvent evaporates. Typically, millimolar solutions are used for this method, which therefore only need small quantities of substance. The solvent has to be chosen not to adsorb on the substrate and leave enough time (prior to evaporation) to allow molecular organization. This method leads to reproducible results, also for supermolecules. Solution casting on a heated substrate is often referred to as *sizzling* [137].

Spin coating (Fig. 10.24i) is a similar technique, but applied to a rotating substrate. The centrifugal force stretches the molecules in radial direction. This method is particularly useful to study polymers. The parameters to be optimized are the revolutions per minute and the concentration.

58) Typical STM substrates for ambient conditions are: graphite (HOPG), Pt(100), Au(111); for use in ultrahigh vacuum: Cu(111), Ag(001).

59) The containers of the molecules are directly or indirectly heated and manifold forms are employed: crucibles, boots, coils, ...

60) Here small heated pockets with a hole are used to produce a molecular beam.

61) In order to use electro-spray on a ultrahigh vacuum chamber multistage differential pumping has to be employed. The pressure inside an UHV chamber is a factor 1000 to 10000 lower than in a conventional mass spectrometer.

10.4 What Results have been Obtained in the Field of Supramolecular Chemistry?

Here, further exemplary results are given to illustrate the diverse approaches to study supramolecular chemistry with scanning probe techniques. As mentioned previously, one way to study molecules is to build self-organized or self-assembled monolayers[62] [138], but also single molecules or supermolecules as well as supramolecular assemblies were studied in the past. Only some of many examples can be mentioned here.

62) Two-dimensional systems possess a unique topological ordering – not found in either three- or one-dimensional systems.

10.4.1
Coronenes, Crown ethers, Cryptands, Macrocycles, Squares, Rectangles

Coronenes belong to the category of flat π-orbital rich molecules, which self-organize well on surfaces. Some interesting supramolecular assemblies have been initiated around coronenes, and studied with STM. Coronene, thermally evaporated [139] onto HOPG(0001) surfaces in ultrahigh vacuum, initially forms a closed monolayer. When more molecules are deposited islands build on top (Stranski-Krastanov growth mode). Sub-molecular resolution images could be obtained of the monolayer between the islands with STM. However, it has not been possible to determine the boundaries of a molecule.

Heckl and co-workers [140] incorporated coronene molecules into a molecular network of trimesic acid molecules. Some of the coronene molecules apparently rotated and single ones could be removed with an STM tip operated at the liquid–solid interface. The authors report that coronone molecules alone do not self-organize in heptanoic acid on graphite – the support of the trimesic acid molecule network is necessary.

For the organization of coronene in heptanoic acid, the substrate material and/or preparation procedure apparently matters: Hipps and co-workers [141] investigated supramolecular assemblies of coronene with alkaline acids (hexanoic, heptanoic and octanoic) with STM also at the liquid–solid interface. While the dipole-dipole interactions produce bi-layers of one coronene molecule, which are surrounded by 12 acid molecules in the case of hexanoic and heptanoic acid, in octanoic acid, on the other hand, no solvent molecules are present in the coronene monolayer.

Crown ethers: Due to their inclusion capability, crown ether molecules are interesting hosts. Some detailed insights into the adsorption of crown ethers on surfaces as well as its inclusion of hosts have been already obtained in STM studies.

Phthalocyanines functionalized with four benzo*crown ether* rings disubstituted with enantionmerically pure (*S*)-3,7-dimethyloctyl chains have been studied at the gel-graphite interface with STM [142]. The molecules arrange "face-on" (in hexagonal closed packing). In co-existence to this surface arrangement also π–π stacked lamellae (molecules adsorbed "edge-on" graphite) were characterized. By appropriate tuning of the tunneling parameters (raising the tunneling current and decreasing the voltage) structural rearrangement from the face-on to the "edge-on" lamellae structure can be induced.

Adlayers of metal free and preformed potassium complexes of dibenzo-18-*crown-6-ether* molecules (organized on a Au(111) surface under potential control) were in-situ imaged with STM [143]. The potassium ion appeared as ball shaped protrusion in the inclusion complex.

Itaya and co-workers [144] self-organized 15-crown-5-ether-substituted cobalt(II) phthalocyanines by immersing Au(111) or Au(100) substrates into benzene-ethanol (9:1 v/v) solutions of the molecules. After transferring into aqueous $HClO_4$ solution the authors studied the inclusion of Ca^{2+} with STM (Ca concentration 1 mM). Interestingly, the crystalline phase of the substrate mattered: Calcium was only encapsulated in the molecules on the Au(111) surface and not on the Au(100) surface.

Single tetralactam *macrocycles* on Au(111) surfaces apparently only build disordered closed films, and clustering at step edges and surface defect sites occurs [145].

Monolayers of cyclothiophene *macrocycles* have been observed to form a complex with C_{60} molecules, when studied with STM at a 1,2,4-trichlorobenzene solution/graphite interface [13].

Cryptants are three dimensional "crown ether" like molecules suited for inclusion complexes. Stievenard and co-workers [146] made an attempt to study nanoclusters of iso-hexa-imino cryptand molecules deposited by solution casting in chloroform on HOPG surfaces with STM and infrared spectroscopy.

Square and rectangular like metalo-supramolecular macrocycles are widely studied in supra molecular chemistry; some can be imaged with STM. If a molecular square organizes with "edge-on" or "face-on" to a substrate delicately depends on the balance between the molecule-molecule and molecule-substrate interaction forces. On a graphite surface in a dry environment (prepared with solution casting) supramolecular $[Pt(dppp)(4,4'\text{-bpy})]_4\text{-}(CF_3SO_3)_8$ (bpy = bipyridine) squares (Fig. 10.25a) form a lamellar, stacked structure which can be imaged with STM in a high-lateral resolution [147]. On a charged copper surface in an electrochemical STM cell, Broekmann and co-workers [147] have constructed a template of chloride ions on top of the Cu surface thereby forcing $[Pt(\text{ethylenediamnene})(4,4'\text{-bpy})]_4(NO_3)_8$ macrocycles (Fig. 10.25b) to lie flat.

Metalo-supramolecular rectangles cyclobis[(1,8 bis(*trans*-$Pt(PEt_3)_2$)anthracene)-(1,4′-bis(4-ethynylpyridyl)benzene]$(PF_6)_4$ (Fig. 10.25c) self-organize on graphite and on an Au(111) surface in an electrochemical cell. It has been observed that single rectangles stand "edge-on" (with their long edge) on the underlying HOPG surface, while on a Au(111) surface, the macrocycles lie flat ("face-on") [148].

Supramolecular rectangles (1,8-Bis(*trans*-$Pt(PEt_3)_2$)anthracene)(4,4′-bpy)]$_2(PF_6)_4$ (Fig. 10.25d) and (the above studied) [147] squares $[Pt(dppp)(4,4'\text{-bpy})]_4(CF_3SO_3)_8$ (Fig. 10.25a) form adlayers on Au(111) surfaces with the rectangles/squares lying flat – the squares can be studied in submolecular resolution [149].

Obeying a topologically similar structure, calixarenes and cyclodextrines are generally smaller and show a greater shape persistency; they are discussed subsequently.

Fig. 10.25. Structural formulae of the (a) $[Pt(dppp)(4,4'\text{-bpy})]_4\text{-}(CF_3SO_3)_8$ (bpy = bipyridine) [147]; (b) [Pt(ethylenediamnene)-(4,4′-bpy)]$_4(NO_3)_8$ [147]; (c) cyclobis[(1,8bis(trans-$Pt(PEt_3)_2$)anthracene)-(1,4′-bis(4-ethynylpyridyl) benzene] $(PF_6)_4$ [148]; (d) (1,8-Bis(trans-$Pt(PEt_3)_2$)anthracene)(4,4′-bpy)] 2$(PF_6)_4$ [149]; (e) porphyrin boxes (cyclic tetramers of zinc porphyrin-5-tetrapyridylporphyrin) [160].

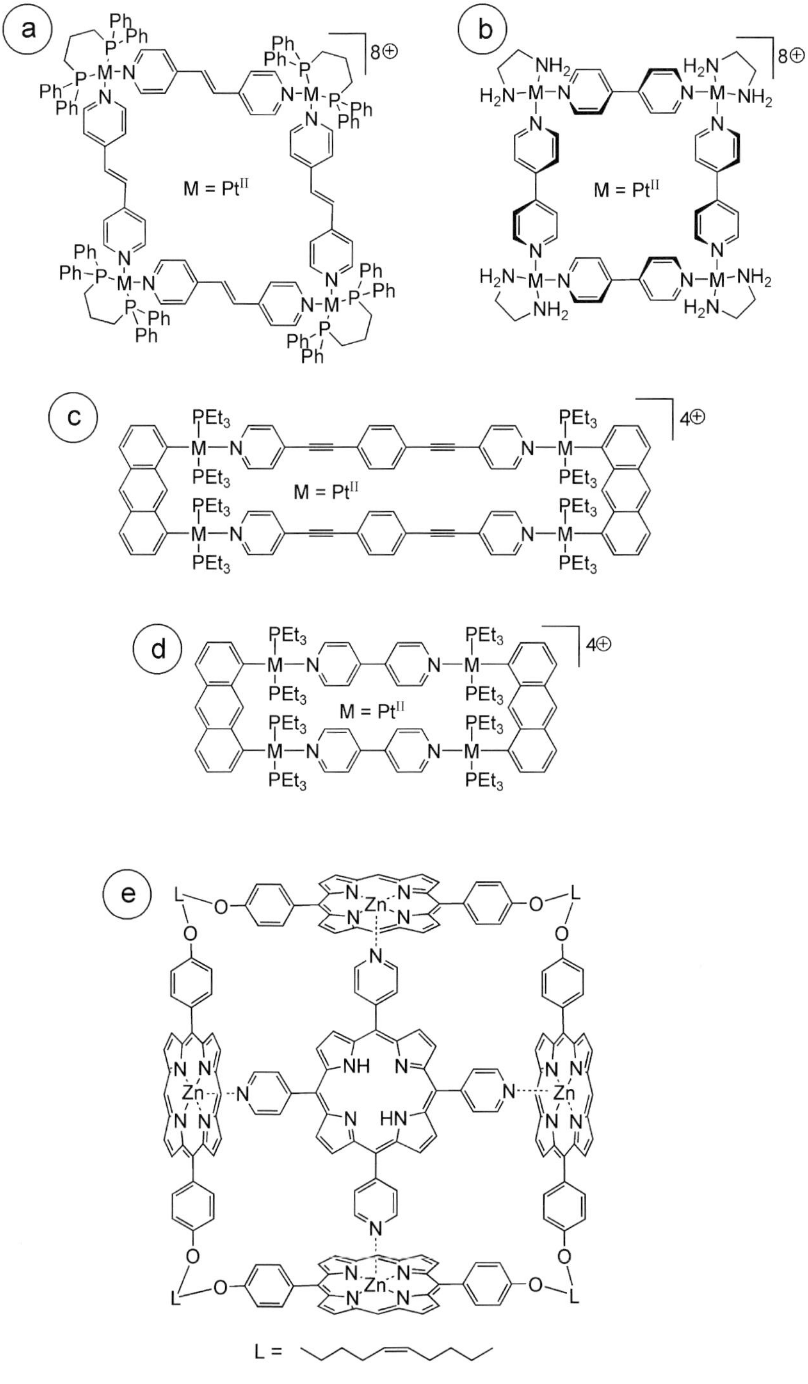

a

M = Pt^{II}

b

M = Pt^{II}

c

M = Pt^{II}

d

M = Pt^{II}

e

L =

10.4.2
Calixarenes, Cyclodextrins, Molecular Sieves and Boxes

The potential applications of *calixarenes* range from molecular hosts for sensor techniques and medical diagnostics, use in decontamination of wastewater, construction of artificial enzymes, new materials for nonlinear optics to sieve membranes with molecular pores.

Early reports on AFM studies on calix[4]resorcinarene and calix[4]arene sulfide molecules [150, 151], suggested, that the calixarenes self-organize in monolayers on Au(111) surfaces, however, they did not identify a molecular ordering.

Bai and co-workes reported a series of STM studies on various types of calixarenes on graphite and Au(111) surfaces. Initially, they included *p-tert*-butylcalix[8]-arene in between domains of octa-alkoxyl-substituted phthalocyanine ($PcOC_8$) monolayers[63] on graphite [152]. The adsorption[64] of calix[4]arene (lower rim: OH) on (111) facets of Au-beads[65] lead to an "edge-on" geometry of calixarene dimers with the lower rim pointing towards lower rim (Fig. 10.26a) [153].

On the other hand, deposition of calix[4]arene (lower rim: propyl, upper rim: COOH) again on (111) facets of Au-beads – this time under potential control – resulted in presumably "face-up" (upright) adsorbed monolayers (Fig. 10.26b) [154].

Calix[8]arene OBOCMC8 ($C_{104}H_{128}O_{24}$) and the inclusion complex C60/OBOCMC8, both adsorbed[66] "face-up" on the (111) facets of Au-beads (Fig. 10.26c) [155].

Langlais et al. [156] deposited calixarenes on (1×2) reconstructed surfaces [60] of Au(110) by organic molecular beam epitaxy (OMBE) in ultrahigh vacuum. The molecules interact so strongly with the substrate that the original (1×2) missing row structure of Au(110) transforms into a (1×3) structure with troughs of [111]-facets. The calix[4]arenes fill the troughs in the reconstruction with two opposite phenols groups bound to the [111]-facets.

The above listed examples illustrate that calixarene adsorption and organization in monolayers not only depends on the type of surface, but also on the preparation and measurement conditions. The cyclodextrins discussed below even form nanotube-like chains.

Single molecule force spectroscopy was used to measure the inclusion of a guest into a calixarene host molecule, which served as example in the paragraph on single molecule force spectroscopy in this chapter (see Fig. 10.23) [96].

63) both molecules (in 1:1 ratio) were deposited via solution casting in 1mM toluene.

64) in liquid STM in 0.1 M $HClO_4$; The tips were prepared from electrochemically etched W wire (0.25 mm in diameter) in 0.6 M KOH. The side wall of the tip was sealed with transparent nail polish to minimize the faradaic current.

65) Crystallization of a molten ball formed at the end of a pure Au wire in a hydrogen-oxygen flame formed the single-crystal beads of Au; (111) facets on them were used for the STM studies.

66) A saturated solution of OBOCMC8 and C60/OBOCMC8 was prepared in ultra pure $HClO_4$ solution (0.1 M) and deionized water. The authors deaerated the solutions with high-purity nitrogen before carrying out the experiments.

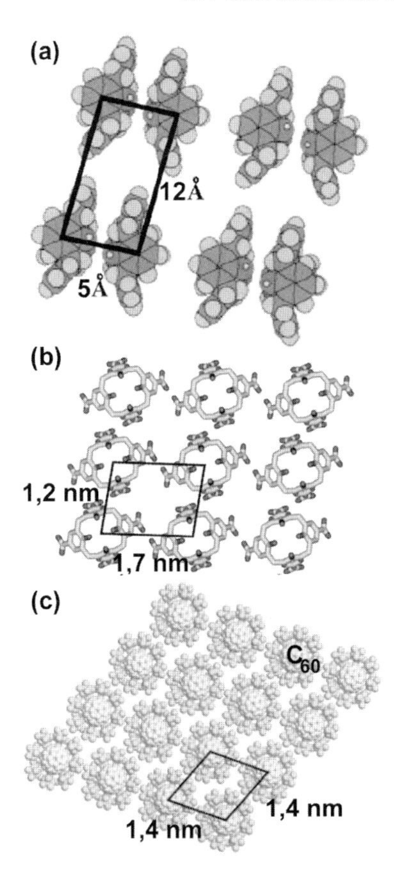

Fig. 10.26. Proposed models for (a) calix[4]arene (lower rim: OH) dimers in "edge on" – with the lower rim pointing towards lower rim – adsorption. (Reprinted from [153], Copyright (2002), with permission from Elsevier); (b) calix[4]arene (lower rim: propyl, upper rim: COOH) in "face up" (upright) adsorption. (Reprinted in parts with permission from [154]. Copyright 2003 American Chemical Society); (c) C60/ Calix[8]arene OBOCMC8 inclusion complexes again adsorbed upright standing. (Reprinted in parts from [155] with permission of the authors and Wiley, VCH, copyright 2003). All molecules imaged at the solution interface to the (111) facets of Au-beads.

Cyclodextrin is a well-known non-aromatic host molecule, which serves not only in host-guest chemistry but also in supramolecular chemistry as a component of the topochemical molecular architecture along with rotaxane and polyrotaxanes.

67) Au(111) facets formed on a single crystal bead were produced by annealing Au wire in a H_2–O_2 flame to form a droplet, which was quenched in pure water saturated with H_2. The bead was transferred into a STM cell filled with 10–mM $NaClO_4$ aqueous solution containing β-cyclodextrin in a typically 2.0 μM concentration. Order was obtained by potential control (see text).

The formation of nanotube like cyclodextrin chains as a function of the applied potential has been studied by Kunitake and co-workers with STM[67] [157]. Up to a potential of −0.6 V relative to a saturated calomel electrode (SCE) the chains do not adsorb, between −0.6 and −0.2 V the nanotube like cyclodextrin chains are formed, above −0.2 V only disordered adlayers are observed.

An example of a STM manipulation on a β-cyclodextrin necklace has been given above (see Fig. 10.13) [75]. In further studies the same group further clarified the intramolecular structure of the cyclodextrin polymer necklace[68] with STM [158]. "Head-to-head" ("tail-to-tail") to "head-to-tail" arrangement of the cyclodextrin molecules were observed in the ratio (8:2).

Carbon *molecular sieves* have been prepared of active carbon precursors with chemical vapor deposited (CVD) amorphous carbon. Villar-Rodil and co-workers [159] employed STM to visualize the changes brought about by the CVD treatment on the pore mouth structure. They used STM as extremely surface-sensitive technique, since the CVD treatments are supposed to modify only the pore entrances.

Self-organized porphyrin *boxes* (cyclic tetramers of zinc porphyrin-5-tetrapyridylporphyrin) (Fig. 10.25e) have been imaged with STM at the liquid–solid interface between Phenyloctane-$CHCl_3$ and HOPG. A porphyrin box appeared as single protrusion in a regular pattern with four-fold symmetry. Apparently, the Zn-porphyrin located in the cyclic ring adsorb "edge-on" on the graphite in optimal interaction with Zn-porphyrins of neighboring porphyrin boxes leaving an open "face-on"-adsorbed tetrapyridylprophyrin template [160].

The multifold application of porphyrins in supramolecular chemistry suggests to a closer look at the SPM studies of this molecule, before more complex interconnected supermolecules are discussed.

10.4.3
Porphyrins and Phorphyrin Oligomers

A STM study [108] on vacuum deposited Cu-tetra[3,5 di-*t*-butylphenyl] *porphyrin* molecules on Au(110), Ag(110) and Cu(100) served as example in the discussion previously on differences between AFM and STM. Depending on the substrate the porphyrin molecules took different conformations, however, solely the legs of this porphyrin molecule have been visible in the STM measurements.

Itaya and co-workers investigated 5,10,15,20-tetrakis(*N*-methylpyridinium-4-yl)-21*H*,23*H*-porphine (TMPyP) on bare Au(111), Iodine-Au(111) [161] and Sulfur-Au(111) [162] with STM[69] under potential control.

68) α-cyclodextrins (CyD) in 0.1 M NaOH were cast dropwise onto freshly cleaved MoS_2 substrates and dried in air at room temperature, and then STM was performed under ambient conditions. The concentration of CyD necklaces was sufficiently diluted to enable observation of an isolated chain of α-CyDs.

69) Also here, the tunnelling tip was prepared from an electrochemically etched W wire with its side wall sealed with transparent nail polish. The tip was previously soaked in the electrolyte solution for several hours to remove soluble contaminants before STM measurements started.

On bare Au(111) [161]: after injection of a dilute aqueous solution of TMPyP directly into the STM cell under potential control at 0.8 V (versus a reversible hydrogen electrode (RHE)) immediately spots appeared and covered the surface.

On Iodine-Au(111) [161]: a centered rectangular $c(p \times \sqrt{3}R\text{-}30°)$ phase and a rotated hexagonal phase has been found depending on the potential.[70] Again atomic resolution was achieved on the iodine adlayer, and then TMPyP was injected.[71] With time the iodine adlayer blurred and adsorbed TMPyP molecules became visible in STM images, in several ordered structures depending on the applied potential.

On S-Au(111) [162]: The STM image indicates that the sulphur adlayer (without TMPyP) obeys a $(\sqrt{3} \times \sqrt{3})$R30° structure and after injection of TmPyP shows a highly ordered adlayer[72] of TMPyP molecules showing a reduced mobility compared to on iodine-Au(111).

Bai and co-workers [163] have investigated hydrogen bonded networks of 5,10,15,20-tetrakis (4-carboxyphenyl)-21H,23H-porphyrin (TCPP) on HOPG[73] with STM. TCPP molecules alone did not lead to an observable molecular substructure, however, when co-adsorbed with stearic acid sub-molecular resolution could be obtained in the domains of the 2-D networks, the structure of which differed from the structure of 3-D bulk crystals of TCPP.

Shimizu and co-workers [164] have achieved molecular resolution on 5,10,15,20-tetra(4-octacyloxyphenyl)porphyrins (C_{18}OPP) and Rh(C_{18}OPP) mixtures (ratio 1:9) with STM at the liquid–solid interface of phenyloctane and graphite.

Jung and co-workers investigated both (a) the interaction of diporphyrins (3,5-di(tert-butyl)phenyl substituents) with C_{60} molecules on Ag(111) surfaces [165]; and (b) single copper-octaethyl porphyrins on top of thin NaCl islands on metals [166] with STM at room temperature in ultrahigh vacuum.[74] In the latter case, the electronic structure of the porphyrin was different on the metal surfaces when compared to that on thin layers of NaCl.

70) The centred rectangular $c(p \times \sqrt{3}R\text{-}30°)$ phase was found with in situ STM and ex situ with LEED in 0.1 M $HClO_4$ (without TMPyP) in the potential range between 0.5 and 1.2 V vs. RHE. The rotated hexagonal phase appeared at potentials more positive than 1.3 V vs. RHE.

71) After achieving atomic resolution on the iodine adlayer, again TMPyP was injected under potential control. Within the first few minutes no significant changes were observed with only the iodine adlayer structure being resolved (more see text).

72) The $(\sqrt{3} \times \sqrt{3})$R30° structure is observed in a potential range between 0.2 and 1.1 V vs. RHE. After observation of the S adlayer, again, a few droplets of TMPyP solution were added to 0.01 M $HClO_4$. Approximately 5–10 min after the TMPyP solution was added, small ordered domains with a square shape were found on atomically flat terraces. The domains increased in size with increasing immersion time. After 30 min, the S-Au(111) surface is almost covered by a highly ordered TMPyP adlayer (still with several defects of the molecular size). The surface mobility of the adsorbed TMPyP molecules is reported to be slower than on the I-Au(111).

73) The adlayers TCPP:stearic acid (1:3) were prepared by solution casting in toluene (<1 mM concentration) and investigated under ambient conditions.

74) The molecules were deposited with OMBE, NaCl was sublimed from a boron-nitride crucible.

Tanaka and co-workers have been able to bring (a) single molecule sheet like porphyrin oligomers ($C_{1244}H_{1350}N_{84}Ni_{20}O_{88}$) [167] and (b) porphyrine macrocycles (ferrocine bridged trisporphyrin – 10mer) [168] into ultrahigh vacuum by pulse injection on Cu(100) and Au(111), respectively. The macrocyle and the oligomer could be imaged in sub-molecular resolution with STM (the macrocyle at 77 K).

As mentioned, studies on complex interconnected supermolecules are now exemplarily discussed.

10.4.4 Complex Interconnected Supermolecules: Rotaxanes and Catenanes

Expected limitations of the miniaturization of silicon based electronics (top-down approach) have led to various attempts to build electronic devices with components of molecular scale (in a bottom-up approach) [169], in particular with complex interconnected supermolecules such as rotaxanes or catenanes [170, 171].

Stoddart and co-workers have developed molecular switch tunnel junctions [172] based on a [2]*rotaxane*, sandwiched between silicon and metallic electrodes. The rotaxane bears a cyclophane that shuttles along the molecular string toward the electrode and back again driven by an electrochemical translation. They used electrochemical measurements at various temperatures [173] to quantify the switching process of molecules not only in solution, but also in self-assembled monolayers and in a polymer electrolyte gel. Independent of the environment (solution, self-assembled monolayer or solid-state polymer gel), but also of the molecular structure – rotaxane or *catenane* – a single and generic switching mechanism is observed for all bistable molecules [173].

Rabe and co-workers [174] have self-organized and imaged [2]catenanes consisting of two 87-membered rings on a graphite surface at the liquid–solid interface. Not all types of catenanes self-organized; it was important, that the catenanes had enough flexibility for adopting a conformation, allowing interaction with the substrate and, on the other hand, leaving enough confinement, so that a loss in entropy upon assembling on a surface will not override the gain in enthalpy.

10.4.5 Supramolecular Assemblies, Grids, Arrays, Chains

M. Möller and co-workers [175] studied ordered arrays of [2×2]-*grid*-type FeII complexes with STM at the liquid–solid interface of 1,2,4-trichlorobenzene (saturated solution) and graphite (HOPG grade). They observed "face-on" (Figs. 10.27a and 10.27b) and "edge-on" adsorption of the grid and claim, that particularly the alkane-chain substitution enhanced the organization on the surface and thus improved the resolution of the STM images.

A diluted solution of [2×2] grid-type-4,6-bis(2′,2″-bipyridyl-6-yl)pyrimidine-Co(II) complexes (solvent: acetone) has been cast on graphite (HOPG-grade) and investigated again by M. Möller and co-workers with STM [176]. The authors report that neither the Langmuir-Blodgett technique nor assembly at the air-water

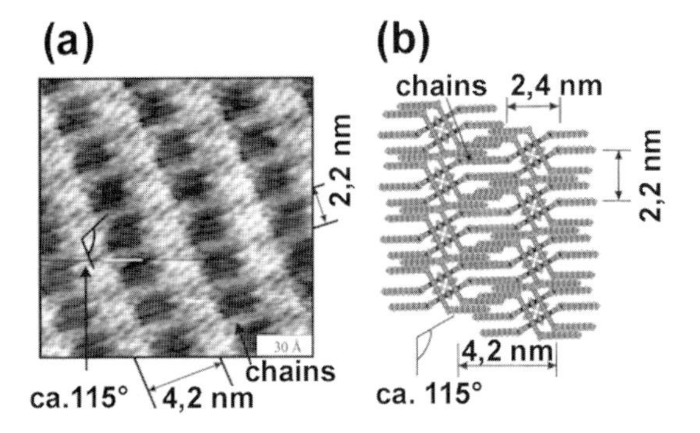

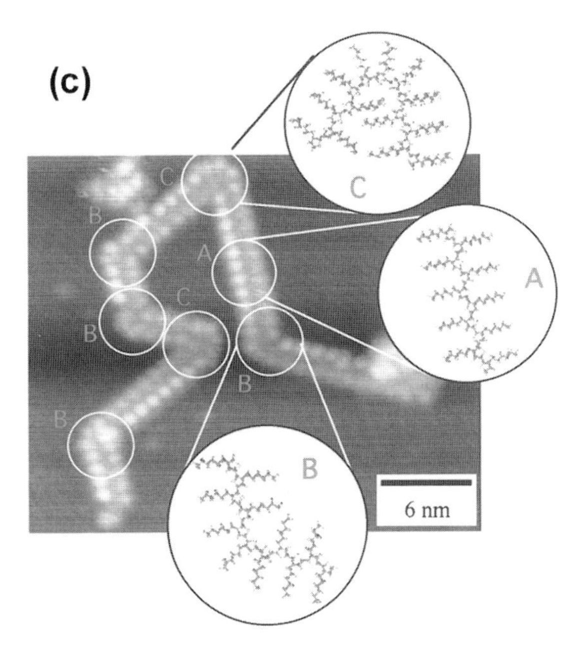

Fig. 10.27. (a) STM image [175] of an array of [2×2]-grid-type FeII complexes (at the liquid-solid interface of 1,2,4-trichlorobenzene and graphite); (b) model of the structure: marked with "chains" are alkane chains lying on top of each other and thus leading to higher contrast in the STM image. (Reprinted in parts from [175] with permission of Wiley, VCH); (c) low temperature STM image [178] of a PHT polymer chain on Cu(111) (deposited by pulse injection into the UHV). (Reproduced from [178] with permission of JCS, copyright 2002).

interface lead to self-organized grids. Rather preformed single grids had to carefully be separated by a equilibrium sedimentation process in an analytical ultracentrifuge (to separate grid clusters, which are predominantly observed) and applied in dilute solution. Both solution casting (applying a droplet and the solvent evaporates) as well as dipping into a dilute solution did produce ordered arrays of the preformed grids. Moreover, the authors extracted single grids by applying a voltage pulse of −0.5 V with the STM tip.

Lin and co-workers [177] have studied self-constructing *arrays* of individual Fe-1,2,4-benzenetricarboxylic acid (Fe-tmla) complexes on a Cu(111) substrate, produced by OMBE deposition of precursor tmla adlayers and iron atoms in ultrahigh vacuum. These self-constructed metal-organic arrays (of four tmla molecules per one Fe atom) form large homochiral monolayers (i.e. both *S* and *R* chirality are found) extending laterally up to 200 nm.

Tanaka and co-workers [178] have introduced poly(3-hexylthiophene-2,5-diyl) (PHT) into ultrahigh vacuum by pulse injection onto Cu(111) surfaces. With low temperature STM (92K) they could observe the polymer chain with high resolution and study details of curving and kinks in the chain (Fig. 10.27c).

These examples have been chosen to illustrate just how manifold the applications of scanning probe techniques are to the field of supramolecular chemistry. The increasing number of publications each year leads to the conclusion that we are probably only at the beginning of many more interesting investigations.

Acknowledgements

I thank Peter Haier for his help with the extensive work on the graphics. Special thanks to Carsten Rohr for his help in locating literature examples to be quoted and referred to in this chapter. I would like to acknowledge Urs Hubler, Regina Hoffmann, Christian Loppacher-Voirol, Lukas Eng, Ernst Meyer, Thorsten Hugel, Kathrin Gruber, Dirk Grupe and again Carsten Rohr for fruitful discussion on sections of this manuscript. I am indebted to Catherine Housecroft and Tom Sobey for proof-reading the manuscript. Last but not least, I would like to thank Christoph Schalley for his invitation to participate in the book and for his useful comments on the manuscript.

References

1 G. Binnig, H. Rohrer, *Rev. of Mod. Phys.*, **1999**, 71, S324–330.
2 B. Bhushan, H. Fuchs, S. Hosaka, *Applied Scanning Probe Methods I–IV*, Springer, Berlin, Heidelberg, 2004–06.
3 E. Meyer, H. Hug, R. Bennewitz, *Scanning Probe Microscopy: the Lab on a Tip*, Springer, Berlin, Heidelberg 2004.
4 A. de Stefanis, A. Tomlinson, *Scanning Probe Microscopies: From Surface Structure to Nano-scale Engineering*, Trans Tech Publications Inc., Uetikon-Zürich, 2001.
5 D. Bonnell, B. Huey, J. Tersoff, R. Hamers, D. Padowitz, W. Unertle, G. Rohrer, R. Smith, S. Kalinin, W. Kaiser, S. Lindsay, N. Burnham, R. Colton, D. Higgins, E. Mei, A. Bard, F. Fan, *Scanning Probe Microscopy and Spectroscopy: Theory, Techniques, and Applications*, Wiley-VCH, Weinheim, New York, 2001.
6 C. Bai, *Scanning Tunneling Microscopy and Its Application*, Springer, Berlin, Heidelberg 2000.

7 T. Sakurai, Y. Watanabe, *Advances in Scanning Probe Microscopy*, Springer, Berlin, Heidelberg 2000.

8 R. Wiesendanger, *Scanning Probe Microscopy and Spectroscopy: Methods and Applications*, Cambridge University Press, Cambridge 1998.

9 R. Wiesendanger, H.-J. Güntherodt, (Eds.), *Scanning Tunneling Microscopy I, II, III*, Springer, Berlin, Heidelberg, 1994–96.

10 O. Marti (Ed.), *STM and SFM in Biology*, Academic Press, Boston 1993.

11 J. Stroscio, W. J. Kaiser (Eds.), *Scanning Tunneling Microscopy*, Academic Press, Bosten, 1993.

12 C. J. Chen, *Introduction to Scanning Tunneling Microscopy*, Oxford University Press, Oxford, New York 1993.

13 E. Mena-Osteritz, P. Bäuerle, *Adv. Mat.* **2006**, 18, 447–451.

14 L. Piot, A. Marchenko, J. Wu, K. Müllen, D. Fichou, *J. Am. Chem. Soc.* **2005**, 127, 16245–16250.

15 B. A. Hermann, *Self Assembled Monolayers* in the *Encyclopedia of Supramolecular Chemistry*, J. L. Atwood, J. W. Steed (Eds.), Marcel Dekker, New York, webpublished July 2006.

16 P. Samori, *J. Mat. Chem.* **2004**, 14, 1353–1366.

17 R. Czajka, L. Jurczyszyn, H. Rafii-Tabar, *Progress in Surface Science*, **1998**, 59, 13–23.

18 W.-D. Schneider, *Surf. Sci.* **2002**, 514, 74–83.

19 R. M. Tromp, *J. Phys.: Cond. Matter* **1989**, 1, 10211–10228.

20 S.-W. Hla, *J. Vac. Sci. Technol.* B **2005**, 23, 1351–1360.

21 K.-F. Braun, S.-W. Hla, *Properties of Single Molecules: Manipulation, Dissociation and Synthesis with the Scanning Tunnelling Microscope* in *Properties of Single Molecules on Crystal Surfaces*, F. Rosei (Ed.), Imperial College Press/American Scientific Publishers, London 2004.

22 F. Moreso, G. Meyer, H. Tang, C. Joachim, K.-J. Rieder, *J. of Elec. Spect. and Rel. Phen.* **2003**, 129, 149–155.

23 K. Morgenstern, *Phys. Stat. Sol.* B **2005**, 242, 773.

24 R. Curtis, T. Mitsui, E. Ganz, *Rev. Sci. Instr.* **1997**, 68, 2790–2796.

25 X. F. Wang, K. S. Ryu, D. A. Bullen, J. Zou, H. Zhang, C. A. Mirkin, C. Liu, *Langmuir* **2003**, 19, 8951–8955.

26 G.-Y. Liu, S. Xu, Y. Qian, *Acc. Chem. Res.* **2000**, 33, 457–466.

27 S. De Feyter, F. C. De Schryver, *J. Phys. Chem. B*, **2005**, 109, 4290–4302.

28 D. M. Kolb, F. C. Simeone, *Electrochimica Acta*, **2005**, 50, 2989–2996.

29 M. de Wild, S. Berner, H. Suzuki, L. Ramoino, A. Baratoff, T. A. Jung, Molecular Assembly and Self-Assembly: Molecular Nanoscience for Future, Molecular Electronics III *Ann. NY Acad. Sci.* **2003**, 1006, 291–305.

30 S. I. Park, C. F. Quate, *Rev. Sci. Instrum.* **1987**, 58, 2010–2017.

31 D. J. Müller, M. Ammrein, A. Engel, *J. Struct. Biol.* **1997**, 119, 172–188.

32 H. Tokumoto, M. Iwatsuki, Jpn. *J. Appl. Phys.* **1993**, 32, 1368–1378.

33 J. G. Hou, Y. Jinlong, W. Haiqian, L. Qunxiang, Z. Changgan, Y. Lanfeng, W. Bing, D. M. Chen, Z. Qingshi, *Nature* **2001**, 409, 304–305.

34 G. Meyer, *Rev. Sci. Instrum.* **1996**, 67, 2960.

35 M. Bott, T. Michely, and G. Comsa, *Rev. Sci. Instrum.* **1995**, 66, 1995.

36 S. Horch, P. Zeppenfeld, R. David, and G. Comsa, *Rev. Sci. Instrum.* **1994**, 65, 3204.

37 A. Bouhelier, *Micros. Res. & Techn.* **2006**, 69 (7) in print.

38 S. Grafström, *J. Appl. Phys.* **2002**, 91, 1717-1754.

39 M. Sakurai, C. Thirstrup, M. Aono, *Appl. Phys. A* **2005**, 80, 1153–1160.

40 J. Wintterlin, *Chaos*, **2002**, 12, 108–117.

41 T. Belser, M. Stöhr, A. Pfaltz, *J. Am. Chem. Soc.* **2005**, 127, 8720–8731.

42 C. Sachs, M. Hildebrand, S. Völkening, J. Wintterlin, G. Ertl, *Science* **2001**, 293, 1635.

43 T. Milic, N. Chi, D. Yablon, G. Flynn, J. Batteas, C. M. Drain, *Angew. Chem. Int. Ed.* **2002**, 41, 2117–2119.

44 J. Zhang, M. Grubb, A. G. Hansen, A. M. Kuznetsov, A. Boisen, H. Wackerbarth, J. Ulstrup, *J. Phys. Condens. Matter* **2003**, 15, S1873–S1890.
45 F. Rosei, M. Schunack, Y. Naitoh, P. Jiang, A. Gourdon, E. Laegsgaard, I. Stensgaard, C. Joachim, F. Besenbacher, *Prog. in Surf. Sci.* **2003**, 71, 95–146.
46 S. De Feyter, F. C. De Schryver, *Supermicroscopy* in the *Encyclopedia of Supramolecular Chemistry*, J. L. Atwood, J. W. Steed (Eds.), Marcel Dekker, New York, **2004**, L1394–L1400.
47 L. C. Pinheiro, *Scanning Tunneling Microscopy* in the *Encyclopedia of Supramolecular Chemistry*, J. L. Atwood, J. W. Steed (Eds.), Marcel Dekker, New York, **2004**, 1202–1208.
48 J. H. Ferris, J. G. Kushmerick, J. A. Johnson, M. G. Yoshikawa Youngquist, R. B. Kessinger, H. F. Kingsbury, and P. S. Weiss, *Rev. Sci. Instrum.* **1998**, 89, 2691.
49 S. I. Park, C. F. Quate, *Rev. Sci. Instrum.* **1987**, 58, 2004–2009.
50 D. W. Pohl, *IBM J. Res. Develop.* **1986**, 30, 417–437.
51 S. Schintke, W.-D. Schneider, *J. of Phys.: Condens. Matt.* **2004**, 16, R49–R81.
52 G. Binnig, H. Rohrer, C. Gerber, E. Weibel, *Phys. Rev. Lett*, **1982**, 49, 57–61.
53 G. Binnig, H. Rohrer, *Helv. Phys. Acta* **1982**, 55, 726–735.
54 G. Binnig, H. Rohrer, *Scanning Tunneling Microscopy – from Birth to Adolescence* (Nobel Lecture) in Les Prix Nobel (The Nobel Foundation, Stockholm) 1986, 85; reprinted in Rev. Mod. Phys. 1987, 59, 615.
55 H.-P. Rust, J. Buisset, E. K. Schweizer, L. Cramer, *Rev. Sci. Instrum.* **1997**, 68, 129.
56 J. Frohn, J. F. Wolf, K. Besocke, M. Teske, *Rev. Sci. Instrum.* **1989**, 60, 1200.
57 G. W. Stupian, M. S. Leung, *Rev. Sci. Instrum.* **1988**, 60, 181.
58 D. W. Pohl, *Rev. Sci. Instrum.* **1987**, 58, 54–57.
59 P. Samorí, N. Severin, C. D. Simpson, K. Müllen, J. P. Rabe, *J. Am. Chem. Soc.* **2002**, 124, 9454–9457.
60 D. E. Hooks, T. Fritz, M. D. Ward, *Adv. Mat.* **2001**, 13, 227–241.
61 J. Repp, G. Meyer, M. Stojkovic, A. Gourdan, C. Joachim, *Phys. Rev. Lett.* **2005**, 94, 026803.
62 W. Ho, *J. Chem. Phys.* **2002**, 117, 11033–11061.
63 S.-W. Hla, K.-H. Rieder, *Annu. Rev. Phys. Chem.* **2003**, 54, 307–30.
64 S.-W. Hla, *Superlattices and Microstructures*, **2001**, 31, 63–72.
65 G. Meyer, L. Bartels, K.-H. Rieder, *Superlattices and Microstructures*, **1999**, 25, 463–471.
66 H. Uji-i, S. Nishio, H. Fukumura, *Chem. Phys. Lett.* **2005**, 408, 112–117.
67 S. Sek, K. Swiatek, A. Misicka, *J. Phys. Chem. Lett.* **2005**, 109, 23121–23124.
68 G. V. Nazin, X. H. Qiu, and W. Ho, *Science* **2003**, 302, 77–81.
69 R. J. Hamers, *STM on semiconductors* in *Scanning Tunneling Microscopy* I, R. Wiesendanger, H.-J. Güntherodt, (Eds.), Springer, Berlin, Heidelberg, 1994, 86.
70 D. M. Eigler, E. K. Schweizer, *Nature* **1990**, 344, 524.
71 J. A. Stroscio, D. M. Eigler, *Science* **1991**, 254, 1319.
72 T. W. Fishlock, A. Oral, R. G. Egdell, J. B. Pethica, *Nature* **2000**, 404, 743.
73 L. Gross, K.-H. Rieder, F. Moresco, S. M. Stojkovic, A. Gourdan, C. Joachim, *Nature Materials* **2005**, 4, 892–895.
74 R. Yamachika, M. Grobis, A. Wachowiak, M. F. Crommie, *Science* **2004**, 304, 281–284.
75 H. Shigekawa, K. Miyake, J. Sumaoka, A. Harada, M. Komiyama, *J. Am. Chem. Soc.* **2000**, 122, 5411–5412.
76 T. T. Tsong, C. S. Chang, W. B. Su, H. N. Lin, T. C. Chang, R. L. Lo, *Chinese J. of Phys.* **1994**, 32, 667–684.
77 S.-W. Hla, K.-F. Braun, V. Iancu, A. Deshpande, *Nano Letters* **2004**, 4, 1997–2001.

78 M. Tsukada, K. Kobayashi, N. Isshiki, S. Watanabe, H. Kageshima, T. Schimizu, The role of the tip atomic and electronic structure in scanning tunneling microcopy and spectroscopy in *Scanning Tunneling Microscopy III*, R. Wiesendanger, H.-J. Güntherodt, (Eds.), Springer, Berlin, Heidelberg, 1996, 80.

79 J. A. Nieminen, E. Niemi, K.-J. Rieder, *Surf. Sci.* **2004**, 552, L42–L52.

80 O. Wolter, T. Bayer, J. Greschner, *J. Vac. Sci. & Techn. B: Microelectr. and Nanom. Struct.* **1991**, 9, 1353–1357.

81 G. Meyer, N. M. Amer, *Appl. Phys. Lett.* **1990**, 57, 2089–2091.

82 T. Hugel, M. Seitz, *Macromol. Rapid Commun.* **2001**, 22, 989–1016.

83 J. Israelachvili, *Intermolecular and Surface Forces*, Academic Press (1985–2004).

84 F. J. Giessibl, H. Bielefeldt, *Phys. Rev. B: Cond. Matt. a. Mate. Phys.* **2000**, 61, 9968–9971.

85 U. Dürig, *Appl. Phys. Lett.* **1999**, 75, 433–435.

86 E. Meyer, R. M. Overney, K. Dransfeld, T. Gyalog, *Nanoscience-Friction and Rheology on the Nanometer Scale*, World Scientific, Singapore 1998.

87 R. M. Overney, E. Meyer, J. Frommer, D. Brodbeck, R. Lüthi, L. Howald, H.-J. Güntherodt, M. Fujihira, H. Takano, Y. Gotoh, *Nature* **1992**, 359, 133.

88 G. J. Leggett, *Anal. Chim. Acta* **2003**, 479, 17–38.

89 E. Gnecco, R. Bennewitz, T. Gyalog, E. Meyer, *J. Phys.: Condens. Matter* **2001**, 13, R619–R642.

90 T. Fukuma, M. Kimura, K. Kobayashi, K. Matsushige, H. Yamada, *Rev. Sci. Instr.* **2005**, 76, 053704 1–8.

91 H. Schönherr, V. Paraschiv, S. Zapotoczny, M. Crego-Calama, P. Timmerman, C. W. Frank, G. J. Vancso, D. N. Reinhoudt, *Proc. Natl. Acad. Sci. USA* **2002**, 99, 5024–5027.

92 H. Clausen-Schaumann, M. Seitz, R. Krautbauer, H. E. Gaub, *Curr. Opin. Chem. Bio.* **2000**, 4, 524–530.

93 Janshoff, M. Neitzert, Y. Oberdörfer, H. Fuchs, *Angew. Chem. Int. Ed.* **2000**, 39, 3212–3237.

94 E. Evans, *Annu. Rev. Biophys. a. Biomol. Struc.* **2001**, 3, 105–108.

95 M. Lantz, H. J. Hug, R. Hoffmann, P. J. A. van Schendel, P. Kappenberger, S. Martin, A. Baratoff, and H.-J. Güntherodt, *Science* **2001**, 291, 2580–2583.

96 R. Eckel, R. Ros, B. Decker, J. Mattay, D. Anselmetti, *Angew. Chem. Int. Ed.* **2005**, 44, 484–488.

97 J. Barth, *Surf. Sci. Rep.* **2000**, 40, 75–149.

98 F. Besenbacher, *Rep. Prog. Phys.* **1996**, 59, 1737–1802.

99 S. A. Burke, J. M. Mativetsky, R. Hoffmann, P. Grütter, *Phys. Rev. Lett.* **2005**, 94, 096102.

100 T. Kunstmann, A. Schlarb, M. Fendrich, T. Wagner, R. Möller, *Phys. Rev. B* **2005**, 71, R121403.

101 H. G. Hansma, K. J. Kim, D. E. Laney, R. A. Garcia, M. Argaman, M. J. Allen, S. M. Parsons, *J. Struct. Biol.* **1997**, 119, 99–108.

102 J. Fritz, D. Anselmetti, J. Jarchow, X. Fernàndez-Busquets, *J. Struct. Biol.* **1997**, 119, 165–171.

103 H. Tanaka, T. Kawai, *Surf. Sci.* **2003**, 539, L531–L536.

104 B. Gotsmann, C. Schmidt, C. Seidel, H. Fuchs, *Eur. Phys. J. B* **1998**, 4, 267–268.

105 T. Uchihashi, T. Ishida, M. Komiyama, M. Ashino, Y. Sugawara, W. Mizutani, K. Yokoyama, S. Morita, H. Tokumoto, M. Ishikawa, Appl. *Surf. Sci.* **2000**, 157, 244–250.

106 A. Sasahara, H. Uetsuka, H. Onishi, *Surf. Sci.* **2001**, 481, L437–L442.

107 Ch. Loppacher, M. Guggisberg, O. Pfeiffer, E. Meyer, M. Bammerlin, R. Lüthi, R. Schlittler, J. K. Gimzewski, H. Tang, C. Joachim, *Phys. Rev. Lett.* **2003**, 60, 066107, 1–4.

108 T. A. Jung, R. R. Schlittler, J. K. Gimzewski, *Nature* **1997**, 386, 696–698.

109 K. Morgenstern, S. W. Hla, K.-H. Rieder, *Surf. Sci.* **2003**, 523, 141–150.

110 Q. Chen, T. Rada, A. McDowall, N. V. Richardson, *Chem. Mater.* **2002**, 14, 743–749.
111 M. E. Stawasz, D. L. Sampson, B. A. Parkinson, *Langmuir* **2000**, 16, 2326–2342.
112 L. C. Giancarlo, G. W. Flynn, *Acc. Chem. Res.* **2000**, 33, 491–501.
113 F. Schreiber, *Progr. in Surf. Sci.* **2000**, 65, 151–256.
114 T. Fukuma, K. Kobayashi, T. Horiuchi, H. Yamada, K. Matsushige, *Appl. Phys. A* **2000**, 72, 1–4.
115 L. J. Scherer, L. Merz, E. C. Constable, C. E. Housecroft, M. Neuburger, B. A. Hermann, *J. Am. Chem. Soc.* **2005**, 127, 4033–4041.
116 B. A. Hermann, L. Scherer, E. C. Constable, C. E. Housecroft, *Adv. Func. Mat.* **2006**, 16, 221–235.
117 A. Sasahara, H. Uetsuka, H. Onishi, *J. Phys. Chem. B* **2001**, 105, 1–4.
118 G. Flynn, *Proc. Scanning: J. Scan. Microsc.* **2001**, 23, 79–80.
119 L. Pérez-Garcia, D. Amabillino, *Chem. Soc. Rev.* **2002**, 31, 342–356.
120 V. Humblot, S. M. Barlow, R. Raval, *Progr. in Surf. Sci.* **2004**, 76, 1–19.
121 S. Stephanow, N. Lin, F. Vidal, A. Landa, M. Ruben, J. V. Barth, K. Kern, *Nanoletters* **2005**, 5, 901–904.
122 L. Merz, H.-J. Güntherodt, L. J. Scherer, E. C. Constable, C. E. Housecroft, M. Neuburger, B. A. Hermann, *Chem. Eur. J.* **2005**, 11, 2307–2318.
123 J. K. Gimzewski, C. Joachim, R. R. Schlittler, V. Langlais, H. Tang, I. Johannsen, *Science* **1998**, 281, 531.
124 M. Stöhr, Th. Wagner, M. Gabriel, B. Weyers, R. Möller, *Phys. Rev. B* **2001**, 65, 033404.
125 Z. J. Donhauser, B. A. Mantooth, K. F. Kelly, L. A. Bumm, J. D. Monnell, J. J. Stapleton, D. W. Price Jr., A. M. Rawlett, D. L. Allara, J. M. Tour, P. S. Weiss, *Science* **2001**, 292, 2303–2307.
126 J. Henzl, M. Mehlhorn, H. Gawronski, K.-H. Rieder, K. Morgenstern, *Angew. Chem. Int. Ed..* **2006**, 45, 603.
127 C. Cai, M. Bösch, C. Bossard, B. Müller, Y. Tao, A. Kündig, J. Weckesser, J. V. Barth, L. Bürgi, O. Jeandupeux, M. Kiy, I. Biaggio, I. Liakatas, K. Kern, P. Günter, *Proc. ACS Symposium* **1999**, 3, 15–23.
128 R. J. Jackman, J. L. Wilbur, G. M. Whitesides, *Science* **1995**, 269, 664–666.
129 L. M. Demers, D. S. Ginger, S.-J. Park, Z. Li, S.-W. Chung, C. A. Mirkin, *Science* **2002**, 296, 1836–1838.
130 S. Hong, J. Zhu, C. A. Mirkin, *Science* **1999**, 286, 523–525.
131 P. M. Albrecht, J. W. Lyding, *Appl. Phys. Lett.* **2003**, 83, 5029–5031.
132 A. J. Bard, M. V. Mirkin (Eds.), *Scanning Electrochemical Microscopy*, Marcel Dekker, New York, 2001.
133 E. Meyer, R. M. Overney, D. Brodbeck, L. Howald, R. Lüthi, J. Frommer, H.-J. Güntherodt, *Phys. Rev. Lett.* **1992**, 69, 1777.
134 O. Salata, *Curr. Nanosci.* **2005**, 1, 25–33.
135 S. Rauschenbach, F. L. Stadler, E. Lunedei, N. Malinowski, S. Koltsov, G. Costantini, K. Kern, *Small* **2006**, 2, 540–547.
136 H. Tanaka, C. Hamai, T. Kanno, T. Kawai, *Surf. Sci.* **1999**, 432, L611–L616.
137 W. M. Heckl, D. P. E. Smith, G. Binnig, H. Klagges, T. W. Hänsch, J. Maddocks, *Proc. Natl. Acad. Sci. USA* **1991**, 88, 8003.
138 S. K. Sinha (Ed.), *Ordering in Two Dimensions*, Elsevier North Holland, Amsterdam, 1980.
139 M. Lackinger, S. Griessl, W. M. Heckl, M. Hietschold, *Anal. Bioanal. Chem.* **2002**, 374, 685–68.
140 S. J. H. Griessl, M. Lackinger, F. Jamitzky, T. Markert, M. Hietschold, W. M. Heckl, *Langmuir* **2004**, 20, 9403–9407.
141 B. J. Gyarfas, B. Wiggins, M. Zosel, and K. W. Hipps, *Langmuir* **2005**, 21, 919–923.
142 P. Samorí, H. Engelkamp, P. de Witte, A. E. Rowan, R. J. M. Nolte, J. P. Rabe, *Angew. Chem. Int. Ed.* **2001**, 40, 2348–2350.

143 A. Ohira, M. Sakata, C. Hirayama, M. Kunitake, *Org. Biomol. Chem.* **2003**, 1, 251–253.
144 S. Yoshimoto, K. Suto, A. Tada, N. Kobayashi, K. Itaya, *J. Am. Chem. Soc.* **2004**, 126, 8020–8027.
145 I. Kossev, S. Fahrenholz, A. Görling, W. Hieringer, C. A. Schalley, M. Sokolowski, *Synthetic Metals* **2004**, 147, 159–164.
146 L. Markey, D. Stievenard, A. Devos, M. Lannoo, F. Demol, M. de Backer, *Supramol. Sci.* **1997**, 4, 315–319.
147 C. Safarowsky, L. Merz, A. Rang, P. Broekmann, B. A. Hermann, C. A. Schalley, *Angew. Chem. Int. Ed.* **2004**, 43, 1291–1294.
148 J.-R. Gong, L.-J. Wan, Q.-H. Yuan, C.-L. Bai, H. Jude, P. J. Stang, *Proc. Natl. Acad. Sci. USA* **2005**, 102, 971–974.
149 Q.-H. Yuan, L.-J. Wan, H. Jude, P. J. Stang, *J. Am. Chem. Soc.* **2005**, 127, 16279–16286.
150 H. Schönher, G. J. Vancso, B.-H. Huisman, F. C. J. M. Van Veggel, D. N. Reinhoudt, *Langmuir* **1999**, 15, 5541.
151 H. Schönher, G. J. Vancso, B.-H. Huisman, F. C. J. M. Van Veggel, D. N. Reinhoudt, *Langmuir* **1997**, 13, 1567.
152 Y.-M. Liu, S. Lei, S. Yin, S. Xu, Q.-Y. Zheng, Q. Zeng, C. Wang, L.-J. Wan, C.-L. Bai, *J. Phys. Chem. B* **2002**, 106, 12569–12574.
153 G.-B. Pan, L.-J. Wan, Q.-Y. Zheng, C.-L. Bai, K. Itaya, *Chem. Phys. Lett.* **2002**, 359, 83–88.
154 G.-B. Pan, J.-H. Bu, D. Wang, Y.-M. Liu, L.-J. Wan, Q.-Y. Zheng, C.-L. Bai, *J. Phys. Chem. B* **2003**, 107, 13111–13116.
155 G.-B. Pan, Y.-M. Liu, H.-M. Zhang, L.-J. Wan, Q.-Y. Zheng, C.-L. Bai, *Angew. Chem. Int. Ed.* **2003**, 42, 2747–2751.
156 V. A. Langlais, Y. Gauthier, H. Belkhir, O. Maresca, *Phys. Rev. B*, **2005**, 72, 085444.
157 A. Ohira, T. Ishizaki, M. Sakata, I. Taniguchi, C. Hirayama, M. Kunitake, *Colloids and Surfaces A: Physicochemical and Engineering Aspects* **2000**, 169, 27–33.
158 K. Miyake, S. Yasuda, A. Harada, J. Sumaoka, M. Komiyama, H. Shigekawa, *J. Am. Chem. Soc.* **2003**, 125, 5080–5085.
159 J. I. Paredes, S. Villar-Rodil, A. Martínez-Alonso, J. M. D. Tascón, *J. Mater. Chem.* **2003**, 13, 1513–1516.
160 P. C. M. van Gerven, J. A. A. W. Elemans, J. W. Gerritsen, S. Speller, R. J. M. Nolte, A. E. Rowan, *Chem. Commun.* **2005**, 3535–3537.
161 M. Kunitake, U. Akiba, N. Batina, K. Itaya, *Langmuir* **1997**, 13, 1607–1615.
162 L.-J. Wan, S. Shundo, J. Inukai, K. Itaya, *Langmuir* **2000**, 16, 2164–2168.
163 S. B. Lei, C. Wang, S. X. Yin, H. N. Wang, F. Xi, H. W. Liu, B. Xu, L. J. Wan, C. L. Bai, *J. Phys. Chem. B* **2001**, 105, 10838–10841.
164 T. Ikeda, M. Asakawa, M. Goto, K. Miyake, T. Ishida, T. Shimizu, *Langmuir* **2004**, 20, 5454–5459.
165 D. Bonifazi, H. Spillmann, A. Kiebele, M. de Wild, P. Seiler, F. Cheng, H.-J. Güntherodt, T. Jung, F. Diederich, *Angew. Chem. Int. Ed.* **2004**, 43, 4759–4763.
166 L. Ramoino, M. von Arx, S. Schintke, A. Baratoff, H.-J. Güntherodt, T. A. Jung, *Chem. Phys. Lett.* **2006**, 417, 22–27.
167 K.-I. Sugiura, H. Tanaka, T. Matsumoto, T. Kawai, Y. Sakata, *Chem. Lett.* **1999**, 1193.
168 O. Shoji, H. Tanaka, T. Kawai, Y. Kobuke, *J. Am. Chem. Soc.* **2005**, 127, 8598–8599.
169 R. M. Metzger, *Unimolecular Electronics and unimolecular rectifiers* in the *Encyclopedia of Supramolecular Chemistry*, J. L. Atwood, J. W. Steed (Eds.), Marcel Dekker, New York, 2004.
170 M. A. Reed, J. Tour, *Scientific American* **2000**, 282, 86–93.
171 J. R. Heath, M. A. Ratner, *Physics Today* **2003**, 56, 43–49.
172 A. H. Flood, J. F. Stoddart, D. W. Steuermann, J. R. Heath, *Science* **2004**, 306, 2055–2056.

173 A. H. Flood, A. J. Peters, S. A. Vignon, D. W. Steuermann, H.-R. Tseng, S. Kang, J. R. Heath, J. F. Stoddart, *Chem. Eur. J.* **2004**, 10, 6558–6564.
174 P. Samori, F. Jäckel, Ö. Ünsal, A. Godt, J. Rabe, *ChemPhysChem* **2001**, 7, 461–464.
175 A. Mourran, U. Ziener, M. Möller, E. Breuning, M. Ohkita, J.-M. Lehn, *Eur. J. Inorg. Chem.* **2005**, 2641–2647.
176 A. Semenov, J. P. Spatz, M. Möller, J.-M. Lehn, B. Sell, D. Schubert, C. H. Weidl, U. S. Schubert, *Angew. Chem. Int. Ed.* **1999**, 38, 2547–2550.
177 A. Dimitriev, H. Spillmann, M. Lingenfelder, N. Lin, J. V. Barth, K. Kern, *Langmuir* **2004**, 20, 4799–4801.
178 H. Kasai, H. Tanaka, S. Okada, H. Oikawa, T. Kawai, H. Nakanishi, *Chem. Lett.* **2002**, 696.

11
The Characterization of Synthetic Ion Channels and Pores

Stefan Matile and Naomi Sakai

11.1
Introduction

"Membrane transport is the quintessential supramolecular function." The objective of this chapter is not to defend this statement by one of the pioneers of the field [1] but to provide a supramolecular chemist newly entering the field with an introductory guideline how to measure this function. Emphasis will be on transport by synthetic ion channels and pores [2], summarizing lessons learned during years of research on the topic. The term "synthetic" applies to ion channels and pores that are constructed from scratch with scaffolds that are not known to occur in nature; this covers neither the organic synthesis of biological scaffolds (e.g. α-helices) nor chemically or biotechnologically modified biological channels and pores.

We use the terms ion channels and pores for the transport of inorganics and organics, respectively. Ion channels and pores can mediate this transport across lipid bilayers without moving themselves and without disturbing the suprastructure of the membrane (Fig. 11.1a). Transporters that shuttle across the bilayer during transport are named ion carriers (Fig. 11.1b). Detergents differ from ion channels and pores because they disturb the suprastructure of the bilayer membrane. Differentiation between these modes of transport can be difficult because of analytical challenges as well as because they depend not only on suprastructures but also on experimental conditions (e.g. carriers and detergents can behave like ion channels and pores under appropriate conditions, and *vice versa*) [2]. Synthetic ion carriers have been studied in detail for decades, whereas the design of supramolecules that operate by the more complex endovesiculation (Fig. 11.1c) or fusion (Fig. 11.1d) remains as a future challenge. Mainly because of the enormous progress made in the synthesis of complex supramolecules, synthetic ion channels and pores can be considered as the topic of current interest in the field.

The structural classification of synthetic ion channels and pores differentiates between (macro)molecules and different classes of supramolecules [2]. Unimolecular synthetic ion channels and pores are usually hollow, often helical (macro)molecules that are long enough to span common lipid bilayer membranes (2–4 nm).

Analytical Methods in Supramolecular Chemistry. Edited by Christoph Schalley

ISBN: 978-3-527-31505-5

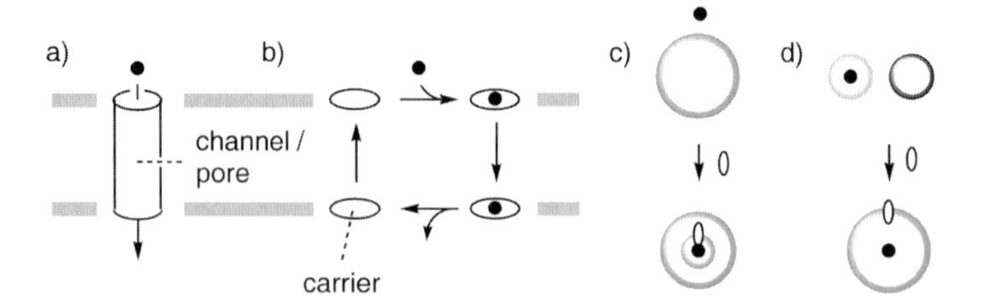

Fig. 11.1. Transport of ions and molecules (filled circles) across lipid bilayer membranes (grey) by (a) ion channels and pores; (b) carriers; (c) endovesiculators and (d) fusogens (empty black symbols).

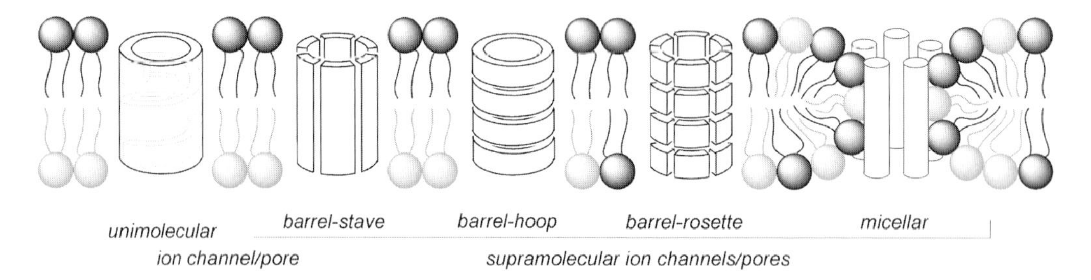

Fig. 11.2. Structural classification of synthetic ion channels and pores.

Virtual vertical cutting of this unimolecular "barrel" gives the "barrel-stave", horizontal cutting the "barrel-hoop", horizontal and vertical cutting the "barrel-rosette" motif (Fig. 11.2). More complex motifs that include modification of the lipid bilayer are summarized as micellar pores.

The barrel-stave architecture is a classic for both biological and synthetic ion channels and pores (Fig. 11.3). Whereas barrel-hoop motifs have received considerable attention, barrel-rosette ion channels and pores are just beginning to emerge. The more complex micellar (or "toroidal") ion channels and pores are, on the one hand, different from detergents because the micellar defects introduced into the lipid bilayer are only transient. Micellar pores differ, on the other hand, from membrane-spanning (i.e. "transmembrane") barrel-stave pores because (a) they disturb the bilayer suprastructure and (b) they always remain at the membrane–water interface. A representative synthetic barrel-stave pore is shown in Fig. 11.3 [3, 4]; comprehensive collections of recently created structures can be found in pertinent reviews [2].

11.2 Methods

Two main methods exist to characterize synthetic ion channels and pores [2]. Conductance experiments in planar or "black" lipid membranes (BLMs) reveal the

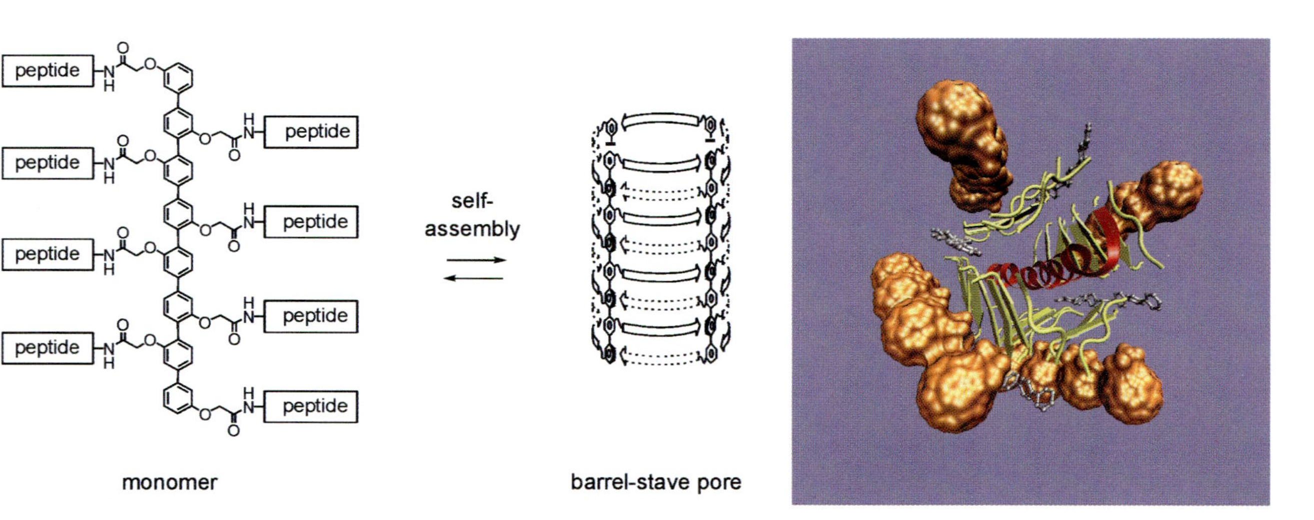

Fig. 11.3. A representative synthetic barrel-stave pore self-assembled from four p-octiphenyl monomers (left). Molecular model with p-octiphenyl staves in grey, β-sheet hoops in yellow, external fullerene ligands in gold and an internal α-helix blocker in red (right, adapted from Ref. 4).

electric properties of the hollow interior of an ion channel or a pore [5]. Fluorescence spectroscopy with labeled, usually large, unilamellar vesicles (LUVs) can provide information on influx and efflux of ions and molecules through ion channels and pores as well as on their suprastructures, depending on the fluorescent probes used [3, 6–11]. For a supramolecular chemist entering the field, LUVs may be preferable because they do not require specialized equipment or expertise, provide a rapid overview on activities and are sufficient to identify most of the relevant characteristics described in Section 11.3. BLM conductance experiments require skilled hands and are time consuming. Moreover, the frequent occurrence of artifacts requires insightful controls and cautious data interpretation. However, BLM experiments are needed to prove the presence of an ion channel or a pore and can provide deeper mechanistic insights than LUVs. In the following, the basic principles of both methods are briefly introduced before the discussion of their application to characterize synthetic ion channels and pores in Section 11.3.

11.2.1 Planar Bilayer Conductance

A conventional set-up to characterize synthetic ion channels and pores in BLMs uses two chambers that are filled with buffer and named *cis* and *trans* (Fig. 11.4a) [5]. Here, the *cis* chamber is arbitrarily connected to the input electrode and the *trans* chamber to a reference electrode. The BLM is formed in a tiny hole (micro-

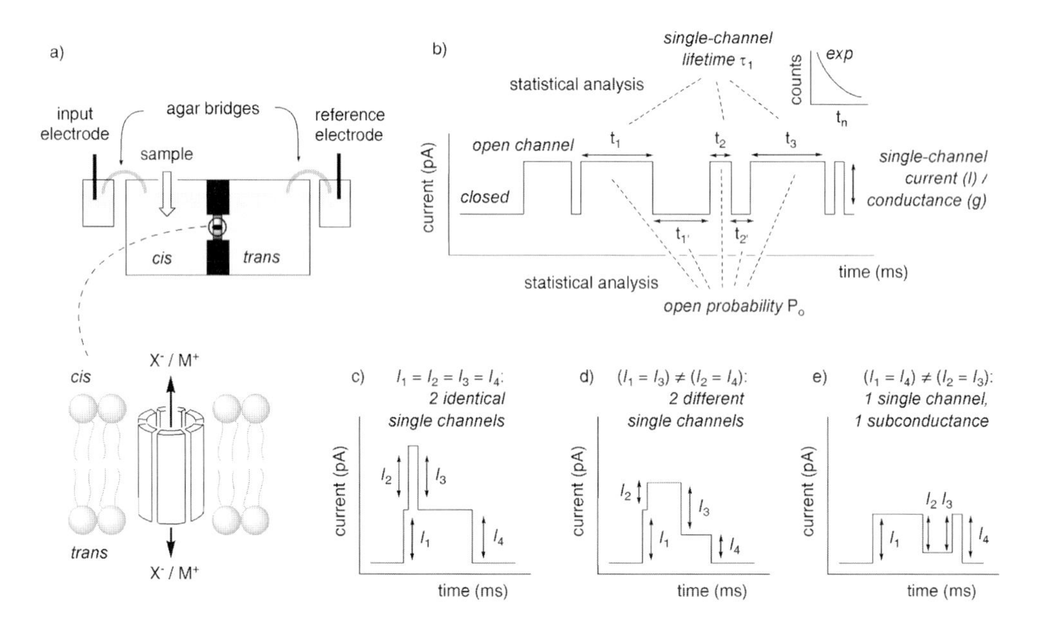

Fig. 11.4. Standard configuration of a BLM experiment (a) and analysis of single-channel recordings (b–e).

meters) between *cis* and *trans* chamber. The BLM acts as an insulator. Without ion channels or pores, no current can flow between the two electrodes. In other words, phase transfer of anions X^- and cations M^+ in the buffer across the hydrophobic core of the lipid bilayer is not possible.

Synthetic ion channels and pores can be added to either *cis* or *trans* chamber. Repulsive voltage can then be applied to drive them into the BLM between the chambers. Although not desirable and more demanding to reproduce, poor partitioning can be bypassed by premixing of channels/pores and lipids before BLM assembly in the workstation (compare Section 11.4.1). Anyway, once put in place, the current flowing through the synthetic ion channel or pore in the BLM is then recorded as a function of time. It provides insights on the electric properties of the interior of the functional supramolecule.

In multichannel experiments, "macroscopic" currents flowing through an ensemble of synthetic ion channels and pores in the BLM are studied. The fast ion flux through ion channels and pores also allows for the resolution of currents flowing through single functional supramolecules. Because transport by ion carriers is too slow for this single-molecule resolution and detergents simply break the BLM, the appearance of single-molecule currents is often considered as the ultimate experimental evidence for the existence of an ion channel or a pore. However, single-channel currents are not a prerequisite of ion channels [5].

Single-channel currents appear and disappear in a stochastic manner (Fig. 11.4b–e). It is important to realize that these typical on-off transitions between open and closed channels are conventional single-molecule phenomena that are, however, invisible in the average of the ensemble, that is, on the macroscopic or "multichannel" level. They should not be confused with the macroscopic modes of gating discussed in Section 11.3.

The observation of two or more single-molecule transitions of identical magnitude identifies a single ion channel or pore (Fig. 11.4c). Asymmetric repetition or two or more current levels of different magnitude demonstrates the co-existence of two or more single ion channels or pores of different active structure (Fig. 11.4d). This sample heterogeneity is more frequently observed with supramolecular than with unimolecular ion channels and pores. Time-dependent sample heterogeneity can indicate the presence of one or several detectable intermediates during formation of the final pore (as in Engelman's "two-state" model [12]) [9, 12–14]. Symmetric repetition of two or more current levels of different magnitude indicates the existence of two or more subconductance levels of a single ion channel or pore (Fig. 11.4e). Despite attractive possibilities including conformational changes, counterion effects (Section 11.3.5) and molecular recognition (Section 11.3.6), it is usually not possible to identify the structural origin of subconductance levels.

Key parameters obtained from single-channel currents are the conductance g, the lifetime τ_1 and the open probability P_o of the single ion channel or pore (Fig. 11.4b). The conductance g informs on the ability of channels and pores to transport ions and therefore often relates to the inner channel diameter (Section 11.3.3). The conductance g increases usually with increasing ionic strength toward saturation at g_{MAX} (Section 11.3.6). The lifetime is the "average" time a single channel remains open. It can be obtained from the exponential decay in the corresponding histogram. The lifetime corresponds to the kinetic stability of the functional supramolecule. Single-channel lifetimes are often in the range of milliseconds but can vary from resolution limit around one millisecond to more than minutes. Taking the silent periods also into account, the open probability P_o, that is the probability to observe an open single channel, reveals the thermodynamic stability of the functional supramolecule.

Failure to detect single-channel currents does not imply that the synthetic ion channel or pore does not exist. Moreover, occasional detection of single-channel currents without or even with reasonable reproducibility does not always imply that the synthetic ion channel or pore exists (single-molecule experiments are sensitive to artifacts, detergents like triton X-100 can produce "single-channel currents" [2]; backup from vesicle flux experiments is always desirable). Sources of failure to detect single-channel currents for existing synthetic ion channels or pores include incorrect lipid composition of the BLM (many variables), insufficient partitioning into the BLM (small surface as well as often too hydrophobic synthetic ion channels and pores, see Section 11.4.1), inactivity at the high ionic strength usually used for good signal-to-noise ratio with large currents, or simply transport rate. Only limited information can be obtained from single-channel BLM experiments for very labile (lifetimes below 1 ms) but also for the arguably most desirable inert synthetic ion channels and pores with "infinite" lifetime [2, 9].

11.2.2
Fluorescence Spectroscopy with Labeled Vesicles

A conventional set-up to characterize synthetic ion channels and pores by fluorescence spectroscopy in vesicles uses large unilamellar vesicles (LUVs) that are labeled with one or more internal (P_I), external (P_E) or membrane-bound (P_M) fluorescent probes (Fig. 11.5a). Internal probes such as HPTS, ANTS or CF are most common to determine activity, external probes are used to reveal specific characteristics (e.g. the voltage-sensitive safranin O, Section 11.3.4), membrane-bound probes are mainly but not exclusively used for structural studies (Section 11.4).

To determine activity, the change in fluorescence emission of internal probes in response to the addition of synthetic ion channels or pores to the LUVs is usually measured as a function of time at different concentrations (Fig. 11.5b). To calibrate the dose response, the emission I_∞ of the free fluorophore is determined at the end of each experiment by, e.g. the addition of a detergent like triton X-100. The minimal and maximal detectable activities I_{MIN} and I_{MAX} in the calibrated curves are then used to recalibrate for a fractional activity Y, and a plot of Y as a function of the concentration of the channel or pore yields the EC_{50}, that is the effective channel concentration needed to observe $Y = 0.5$ (Section 11.3.2). EC_{50} values depend on many parameters and can be further generalized (e.g. the dependence on LUV concentration reveals the $EC_{50(MIN)}$ at low, together with the maximal lipid/pore ratio at high, LUV concentration, Section 11.3) [15]. With respect to BLMs,

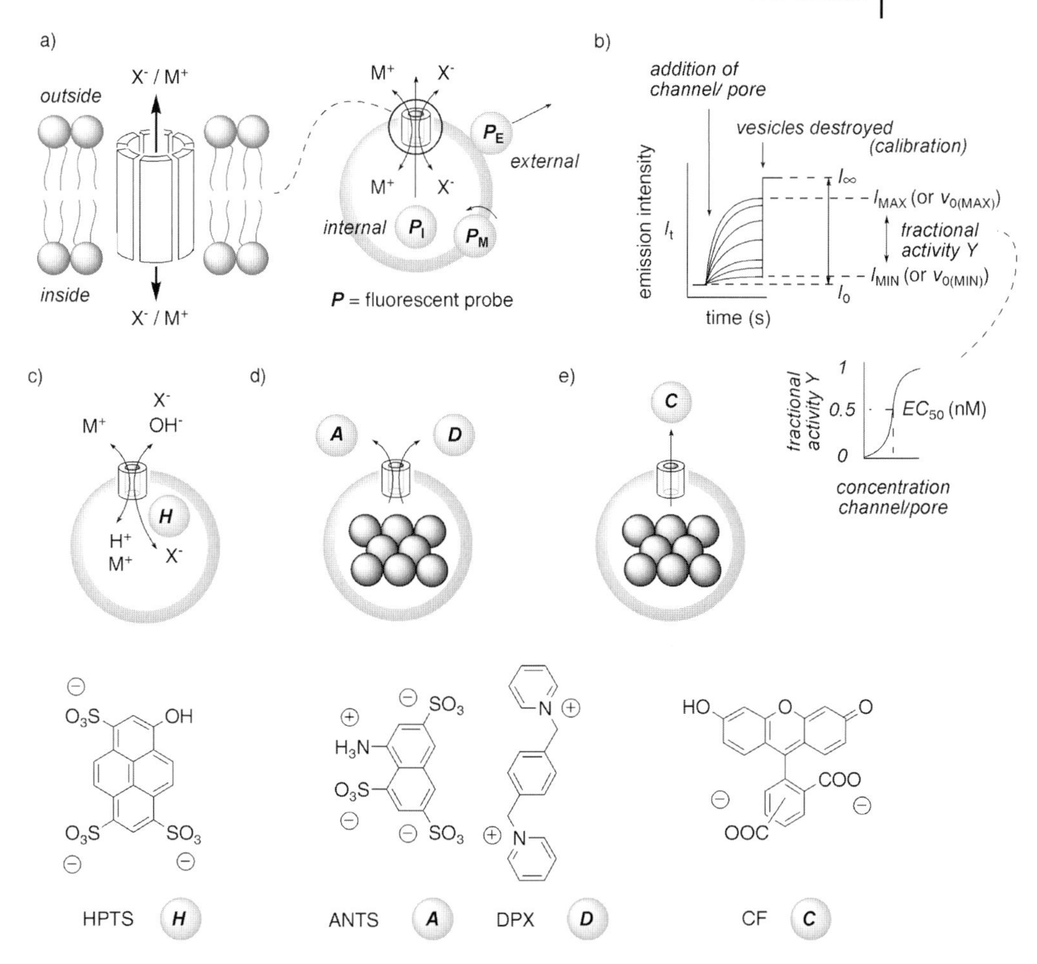

Fig. 11.5. Standard configuration (a) and analysis (b) of vesicle flux experiments with details on (c) the HPTS, (d) the ANTS/DPX and (e) the CF assay.

the EC_{50} reports a sum of contributions including open probability P_o and conductance g, with contributions from P_o being usually most important. For updated original procedures, see, e.g. Refs [3] or [10].

As in BLMs, failure to detect activity in LUVs does not imply that the synthetic ion channel or pore does not exist. *As a general rule, the difficulty of identifying activity in LUVs increases with increasing selectivity of either pore or assay* (for the unrelated partitioning problem, see 4.1) [3, 7–11]. In other words and similar to the situation in BLMs, the most interesting synthetic ion channels or pores are the most difficult ones to detect. The HPTS assay is arguably the most useful assay to begin with because of its poor selectivity (Fig. 11.5c) [9–11].

HPTS is a pH-sensitive fluorophore ($pK_a \sim 7.3$) [6]. The opposite pH sensitivity of the two excitation maxima permits the ratiometric (i.e. unambiguous) detection of pH changes in double-channel fluorescence measurements. The activity of synthetic ion channels is determined in the HPTS assay by following the collapse of an applied pH gradient. In response to an external base pulse, a synthetic ion channel can accelerate intravesicular pH increase by facilitating either proton efflux or OH^- influx (Fig. 11.5c). These transmembrane charge translocations require compensation by either cation influx for proton efflux or anion efflux for OH^- influx, i.e. cation or anion antiport (Fig. 11.5a). Unidirectional ion pair movement is osmotically disfavored (i.e. OH^-/M^+ or X^-/H^+ symport). HPTS efflux is possible with pores only (compare Fig. 11.5b/c). Modified HPTS assays to detect endovesiculation (Fig. 11.1c) [16], artificial photosynthesis [17] and catalysis by pores [18] exist.

In the CF assay, LUVs are loaded with CF at concentrations high enough for self-quenching (Fig. 11.5e) [3]. CF efflux through large enough pores results in fluorophore dilution, and the disappearance of self-quenching is detected as an increase in CF emission (Fig. 11.5b). In the ANTS/DPX assay, the fluorophore ANTS is entrapped in LUVs together with its quencher DPX (Fig. 11.5d) [19]. Efflux of either the anionic ANTS or the cationic DPX through a pore can then be monitored as an increase in ANTS emission. Different to the HPTS assay, both ANTS/DPX and CF assay are sensitive to the inner diameter of synthetic ion channels (not detected) or pores (detected, see Section 11.3.3). Different to ANTS/DPX assay, the CF assay is also sensitive to pH and to the ion selectivity of the pore (Sections 11.3.1 and 11.3.5). Many other fluorescent probes exist, selection is often determined by cost [6]. The use of some more specific probes is introduced below (Sections 11.3.4 and 11.4).

11.2.3 Miscellaneous

Because of their high development, sensitivity and broad applicability, fluorescence spectroscopy with labeled LUVs and BLM conductance experiments are the two methods of choice to study synthetic ion channels and pores. However, several other methods exist. ^{23}Na NMR spectroscopy in vesicles is often used by supramolecular chemists [2]. In this assay, a paramagnetic shift reagent that cannot cross the membrane is added either inside or outside to separate the chemical shifts of intra- and intervesicular Na^+. Sodium flux through synthetic ion channels is then detectable by line-width analysis or peak integration. Whether this method is useful to determine specific characteristics of synthetic ion channels has hardly been explored (Section 11.3). Pores that are large enough to transport the shift reagent (usually dyprosium triphosphate) cannot be studied.

Ion-selective electrodes are another example for interesting methods to analyze synthetic ion channels and pores in LUVs that are arguably less developed and less frequently used. Conductance experiments in supported lipid bilayer membranes may be mentioned as well [20].

11.3 Characteristics

The activity of ion channels and pores is coupled to and, therefore, controlled by many different parameters. This includes membrane composition, polarization and stress, molecular recognition, molecular transformation, light, and so on. The intrinsic relativity of their "activity" makes absolute comparisons between synthetic ion channels and pores often difficult but also less important. Because of their potential for practical applications, it is indeed exactly the creation of "smart" synthetic ion channels and pores with high selectivity and sensitivity that is attracting more and more scientific attention. Highest current and future importance in the field concerns the rational design of synthetic ion channels and pores that respond to specific chemical or physical stimulation [2, 3, 9–11, 15]. In other words, the relevant question concerning the activity of synthetic ion channels and pores is usually not "how active?" but rather "which activity?" This chapter introduces some analytical methods available to determine pertinent aspects of the activities of synthetic ion channels and pores.

11.3.1 pH Gating

Determination of the pH profile of a synthetic ion channel or pore early on is important to focus the further, often time consuming in-depth characterization on the pH of relevant activity [19, 21]. pH sensitivity is usually introduced with acidic or basic functional groups with a pK_a near the pH of interest. Synthetic ion channels and pores with multiple acids or bases in the ion-conducting pathway exhibit parabolic pH profiles (Fig. 11.6). With fractional activity normalized between minima and maxima ($Y = 0$ at pH_{MIN}, $Y = 1$ at pH_{MAX}), characteristics such as pH_{MAX},

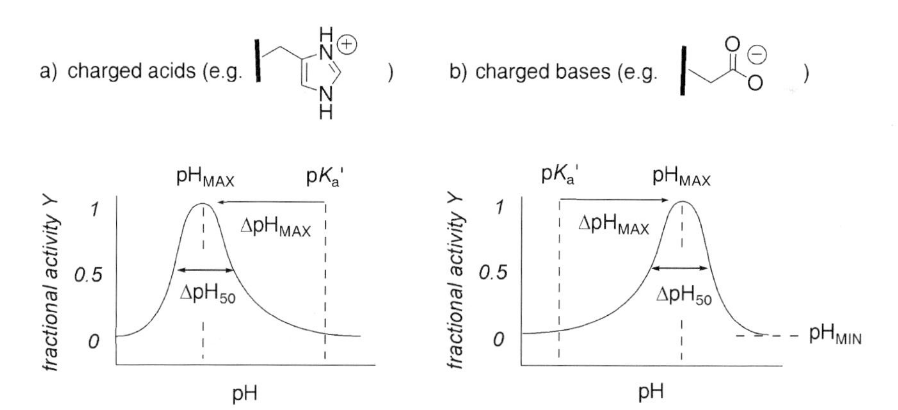

Fig. 11.6. pH profiles summarize the pH dependence of synthetic ion channels and pores with multiple charged acids (a) or bases (b) with intrinsic pK_a'; characteristics are the maximal pH_{MAX}, the effective pH_{50}, ΔpH_{50} for parabolic profiles, and ΔpH_{MAX}.

ΔpH_{50} and ΔpH_{MAX} can be defined (see Fig. 11.5 for Y). ΔpH_{50} is the difference between two effective pH_{50} values at $Y = 0.5$. It depends on the number n of acids/bases in the channel/pore and reflects the cooperativity of pH gating. ΔpH_{MAX}, the shift of pH_{MAX} from the pH corresponding to the intrinsic pK_a', is positive for charged bases and negative for charged acids. It depends on the number n of acids or bases in the channel/pore and reflects the magnitude of proximity effects. These proximity effects on the $pK_a(i)$ of n individual acids or bases described in Eq. (11.1) are thought to account for the parabolic pH profiles of synthetic ion channels and pores with multiple acids/bases.

$$pK_a(i) = pK_a' + \gamma \log(\Omega_{i-1}/\Omega_i) + \gamma(i-1)G_r/2.303RT \tag{11.1}$$

pK_a' = intrinsic pK_a; $\log(\Omega_{i-1}/\Omega_i)$ = probability of proton transfer with $\Omega_i = n!/(n-i)!\ i!$ possibilities to add or remove $1 \le i \le n$ protons for n sites; $\gamma = +1$ for charged bases, $\gamma = -1$ for charged acids; G_r = average electrostatic repulsion between two charges ($G_r \propto r^{-2}$, etc).

Application of Eq. 11.1 to parabolic pH profiles of synthetic ion channels and pores reveals that highest activity occurs at partial (de)protonation of bases (acids). This trend is generalized in the internal charge repulsion (*ICR*) model, stating that intermediate *ICR* is required for maximal activity, whereas undercharged hollow supramolecules tend to "implode" and overcharged ones to "explode" [19]. Both ionic strength (see Section 11.3.6) and lateral pressure from the surrounding membrane ("external membrane pressure", *EMP*) influence the parabolic pH profile of *ICR*-compatible synthetic ion channels and pores [22].

The determination of pH profiles is no problem with the ANTS/DPX assay but more difficult with the pH-sensitive CF and HPTS assays [19, 21]. Determination in BLMs poses no problems and ^{23}Na NMR spectroscopy should be possible as well. The determination of the ionic strength dependence in BLMs is routine and gives important information on the affinity of the ions to the channels (Section 11.3.6). However, the low conductance often observed at low ionic strength can make the experiments difficult. For vesicle experiments with varied external ionic strength, osmotic stress can be avoided with high internal ionic strength and compensation of external under-pressure using sucrose [15, 18].

11.3.2
Concentration Dependence

Determination of the dependence of the fractional activity on the concentration of the monomer forming the uni- or supramolecular active structure is required to obtain the EC_{50} (Section 11.2.2). The shape of the c_M profile contains further information on the thermodynamics and, in some cases, the stoichiometry of the self-assembly of supramolecular ion channels and pores (Fig. 11.7) [21].

Class I channels and pores follow the Hill Eq. (11.2) applied to self-assembly (Fig. 11.7a)

$$\log Y = n \log c_M - n \log K_D \tag{11.2}$$

Y = fractional activity (Fig. 11.5); c_M = monomer concentration; n = Hill coefficient; K_D = dissociation constant of the supramolecule.

The Hill coefficient n obtained from the curve fit of the c_M profile of Class I channels and pores (Fig. 11.7a) corresponds to the number of monomers in the active supramolecule (if self-assembly indeed occurs from an excess monomer in solution. With self-assembly from excess dimer, the number of monomers per active supramolecule is $2n$, and so on). The compatibility with the Hill equation further demonstrates the presence of excess monomer besides a small population of active supramolecule. The presence of excess monomer, in turn, reveals that the self-assembly of the channel or pore is an endergonic process. Structural studies of unstable $n > 1$ supramolecules at concentrations near the EC_{50} by conventional methods are therefore meaningless. For example, NMR or IR measurements will report on the inactive monomers, whereas the unstable active structure of Class I channels and pores is "invisible" (see Section 11.4 for methods to selectively detect and study minority populations of active supramolecules). In BLMs, the thermodynamic instability of Class I channels and pores is expressed in low open probabilities P_o (Fig. 11.4). The $n > 1$ of Class I channels and pores is unrelated to the kinetic stability expressed in short lifetimes for labile Class IA and long lifetimes for inert Class IB supramolecules.

The self-assembly of Class II channels and pores is exergonic. Their c_M profile is, therefore, incompatible with the Hill equation, revealing either $n = 1$ (Fig. 11.7b)

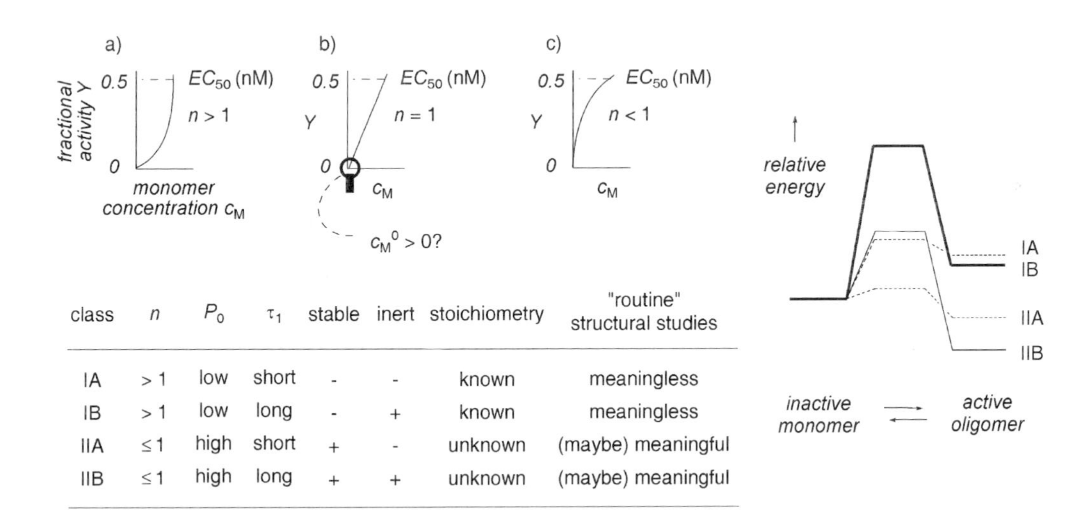

class	n	P_o	τ_1	stable	inert	stoichiometry	"routine" structural studies
IA	> 1	low	short	-	-	known	meaningless
IB	> 1	low	long	-	+	known	meaningless
IIA	≤ 1	high	short	+	-	unknown	(maybe) meaningful
IIB	≤ 1	high	long	+	+	unknown	(maybe) meaningful

Fig. 11.7. c_M Profiles, describing the dependence of synthetic ion channels and pores on the monomer concentration, can differentiate either between unimolecular (b and c, $c_M{}^0 = 0$) and supramolecular (a–c, $c_M{}^0 \geq 0$) or between unstable (class I, a) and stable (class II, b and c) active suprastructures and reveal their stoichiometry (n, class I only), whereas BLM-lifetimes (τ_1) are needed to differentiate between labile (class A) and inert (class B) functional supramolecules.

or $n < 1$ (Fig. 11.7c), depending on their frequent tendency to precipitate rather than to partition into the membrane at increasing concentration. Both labile (IIA) and inert (IIB) Class II channels and pores are, in principle, compatible with structural studies by non-selective routine methods [10, 21]. The Hill coefficient, however, reveals neither the number of monomers per active supramolecule nor the supramolecular nature of the channel or pore as such. A linear c_M profile that intercepts the x axis above 0 ($c_M{}^0 > 0$, $n = 1$) can demonstrate the latter, revealing a less stable active supramolecule (intermediate P_O) with a K_D slightly below the concentration range accessible with the c_M profile (Fig. 11.7b). In other words, $n \leq 1$ profiles do not identify unimolecular ion channels and pores. In practice, unstable ion channels and pores ($n > 1$) are often preferable over stable ones ($n \leq 1$) to avoid precipitation of the hydrophobic prepores from the media [21, 23].

It is important to note that Hill analysis of supramolecular synthetic ion channels and pores is not fully developed and deserves appropriate caution. The usefulness of chemical or thermal denaturation [12–14, 24] to temporarily transform exergonic Class II into endergonic Class I supramolecules for clear-cut demonstration of their true Hill coefficient as well as their supramolecular nature, for example, remains to be explored.

11.3.3
Size Selectivity

The relation between the inner diameter d and the conductance g of synthetic ion channels and pores is described in the Hill equation [5].

$$1/g = l\rho/[\pi(d/2)^2] + \rho/d \tag{11.3}$$

g = conductance of single channels/pores (Fig. 11.4); ρ = resistivity of the recording solution (experimental value); l = length of the channel/pore.

This model is quite unproblematic for synthetic pores with large inner diameters. The meaningfulness of the approximation of channels/pores as electrolyte filled cylinders of known length depends from case to case. However, it decreases generally with decreasing diameter. Significant underestimates are obtained when restricted mobility in the confined interior of small ion channels reduces their conductance. This is particularly pronounced with saturation behavior from electrolyte binding within the ion channel at high ionic strength (Section 11.3.6). However, Sansom's empirical correction factors are available to correct for increasing underestimates with decreasing conductance [25].

Supramolecular chemists without BLM-workstation can approximate the inner diameter of synthetic ion channels and pores by size exclusion experiments in LUVs. Differences in activity between the HPTS assay – compatible with all diameters – and the ANTS/DPX assay reserved for pores with diameter larger than ~5 Å and the CF assay for pores with larger than ~10 Å can differentiate between ion channels and pores (Fig. 11.5). Larger fluorescent probes like CF-dextrans are available to identify giant pores or defects [26]. However, we caution that even probes as large as CF can possibly be translocated by a carrier mechanism rather than through a large pore (see Section 11.1) [27]. The combination and comparison of data on diameter from BLMs, LUVs and molecular models is ideal [3, 10, 24]. With synthetic ion channels and pores with internal active sites for blockage, size dependence experiments on molecular recognition can qualitatively complement above insights from size exclusion experiments on molecular translocation [2–4].

11.3.4 Voltage Gating

The dependence of the activity of ohmic synthetic ion channels and pores on the membrane potential, i.e. the voltage V applied to BLMs or the polarization of LUVs, follows Ohm's law

$$I = gV \tag{11.4}$$

The slope of the IV profile of ohmic ion channels and pores gives their conductance g (Fig. 11.4, 8c, dotted). Non-ohmic ion channels are ion channels that violate Eq. (11.4) (Fig. 11.8c, solid). Because of importance in biology and materials science, the creation of non-ohmic ion channels and pore has attracted considerable interest in supramolecular chemistry [2, 9]. The key parameter characterizing non-ohmic behavior is the gating charge z_g [5, 9].

$$Y \propto c_M{}^n \exp(z_g eV/kT) \tag{11.5}$$

Y = fractional activity; c_M = monomer concentration; n = Hill coefficient (Eq. 2); e = elementary charge; k = Boltzmann constant; T = absolute temperature.

For ohmic channels and pores, $z_g = 0$. Formally, the gating charge z_g is the charge that has to be translocated across the bilayer to open the ion channel or pore. In single-channel BLM-experiments, z_g is determined from the change in open probability P_o with applied voltage [9]. However, macroscopic I–V profiles also produce correct z_g values [9]. This is the case because contributions from changes in single-channel conductance with voltage (i.e. rectification) [28] are usually negligible compared to changes in P_o (i.e. number and stability of open channels).

Supramolecular chemists without BLM-workstation can determine gating charges in polarized vesicles (Fig. 11.8), using the HPTS assay for synthetic ion channels and the ANTS/DPX assay for synthetic pores (Fig. 11.5) [7]. To polarize vesicles, an inside-negative Nernst potential is applied with a potassium gradient (Eq. 11.6), osmotically balanced with sodium, coupled with the potassium carrier valinomycin at intermediate concentrations sufficient for rapid potential buildup without immediate collapse, and monitored by an emission increase of the externally added probe safranin O (Fig. 11.8a and b).

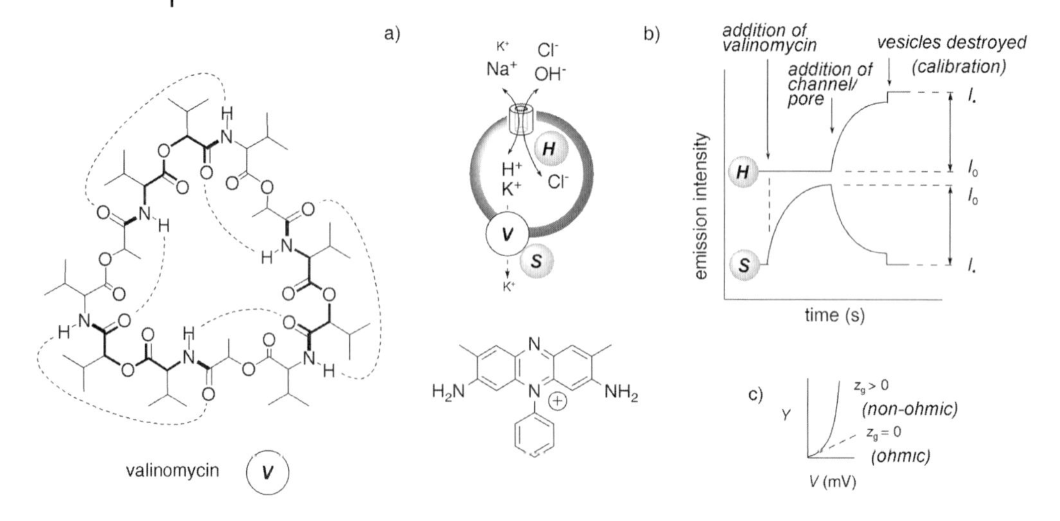

Fig. 11.8. Gating charges z_g determined for non-ohmic (i.e. voltage-gated) synthetic ion channels or pores (c) in valinomycin polarized, doubly-labeled LUVs (a) with internal HPTS to measure changes in pH and external Safranin O to measure depolarization (b) (compare Fig. 11.4 for data analysis).

$$V = RT/zF \ln(\text{external } [K^+]/\text{internal } [K^+]) \tag{11.6}$$

V = Nernst potential; R = gas constant; T = absolute temperature; $z = +1$ (charge of K^+); F = Faraday constant.

In a pH gradient, the activity of added synthetic ion channels and pores can then be observed in the HPTS channel as increase and in the safranin O channel as decrease of fluorophore emission (Fig. 11.8b). Changes in activity at different Nernst potentials give then the gating charge z_g (Fig. 11.8c). Gating charges from BLMs and LUVs are comparable [9]. "Reversed" non-ohmic behavior, with inactivation at high voltage is quite common and characteristic for highly symmetric, ohmic ion channels and pore like β-barrels [9, 29]. Hysteresis in *IV* curves can imply voltage-sensitive formation of voltage-insensitive and highly inert ion channels and pores [9], presumably in a "two-state" or multistate process [9, 12–14].

11.3.5 Ion Selectivity

The anion/cation selectivity describes the overall preference of synthetic ion channels or pores for either anion or cations [5, 8]. The anion/cation selectivity can be determined in BLMs using salt concentration gradients between *cis* and *trans* chamber (Fig. 11.9A and Fig. 11.4). With salt gradients, a current will flow without applied voltage. The direction of this zero current reveals the preferred movement of K^+ or Cl^- through the synthetic ion channel or pore. The voltage required to cancel this zero current is the reversal potential V_r. Calculation of the permeability ratios with the Goldman–Hodgkin–Katz (GHK) *voltage* Eq. (11.7) from the reversal potential V_r reveals anion/cation selectivity quantitatively.

$$P_{A^-}/P_{M^+} = [a_M cis - a_M trans \exp(-V_r F/RT)]/[a_A cis \exp(-V_r F/RT) - a_A trans] \tag{11.7}$$

P_{A^-}/P_{M^+} = anion/cation permeability ratio; $a_M cis$ = cation activity in *cis* chamber; V_r = reversal potential; R = gas constant; T = temperature; F = Faraday constant.

As discussed above, synthetic ion channels added to LUVs loaded with the pH-sensitive fluorophore HPTS and exposed to a pH gradient mediate the collapse of the latter by either H^+/K^+ or OH^-/Cl^- antiport (Figs. 9B and 5). Sensitivity of the measured rate (e) to external cation (f) but not anion exchange (g) identifies cation selectivity (Fig. 11.9h) [30]; sensitivity to external anion but not cation exchange identifies anion selectivity in LUVs [9, 10]. Anion/cation selectivity of synthetic ion channels determined in BLMs and LUVs are comparable [9, 24]. Whereas the determination of anion/cation selectivities of synthetic pores in BLMs is as with synthetic ion channels and unproblematic, the HPTS assay is not applicable for this purpose. Indications on anion/cation selectivities of synthetic pores in LUVs can be obtained from comparison of CF and ANTS/DPX assays, because the CF assay reports activity for anion selective pores only [8].

The selectivity sequence describes the preference of a cation channel between different cations and an anion channel between different anions [8, 30–36]. In BLMs and LUVs, selectivity sequences are determined like the anion/cation selectivity (Fig. 11.9). In BLMs, the gradient between *cis* and *trans* chamber is not established with different concentrations of the same salt but between identical concentrations of different salts. In LUVs, the rate of the increase in HPTS emission in the presence of different external cations M^{n+} or anions A^{n-} are compared to determine cation and anion selectivity sequences, respectively [10, 30].

Many theories, variations and refinements on ion selectivity exist [5, 31–36]. Arguably, the selectivity sequences of cation channels are best described as Eisenman topologies, where ion selectivities (i.e. respective permeability ratios, conductances or fractional activities) are plotted as a function of the reciprocal radius of the cation (Fig. 11.9C) [31]. These eleven topologies cover sequences between the one completely determined by dehydration penalty (Eisenman I, $Cs^+ > Rb^+ > K^+ > Na^+ > Li^+$) and the one completely determined by cation binding in the channel (Eisenman XI, $Li^+ > Na^+ > K^+ > Rb^+ > Cs^+$). Eisenman I topologies are the least interesting ones because they do not need refined synthetic ion channels or pores to occur. Eisenman IV topologies are arguably the most interesting ones because of the importance of potassium selectivity in neurons and beyond ($K^+ > Rb^+ > Cs^+ > Na^+ > Li^+$) [8, 30, 31].

Anion selectivity topologies beyond the seven halide sequences are less common because of possible mismatches of size and dehydration energy (Fig. 11.9D) [33–35]. Because of its frequent occurrence, complete dependence on dehydration en-

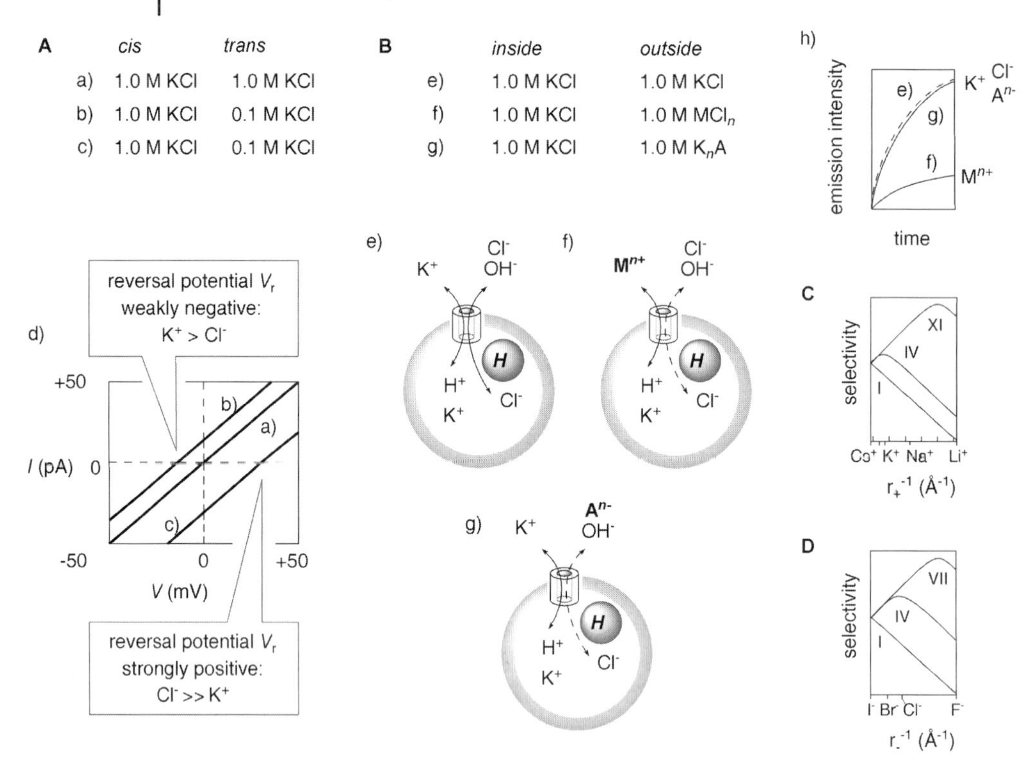

Fig. 11.9. Ion selectivity in (A) BLMs and (B) LUVs. (A) $V_r \neq 0$ mV in current-voltage profiles (d) with cis-trans salt gradients (b, c) reveal ion selectivity in BLMs according to the GHK equation (11.7) [8]. (B) Sensitivity to external cation (e, f) and insensitivity to external anion exchange (e, g) in the HPTS assay (Fig. 11.5) reveals cation selectivity in LUVs (h). (C) Cation selectivity sequences can be classified as Eisenman topologies I–XI, (D) anion selectivity sequences as halide topologies I–VII, halide sequence I is often referred to as Hofmeister series, sequences IV are relevant concerning biological potassium and chloride channels.

ergy (i.e. halide topology I) is historically also referred to as Hofmeister or lyotropic series, with hydrophobic anions named chaotropes (structure breakers) and hydrophilic anions named kosmotropes (structure makers) [35]. Chloride selectivity as in topology IV is interesting because of its medicinal relevance (e.g. channel replacement therapy for cystic fibrosis treatment) [2, 23].

Note that selectivity sequences derived from permeability ratios and conductances can differ significantly [31–36]. Because there is no current at the reversal potential, permeability ratios describe the ability of ions to enter a synthetic channel or pore. Permeability ratios, therefore, relate directly to the change in energy from ion stabilization by hydration to ion stabilization by its interaction with the channel referred to in the Eisenman theory. The respective conductance with different permeant ions, that is the slope of the *IV* profile, reveals the ability of an ion to move through the channel and be released at the other side. Permeant ions that are recognized within synthetic ion channels or pores can become blockers at high concentration (Section 11.3.6).

More difficult in BLMs, refined HPTS assays exist to address the special cases of selective transport of protons [11] and electrons [17] in LUVs. In the conventional HPTS assay (Fig. 11.5c), the apparent activity of proton channels *decreases* with increasing proton selectivity because the rate of the disfavored cation (M^+) influx influences the detected velocity more than the favored proton efflux. Disfavored potassium influx can, however, be accelerated with the potassium carrier valinomycin (Fig. 11.8). Increasing activity in the presence of valinomycin identifies proton channels with $H^+ > K^+$ selectivity being at least as high as the maximal measurable increase (in unpolarized LUVs of course, compare Section 11.3.4). Important controls include evidence for low enough valinomycin concentrations to exclude activity without the proton channel (due to disfavored H^+ efflux). The proton carrier FCCP is often used as complementary additive to confirm $M^+ > H^+$ selectivity (e.g. amphotericin B).

As for selective electron transport in LUVs, many assays have been developed early on in studies directed toward artificial photosynthesis [37]. An elegant recent example applies the HPTS assay to active transport [17]. Namely, photoinduced electron transfer is detected as internal pH increase due to proton consumption during the reduction of a water-soluble quinone trapped together with HPTS within LUVs.

11.3.6 Blockage and Ligand Gating

Strong binding of specific ions can lead not only to selectivity but also to blockage of synthetic ion channels. In BLMs, blockage by permeant ions is revealed by saturation behavior of the conductance g with increasing activity a_{ION} of the ion of interest (Fig. 11.10D) [33]. Fit of the salt profile to Eq. (11.8) gives the inhibitory concentration IC_{50} (or the apparent dissociation constant K_D) and the maximal conductance g_{MAX} for a permeant ion blocker.

$$g = g_{MAX} a_{ION} / IC_{50} + a_{ION} \qquad (11.8)$$

Note that the Hill model to measure inner diameters of channels and pores is obviously incompatible with blockage by permeant ions (Section 11.3.3). Multiple binding to line up several permeant ion blockers in a single transmembrane file for selective ion hopping through the channel can be considered as elegant solution for ion selectivity without blockage [38]. This same supramolecular solution of the dilemma of "selectivity without loss in speed" is applicable to the selective transport of protons on the one hand [11] and that of molecules through large synthetic multifunctional pores on the other) [39]. Anomalous mole fraction behavior can be used to imply multiple electrolyte binding [6, 33–35].

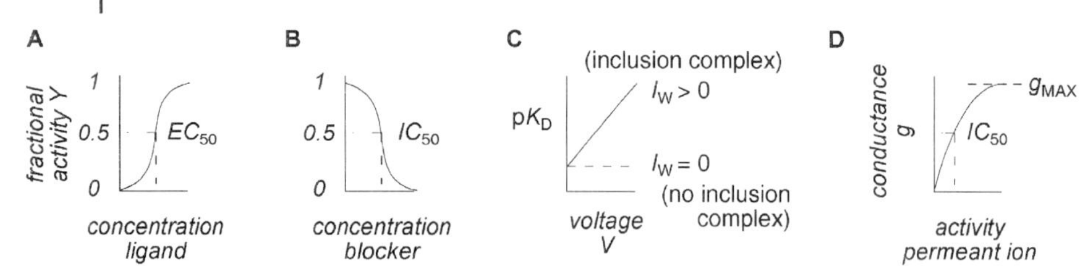

Fig. 11.10. Synthetic multifunctional ion channels and pores either (A) open or (B) close in response to chemical stimulation. Dose response curves for (A) ligand gating or (B) blockage are characterized by effective (EC_{50}) or inhibitory concentrations (IC_{50}) and Hill coefficients (n). (C) Woodhull analysis of the voltage dependence of blockage reveals the depth of molecular recognition. (D) The special case of blockage by permeant electrolytes is described by IC_{50} and g_{MAX} (the maximal conductance).

Multiply-charged synthetic ion channels and pores can scavenge and permanently immobilize counterions [21, 40]. Counterion immobilization can result in significantly reduced single-pore conductance. This translates, according to the Hill model (Section 11.3.3), to a meaningful shortening of the inner pore diameter by internal counterion scavenging. Other significant effects of counterion immobilization on ion channel and pore characteristics include increased single-pore lifetimes (i.e. increased kinetic pore stability, Section 11.2.1) and inversion of anion/cation selectivity (Section 11.3.5) [40].

Molecular recognition by synthetic ion channels and pores is the key to multifunctionality and its many diverse applications to sensing and catalysis [41, 42]. The response to chemical stimulation is ligand gating or blockage (other, in equally imperfect terms: "opening" or "closing", "activation" or "inactivation"). Ligand gating and blockage can be readily determined in BLMs, LUVs and other methods like supported bilayers as change in fractional activity Y in response to chemical stimulation with a guest molecule, either blocker or ligand. Dose response curves can be fitted to the Hill equation (11.9) applied to molecular recognition (Figure 11.10A and B)

$$\log Y = n \log c_{\text{GUEST}} - n \log EC_{50} \tag{11.9}$$

Y = fractional activity (Fig. 11.5); c_{GUEST} = guest concentration (ligand or blocker); n = Hill coefficient; EC_{50} (effective ligand concentration; or IC_{50} for blockage, or apparent K_D = dissociation constant of the host–guest complex).

Effective and inhibitory concentrations correspond to apparent dissociation constants of the host–guest complexes. These values can be quite different from the true K_D values. Particularly cautious interpretation is recommended for stoichiometric binding, molecular recognition may actually be much better than it appears in these cases [43–45]. Quantitative correlation of the values of IC_{50} and KD [44] is not common in the field of synthetic ion channels and pores. The Hill coefficient n

can reflect the number of ligands or blockers needed for function [45]. However, many parameters can further influence Hill coefficients, such as self-assembly of both ligands and ion channels/pores (compare Section 11.3.2) [21, 24].

As with synthetic ion channels and pores as such (Sections 11.3.1 to 11.3.5), molecular recognition by synthetic ion channels and pores [2–4, 41, 42] depends on many parameters such as pH, [46] ionic strength [15, 22], self-assembly [21], voltage [47], topological matching [46, 47], ion selectivity [48], and so on. The voltage dependence (V) of molecular recognition by synthetic ion channels and pores (K_D) described in the Woodhull equation (11.10) is arguably the most interesting parameter [5, 47, 49].

$$pK_D = pK_D(0\ \text{mV}) + (l_w z_{GUEST} FV)/(2.303 lRT) \tag{11.10}$$

$pK_D = -\log K_D$; l_w = Woodhull distance from channel/pore entrance to active site; z_{GUEST} = guest charge; F = Faraday constant; V = voltage applied to the side of guest addition; l = length of ion channel/pore; $l_w/l = \delta$ = electric distance; R = gas constant; T = absolute temperature.

Woodhull analysis is possible in BLMs and polarized LUVs with results being at least qualitatively comparable [47]. The key information obtained is the Woodhull distance l_w from channel/pore entrance to active site. Mechanistic and structural insights accessible with Woodhull analysis include evaluation as to whether or not molecular recognition by synthetic ion channels and pores really takes place in the membrane, discrimination between voltage-sensitive inclusion complexes and voltage-insensitive peripheral association, and measurement of the depth of guest inclusion (see Section 11.4.4) [47].

Synthetic ion channels and pores that open and close in response to molecular recognition are detectors of all chemical reactions as long as substrate and product recognition are not the same [3, 15, 41, 42, 50]. For instance, the conversion of good substrate blockers into poor product blockers can be detected as pore opening (Fig. 11.11A). Pore closing is observed with products as blockers (Fig. 11.11B), whereas pore opening is observed with products as activators (Fig. 11.11C). The combination of synthetic multifunctional pores as general optical transducers of

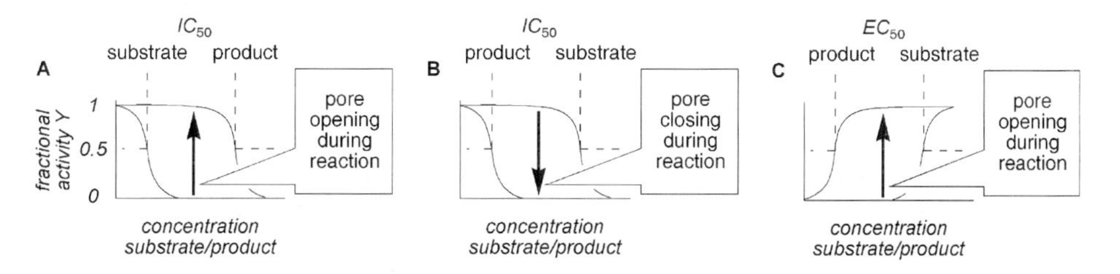

Fig. 11.11. Detection of chemical reactions with synthetic multifunctional pores that operate by blockage (A, B) or ligand gating (C): Pore opening during reaction occurs with substrate as blockers (A) and products as ligands (C), pore closing during reaction with products as blockers (B).

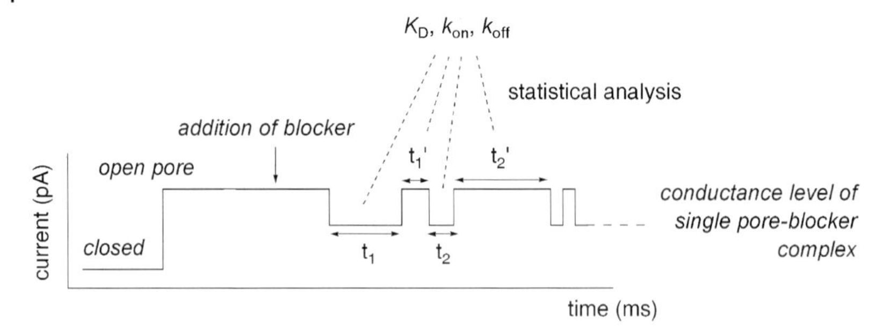

Fig. 11.12. Stochastic detection of molecular recognition in single-molecule measurements in BLMs. Binding of a blocker reduces the conductance of a single open pore (compare Section 11.2.1). Statistical analysis of lifetime and probability to observe the conductance level of single pore-blocker complexes reveals both thermodynamic and kinetic data.

chemical reactions with enzymes as specific signal generators provides access to multicomponent sensing in complex matrixes (e.g. sugar sensing in soft drinks) [15]. *As "universal" detectors and sensors, synthetic ion channels and pores emerge as powerful tools of a new method in (rather than being a target of) supramolecular analytic chemistry.*

The possibility of stochastic detection of single analytes in single-pore measurements in BLMs has been explored extensively with biological and bioengineered pores [51–53], although not yet realized with synthetic ion channels and pores. In brief, the conductance of single pore-blocker complexes, if detectable, may differ from that of the single open pore (Fig. 11.12). With inert single pores that do not frequently open and close by themselves (Fig. 11.4), the entering and leaving of single blockers can possibly be observed in a stochastic manner. Statistical analysis of the obtained "on-off" transitions for guest recognition gives both thermodynamic and kinetic data for the pore-blocker complex (analogous to the insights on pores as such, available from statistical analysis of open-close transitions, see Section 11.2.1).

This stochastic detection of single pore-blocker complexes is attractive for several reasons. Different conductances and lifetimes for different pore-blocker complexes allows, in principle, for "fingerprint-type" sensing [51]. The possibility of single gene sequencing with single pores has, for example, been much discussed although so far not realized [52]. Different conductances for single pore–substrate and pore–product complexes opens appealing perspectives for the detection of reactions, including single reactive intermediates [53]. Catalytic activity has, however, so far been reported only for synthetic pores (Section 11.3.7) [54, 18].

11.3.7 Miscellaneous

A practically unexplored application of molecular recognition by synthetic multifunctional ion channels and pores is catalysis [54]. An HPTS-based assay is being developed to analyze the voltage dependence of Michaelis–Menton kinetics in LUVs (similar to Woodhull analysis of voltage-sensitive molecular recognition, see Section 11.3.6) [18]. The obtained key characteristics – analogous to Woodhull distances (l_w, Section 11.3.6) or gating charges (z_g, Section 11.3.4) – are "steering factors" that quantify remote control by membrane potentials to guide substrates into and products out of the catalytic pore.

An enormously important and diverse aspect of molecular recognition by synthetic ion channels and pores concerns specific interactions with the surrounding membrane. Only *membrane potentials* have been considered so far (Sections 11.3.4 and 11.3.6). The recognition of *surface potentials* by synthetic ion channels and pores has attracted much interest for the development of new antimicrobials [2]. The dependence on surface potentials (simply speaking, the summed charges of the lipid headgroups in the bilayer) is best expressed in $Y - \Psi_0$ profiles; surface potentials Ψ_0 can be obtained from the Gouy–Chapman theory [9]. Sensitivity of synthetic ion channels and pores to *bilayer fluidity* is another interesting topic. Fractional activity in dipalmitoyl phosphatidylcholine (DPPC) LUVs as a function of temperature is often measured for this purpose because of the convenient phase transition of the neutral DPPC bilayers (41 °C). Beware of the often-heard notion that increasing activity with decreasing fluidity implies a channel and increasing activity with increasing fluidity a carrier mechanism (Fig. 11.1A and B): Contributions from other parameters, particularly partitioning, account usually for changes in activity with bilayer fluidity [22, 55]. These effects can be clarified with reasonably straightforward and reliable structural studies (Section 11.4.1).

Dependence of the activity on *bilayer thickness* is sometimes examined to prove transmembrane orientation of synthetic ion channels and pores. Parabolic $Y - n$ profiles are unavoidable to justify this conclusion (n = number of methylenes in the alkyl tails of lipids; varied to vary bilayer thickness) [56]. Meaningful analysis of the somewhat related dependence of activity on *bilayer stress* requires osmolarity profiles [57, 58]. Dependence of activity of *bilayer composition* is a vast topic that interconnects with several of the above parameters; sterol sensitivity with emphasis on ergosterol and cholersterol has attracted much attention for the development of new fungicides [59], targeted pore formation with lipid II and lipid A is thought to account for the antibacterial activities of nisin [60] and cationic amphiphilic steroids [61] respectively [2]. Simple dose response curves with Hill analysis are applicable to this topic (Section 11.3.6). *Bilayer heterogeneity* including sterol sensitivity or the trendy "rafts" is in general a topic where synthetic ion channels and pores have been suggested to possibly be of some use as analytic tools [2, 62]. *Flip-flop*, finally, is a central topic in mechanistic and structural studies on synthetic ion channels and pores [63]. Flip-flop describes the reversible vertical motion of lipids from one leaflet to the other leaflet of a lipid bilayer. In intact membranes, flip-flop is slow. It can be accelerated by biological and synthetic flippases, a special class of membrane transporters related to ion carriers [63]. Synthetic ion channels and pores with flippase activity are likely to affect the suprastructure of the surrounding bilayer, i.e. have micellar active structure (Fig. 11.2) [64]. Routine flippase assays with fluorescently labeled lipids can be found in the literature [48, 63, 64].

11.4
Structural Studies

In research focusing on the creation of function, experimental evidence for this desired function is naturally all that really matters (Section 11.3) [4]. The intrinsically lower significance on the one hand and the complexity of the topic on the other can complicate research focusing on structural studies. The comments made in the introduction to functional studies (see Section 11.3.0) are even more important with regard to structural studies: It is extremely important to ask meaningful questions and to work under conditions that are relevant for function. Beware of high-resolution NMR insights or crystal structures that have nothing to do with the active structure of synthetic ion channels and pores.

As target molecules of multistep organic synthesis, the molecular structure of the monomers that form synthetic ion channels and pores is always known from routine analytical methods. Structural studies, therefore, ask the question how these monomers form active ion channels and pore, i.e. focus on conformational or supramolecular analysis. Being the outcome of rational design based on known structural information, this question often comes down to the not very inspired question whether or not synthetic ion channels and pores that are already known to function as expected also form the expected active structure. Added to a lipid bilayer, a monomer of known structure can avoid the membrane (Fig. 11.13, Aa) or accumulate in transmembrane orientation (Figure 11.13, Ab), in the middle of the

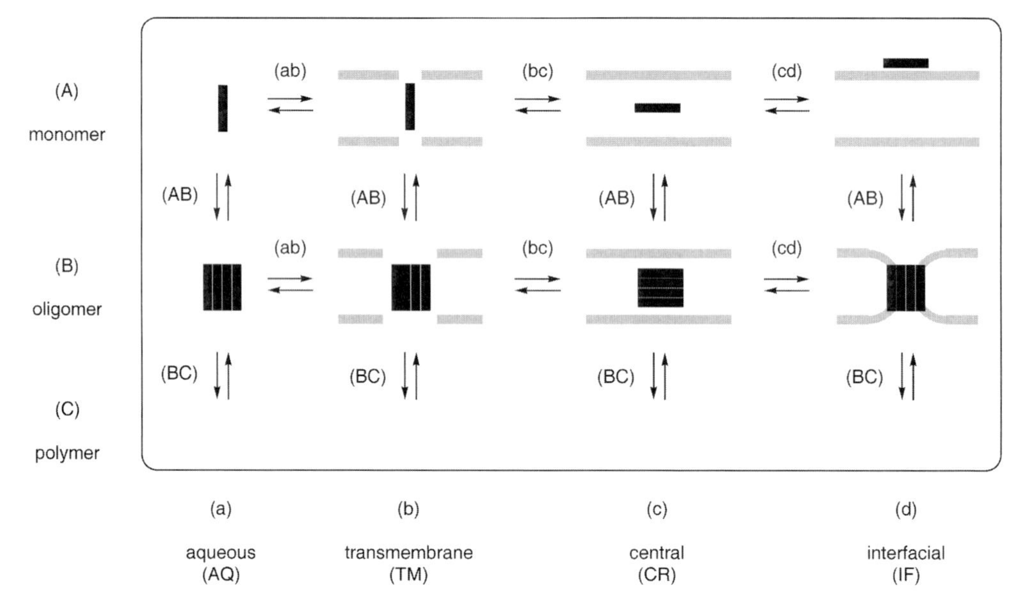

Fig. 11.13. Dynamic supramolecular polymorphism of synthetic ion channels and pores including unimolecular (Ab), barrel-stave (Bb) and micellar motifs (Bd; compare Fig. 11.2, text and refs [42] and [4]).

membrane (Fig. 11.13, Ac) or at the membrane–water interface (Fig. 11.13, Ad). Alternatively, the monomers can self-assemble – in the water (Fig. 11.13, a) or in the membrane (Fig. 11.13, b–d) – into oligomers like barrel-stave, barrel-hoop, barrel-rosette supramolecules (Fig. 11.13B, see Fig. 11.2). Continuing self-assembly into supramolecular polymers such as vesicles, fibers or microcrystals is known as often undesired high-concentration side-effect in water [42] but unknown as a productive route to ion channels and pores in lipid bilayer membranes (Fig. 11.13C) [21, 23].

In principle, all the suprastructures shown in a still highly simplified manner in Fig. 11.13 are in equilibria that depend on many parameters such as pH (Section 11.3.1), concentration (Section 11.3.2), ionic strength (Section 11.3.1), nature of the bilayer (Sections 11.3.4 and 11.3.7), and so on. This dynamic supramolecular polymorphism restricts meaningful structural studies to conditions that are relevant for function but often incompatible with routine analytical methods (e.g. nanomolar to low micromolar concentrations in lipid bilayer membrane. The fact that the active conformers or supramolecules are often not the thermodynamically dominant ones [21] (Section 11.3.2) calls for additional caution as well as selective methods of detection). *As a general rule, the complexity of the supramolecular polymorphism of synthetic ion channels and pores decreases with increasing complexity (size) of the monomer* (in other words, synthetic efforts are often worthwhile [2]; compare Fig. 11.2).

Consideration of the supramolecular polymorphism of synthetic ion channels and pores reveals the relevant questions for structural studies: Where do the monomers act (Fig. 11.13b–d), how many of them are needed (Fig. 11.13A–C), and where and how can they be regulated? In the following, analytical methods available to address these questions will be briefly mentioned but not introduced in detail (compare, e.g. ref [4]).

11.4.1
Binding to the Bilayer

One of the general methods to detect the binding to lipid bilayer membranes quantitatively is equilibrium dialysis (Fig. 11.13a,b). The key parameter, the partition coefficient K_x, is described in Eq. (11.11) [24, 65].

$$I = I_{MIN} + (I_{MAX} - I_{MIN})/(1 + [H_2O]/K_x[\text{lipid}]) \qquad (11.11)$$

I = response, I_{MIN} = I without lipids, I_{MAX} = I at saturation, $[H_2O]$ = concentration of water (55.3 M).

The binding of synthetic ion channels and pores to lipid bilayer membranes often causes a change in intra- or intermolecular self-organization that is visible in sufficiently sensitive methods such as fluorescence (e.g. tryptophan emission) [14] or circular dichroism spectroscopy and can be used to determine the partition coefficient. Convenient methods of detection under relevant conditions are fluorescence resonance energy transfer (FRET) or fluorescence depth quenching (FDQ) [3, 4, 6]. Many fluorescent probes for the labeling of both synthetic ion channels/

pores and lipid bilayer membranes exist. Isothermal calorimetry (ITC) can be used to gain further insight into the thermodynamics of channel–lipid interactions.

The reversibility of binding to the bilayer can be measured in a so-called "hopping experiment" [3]. In this assay on intervesicular transfer, the activity of synthetic ion channels or pores is measured in LUVs as described (Section 11.2.2). Then, fresh LUVs loaded with fluorescent probes are added for a second time. Inactivity in this second round demonstrates irreversible, activity reversible binding of synthetic ion channels or pores to the bilayer membrane of the initially added LUVs.

Partition coefficients depend on many parameters including pH, ionic strength, counterions [27] but also bilayer composition, polarization [9], surface potential or fluidity [22, 55] (Section 11.3.7). Needless to explain but important to reiterate that general extrapolation of partition coefficients from bulk membranes (water–chloroform or water–octanol extractions, U-tube experiments) is not possible. Poor partitioning is a frequently underrecognized limiting factor with synthetic ion channels and pores that are too hydrophobic. Varied, at best multicomponent, stock solutions can be a useful and simple solution to solve this problem (i.e. to minimize competing precipitation as supramolecular polymers, Fig. 11.13, BC vs ab). Innovative design strategies of synthetic ion channels and pores with "hyodrophilic–lipophilic switches" to control and maximize partitioning are desirable [21].

11.4.2 Location in the Bilayer

The method of choice to determine, under meaningful conditions, the location of synthetic ion channels and pores in bilayer membranes is fluorescence depth quenching (FDQ) [4, 11]. In this well developed but costly analysis, the position of a quencher in the lipid bilayer is varied systematically. Analysis of the dependence of the efficiency to quench a fluorescent synthetic ion channel or pore on the position of the quencher reveals transmembrane, central or interfacial location (Fig. 11.13b–d).

Evaluation of the meaningfulness of results from less sensitive and less selective methods needs additional attention (e.g. attenuated total reflectance infrared (ATR–IR) or solid-state NMR spectrometry) [2]. Indirect insights from functional studies include support for transmembrane orientation (Fig. 11.13b) from parabolic dependence of the activity of synthetic ion channel or pore on bilayer thickness (Section 11.3.7) [56] and other readouts in support of operational hydrophobic matching. Flippase activity may provide some support for interfacial location (Section 11.3.7, Fig. 11.13d) [61, 62].

11.4.3 Self-Assembly

The method of choice to determine whether or not synthetic ion channels and pores act as monomers or supramolecular oligomers (Fig. 11.13, A vs B) is Hill analysis of the dependence of activity on the monomer concentration (Section 11.3.2) [21]. However, whereas nonlinear concentration dependence is solid evidence for the endergonic self-assembly of n (or more) monomers into active supramolecules, linear concentration dependence does not confirm monomeric active structures. Meaningful and noninvasive methods to differentiate between exergonic self-assembly and unimolecular self-organization are needed (Hill plots at chemical denaturation) [66].

c_M profiles from structural studies by circular dichroism or fluorescence spectroscopy can be readily validated by comparison with results from function. Several labeling methods exist to study self-assembly under meaningful high-dilution conditions (FRET [6]; ECCD, exciton-coupled circular dichroism [67]). Validated c_M profiles from structural studies or from other functions such as catalysis can be used to dissect self-assembly in water with self-assembly in the bilayer membrane (Fig. 11.13B, a vs b) [21].

Parallel and antiparallel self-assembly can be differentiated by voltage dependence: Parallel self-assembly of asymmetric monomers gives voltage-sensitive, antiparallel self-assembly voltage-insensitive synthetic ion channels and pores [9]. Other indirect evidence from function such as inner diameters from Hill analysis of single-channel conductance (Section 11.3.3) or other size exclusion experiments is often used to support indications for supramolecular active structures; molecular modeling can be of help as well [3, 4, 10].

11.4.4
Molecular Recognition

Whereas functional studies on molecular recognition by synthetic ion channels and pores are straightforward (Sections 11.3.6 and 11.3.7), structural studies can be troublesome, naturally even more difficult than the structural studies of synthetic ion channels and pores as such (Sections 11.4.1 to 11.4.3) [4]. In brief, the meaningful questions remain similar: Where do the ligands and blockers act, how many of them are needed, and so on. Some methods described for ion channels and pores as such can be adapted to the study of molecular recognition. Most importantly, FDQ and also FRET can reveal whether host–guest complexes are formed in the membrane or in the water. Woodhull analysis is a unique method to measure the depth of molecular recognition within inclusion complexes [47]. Pseudo-rotaxanes with polymer blockers moving through pores can be large enough for structural and mechanistic studies by atomic force microscopy (AFM) [68].

11.5
Concluding Remarks

The objective of this contribution is to equip the supramolecular chemist entering the field with an introduction to the characterization of synthetic ion channels and pores. Emphasis is on fundamental functional characteristics such as ion selectivity, voltage gating, ligand gating, blockage, and so on. A brief introduction to the most common methods available to measure these activities is included at the beginning, and an introductory discussion of meaningful questions that can be addressed in structural studies at the end.

The material discussed is only the "tip of the iceberg." There is an extensive, highly developed analytical biophysics literature available on each topic described. Recommended leading references can be found in the bibliography, often introduced via an illustrative and simplifying application to synthetic ion channels and pores from our group. Arguably one of the biggest challenges faced by a supramolecular chemist entering the field is to find the right balance between synthetic creativity and analytical depth, that is not to get lost in biophysics without being too superficial or even incorrect (and to hope for referees with an understanding of this balance). The introduction is, therefore, highly simplified and to a good part subjective, building on our own experience of years of research on the topic. It should, hopefully, suffice to serve as dependable guide to a well-balanced characterization of an exciting new functional supramolecule.

Acknowledgement

We thank J. Mareda and G. Bollot for the preparation of Fig. 11.3 and the Swiss NSF for financial support.

References

1 Fyles, T. M. *Membrane Transport*, 13th International Symposium on Supramolecular Chemistry, Notre Dame, IL, 2004.
2 (a) Matile, S.; Som, A.; Sordé, N. *Tetrahedron* **2004**, 60, 6405–6435, and references therein. (b) Sisson, A. L.; Shah, M. R.; Bhosale, S.; Matile, S. *Chem. Soc. Rev.*, in press.
3 Gorteau, V.; Perret, F.; Bollot, G.; Mareda, J.; Lazar, A. N.; Coleman, A. W.; Tran, D.-H.; Sakai, N.; Matile, S. *J. Am. Chem. Soc.* **2004**, 126, 13592–13593.
4 Baudry, Y.; Bollot, G.; Gorteau, V.; Litvinchuk, S.; Mareda, J.; Nishihara, M.; Pasini, D.; Perret, F.; Ronan, D.; Sakai, N.; Shah, M. R.; Som, A.; Sordé, N.; Talukdar, P.; Tran, D.-H.; Matile, S. *Adv. Funct. Mater.* **2006**, 16, 169–179.
5 Hill, B. *Ionic Channels of Excitable Membranes*, 3rd Edn, Sinauer, Sunderland, MA, 2001.
6 Haugland, R. P.; Spence, M. T. Z.; Johnson, I.; Basey, A. *The Handbook. A Guide to Fluorescent Probes and Labeling Technologies*, 10th Edn, Molecular Probes, Eugene. OR, 2005.
7 Sakai, N.; Matile, S. *Chem. Biodiv.* **2004**, 1, 28–43.
8 Sakai, N.; Matile, S. *J. Phys. Org. Chem.* in press.
9 Sakai, N.; Houdebert, D.; Matile, S. *Chem. Eur. J.* **2003**, 9, 223–232.
10 Talukdar, P.; Bollot, G.; Mareda, J.; Sakai, N.; Matile, S. *J. Am. Chem. Soc.* **2005**, 127, 6528–6529.
11 Weiss, L. A.; Sakai, N.; Ghebremariam, B.; Ni, C.; Matile, S. *J. Am. Chem. Soc.* **1997**, 119, 12142–12149.

12 Popot, J.; Engelman, D. *Biochemistry* **1990**, 29, 4031–4037.
13 Bowie, J. U. *Proc. Natl. Acad. Sci. USA* **2004**, 101, 3995–3996.
14 Hong, H.; Tamm, L. K. *Proc. Natl. Acad. Sci. USA* **2004**, 101, 4065–4070.
15 Litvinchuk, S.; Sordé, N.; Matile, S. *J. Am. Chem. Soc.* **2005**, 127, 9316–9317.
16 Matsuo, H.; Chevallier, J.; Vilbois, F.; Sadoul, R.; Fauré, J.; Matile, S.; Sartori Blanc, N.; Dubochet, J.; Gruenberg, J. *Science* **2004**, 303, 531–534.
17 Steinberg-Yfrach, G.; Liddell, P. A.; Hung, S.-C.; Moore, A. L.; Gust, D.; Moore, T. A. *Nature* **1997**, 385, 239–241.
18 Sakai, N.; Sordé, N.; Matile, S. *J. Am. Chem. Soc.* **2003**, 125, 7776–7777.
19 Baumeister, B.; Som, A.; Das, G.; Sakai, N.; Vilbois, F.; Gerard, D.; Shahi, S. P.; Matile, S. *Helv. Chim. Acta* **2002**, 85, 2740–2753.
20 Terrettaz, S.; Ulrich, W.-P.; Guerrini, R.; Verdini, A.; Vogel, H. *Angew. Chem. Int. Ed.* **2001**, 40, 1740–1743.
21 Litvinchuk, S.; Bollot, G.; Mareda, J.; Som, A.; Ronan, D.; Shah, M. R.; Perrottet, P.; Sakai, N.; Matile, S. *J. Am. Chem. Soc.* **2004**, 126, 10067–10075.
22 Som, A.; Matile, S. *Chem. Biodiv.* **2005**, 2, 717–729.
23 Broughman, J. R.; Shank, L. P.; Takeguchi, W.; Schultz, B. D.; Iwamoto, T.; Mitchell, K. E.; Tomich, J. M. *Biochemistry* **2002**, 41, 7350–7358.
24 Talukdar, P.; Bollot, G.; Mareda, J.; Sakai, N.; Matile, S. *Chem. Eur. J.* **2005**, 11, 6525–6532.
25 Smart, O. S.; Breed, J.; Smith, G. R.; Sansom, M. S. P. *Biophys. J.* **1997**, 72, 1109–1126.
26 Gorteau, V.; Bollot, G.; Mareda, J.; Pasini, D.; Tran, D.-H.; Lazar, A. N.; Coleman, A. W.; Sakai, N.; Matile, S. *Bioorg. Med. Chem.* **2005**, 13, 5171–5180.
27 Perret, F.; Nishihara, M.; Takeuchi, T.; Futaki, S.; Lazar, A. N.; Coleman, A. W.; Sakai, N.; Matile, S. *J. Am. Chem. Soc.* **2005**, 127, 1114–1115.
28 Goto, C.; Yamamura, M.; Satake, A.; Kobuke, Y. *J. Am. Chem. Soc.* **2001**, 123, 12152–12159.
29 Bainbridge, G.; Gokce, I.; Lakay, J. H. *FEBS Lett.* **1998**, 431, 305–308.
30 Tedesco, M. M.; Ghebremariam, B.; Sakai, N.; Matile, S. *Angew. Chem. Int. Ed.* **1999**, 38, 540–543.
31 Eisenman, G.; Horn, R. *J. Membrane Biol.* **1983**, 76, 197–225.
32 Wright, E. M.; Diamond, J. M. *Physiol. Rev.* **1977**, 57, 109–156.
33 Qu, Z.; Hartzell, H. C. J. *Gen. Physiol.* **2000**, 116, 825–884.
34 Hartzell, C.; Putzier, I.; Arreola, *J. Annu. Rev. Physiol.* **2005**, 67, 719–758.
35 Lindsdell, P. *J. Physiol.* **2001**, 531, 51–66.
36 Dawson, D. C.; Smith, S. S.; Mansoura, M. K. *Physiol. Rev.* **1999**, 79, 47–75.
37 Robinson, J. N.; Cole-Hamilton, D. J. *Chem. Soc. Rev.* **1991**, 20, 49–94.
38 Doyle, D. A.; Cabral, J. M.; Pfuetzner, R. A.; Kuo, A.; Gulbis, J. M.; Cohen, S. L.; Chait, B. T.; MacKinnon, R. *Science* **1998**, 280, 69–77.
39 Ronan, D.; Sordé, N.; Matile, S. *J. Phys. Org. Chem.* **2004**, 17, 978–982.
40 Sakai, N.; Sordé, N.; Das, G.; Perrottet, P.; Gerard, D.; Matile, S. *Org. Biomol. Chem.* **2003**, 1, 1226–1231.
41 Sakai, N.; Mareda, J.; Matile, S. *Acc. Chem. Res.* **2005**, 38, 79–87.
42 Sakai, N.; Matile, S. *Chem. Commun.* **2003**, 2514–2523.
43 Straus, O. H.; Goldstein, A. *J. Gen. Physiol.* **1943**, 26, 559–585.
44 Cheng, Y.; Prusoff, W. H. *Biochem. Pharmacol.* **1973**, 22, 3099–3108.
45 Connors, K. A. *Binding Constants*, John Wiley & Sons, New York, 1987.
46 Litvinchuk, S.; Matile, S. *Supramol. Chem.* **2005**, 17, 135–139.
47 Baudry, Y.; Pasini, D.; Nishihara, M.; Sakai, N.; Matile, S. *Chem. Commun.* **2005**, 40, 4798–4800.

48 Das, G.; Onouchi, H.; Yashima, E.; Sakai, N.; Matile, S. *ChemBioChem* **2002**, 3, 1089–1096.
49 Woodhull, A. M. *J. Gen. Physiol.* **1973**, 61, 687–708.
50 Das, G.; Talukdar, P.; Matile, S. *Science* **2002**, 298, 1600–1602.
51 Gu, L. Q.; Braha, O.; Conlan, S.; Cheley, S.; Bayley, H. *Nature* **1999**, 398, 686–690.
52 Deamer, D. W.; Branton, D. *Acc. Chem. Res.* **2002**, 35, 817–825.
53 Luchian, T.; Shin, S.-H.; Bayley, H. *Angew. Chem. Int. Ed.* **2003**, 42, 1926–1929.
54 Baumeister, B.; Sakai, N.; Matile, S. *Org. Lett.* **2001**, 3, 4229–4232.
55 Otto, S.; Osifchin, M.; Regen, S. L. *J. Am. Chem. Soc.* **1999**, 121, 10440–10441.
56 Weber, M. E.; Schlesinger, P. H.; Gokel, G. W. *J. Am. Chem. Soc.* **2005**, 127, 636–642.
57 Benachir, T.; Lafleur, M. *Biophys. J.* **1996**, 70, 831–840.
58 Sakai, N.; Matile, S. *Chirality* **2003**, 15, 766–771.
59 Matsumori, N.; Sawada, Y.; Murata, M. *J. Am. Chem. Soc.* **2005**, 127, 10667–10675.
60 Hasper, H. E.; de Kruijff, B.; Breukink, E. *Biochemistry* **2004**, 43, 11567–11575.
61 Ding, B.; Yin, N.; Liu, Y.; Cardenas-Garcia, J.; Evanson, R.; Orsak, T.; Fan, M.; Turin, G.; Savage, P. B. *J. Am. Chem. Soc.* **2004**, 126, 13642–13648.
62 Otto, S.; Janout, V.; DiGiorgo, A. F.; Young, M.; Regen, S. L. *J. Am. Chem. Soc.* **2000**, 122, 1200–1204.
63 Smith, B. D.; Lambert, T. N. *Chem. Commun.* **2003**, 2261–2268.
64 Matsuzaki, K.; Murase, O.; Fujii, N.; Miyajima, K. *Biochemistry* **1996**, 35, 11361–11368.
65 White, S. H.; Wimley, W. C.; Ladokhin, A. S.; Hristova, K. *Methods Enzymol.* **1998**, 295, 62–87.
66 Bhosale, S.; Matile, S. *Chirality* **2006**, *18*, 849–856.
67 Nakanishi, K.; Berova, N. *The Exciton Chirality Method.* In *Circular Dichroism – Principles and Applications*, Nakanishi, K.; Berova, N.; Woody, R. W., Eds. VCH, Weinheim, Germany, **1994**, 361–398.
68 Kumaki, J.; Yashima, E.; Bollot, G.; Mareda, J.; Litvinchuk, S.; Matile, S. *Angew. Chem. Int. Ed.* **2005**, 44, 6154–6157.

12
Theoretical Methods for Supramolecular Chemistry[1)]

Barbara Kirchner and Markus Reiher

12.1
Introduction

Supramolecular chemistry [1, 2] comprises chemical processes of nanometer-sized molecules often interacting via many, comparatively weak, non-bonding contacts. An important and generic chemical process in supramolecular chemistry is molecular recognition in host–guest complexes[2)], which is, for instance, the basis of protein–substrate interactions in biochemistry. In general, reactions, in which a host or template takes care of a spatial pre-organization of a reactant prepared for a well-defined chemical reaction or molecular motion, play an important role in chemistry [3–11]. The number of template-assisted reactions known is steadily increasing though most of them are usually discovered by serendipity. Apart from such accidentally discovered (and often *a posteriori* recognized) template-assisted processes, the various reactants can often hardly be classified according to a unifying scheme because such reactions can be found in any branch of chemistry. It is therefore not surprising that a generally applicable model of host- or template-assisted reactions has not been developed yet. We should emphasize that a (thermochemical) model for intermolecular host–guest interactions and the theoretical means to quantitatively assess the static and dynamic processes involved in this model should be understood as the straightforward approach to theoretical supramolecular chemistry. The quantitative description can then be transfered to other aspects of supramolecular chemistry like molecular switches and motors [12] in a straightforward fashion.

The large number of atoms involved in supramolecular assemblies require sophisticated analytical methods of experiment and theory to selectively extract relevant information for the chemistry of a particular supramolecular aggregate.

1) II. Part of a series of papers on the theory of template-assisted chemical processes (see for part I: J. Am. Chem. Soc. *127* **2005**, 8748–8756).

2) In the following, the discussion of host–guest, template–substrate, or template–guest interactions shall be understood as a typical model process occuring in supramolecular assemblies.

Analytical Methods in Supramolecular Chemistry. Edited by Christoph Schalley

ISBN: 978-3-527-31505-5

In principle, theoretical methods provide – by construction – an excellent space (and time) resolution since the molecular structure is an essential ingredient for any kind of more approximate modeling. In practice, however, accuracy is lost because the description of the intermolecular forces is approximate or when there are a limited number of time intervals in a simulation. Nonetheless, theoretical approaches can provide detailed information often not accessible by experiment.

Naturally, quantum chemical approaches to molecular recognition are usually employed for selected systems since the complexity of these systems requires a system-specific analysis which makes it difficult to extract results of general validity; for examples, see Refs. [13–16] for studies of molecular tweezers. Further examples are mentioned in a review article by Schatz considering *ab initio* calculations on calixarenes and calixarene complexes [17]. Schatz concludes that although the systems are quite big, useful contributions have been made by *ab intio* calculations. However, a general model is needed in order to make host–guest processes and template-assisted reactions accessible to a comparison of quantitative measurements and calculations, which may finally provide the basis for rational host design and for the prediction of template effects (compare the recent attempt by Hunter [18]).

In principle, the interactions between host and guest may be modeled by suitably constructed empirical potentials, which are widely used in traditional molecular mechanics. A different approach to intermolecular interactions is possible through *first-principles* or *ab initio* theory, which is based on the fundamental principles of quantum mechanics and does not require any input of experimental data[3]. The broad and intellectually appealing idea of this reductionistic *first-principles* approach is to describe the motion and interaction of elementary particles, i.e. electrons and atomic nuclei. This approach is based on the laws of quantum mechanics such that any molecular assembly can be treated with the same methods and *without* additional adaptions of potential energy terms using empirical data from experiment. However, this elegant approach is associated with two drawbacks, which are the non-decomposable nature of a highly correlated electronic system (i.e. of the molecule or molecular assembly) and the very high computational costs. Hence, decomposition schemes – even if they are not in the spirit of the *first-principles* methods and in contradiction with holistic quantum mechanics – need to be developed to enable qualitative understanding. Also, fast *first-principles* electronic structure methods need to be devised in order to make such calculations feasible for large supramolecular aggregates.

In addition, molecular recognition of a host and its guest is a typical dynamic process running through many local minimum structures until host and guest are

3) Note that we use the terms *first-principles* and *ab initio* synonymously. Note also that density functional theory (DFT) is by construction a *first-principles* or *ab initio* theory. Of course, the exact exchange–correlation functional is not known and one uses approximate functionals instead. For this reason, some people tend to emphasize this *semi-empirical* nature of contemporary DFT. However, the number of parameters in present-day density functionals is usually embarassingly small while their range of applicability is enormous.

locked in a somewhat final configuration, namely in the host–guest complex. The explicit calculation of these dynamics provides an overwhelming amount of data, which needs to be evaluated statistically at the cost of a loss of information on relevant individual structures. It might, however, be possible in certain cases to map these complex dynamics onto relevant static structures, which are stationary points on the Born–Oppenheimer energy hypersurface. That this might be indeed possible was confirmed, for instance, in the case of oligopeptide structures in solution by van Gunsteren and co-workers (see sect. 12.5 below for references).

This account presents an introduction to the basic principles of quantum chemical methods and molecular dynamics simulations. Their applicability for the description of well-known intermolecular interaction patterns like hydrogen bonds or π–π stacking is discussed. Decomposition schemes are mandatory for a chemical understanding of host–guest complexes with a large number of attractive contacts. We propose an extension of a method for the calculation of hydrogen bond energies based on concepts known from ESCA spectrocsopy [19] (electron spectroscopy for chemical analysis) and relate this method to comparable approaches. Moreover, in an effort to arrive at a general classification scheme for many different host–guest processes and template-assisted reactions we propose a generalized thermochemical scheme in close analogy to classical theories of chemical kinetics. As mentioned above, due to the complexity of these processes theoretical studies are usually restricted to special cases. A unifying view of the many different processes of template-assisted reactions and host–guest recognition has not yet been attempted. Accordingly, we first propose a general mechanism of template-assisted reactions, which also allows us to clarify the relation to related concepts (e.g. to catalysis). Although elementary in nature, these clarifications appear to be needed for a rational design of supramolecular processes. It can be expected that a general mechanism of template-assisted reactions and of host–guest processes then permits a reasonable decomposition of the involved free enthalpy contributions to chemically meaningful thermochemical quantities of elementary reactions. Within the proposed reaction scheme, it might be possible to derive thermodynamic and kinetic boundary conditions for the rational design of template-assisted chemical transformations. We call this model the thermochemical theory of template-assisted elementary reactions – in short: the TAR model.

This work is organized as follows: Section 12.2 provides an introduction to the basic principles of quantum chemical and molecular dynamics methods. The focus is on the value of these methods for the description of supramolecular assemblies and processes. Section 12.3 discusses standard classifications for inter-molecular interactions, where the phenomenologically observed interaction patterns are arranged into a small number of classes. In Section 12.4, we present selected decomposition schemes (mainly from our own work) and finally present in Section 12.5 the generalized thermochemical TAR scheme for the static description of host–guest processes and of template-assisted reactions. Since a large number of studies deal with theoretical methods as a tool to complement the experimental data, we can only give a limited insight into the theoretical treatment of supermolecules that have appeared in literature. For this reason, the presentation of each approach is far from being complete. Thus, we can hardly overemphasize that a complete coverage of all relevant literature, i.e. an inclusion of all references dealing with the theoretical treatment of supramolecular assemblies – and also with theoretical and computational methods in general – is impossible owing to the width of the field. It was therefore only possible to include selected references as well as exemplary, mainly recent studies which highlight particular theoretical approaches. We kindly ask the reader who is particularly interested in certain topics to also check the references in the literature cited here. For a collection of review articles of early computational approaches to supramolecular chemistry we refer the reader to the excellent and extensive book by Wipff [20].

12.2
A Survey of Theoretical Methods

In this section we present a brief overview of theoretical methods that can be employed to obtain structural and energetic data on supramolecular assemblies. As pointed out in the introduction, quantum chemical *first-principles* methods are particularly appealing because all interactions between individual atoms and functional groups of host and guest are reduced to the interactions of elementary particles. In principle, these methods can provide an optimum access to space and time resolution of the molecular processes involved. For computational, but also for conceptual, reasons it can be advantageous to map these first-principles results onto empirical force fields, which describe the interactions between sites. These sites can consist of atoms or functional groups. Simple force field expressions contain pair interaction terms like electrostatic Coulombic monopole–monopole $1/r$ terms or Lennard–Jones 6–12-potentials containing attractive $-1/r^6$ and repulsive $+1/r^{12}$ terms. In practice, these (classical) force-fields are often constructed intuitively and pragmatically or through a coarse-graining ansatz (see Ref. [21] for tutorial reviews). Hence, their validity needs to be thoroughly investigated (compare the evaluation of force fields for the description of aromatic interactions of supramolecular model systems in Ref. [22]).

Classical force-fields, which are elements of an approach based on Newtonian mechanics, are very successful in polymer and biochemistry. They have been extensively applied in docking studies (see Ref. [23] for a very recent example), which cannot be reviewed here in detail. However, force fields have also been applied in other areas of chemistry and naturally to supramolecular chemistry. For instance, Müller and collaborators derived a force-field especially addressed to the inorganic host–guest species of polyoxovandates [24]. Polyoxometalates are an example for a class of compounds whose sizes range from small clusters to huge systems (cf. the beautiful wheel- and sphere-like polyoxomolybdates synthesized by Müller and collaborators within the last decade [25–28]). While the huge systems are computationally accessible only by force-field methods, their smaller relatives can be treated explicitly with *first-principles* methods. Rohmer and coworkers reviewed structures and reactivity of polyoxometalates [29]. They note that although a lot of empirical knowledge was gathered [30–32], only little effort [33, 34] was devoted to modeling the formation of such molecules by quantum chemical *first-principles* techniques. The authors investigated quantum-chemically calculated molecular electrostatic potentials V_{ESP},

$$V_{\text{ESP}}(\mathbf{r}) = \sum_A \frac{Z_A e}{|\mathbf{R}_A - \mathbf{r}|} + \int \mathrm{d}^3 r' \frac{\rho(\mathbf{r}')}{|\mathbf{r}' - \mathbf{r}|}, \tag{12.1}$$

in order to gain insight into the host–guest interaction energy (see, e.g. Ref. [35] for a discussion of the V_{ESP} in chemical applications; units are chosen such that the pre-factor $1/(4\pi\varepsilon_0)$ can be suppressed). The first term in Eq. (12.1) describes the sum of the point charge potentials of the atomic nuclei A with nuclear charge number Z_A and resulting charge $Z_A e$ at position $\mathbf{R}_A$, while the second refers to the contribution of the electronic density ρ at distance $\mathbf{r}$.

The electrostatic potential is commonly used as a means to understand possible channels for a substrate to approach a host on the basis of electrostatic interactions. As an example Fig. 12.1 depicts the electrostatic potential of a biomimetic arginine fork studied by quantum chemical methods [14, 36]. This artifical argi-

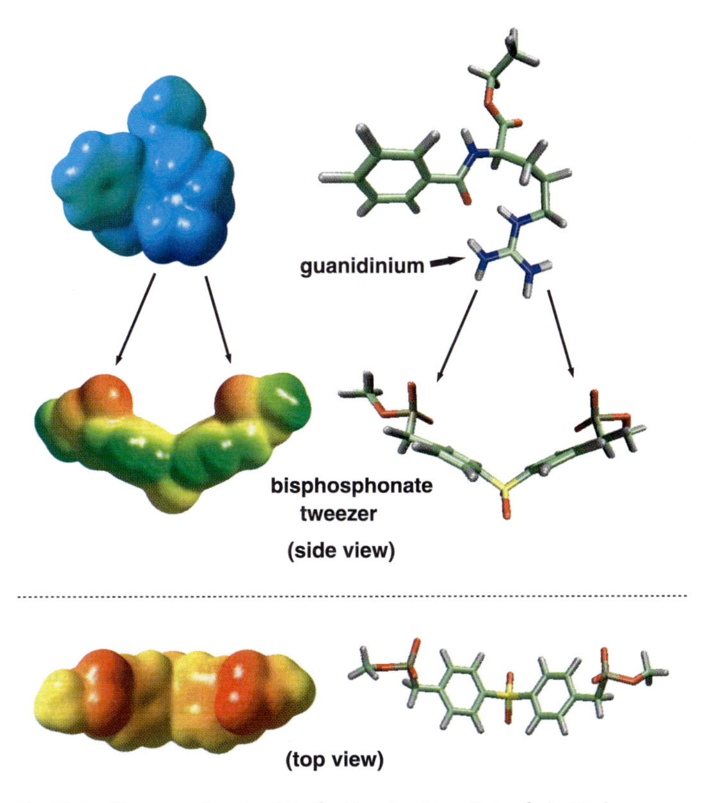

Fig. 12.1. Electrostatic potential of a biomimetic arginine fork. (Red color: negatively charged bisphosphonate tweezer; blue color: positively charged guanidinium host; structures taken from Ref. 14.)

nine receptor molecule was developed by Schrader and coworkers [37, 38]. The receptor complex in Fig. 12.1 consists of a bisphosphonate tweezer that clamps a guanidinium group of the guest (see also Refs. [39–42] for comparable bisphosphonate receptor molecules by Schrader et al.). Structural and energetic information about this system has also been obtained from experimental NMR investigations and in parts from molecular mechanics calculations [38, 43]. The complex is designed to imitate the arginine–phosphonate diester interaction of the arginine fork [44], which is a key element in RNA–protein recognition. The relevance of understanding the interaction modes of such arginine/guanidinium moieties with phosph[on]ate or sulf[on]ate residues has been recently stressed [45]. The electrostatic potential can give a first glance on these modes. However, the detailed interactions are best studied via *first-principles* methods to be introduced in the next section.

Coming back to the case of polyoxometalates, Rohmer et al. [29] found that electronically inverse host anions are formed in solution by means of a template mechanism which tends to maximize the electrostatic potential at the place of the guest anion. These authors also provided a correlation between the topology of the host and its molecular electrostatic potential, which explains on the basis of simple geometric considerations the difference between electronically normal and electronically inverse hosts. With the aid of this correlation it can be shown that the host cage tends to adapt not only to the shape of the guest molecule, but also to its electrostatic potential distribution [29].

12.2.1
First-principles Methods

Quantum chemical *first-principles* methods are based on the fundamental principles of quantum mechanics. Despite the complicated nature of these methods, we try to give in this section a brief but comprehensive formal description of the most important ones. Technical and practical aspects are left aside as this scientific account is not intended to be a manual for their usage. We may refer the reader interested in these aspects to the excellent general books on computational chemistry by Cramer [46] and by Jensen [47] or to the classic text [48].

The central physical quantity of interest for chemistry and supramolecular chemistry is the total electronic energy E_{el} of a molecular system consisting of N electrons and M atomic nuclei. This energy is the solution of the time-independent electronic Schrödinger equation,

$$\hat{H}_{el}\Psi_{el} = E_{el}\Psi_{el}, \tag{12.2}$$

where Ψ_{el} denotes the total electronic wave function and $\hat{H}_{el}$ is the many-electron Hamiltonian operator,

$$\hat{H}_{el} = \sum_{e=1}^{N} \hat{t}_e + \sum_{e=1}^{N}\sum_{K=1}^{M} \hat{v}_{eK} + \sum_{e=1}^{N}\sum_{e'>e}^{N} \hat{v}_{ee'} + \sum_{K=1}^{M}\sum_{K'>K}^{M} \hat{v}_{KK'}, \tag{12.3}$$

which contains all Coulombic pair interaction energy operators $\hat{v}_{ij} = q_i q_j / |\mathbf{r}_i - \mathbf{r}_j|$ (again, the $1/(4\pi\varepsilon_0)$ factors have been removed through a proper definition of units) of two elementary particles i and j (q_i, q_j are their charges and $|\mathbf{r}_i - \mathbf{r}_j|$ is their distance) and their individual kinetic energy operators $\hat{t}_e$, which are in the most simple, non-relativistic form equal to $-\hbar^2/(2m_e)\nabla_e^2 = -\hbar^2/(2m_e) \cdot (\partial^2/\partial x_e^2 + \partial^2/\partial y_e^2 + \partial^2/\partial z_e^2)$. The solution of Eq. (12.2) provides the wave function Ψ_{el} and the energy E_{el}. However, the solution is complicated. Decades of method development in quantum chemistry provided a diverse tool box of methods [49–52]. All of these methods start with an ansatz to approximate the electronic wave function Ψ_{el}.

The most simple ansatz Φ_0 for the complicated function $\Psi_{el} = \Psi_{el}(\mathbf{r}_1, \mathbf{r}_2, \ldots \mathbf{r}_N)$ which depends on $3N$ variables, namely, on the coordinates of all electrons, is an antisymmetrized product of one-electron functions, i.e. of molecular (spin) orbitals ψ_i,

$$\Psi_{el} \approx \Phi_0 = \hat{\mathscr{A}} \psi_1(\mathbf{r}_1), \psi_2(\mathbf{r}_2), \ldots \psi_N(\mathbf{r}_N), \tag{12.4}$$

where the spin variables have been omitted for the sake of brevity. This approximation Φ_0 to Ψ_{el} is called a Slater determinant and represents the basis of the Hartree–Fock model [53]. The antisymmetrizer $\hat{\mathscr{A}}$ ensures that the Pauli principle is fulfilled [54]. In order to calculate the electronic energy E_{el}, the orbitals ψ_i need to be determined. By construction, they are the solution functions of self-consistent field (SCF) equations which read for each orbital ψ_i:

$$\left[-\frac{\hbar^2}{2m_e} \nabla_i^2 + \sum_{K=1}^{M} \hat{v}_{eK} + \hat{v}_C(\{\psi_j\}) + \hat{v}_x(\{\psi_j\}) \right] \psi_i(\mathbf{r}_i) = \varepsilon_i \psi_i(\mathbf{r}_i). \tag{12.5}$$

The sum of all operators in brackets on the left hand side is called the Fock operator $\hat{f}$. The Coulomb and exchange operators, $\hat{v}_C(\{\psi_j\})$ and $\hat{v}_x(\{\psi_j\})$, respectively, result from the electron–electron interactions terms $\hat{v}_{ee'}$ in the Hamiltonian of Eq. (12.3). Advanced readers may notice that these equations are the canonical Hartree–Fock equations, which yield canonical orbitals and corresponding orbital energies ε_i. The subscript i at the electronic coordinate $\mathbf{r}_i$ is not necessary but has been kept for the sake of consistency in this very brief presentation. Once the SCF equations are solved, which can be routinely accomplished with quantum chemistry program packages (such as Gaussian [55], Turbomole [56] or Molpro [57] – see Ref. [58] for a list of programs) the electronic energy of a molecule can be calculated,

$$E_{el} \approx E_{el}^{\mathrm{HF}} = \frac{1}{2} \sum_{i=1}^{N} \left[\varepsilon_i + \int d^3 r \psi_i^\star(\mathbf{r}) \left(-\frac{\hbar^2}{2m_e} \nabla_i^2 + \sum_{K=1}^{M} \hat{v}_{eK} \right) \psi_i(\mathbf{r}) \right] + \sum_{K=1}^{M} \sum_{K'>K}^{M} \hat{v}_{KK'}. \tag{12.6}$$

It was soon realized that the error introduced by the Hartree–Fock model, which is the so-called correlation energy $\Delta E = E_{el} - E_{el}^{\mathrm{HF}}$, is small for closed-shell systems (of the order of a few percent) but decisive for chemical reaction energetics. Moreover, weak interactions of the van der Waals type cannot be described with such a single-determinant Hartree–Fock model. Consequently, Hartree–Fock calculations on supramolecular assemblies, whose interaction is governed by weak dispersion forces, cannot provide accurate quantitative results and are likely to yield a wrong qualitative picture. However, in certain cases Hartree–Fock results may be of value. For instance, Houk and coworker have investigated the role of [C–H $\cdots$ O] interactions in supramolecular complexes by means of dynamic and static Hartree–Fock calculations [59].

In order to improve on the Hartree–Fock model, the use of perturbation theory is common. The first energy correction is obtained at second order and the corresponding method is calles second-order Møller–Plesset perturbation theory (MP2). MP2 calculations provide a first estimate for the correlation energy ΔE, which turned out to be also useful for estimates of the interaction energy in cases dominated by dispersive interactions (see the next section for an overview on how interaction energies can be calculated from total electronic energy estimates).

Before we proceed to describe the basic ideas of post-Hartree–Fock methods, we should briefly recall that the orbitals ψ_i in Eq. (12.5) are usually expanded in a finite set of known basis functions χ_μ (often denoted linear combination of atomic orbitals – LCAO),

$$\psi_i = \sum_{\mu=1}^{m} c_\mu^{(i)} \chi_\mu \tag{12.7}$$

so that the determination of the orbitals is equivalent to the determination of the expansion coefficients $c_\mu^{(i)}$. For the calculation of these molecular orbital coefficients, the Hartree–Fock Eqs. (12.5) are rewritten in matrix form (Roothaan equations [60]) and can be solved with standard linear algebra techniques on computers. In the Roothaan equations, the Fock operator is then rewritten in matrix form,

$$\hat{f} \rightarrow \mathbf{f} = \{f_{\mu\nu}\} = \left\{ \int \mathrm{d}^3 r \chi_\mu^\star \hat{f} \chi_\nu \right\} \tag{12.8}$$

A natural extension of the Hartree–Fock model is to approximate the electronic wave function Ψ_{el} by an expansion in terms of many Slater determinants Φ_I,

$$\Psi_{el} = \sum_{I=0}^{\infty} C_I \Phi_I, \tag{12.9}$$

which provide a suitable many-electron basis set comparable to the LCAO expansion of orbitals in terms of one-electron basis functions. For an infinite set of one-electron basis functions, the number of Slater determinants is also infinite. Then, Ψ_{el} is represented exactly and the exact electronic energy, which also includes dispersion effects correctly, is obtained. However, this comes with infinite computational costs. Hence, methods needed to be devised, which allow us to approximate the infinite expansion in Eq. (12.9) by a finite series to be as short as possible. A straightforward approach is the employment of truncated configuration interaction (CI) expansions. Note that "(electronic) configuration" refers to the set of molecular orbitals used to construct the corresponding Slater determinant. It is a helpful notation for the construction of the truncated series in a systematic manner and yields a classification scheme of Slater determinants with respect to their degree of "excitation". Excitation does not mean physical excitation of the molecule but merely substitution of orbitals occupied in the Hartree–Fock determinant Φ_0 by virtual, unoccupied orbitals. Within the LCAO representation of molecular orbitals the virtual orbitals are obtained automatically with the solution of the Roothaan equations for the occupied orbitals that enter the Hartree–Fock determinant.

The excitation-classification produces the CI-Singles-(S), CI-Singles-Doubles-(SD), CI-Singles-Doubles-Triples-(SDT), etc. wave functions and corresponding electronic energies E_{el}^{CIS}, E_{el}^{CISD}, E_{el}^{CISDT}, etc. The higher the included excitations are, the more accurate the electronic energy is. However, this does not guarantee that the essential part of the correlation energy is captured in a computationally feasible model like CISD. Improved CI methods utilize a selected set of reference Slater determinants, which are then all included in the excitation process. The preselection of determinants guarantees that the most important determinants are incorporated as basis functions in the CI expansion. The most prominent scheme is the multi-reference doubles-CI method (MRD-CI) by Peyerimhoff and Buenker [61–63].

A conceptually straightforward improvement on the CI approximation is to reoptimize the molecular orbitals for a truncated CI expansion. This approach is called multi-configuration self-consistent field method (MCSCF) and its most prominent variant is the complete active space SCF method (CASSCF) [64]. In the first generation of MCSCF methods [65, 66], the CI coefficients C_I in Eq. (12.9) are calculated from the CI eigenvalue problem for a given set of molecular orbitals, then the orbitals are re-optimized for these CI coefficients. This process is iterated until convergence is reached. Convergence is usually slow. The second generation of MCSCF methods therefore optimizes CI coefficients and orbitals simultaneously where the Hessian of the total energy with respect to the variable parameters (CI coefficients and molecular orbitals) is also utilized [67, 69].

In order to improve on the result of a truncated CI expansion, one may set on top of a CI-type calculation, a perturbation theory calculation to cover the missing small contribution to the correlation energy. Accordingly, CASSCF plus second-order perturbation theory is called CASPT2, CI plus second-order perturbation theory is called MR-MP2, and so on.

The *first-principles* description of supermolecules requires particularly efficient computational methods and algorithms. None of the post-Hartree–Fock methods metioned so far bears this feature but they provide a reliable description of intermolecular forces. The CASSCF approach, for instance, scales exponentially which prohibits its use for large supermolecules. A solution to this problem came from a different direction. This other approach to the electron correlation problem is density functional theory (DFT). The reason for the extensive use of contemporary DFT is its low cost while giving a comparatively good description of the electronic structure.

The basic steps for the derivation of electronic energies within DFT are given in the following: The total ground state energy E_{el} of the interacting electrons of a molecule with fixed nuclei at positions $\{\mathbf{R}_I\}$ can be calculated exactly according to the first Hohenberg–Kohn theorem [70] solely by using the electronic density $\rho(\mathbf{r})$,

$$\rho(\mathbf{r}_1) = N \int d^3r_2 \cdots \int d^3r_N |\Psi_{el}(\mathbf{r}_1, \mathbf{r}_2, \ldots \mathbf{r}_N)|^2. \tag{12.10}$$

Hence, the electronic energy is a functional of the density $E_{el} = E_{el}[\rho(\mathbf{r})]$ instead of a functional of the orbitals $E_{el} = E_{el}[\{\psi_i(\{\mathbf{r}\})]$ (see Eq. (12.6) for the explicit expression in Hartree–Fock theory). However, the formally exact principles of DFT are connected with the drawback that the energy expression as a function of the density is not known. It can be approximated and is usually divided into different physically meaningful contributions (i.e. an electron–electron interaction energy functional is formulated, which is decomposed in a classical Poisson-type electrostatic interaction term and a rest term – the exchange–correlation functional – which is not known).

In practice, almost all DFT calculations today are carried out within Kohn–Sham DFT [71], where the density is expressed by a set of orbitals. These Kohn–Sham orbitals are elements of a Slater determinant, which exactly describes a non-interacting fermionic N-particle substitute system with density ρ^{KS}. The density ρ^{KS} shall now exactly match the electronic density ρ of the molecule under consideration. In this way, the electron correlation problem is moved from the difficulty to construct the electronic wave function Ψ_{el} to finding the unknown exchange–correlation energy functional. For a non-interacting fermionic system, which contains only local potential energy operators in the Hamiltonian, a single Slater determinant represents the exact wave function. The electronic density of a single Slater determinant can easily be computed from Eq. (12.10) to become the sum of the squared molecular orbitals,

$$\rho^{KS}(\mathbf{r}) = \sum_{i=1}^{N} |\psi_i^{KS}(\mathbf{r})|^2, \tag{12.11}$$

(for the sake of brevity, we again neglected spin degrees of freedom and accordingly omitted the occupation numbers so that the sum is taken over N spin orbitals).

For our purposes here, it is only important to know that Kohn–Sham DFT is algorithmically very similar to Hartree–Fock theory. Consequently, it is computationally much cheaper than the post-Hartree–Fock correlation methods described above, but it includes electron correlation effects through an explicit but unknown potential energy term, the exchange–correlation energy functional $E_{xc}[\rho]$. The Kohn–Sham energy E_{el}^{KS} reads,

$$E_{el}^{KS}[\{\psi_i^{KS}\}] = T[\{\psi_i^{KS}\}] + \int d^3r \sum_{K=1}^{M} v_{eK}(\mathbf{r})\rho(\mathbf{r}) + \frac{1}{2}\int d^3r \hat{v}_C(\mathbf{r})\rho(\mathbf{r})$$
$$+ E_{xc}[\rho] + \sum_{K=1}^{M}\sum_{K'>K}^{M} \hat{v}_{KK'}. \tag{12.12}$$

This energy is an explicit functional of a set of the auxiliary functions, namely the Kohn–Sham orbitals $\{\psi_i^{KS}\}$. The first term in the Kohn–Sham functional Eq. (12.12) is the kinetic energy T of the non-interacting reference system,

$$T[\{\psi_i^{KS}\}] = \sum_{i=1}^{N} \int d^3r \psi_i^{KS,\star}(\mathbf{r}) \left[-\frac{\hbar^2}{2m_e}\right] \nabla^2 \psi_i^{KS}(\mathbf{r}), \tag{12.13}$$

which is in form identical to the Hartree–Fock kinetic energy. The $E_{xc}[\rho]$ term in the Kohn–Sham energy functional is the exchange–correlation density functional. The electronic exchange and correlation effects are lumped together and basically define this functional as the remainder of the exact DFT energy and its Kohn–Sham decomposition. In the last term $\sum_{K=1}^{M}\sum_{K'>K}^{M} \hat{v}_{KK'}$ the interaction energies of the bare nuclear charges are added.

As in the case of Hartree–Fock theory, the Kohn–Sham orbitals ψ_i^{KS} are obtained from SCF equations,

$$\left\{-\frac{\hbar^2}{2m_e}\nabla^2 + \sum_{K=1}^{M} \hat{v}_{eK}(\mathbf{r}) + \hat{v}_C[\rho^{KS}(\mathbf{r})] + \frac{\delta E_{xc}[\rho^{KS}]}{\delta \rho^{KS}(\mathbf{r})}\right\} \psi_i^{KS}(\mathbf{r}) = \varepsilon_i^{KS} \psi_i^{KS}(\mathbf{r}), \tag{12.14}$$

which are the canonical Kohn–Sham equations. They ensure that the solution functions ψ_i^{KS} yield a minimum of the Kohn–Sham energy functional E_{el}^{KS}. These equations are one-particle equations involving an effective one-particle Hamiltonian $\hat{h}^{KS}(\mathbf{r})$, which is the collection of operator terms within the braces in Eq. (12.14).

In practice, only approximate expressions are known for the exchange–correlation density functional $E_{xc}[\rho]$ and the most important ones stem from two classes: the 'generalized gradient approximation' (GGA) functionals also called 'gradient-corrected' functionals and their combination with the exact exchange energy expression $\sum_{i=1}^{N} \int \psi_i^{KS,\star} \hat{v}_x \psi_i^{KS}$ known from Hartree–Fock theory denoted hybrid functionals. The approximate nature of these functionals manifests, for example, in the fact that $-1/r^6$ dispersion terms in weak complexes are not described properly. If the exact exchange–correlation functional would be known, every interaction would be described exactly (Hohenberg–Kohn theorem). However, the electronic density of two fragments decays exponentially and the contemporary functionals do not provide a small but important attractive binding energy assigned to a dispersion interaction.

To increase the accuracy of the present-day exchange–correlation functionals for problems of van der Waals type various cooking recipes have been tested: meta-GGAs, which in addition to the density and its gradient include also second derivatives (for examples see Refs. [72, 75]), GGAs in combination with an (empirical) $1/r^6$ van der Waals correction term, the weighted density approximation [76, 77] and time-dependent density functional theory [72]. GGAs and $1/r^6$ van der Waals corrections terms have already been applied to correct the Hartree–Fock energy, see Ahlrichs et al. [78]. Based on Anderson, Langreth and Lundqvist [79] the van der Waals term is corrected for both separated uniform electron gas and separated atoms with an explicit $1/r^6$ term. Others [80–84] include an empirical $1/r^6$ term. Johnson and Becke developed a correction scheme based on the idea of parameterizing the $1/r^6$ contribution generated by the instantaneous dipole moment of the exchange hole [85]. Roethlisberger and coworkers proposed a correction term based on pseudopotentials tailored for molecular dynamics simulations [86, 87]. Another approach (DFT-D) to semi-empirically correct for the lack of weak forces in contemporary DFT was introduced by Grimme [81]. This empirical scheme was successfully applied by Parac et al. who presented a quantum chemical study of host–guest systems with dimethylene-bridged clips and tetramethylene-bridged tweezers as host molecules and six different aliphatic and aromatic substrates as guests [88]. The DFT-D scheme [81] provides accurate geometries, which compare well with crystal structures. By a partioning of the host into molecular fragments it was shown that the binding energy is clearly dominated by the dispersion interactions of the aromatic units of the clip. An energy decomposition analysis of the interaction energies of some tweezer complexes revealed the decisive role of the electrostatic and dispersion contributions for relative stabilities (compare also Section 12.3). With the help of these calculations [88] it became evident that the benzene-spaced tweezer is a better receptor for aliphatic substrates than its naphthalene analog which possesses a better topology for the binding of aromatic substrates.

Apart from these supramolecular approaches to the interaction energy, advanced perturbation theory techniques, which combine symmetry-adapted perturbation theory and DFT and are thereby able to capture dispersion interactions, are also under development [89–91].

12.2.2
The Supramolecular Approach and Total Interaction Energies

In the last section, we introduced methods which can be employed to calculate the total electronic energy E_{el} for a given molecule with fixed nuclei. The stepwise calculation of electronic energies for various nuclear configurations allows us to scan essential parts of the potential energy hypersurface (the so-called Born–Oppenheimer surface) on which nuclear motions and chemical reactions take place. For a simple thermodynamic picture of chemical reactions, it is necessary to localize stationary points on this hypersurface, which are the valleys of reactants and products. Then follows a subtraction of these energies in order to assign the energy difference to the intrinsic (electronic) reaction energy. Vibrational and other corrections can be included afterwards to account for finite-temperature effects through a so-called frequency analysis (see, e.g. [46, 47, 92, 93]).

Interaction energies are thus well-defined, but more than one definition is possible as we shall see in the following. For modeling the interactions of molecules in a supramolecular assembly one may define a chemical reaction like

$$\text{host} + \text{guest} \longrightarrow [\text{host–guest}]_{\text{complex}}$$

with associated (purely electronic) reaction energies calculated via *first-principles* methods. For the energetic classification of the optimized structures, the (standard) expressions for interaction energies can be defined. Intrinsic interactions energy at 0 Kelvin, E_I, can be calculated within the supramolecular approach

$$E_I = E_{el,\,\text{complex}} - \left[E_{el,\,\text{host}} + E_{el,\,\text{guest}}\right]_{\text{frozen}} \tag{12.15}$$

in order to determine the total strength of the various bonding patterns that clamp together the guest and its host. The absolute energies E_{el} are directly computed with quantum chemical methods. The explicit expression for the calculation of the absolute energy depends on the electronic structure method applied (as discussed explicitly in the last section for the Hartree–Fock and Kohn–Sham DFT energy expressions in Eqs. (12.6) and (12.12), respectively).

Interaction energies calculated at *unrelaxed* fragment structures describe the intrinsic interaction strength. These energies neglect intramolecular relaxation effects of host and guest, which would reduce the interaction energy. For comparison of different conformers, one may calculate differences of interaction energies,

$$\Delta E_{I,\,i} = E_{I,\,i} - E_{I,\,\text{ref}} \tag{12.16}$$

with respect to a selected reference structure denoted 'ref', which may be, e.g. the absolute minimum structure in a sequence of configurations under consideration. (Reaction) energies for adiabatic interactions, which also incorporate the structural relaxation of the fragments, can be defined as differences between total electronic energies of minimum structures,

$$D_e = E_{el,\,\text{complex}} - \left[E_{el,\,\text{host}} + E_{el,\,\text{guest}}\right]_{\text{relaxed}} \tag{12.17}$$

or as relative total energies,

$$\Delta D_{e,\,i} = D_{e,\,i} - D_{e,\,\text{ref}} = E_{\text{complex}_i} - E_{\text{ref}} \tag{12.18}$$

if various complexes i shall be compared energetically. For the calculation of the binding energy D_0, the zero-point vibrational energy (ZPE) differences must be added to the total electronic energy differences,

$$D_0 = D_e + \text{ZPE}, \tag{12.19}$$

which allow the calculation of zero-temperature-corrected relative energies ΔD_0,

$$\Delta D_{0,i} = \Delta D_{e,i} + \Delta \text{ZPE}_i, \tag{12.20}$$

for the comparison of different optimized conformer structures.

The quantum chemical standard model for the inclusion of finite-temperature effects, i.e. for the calculations of enthalpies and entropies at a given temperature, comprises simple model assumptions for the molecule under consideration. The central quantity is the molecular partition function, whose translational, vibrational, and rotational parts are calculated for an ideal gas in a box, for vibrations within the harmonic approximation, and for a classical rigid rotor, respectively. Obviously, this standard quantum-chemical model is far too limited to provide reliable enthalpic and entropic data, especially when the interactions are weak. Also note that the entropy calculated within these approximations does not necessarily account for all entropy contributions stemming from the conformational flexibility of, for instance, a host if many energetically close-lying (near degenerate) local minimum structures exist and only one of these minimum structures is taken into account.

To conclude this section, we reprint in Table 12.1 a comparison of interaction energies for a hydrogen-bonded complex, which models interactions within amide ro-

Tab. 12.1. Total interaction energies E_I of the multiply-hydrogen-bonded complex in Fig. 12.2 in kJ mol^{-1}. BHLYP is a hybrid density functional which features 50% admixture of exact Hartree–Fock-type exchange, AM1 and PM3 are semi-empirical models, which are an efficient approximation to the Hartree–Fock method, and CCSD is a coupled-cluster model.

Method	Basis set	D_e
BHLYP	DZ	−67.3
BHYLP	DZP	−54.8
BHYLP	TZP	−43.9
AM1		−31.4
PM3		−12.1
MP2	TZP	−41.8
CCSD	TZP	−39.3

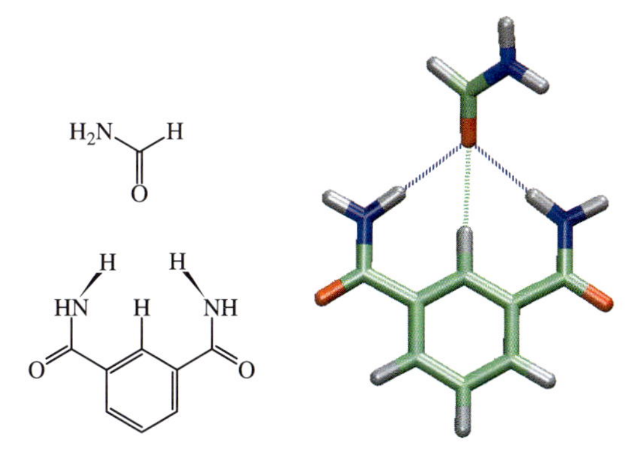

Fig. 12.2. Lewis (left) and optimized structure (right) of a hydrogen-bonded model complex with interaction features as present in rotaxanes.

taxanes (see Fig. 12.2), as calculated with different *first-principles* methods and basis sets taken from Ref. [94]. A typical picture of the range of calculated interaction energies for a multiple-hydrogen-bonded complex emerges from this table.

12.2.3 The Time Dimension: Molecular Dynamics

So far, our discussion has focussed on stationary quantum chemical methods, which yield results for fixed atomic nuclei, i.e. for frozen molecular structures like minimum structures on the Born–Oppenheimer potential energy surface. Processes in supramolecular assemblies usually feature prominent dynamical effects, which can only be captured through explicit molecular dynamics or Monte Carlo simulations [95–98]. Molecular dynamics simulations proved to be a useful tool for studying the detailed microscopic dynamic behavior of many-particle systems as present in physics, chemistry and biology. The aim of molecular dynamics is to study a system by recreating it on the computer as close to nature as possible, i.e. by simulating the dynamics of a system in all microscopic detail over a physical length of time relevant to properties of interest.

For large molecular aggregates only classical mechanics can be applied to describe the nuclear motion of the particles through Newtonian-type equations of motion. This is an approximation, which may only fail if the motion of light particles such as protons needs to be accurately described. If the latter is the case, explicit nuclear quantum dynamics is required. Such extensions that include quantum effects [99, 100] or that incorporate full quantum dynamics have been explored [101].

The equations of motion may be written in various ways. We assume that a system of M particles interacts via a potential V. The particles are described by their positions $\mathbf{R}$ and conjugated momenta $\mathbf{P}$. In the following sets of all positions

$\{\mathbf{R}_1, \mathbf{R}_2, \ldots \mathbf{R}_M\}$ and momenta $\{\mathbf{P}_1, \mathbf{P}_2, \ldots \mathbf{P}_M\}$ are abbreviated as $\mathbf{R}^M$ ($\equiv \{\mathbf{R}_I\}$) and $\mathbf{P}^M$. The equations of motion can be derived from the Euler–Lagrange equations,

$$\frac{d}{dt}\frac{\partial \mathscr{L}}{\partial \dot{\mathbf{R}}_I} - \frac{\partial \mathscr{L}}{\partial \mathbf{R}_I} = 0, \tag{12.21}$$

where the Lagrangian functional $\mathscr{L}$ is defined as kinetic T minus total potential energy V,

$$\mathscr{L}(\mathbf{R}^M, \dot{\mathbf{R}}^M) = T - V = \sum_{I=1}^{M} \frac{1}{2} M_I \dot{\mathbf{R}}_I^2 - V(\mathbf{R}^M), \tag{12.22}$$

where M_I denotes the mass of nucleus I. Alternatively Hamilton's formalism can be applied,

$$\begin{aligned} H(\mathbf{R}^M, \dot{\mathbf{R}}^M) &= \sum_{I=1}^{M} \dot{\mathbf{R}}^M \cdot \mathbf{P}^M - \mathscr{L}(\mathbf{R}^M, \dot{\mathbf{R}}^M) = T + V \\ &= \sum_{I=1}^{M} \frac{\mathbf{P}_I^2}{2M_I} + V(\mathbf{R}^M), \end{aligned} \tag{12.23}$$

with H being Hamilton's function, i.e. the total energy of the system. The force $\mathbf{F}_I$ on particle I is obtained as a partial derivative of the potential V,

$$\mathbf{F}_I(\mathbf{R}^M) = -\frac{\partial V(\mathbf{R}^M)}{\partial \mathbf{R}_I}. \tag{12.24}$$

In practice an analytical function like for example the Lennard–Jones potential is used for the potential, such that the derivative with respect to $\mathbf{R}_I$ can be easily calculated. The equations of motion according to Hamilton's formulation of classical mechanics read [102]

$$\dot{\mathbf{R}}_I = \frac{\partial H}{\partial \mathbf{P}_I} = \frac{\mathbf{P}_I}{M_I}, \tag{12.25}$$

$$\dot{\mathbf{P}}_I = -\frac{\partial H}{\partial \mathbf{R}_I} = -\frac{\partial V(\mathbf{R}^M)}{\partial \mathbf{R}_I} = -\mathbf{F}_I(\mathbf{R}^M). \tag{12.26}$$

From this Newton's second law is obtained by taking the time derivative of the first and equating it to the second part of Hamilton's equations,

$$M_I \ddot{\mathbf{R}}_I = \mathbf{F}_I(\mathbf{R}^M). \tag{12.27}$$

Now numerical integration can be carried out to find the solution, i.e. the trajectory, which represents a collection of sets of nuclear coordinates $\mathbf{R}^M$ for the given time steps. One of the most prominent algorithms for this purpose is due to Störmer and Verlet [95, 96].

In the extended system approach, additional degrees of freedom that control the quantity under consideration can be added to the system [95, 96, 103, 104]. Thereby thermostats or barostats can be simulated [105–107], which provide a quality of modeling thermodynamic processes hardly reached by the stationary quantum chemical methods discussed in the previous two sections.

The description of the forces $\mathbf{F}_I$ is essential for the outcome of simulations [108, 109]. In traditional molecular dynamics simulations, the forces on the particles are obtained as derivatives of the potential, which is usually constructed in a pairwise additive fashion,

$$V(\mathbf{R}^M) = \sum_{I}^{M} \sum_{J>I}^{M} v(\mathbf{R}_{IJ}), \tag{12.28}$$

for the sake of simplicity. Additional many-body correction terms are more difficult to derive and to parametrize in traditional force-field molecular dynamics (compare also Section 12.3.1). One possible solution to this problem is to use polarizable force fields that make up for the neglect of such effects. A second way to circumvent this problem is to use *first-principles* simulations which explicitely include cooperative effects in the description of the system by definition. This is because in *first-principles* simulations the whole electronic structure of the system is known for each nuclear configuration at the corresponding time step (see below).

As an example for a traditional MD study of template-assisted crown ether synthesis we refer to the work by Oh and coworkers [110] who carried out a systematic study of the template effects involved in the synthesis of monobenzo-15-crown-5-ether with the aid of molecular dynamics simulations. They investigated the effect of different metal ions, namely Li^+, Na^+, K^+, Rb^+, and Cs^+. The simulations for each ion at different temperatures show that Li^+ and Na^+ have exerted a markedly favorable effect for keeping the two pertinent atoms of the ether in close proximity. For Li^+ and Na^+ this is independent of temperature, but complexation for all other metals is strongly temperature dependent. The authors also employed Hartree–Fock calculations to determine the minimum energy reaction path of a model complex from which they determined a characteristic distance for the ring being in a closure state [110]. Calculation of the fraction of time in which the ring would stay intact with the help of the closure distance showed that the fraction of this closure-state time decreases in the following order: $Li^+ \approx Na^+ \gg K^+ \approx Rb^+ \approx Cs^+$.

First-principles simulations are techniques that generally employ electronic structure calculations “on the fly”. Since this is a very expensive task in terms of computer time, the electronic structure method is mostly chosen to be density functional theory. Apart from the possibility of propagating classical atomic nuclei on the Born–Oppenheimer potential energy surface represented by the electronic energy $V(\mathbf{R}^M) \equiv E_{el}(\mathbf{R}^M)$, another technique, the Car–Parrinello method, emerged that uses a special trick, namely the extended Lagrangian technique. The basic idea of the Car–Parrinello method is to map a two-component classical/quantum system onto a two component purely classical system. This means that the electrons, which require a quantum mechanical description, are now also treated according to classical mechanics. The extended Lagrangian proposed by Car and Parrinello [111] reads

$$\mathscr{L}_{\mathrm{CP}}[\mathbf{R}^M, \dot{\mathbf{R}}^M, \{\psi_i^{\mathrm{KS}}\}, \{\dot{\psi}_i^{\mathrm{KS}}\}] = \sum_{I=1}^{M} \frac{1}{2} M_I \dot{\mathbf{R}}_I^2 + \sum_i \frac{1}{2} \mu \langle \dot{\psi}_i^{\mathrm{KS}} | \dot{\psi}_i^{\mathrm{KS}} \rangle - E_{el}^{\mathrm{KS}}[\{\psi_i^{\mathrm{KS}}\}; \mathbf{R}^M]. \tag{12.29}$$

Note that the extended Kohn–Sham energy functional E^{KS} is dependent on the orbitals $\{\psi_i^{\mathrm{KS}}\}$ and implicitly on $\mathbf{R}^M$ through the electron–nucleus attraction terms as well as through the nuclear repulsion energy. The corresponding Newtonian equations of motion are derived from the associated Euler–Lagrange Eqs. (12.21) with $\mathscr{L} = \mathscr{L}_{\mathrm{CP}}$ for the nuclear positions and for the orbitals,

$$\frac{d}{dt} \frac{\partial \mathscr{L}_{\mathrm{CP}}}{\partial \dot{\psi}_i^{\mathrm{KS},\star}} = \frac{\partial \mathscr{L}_{\mathrm{CP}}}{\partial \psi_i^{\mathrm{KS}\star}}. \tag{12.30}$$

This yields the Car–Parrinello equations of motion,

$$M_I \ddot{\mathbf{R}}_I(t) = -\frac{\partial E^{KS}}{\partial \mathbf{R}_I} + \sum_{ij} \frac{\partial}{\partial \mathbf{R}_I} \tilde{\varepsilon}_{ij} \int \mathrm{d}^3 r \psi_i^{\mathrm{KS}\star} \psi_i^{\mathrm{KS}}, \tag{12.31}$$

$$\mu \ddot{\psi}_i^{\mathrm{KS}}(t) = -\frac{\delta E^{KS}}{\delta \dot{\psi}_i^{\mathrm{KS}\star}} + \sum_j \tilde{\varepsilon}_{ij} \psi_i^{\mathrm{KS}}, \tag{12.32}$$

for Car–Parrinello molecular dynamics (CPMD) simulations, where μ is a new parameter, namely the so-called fictitious mass or inertia parameter assigned to the orbital degrees of freedom. A broad discussion on how to adjust this mass parameter and a detailed introduction to CPMD can be found, for instance, in Ref [101] – compare also Ref. [112].

In Section 12.2.2 above, we considered the standard model of quantum chemistry for the calculation of thermodynamic functions and their temperature dependence. In the quantum chemical standard model it is particularly difficult to assess the entropy loss stemming from the reduction of configurational degrees of freedom upon recognition of a guest by its host. Loosely speaking, the time dimension in molecular dynamics allows one to use the configurational space scanned by the trajectory for the calculation of entropic contributions although this is also notoriously difficult in MD simulations (see Refs. [113–115] for approaches to estimate entropic contributions in molecular dynamics simulations and Refs. [116, 117] and, in particular, Ref. [118] for applications). Instead of the entropy, it is usually easier to calculate free energies in molecular dynamics simulations (for instance, via thermodynamic integration or umbrella sampling [95]; see also the most recent work by Kästner and Thiel [119]).

12.2.4
A Technical Note: Linear Scaling and Multiscale Modeling

Because of the size of supramolecular assemblies, one aims at the development of *first-principles* methods which scale linearly or sub-linearly with system size for technical reasons. This means that if the size of the supramolecular assembly is doubled this will only require twice as much computer time. The physical reason why one may assume that *first-principles* methods should scale linearly is the fact that systems which show little "electronic coupling" like chains or assemblies of individual molecules or even polypeptide chains should require only an additive amount of computer time upon system enlargement resulting in a linear scaling quantum chemical method. However, standard *first-principles* formalisms do not feature linear scaling. For instance, the calculation of the two-electron Coulomb and exchange integrals resulting from the $\hat{v}_C$ and $\hat{v}_x$ operators in Hartree–Fock theory scales like m^4 with m being the total number of basis functions in the basis set of Eq. (12.7) [53]. This is the computer time-determining step as all other steps require less effort in Hartree–Fock calculations (e.g. the diagonalization of the Fock matrix $\mathbf{f}$ in the Roothaan equations scales like m^3). Consequently, if the size of a supramolecular assembly is doubled, a Hartree–Fock calculation would take about 16 times longer $[(2m)^4 = 2^4 \times m^4 = 16 \times m^4]$. Hence, it is important to introduce technical tricks to reduce the costs of *first-principles* calculations. In this respect, many different technical improvements have been achieved within the last decade and we mention some of them in this sub-section.

Before we list methods which aim at a reduction of the scaling without the introduction of additional approximations, we should mention that very early approaches involved the introduction of crude approximations to make *first-principles* methods computationally feasible. These approaches are the so-called semi-empirical methods. Various integrals occuring in the expression of the Hartree–Fock energy in Eq. (12.6) are then neglected or approximated using data from experiment. Although empirical data are used to parametrize the large number of multi-electron integrals, semi-empirical methods are still quantum mechanical in nature. These approximations limit, however, the precision of semi-empirical methods, particularly when studying molecules that were not present in the initial parameterization procedure. Among the most widely used semiempirical methods are MNDO (Modified Neglect of Differential Overlap) [120], AM1 (Austin Model 1) [121], and PM3 (Parametric Model number 3) [122]. These methods are parametrizations of the Neglect of Diatomic Differential Overlap (NDDO) integral approximation [47].

Owing to the size of supramolecules semi-empirical methods are still in use. For instance, semi-empirical PM3 calculations were carried out for metalacryptands and metalacryptates and with lead as metal the theoretically predicted favorable cryptand formation was subsequently verified experimentally [123]. In another example, the rate of shuttling motions in rotaxanes was examined with semi-empirical AM1 calculation by Ghosh et al. [124] who generated different structures with a Monte Carlo procedure and subsequently optimized low-energy conformers.

A more accurate treatment requires to reduce the size of the system and to use model compounds. E.g. Schalley and coworkers [125, 126] carried out DFT calculations in order to gain insight into the details of the hydrogen bond patterns involved in the formation of mechanically interlocked species such as amide rotaxanes, catenanes, and knots.

A major concern connected with the application of semi-empirical methods to supramolecular problems is that these methods were parameterized to reproduce molecular rather than intermolecular properties. Over the last few years there have been some efforts to improve treatment of the core parameters in semi-empirical methods that play a large role in the nonreproducibility of experimental data [127]. A special problem of semi-empirical methods is that they present an unphysical stabilization effect for short-range $H \cdots H$ interactions: see the analysis of intermolecular $H \cdots H$ interactions in supramolecular chemistry [128] and the history of semi-empirical calculations for this purpose cited therein.

Coming back to linear-scaling *first-principles* calculations without introducing severe approximations, we first note that quantum chemists distinguish two measures of scaling, a formal and an asymptotic scaling (compare also the review by Goedecker [129]). While the first refers to an implementation that ignores natural sparsity, the second applies to large systems where sparsity of matrices can be fully exploited (see, for instance, Refs. [130–136] for examples). With the advent of methods enabling the construction of the Fock matrix to be done with a computational effort that scales linearly with system size, the diagonalization step for solving the Roothaan equations becomes the new computational bottleneck. It is, however, possible to reformulate the SCF problem in terms of a minimization of an energy functional, which depends directly on the density matrix elements, by conjugate gradient methods taking advantage of the fact that the density matrix is sparse for extended systems [137, 138].

The steep fourth-order scaling of the Coulombic term reduces naturally to m^2 in the asymptotic limit, while the scaling of the exchange term is asymptotically linear in insulators. However, linear scaling is reached only for very large systems. Screening and fast multipole techniques have reduced the asymptotic quadratic scaling of the Coulomb term to linear or near-linear ($m \log m$) scaling. Another technique expands the charge distribution into auxiliary basis function and is called density fitting or resolution-of-the-identity technique [139, 140]. Wavefunction-based correlation methods can also be implemented in a linear-scaling fashion using the idea of locally correlating orbitals, which has largely been developed by Werner and collaborators [141–143]. Of course, these techniques apply best when the electronic structure is also local rather than delocalized over a large spatial region (compare a polypeptide versus a graphite-like supermolecule as synthesized in the group of Müllen [144]).

In multiscale modeling approaches the microscopic behavior of a system is linked by a compression operator to the macroscopic state variable. The strategy is then to use the microscopic model to provide necessary information for extracting the macroscale behavior of the system. Such a combined macro-microscale technique is supposed to be much more efficient than solving the full microscopic model in detail (in the volume by Attinger and Koumoutsakos [145] various methods of multiscale modeling to bridge different length and time scales are discussed).

An introduction to coarse graining in the context of molecular dynamics, Monte Carlo, and other techniques to be applied to soft matter was given in a series of lectures published by the John von Neumann Institute for Computing [21]. Such large scale techniques [21, 95, 96] are also applied to supermolecules. For example, Peroukidis et al. [146] introduced a simple molecular theory that relates the self-organization observed in certain systems to their molecular structure. The interactions are modeled by subdividing each molecule into a number of submolecular blocks. Thereby the phase diagram of fullerene-containing liquid crystals could be understood in terms of a simplified coarse-grained model.

Huge systems like DNA are not accessible by traditional *first-principles* algorithms as employed in electronic structure theory. However, *first-principles* electronic structure methods are often needed in order to achieve the necessary level of accuracy for either benchmark calculations that may serve as a reference or in cases where a detailed molecular picture is mandatory. It is therefore desirable to further develop *ab initio* and DFT methods in the context of multiscale modeling [147]. Examples for extended *first-principles* CPMD calculations on electronic and optical properties of DNA and on the reactivity of radical cations can be found in Refs. [148–150].

In a different approach one aims at a combination of fast molecular mechanics (MM) force-fields with a quantum mechanical (QM) description of the relevant spatial region. See Refs. [151–155]) for reviews of these QM/MM techniques.

12.2.5 How to Make the Connection to Experiment?

We have seen in the preceding sections that it is possible to explicitly model the molecular energetics and dynamics of a supramolecular assembly. Here, we shall briefly comment on how to calculate physical quantities that can be directly compared with experimentally measured observables, which is essential if theoretical descriptions and predictions are to be of any meaning. Of course, this endeavor is not special to supramolecular chemistry but is a general task in theoretical and computational chemistry. In supramolecular chemistry, however, new complications arise owing to the many particles involved. Therefore, it is necessary to selectively extract the relevant information needed to describe the features of the supramolecular assembly.

But note that it is not only the proper choice of theoretical methods that determines the accuracy and predictive power of the results obtained. The latter is also largely determined through the molecular model set up. A theoretical description necessarily requires a structure model that should be designed so as to resemble the "real" system as closely as possible. Nevertheless many calculations are carried out on isolated supramolecular assemblies thus neglecting all environmental effects such as solvent effects. Of course, solvent effects can be decisive which has long been known from organic chemistry [156] but the study of the intrinsic effects in an isolated system is possible experimentally in gas-phase experiments like those using a Fourier-Transform Ion Cyclotron Resonance spectrometer (accordingly, technical papers then often missleadingly call the isolated-system calculations as "calculations in the gas-phase"). As a special example for a theoretical study with particular emphasis on the role of a solvent we mention the work by Wipff and coworkers who considered the behavior of host–guest ion complexes at the water/supercritical-CO_2 interface [157]. Their paper focused on the aspect of self-assembly, i.e. at the local increase of concentration at the interface between the two solvents. The authors find two main features: (i) host–guest interactions between macrocyclic molecules and hydrophilic ions allow the transport of the latter from the aqueous to hydrophobic medium, and (ii) formation of supramolecular structures occurs at the interface [157].

It is important to note that solvent molecules can build attractive contacts to supramolecular assemblies of the same strength as within the supramolecular assembly. The appropriate modeling of these effects is then decisive. In principle, two simple approaches are possible: (i) microsolvation where individual solvent molecules are added to the isolated supramolecular assembly and (ii) dielectric continuum models where the solvent is viewed in a static and statistical manner as a dielectric continuum. Roughly speaking, the first approach neglects the fact that individual solvent molecules in optimized structures are not an appropriate model of a fluctuating bulk of solvent molecules while the second approach neglects the fact that classical electrostatic models do not fully capture strong individual and partially covalent contacts of solvent molecules to the supramolecular assembly (e.g. through hydrogen bonds). An optimum choice for modeling solvent interaction is a *first-principles* molecular dynamics approach with periodic boundary conditions as provided by CPMD. For discussion of solvent effects in complex molecular systems with static and dynamic theoretical methods we refer the interested reader to the review in Ref. [158].

The most important information to be extracted from either experiment or theory is *structural* information. In *first-principles* methods this information is readily available. Experimentally, it may be extracted directly from diffraction techniques. While X-ray diffraction provides information on solids, which means on "frozen" structures, other techniques like neutron diffraction can provide statistical information like pair correlation functions that are also easily obtained from a molecular dynamics trajectory.

The next important quantity is the interaction energy, which may be obtained experimentally through the measurement of association or complexation constants K (for instance, via Fourier-Transform Ion Cyclotron Resonance mass spectrometry or Nuclear Magnetic Resonance (NMR) spectroscopy; see also the other chapters in this volume) that are related to the total free enthalpy change ΔG of the host–guest complexation through

$$\Delta G = -RT \ln K. \tag{12.33}$$

Also, various spectroscopic quantities can be calculated in order to test experimental assumptions: Once a structure of a supramolecular assembly has been assumed, optimized or propagated in time, properties like vibrational frequencies, infrared, Raman [93], or Resonance Raman [159] intensities, NMR or EPR parameters can be calculated with *first-principles* methods to be compared with the experimentally measured spectra in order to confirm or reject the structural basis assumed in the interpretation of the experimental spectra. It is impossible to review the work and achievements of theoretical chemistry in this respect. Therefore, we concentrate on selected examples in the following. The interested reader is referred to the book by Kaupp, Bühl and Malkin [160] for the calculation of NMR and ESR parameters and to Refs. [161, 162] for more general discussions of molecular property calculations. NMR parameters are molecular properties probed at atomic nuclei and thus ideal for linear-scaling or empirical approaches. An efficient linear-scaling method for supramolecular systems has been presented recently [163].

Not all molecular properties are, however, of local nature, which lend themselves to efficient computational schemes. Molecular vibrations are typically non-local and delocalized. Nevertheless, it is possible even in such cases to design a tailored quantum chemical method, the Mode-Tracking protocol [164], for the selective calculation of only those vibrations relevant for a certain scientific context. For the selective calculation of various types of vibrational spectra we refer the interested reader to the reviews in Refs. [165, 166]. It should be mentioned that molecular dynamics simulations offer different routes to spectra through autocorrelation functions.

The Mode-Tracking idea originated from the fact that the standard quantum chemical calculation of vibrational spectra within the harmonic approximation requires the calculation of the complete Hessian matrix [92]. The Hessian is the matrix of all second derivatives of the electronic energy E_{el} with respect to the nuclear coordinates $\mathbf{R}^M$. Its calculation gets more computer time demanding the larger the molecule is. However, many if not most vibrations of a supramolecular assembly are of little importance for the function and chemical behavior of this assembly. To provide a solution to the problem of the seemingly inevitable calculation of *all* vibrational modes independent of the size of the system, a formalism was suggested [164], which turns the standard quantum chemical procedure upside down. The central idea of this mode-tracking principle consists of two parts: First, one defines a suitable guess vibration, which maps the scientific issue one is interested in onto one or more selected collective distortions of the equilibrium positions of the atomic nuclei. For example, tracking a stretching vibration simply requires to start with a simple bond elongation. These guess vibrations represent the first approximation for the sought-for normal vibration. Second, the mode-tracking principle requires a refinement step, in which the terms missing in the guess vibrations are constructed automatically until the converged normal mode can be expressed as a superposition of the guess and additional orthogonal basis vectors, which are constructed from the residual vectors. The convergence of these calculations is in general very good as we have shown in an extensive study [167].

As an example, Fig. 12.3 depicts a Zn-porphyrin–potassium-crown-ether compound for which the coordination mode of the cyanide ligand could not be un-

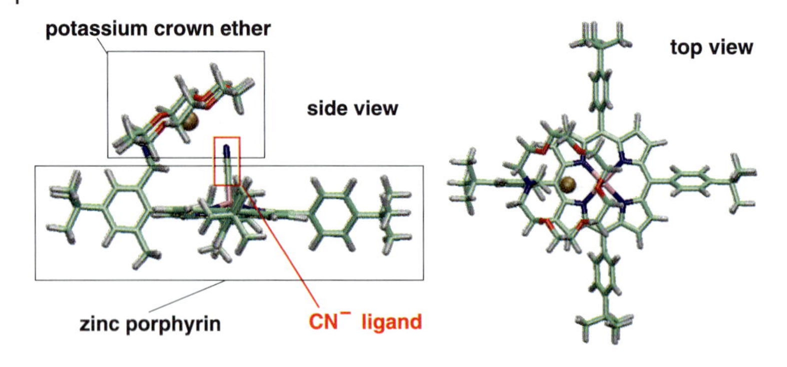

Fig. 12.3. Top and side views of the CN^- isomer of the Zn-porphyrin–potassium-crown-ether compound discussed in the text.

ambiguously identified by X-ray spectroscopy. It was thus uncertain whether CN^- binds via nitrogen (NC isomer) or via carbon (CN isomer) to the Zn atom. In order to investigate this case we ran mode-tracking calculations for the C≡N stretching vibration [168]. Note that it was not necessary to simplify the substitution pattern of the porphyrin in this efficient approach.

The two isomers are energetically separated by 24.2 kJ mol^{-1} (although this is the pure electronic effect, i.e. at 0 Kelvin without zero-point vibrational energy correction, the energy difference is not likely to change upon inclusion of temperature effects on entropy and enthalpy owing to the similar structure of the two isomers). Based on the relative energy of both isomers, the conclusion was drawn that the CN isomer is the one, which has been obtained in experiment [168]. The subsequent mode-tracking calculations converged fast within only two iterations (starting from a pure CN bond elongation as a guess for the stretching mode) to the harmonic wavenumbers. For the CN and the NC isomer, we obtained 2152.8 cm^{-1} and 2133.1 cm^{-1}, respectively [168]. The difference of about 20 cm^{-1} is not significantly large in order to distinguish the two isomers from one another within the quantum chemical methodology employed, as they depend on the harmonic approximation as well as on the density functional and basis set chosen. The experimental IR spectrum shows a very weak peak at 2131 cm^{-1} and it is thus tempting to assume that this stems from the NC isomer [168]. However, one should keep in mind that the calculated frequencies obtained within the harmonic approximation should deviate from experiment. But we might make use of more information obtained in the vibrational spectrum, namely the infrared intensities. The vanishing peak in the experimental IR spectrum is rather unusual for CN^- coordinated to a Zn porphyrin. However, it compares well with the small intensity calculated for the CN isomer. But the intensity calculation for the NC isomer also yields a less intense peak though its intensity is almost twice as big as in the case of the CN isomer. Despite the factor of two, both intensities are small and the energy criterion, which favors the CN over the NC isomer, can be considered decisive.

The Mode-Tracking method has also been successfully employed to understand the intermolecular interactions in methyl lactate clusters produced in a supersonic jet [169] (see Refs. [167, 170–174] for further applications of Mode-Tracking). These clusters are a generic model for intermolecular interactions and chiral recognition. They aggregate via O–H···O–H or O–H···O=C hydrogen bonds. It was shown in Ref. [169] by experiment and MP2 calculations that a tetrameric S_4-symmetric cluster is formed from two R- and two S-units via isolated O–H···C=O hydrogen bonds as the global minimum structure. Enantiopure tetramers prefer cooperative O–H···O–H structures instead. Interestingly, DFT dramatically failed to predict the correct relative energies of various optimized structures – often DFT agrees surprisingly well with MP2 results if molecular aggregates are bound by hydrogen bonds (many references can be found in the literature regarding this issue; see Refs. [36, 175] for examples from our research). The failure of DFT could be due to dispersion forces which may lead to an attraction of the hydrophobic tails of the methyl lactate monomers in the tetramer.

The spectroscopic tool box at our disposal today seems to be infinitely large. As a final example from this tool box, we refer to electronic circular dichroism. Grimme et al. established theoretical tools for the calculations of circular dichroism (CD) spectra of large systems [176]. It is common to employ single structures for such calculations. However, Siering, Grimme and Waldvogel [177] reported a systematic investigation involving combinations of classical molecular dynamics simulations with quantum chemical techniques in order to characterize a chiral receptor that binds prochiral guests. These authors presented the first assignment of highly dynamic enantiofacial discrimination acting on a single heterocyclic substrate with the aid of CD spectroscopy [177].

12.3 Standard Classification of Intermolecular Interactions

Molecular processes in supramolecular assemblies are all governed by the intermolecular interactions of the consistuents of the assembly. The role of these interactions can hardly be overestimated. A main feature of the interactions is that most of them can be reversibly built and broken. Even covalent bonds can exhibit this feature [178]. In a review article by Dance [179] focusing on inorganic intermolecular motifs and their energies, two major topics are the diversity of molecular surfaces and intermolecular interactions (and their energies). The properties and applications of molecular materials are a result or can be understood in terms of the energies of the intermolecular interactions. Dance discusses several definitions of *supramolecular inorganic chemistry*. In some articles supramolecular inorganic chemistry refers to weak intermolecular interactions and their consequences, while in other papers it is connected with the particular compounds exhibiting strong coordinate bonding. Dance proposes as an integrating concept the notion of "controlled assembly of complex matter" [179] and differentiates between "short–strong" and "long–weak" interactions. Spatial regions with long–weak contacts in condensed phase systems define the boundaries of molecules and highlight the difference between inter-molecular and intra-molecular interaction patterns. The absence of molecular boundaries represents structural non-molecularity [179].

Over recent decades, a couple of intermolecular interaction patterns have been identified and refined in order to facilitate understanding of supramolecular processes. The decomposition of a whole interaction pattern into individual contributions is somewhat artificial – only the total interaction energy is well defined – but it is required for our classical, macroscopic view and understanding of these processes. The following list of interaction types contains a few important ones but cannot be considered complete (for instance, magnetic fields or reversibly built and broken covalent bonds are completely neglected):

- *(Electrostatic) monopole attraction and repulsion* of atoms or groups of atoms occur in the presence of local (point) charges. In principle, these classical electrostatic Coulomb interaction are *strong* and *long-ranged*. Chemists tend to identify formal charges on atoms as true point charges. From a quantum chemically point of view, this is in general not true. However, electrostatic monopoles are well-defined in local multipole expansions (see also below).
- *Dipole interactions* are usually weaker than electrostatic monopole interactions but can dominate the intermolecular interactions within a supramolecular assembly. Diederich and coworkers have recently drawn attention onto dipole interactions, and multipolar interactions in general, in such systems based on a statistical analysis of structures [180].
- *Higher and mixed multipole interactions* are also always present and may play a dominant role in the absence of lower multipole moments, especially in the absence of mono- and dipoles. Note also that these electrostatic multipole interactions are purely classic and typical quantum mechanical effects (like Pauli repulsion etc.) are not captured.
- *Hydrogen bonding* [142, 181–186] represents the most famous and most important interaction pattern. It was early recognized [187] that an attractive interaction between hydrogen atoms bound to a given donor atom and an electronegative acceptor atom exists. A standard hydrogen bond can be considered mainly electrostatic in nature though covalent contributions are also present. However, the relative amount of electrostatic contributions depends on the system under consideration but is also dependent on the particular decomposition analysis within which "electrostatic", "dispersion", etc. are defined.

  It is important to note that *weak* hydrogen bonds must not be neglected in multi-valent host–guest interactions as their contribution to the interaction energy can be decisive and can thus determine the structure of a supramolecular assembly (see Refs. [184, 188, 189] for general accounts on weak hydrogen bonds, Ref. [190] for weak C–H $\cdots$ F–C interactions, and Refs. [59, 175] for weak C–H $\cdots$ O= contacts).
- *π–π stacking* denotes an important attractive interaction present when two aromatic systems come into close proximity (see, e.g. Ref. [191]). The exemplary system for π–π stacking is the benzene dimer. The attraction between two benzene

rings is due to dispersion forces. Contemporary DFT methods cannot describe this type of interaction [192, 193] (except if certain precautions are taken as described in Section 12.2). MP2 copes much better with these interactions but tends to overbinding [194–196] (see also the recent accounts in Ref. [197]). A detailed discussion of this dilemma for cyclophanes was presented by Grimme [198].

- *Diffuse dispersive attractions and hydrophobic effects*: One may easily imagine that many different, inhomogeneous electric fields are present in supramolecular assemblies. These fields induce new electric fields, which may also be decomposed into electrostatic multipoles (for instance, dipole moments may be induced depending on the strength of the inducing permanent field and the (local) polarizability of the system, which is the material specific quantity). Interactions based on these effects are usually denoted van der Waals interactions. But note that the induced fields themselves will induce new polarization effects and so on. For a stationary, time-independent equilibrium structure (in the quantum chemical sense) all these effects can in principle be captured at once by the electronic wave function Ψ_{el}.

Especially the weak interaction patterns like van der Waals forces, weak hydrogen bonds (i.e. those with bond energies less than about 10 kJ mol^{-1}) and π–π stackings are often discussed in the light of hydrophobic effects. Such effects are strongly system-dependent, they can hardly be understood in an *ad hoc* fashion, and are thus subject of constant debate (see Refs. [199, 200] for examples and Ref. [201] for a review).

12.3.1
A Complication: Cooperative Effects

The interaction patterns described in the last paragraph are convenient means for a step-wise and additive understanding of molecular recognition processes within supramolecular assemblies. A simple interpretation of cooperative effects would merely consider the action of various attractive interactions at the same time. However, thinking within additive individual contacts silently assumes that the interaction strength is not modulated by the presence of other contacts or even other atoms, which are not involved in such contacts. Thus, the central difficulty, which is associated with the problem that the interactions within a supramolecular assembly cannot be decomposed into single contacts with individual interaction energies, is to account for these cooperative effects.

In general, cooperative or many-body effects can be understood by considering the local polarizability of a given part of the supermolecule. This distributed polarizability is modulated through neighboring atoms and groups of atoms. As a consequence, the interaction strength of the given part is changed. This picture of a modulated local polarizability also lents itself to one possible solution to the problem, which is the use of polarizable force fields in MD simulations (compare, for instance, Refs. [202–213]). However, these polarizable force fields need to be parametrized and cannot be easily obtained for arbitrarily complicated interaction patterns in supramolecular assemblies. By contrast, the *first-principles* methods adjust automatically to a given manifestation of cooperativity defined by a particular nuclear configuration $\mathbf{R}^M$. Hence, wavefunction-based schemes for the assessment of cooperative effects or for the estimation of individual attractive contacts are especially desirable.

The question now arises how all the above-mentioned phenomenologically classified interactions can be quantified. Of course, theory can yield unambiguous results if the additive decomposition of the overall interaction pattern into individual contributions is a suitable approximation in a certain case. It is, however, clear from the outset that many-body effects make a decomposition difficult, although this may be circumvented by a direct reference to the electronic wave function, which automatically adjusts to a given nuclear configuration, i.e. to a given arrangement of atoms. In Ref. [214], for example, an attempt is made to monitor the cooperative action of electrostatics in crown ether hydration via maps of the electrostatic potential.

12.3.2 Distributed Multipoles and Polarizabilities

The description of intermolecular interactions via electrostatic multipoles has a long history and excellent monographs have appeared (see, e.g. Refs. [215, 216]). In principle, one may attempt to decompose a given electronic (and nuclear) charge density distribution in terms of electrostatic multipoles. Naturally, such an expansion is helpful for decomposing and understanding intermolecular interactions at large distance. Leaving the convergence of such a series expansion aside, a single series expansion is of little help when the interactions at shorter distances are to be described since it does not allow for a simple discrimination of different interaction sites. Consequently, *distributed* multipole expansions (see Ref. [215, p. 106] and references cited therein),

$$E_{el.stat.} = \sum_a \sum_b \Bigg[T^{(ab)} q^{(a)} q^{(b)} + \sum_\alpha T_\alpha^{(ab)} \left(q^{(a)} \mu_\alpha^{(b)} - \mu_\alpha^{(a)} q^{(b)} \right) + \sum_{\alpha\beta} T_{\alpha\beta}^{(ab)} \left(\frac{1}{3} q^{(a)} \Theta_{\alpha\beta}^{(b)} - \mu_\alpha^{(a)} \mu_\beta^{(b)} + \frac{1}{3} \Theta_{\alpha\beta}^{(a)} q^{(b)} + \cdots \right) \Bigg] \tag{12.34}$$

where the sums run over sites a interacting with sites b distributed over the supramolecular assembly (with charges q, dipole moments μ_α, quadrupole moments $\Theta_{\alpha\beta}$, etc. where $\alpha, \beta \in \{x, y, z\}$). The distributed multipole expansion needs to fulfill certain constraints (e.g. that they can be mapped on the single multipole expansion mentioned first). It can also easily be written in terms of interaction operators to be added to a Hamiltonian due to the correspondence principle. Accordingly, for the description of induced dipole moments, polarizabilities are usually assigned to the whole molecule while only a distribution scheme may properly account for local redistribution of charges in supramolecular assemblies (the same may be designed for hyperpolarizabilties if nonlinear effects are important).

12.3.3 Local Multipole Expansions in MD Simulations

The idea of distributed dipole moments has also been transferred to the dynamic domain and we shall discuss recent work from our laboratory in this section in more detail. With the help of maximally localized Wannier functions local dipoles and charges on atoms can be derived. The Wannier functions are obtained by Boys' localization scheme [217]. Thus, Wannier orbitals [218] are the condensed phase analogs of localized molecular orbitals known from quantum chemistry. Access to the electronic structure during a CPMD simulation allows the calculation of electronic properties. Through an appropriate unitary transformation $\mathbf{U}$ of the canonical Kohn–Sham orbitals ψ_i^{KS} maximally localized Wannier functions (MLWFs) $w_j(\mathbf{r})$ can be calculated,

$$w_j(\mathbf{r}) = \sum_i U_{ij}\psi_i^{\mathrm{KS}}(\mathbf{r}). \tag{12.35}$$

As proposed by Marzari and Vanderbilt [219], an intuitive solution to the problem of the non-uniqueness of the unitary transformed orbitals is to require that the total spread of the localized function should be minimal. The Marzari–Vanderbilt scheme is based on recent advances in the formulation of a theory of electronic polarization [220, 221]. By analyzing quantities such as changes in the spread (second moment) or the location of the center of charge of the MLWFs, it is possible to learn about the chemical nature of a given system. In particular the charge centers of the MLWFs are of interest, as they provide a classical correspondence to the location of an electron or electron pair.

In condensed phase simulations, the total dipole moment $\mathbf{M}(t)$ of the supercell is often used to calculate the infrared absorption coefficient [222]. However, in electronic structure calculations a straightforward determination of the cell dipole employing charge partition schemes usually fails due to periodic boundary conditions. This problem was solved by the modern theory of polarization [220]. The original approach calculates changes of polarization as a property of the (Berry) phase of the ground state wavefunction using integrals and derivatives of Bloch functions. An equivalent and for the purpose of disordered systems more appropriate real space formulation was developed by Resta [223].

For the analysis of a supramolecular assembly it is most convenient to write the total dipole moment $\mathbf{M}^{el}$ to a good approximation as a sum of individual molecular dipoles μ_I^{el},

$$\mathbf{M}^{el} \approx \sum_I \mu_I^{el}. \tag{12.36}$$

The expectation value $\mathbf{r}_i$ of the position operator for a MLWF i is thus is often called a Wannier function's center (WFC). With this definition the electronic part of the supercell dipole moment reads

$$\mathbf{M}^{el} \approx -2e \sum_i \mathbf{r}_i \tag{12.37}$$

and the electronic part of the molecular dipole moments can be defined as

$$\mu_I^{el} = -2e \sum_{i \in I}^{occ} \mathbf{r}_i, \tag{12.38}$$

where the sum runs over all WFCs associated with the molecule. The dipole time correlation function needed in the calculation of the infrared absorption spectra [222] can be calculated using the molecular dipole moments

$$\langle \mathbf{M}(t) \cdot \mathbf{M}(0) \rangle = \sum_{IJ} \langle \mu_I(t) \cdot \mu_J(0) \rangle, \tag{12.39}$$

(cf. also Section 12.2.5). Analyzing individual terms in the above sum allows a detailed study of the origins of special features in the spectra [224].

We now discuss a recently developed method to derive atomic charges from WFCs [225]. This method is closely related to the D-RESP procedure of the Roethlisberger group [226]. We consider a molecule of M atoms with charges Z_A and atomic positions $\mathbf{R}_A$. The electronic distribution of the molecule is described by n WFCs with charges $-q_w$ at positions $\mathbf{r}_a$. q_w has a value of one for the spin polarized case and a value of two for spin restricted calculations. The electrostatic potential of the molecule derived from the WFCs is defined as

$$V_{\mathrm{ESP}}^{\mathrm{WFC}}(\mathbf{r}) = \sum_A \frac{Z_A e}{|\mathbf{R}_A - \mathbf{r}|} - \sum_a \frac{q_w}{|\mathbf{r}_a - \mathbf{r}|} \tag{12.40}$$

(compare Eq. (12.1)). We then look for a set of charges q_A that reproduce the electrostatic potential of the molecule as closely as possible. The potential is sampled at many positions $\mathbf{r}_i$ outside the molecule. The charges q_A are optimized with a least square fit. The zeroth (total charge) and first (dipole moment) moments of the charge distribution are enforced exactly.

A very important application of the above introduced local dipole analysis scheme will be in the field of supramolecular and template chemistry. Diederich and coworkers [180] note: "For more than two decades supramolecular chemistry

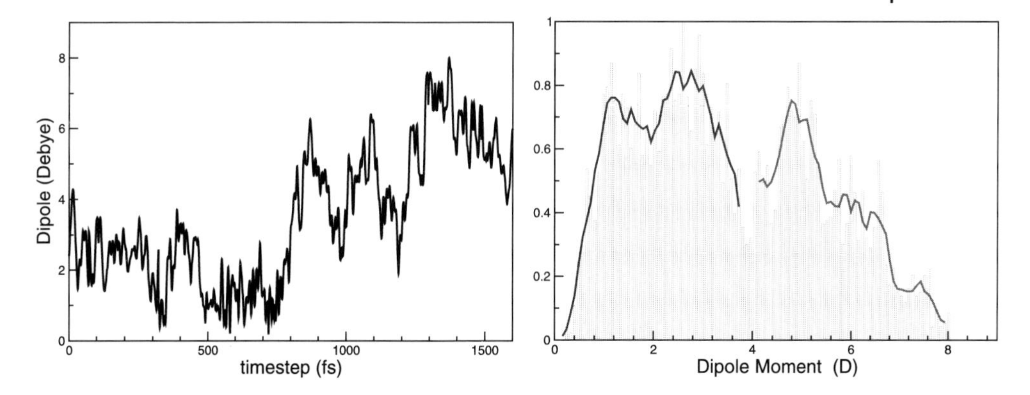

Fig. 12.4. Left: The dipole moment in Debye plotted against the time step. Right: Distribution of dipole moments (left-hand-side region: smaller dipole moments; right-hand-side region: larger dipole moments). Both data sets were calculated from a Car–Parrinello molecular dynamics simulation of an isolated tetralactam macrocycle [227].

has greatly impacted many areas ranging from material science to biomedicine. Intermolecular multipolar interactions have, however, undeservedly been looked at as being too weak, uninteresting, and less important than other nonbonding interactions." and further "Nevertheless, the experimental quantification and the theoretical treatment of these interactions are still underdeveloped and require further elaborate research efforts." Such elaborate research efforts could benefit from the local analysis of the MLWFs. We will show in the following a first attempt of such an analysis applied to a supramolecular system, namely the tetralactam wheel of a rotaxane [227]. We carried out Car–Parrinello simulations with a time step of 0.12 fs. From the total trajectory of 1.5 ps we harvested in every third step the Wannier functions [227].

Figure 12.4 presents the dipole moment along the time step and the distribution of the dipole moment. We observe in the left graph that the dipole drops until the 750th time step is reached. After this the dipole moment suddenly increases. Accordingly, the left panel of Fig. 12.4 shows two remarkable regions: The small dipole region (left hand side) and the large dipole region (right hand side) in the course of the simulation. These changes in dipole moments could influence the intra-supermolecule dynamics of wheel and axle in a rotaxane and thus also the potential shuttling motion of a rotaxane [227]. A closer look at the structure of the macrocycle during the course of a simulation shows one of the amide-groups turned around such that the oxygen atom now points inside the wheel (see Fig. 12.5). At the start of the simulations all oxygen atoms pointed outward the wheel. We may speculate that the motion of the amide bond, which increases the dipole sizably, may induce or increase an important interaction.

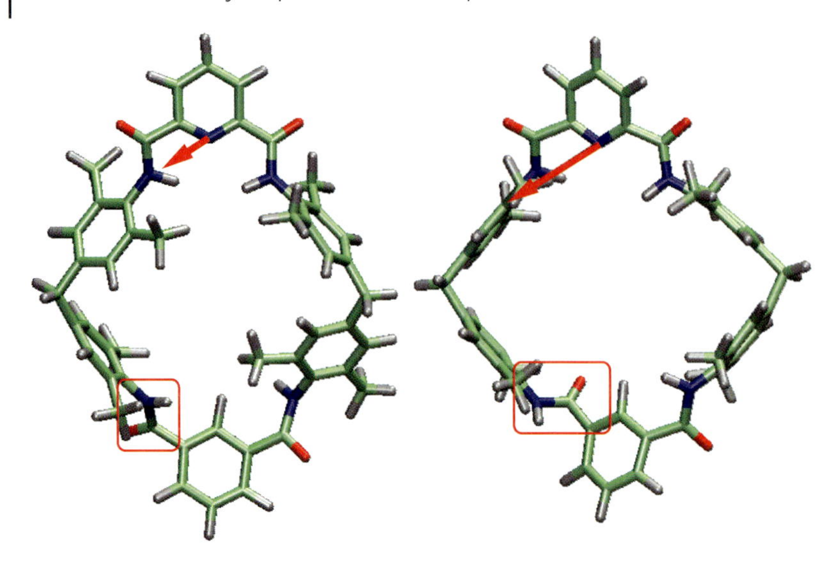

Fig. 12.5. The picture shows the structures of a tetralactam macrocycle. The left structure is from the beginning and the right configuration is from the end of the simulation. The size of the total dipole moment (big red arrow) plotted in each wheel is the average dipole moment.

12.4
Qualitative Understanding and Decomposition Schemes

It is obvious from the discussion in the last sections that the evaluation of distributed multipoles and polarizabilities is not straightforward and also time consuming. In particular, for routine investigations of large extended assemblies it is desirable to have easily obtainable descriptors for local interaction patterns at hand.

Especially molecular dynamics simulations benefit extensively from the interpretation of local interactions [152, 228, 229]. The standard approach for the estimation of local interaction energies in complex aggregates is based on geometric criteria. These solely define the interaction of two fragments of an aggregate on the basis of distances (and occasionally of angles) [228]. It is most desirable to have a single descriptor for the interaction energy at hand. Chandler noted, for example, that attempts towards the quantification of predictions of protein structures with hydrophobic and hydrophilic amino acids "by identifying a single parameter or function that characterizes the strength of hydrophobic interactions have been unsuccessful" [230]. Apart from practical problems with mapping interaction energies onto a single descriptor, we also face fundamental quantum mechanical difficulties: If an aggregate of two subsystems, which interact via more than one site with one another, is decomposed into these two subsystems, the interaction energy for a single attractive site in the aggregate cannot be extracted from the total decomposition energy. A wavefunction-based criterion as opposed to a geometric criterion is desirable because it is sensitive to different environments, in which the hydrogen bond is formed, i.e. it is sensitive to different acceptor atoms, different donor atoms, bifurcated or two-fold hydrogen bonds and other intramolecular or environmental effects (like solvent effects), many-body or cooperative effects. None of this can be detected by standard geometric criteria. First-principles simulations like CPMD allow for new wave-function-based descriptors [231] as the electronic structure is – in addition to the positions of all atomic nuclei involved – available "on the fly". Of course, the above mentioned fundamental problem that the interaction energy is not an observable quantity is in first-principle simulations as apparent as in static calculations. However, the wavefunction naturally tracks all electronic changes in an aggregate. A wavefunction-based descriptor would also be helpful in traditional molecular dynamics because snapshots can be calculated with advanced static quantum chemical methods.

12.4.1
Interaction Energy Decomposition

Decomposition of interaction energies is desired for qualitative chemical analyses of complicated multi-valent interactions in supramolecular aggregates but such a decomposition cannot be uniquely defined within fundamental physical theory. A popular semi-quantitative decomposition method with nice formal features to be mentioned in this context is Weinhold's natural bond orbital (NBO) approach to intermolecular interactions [232, 233]. Comparable is the recently proposed energy decomposition analysis by Mo, Gao and Peyerimhoff [234, 235] which is based on a block-localized wave function. Other energy decomposition schemes proposed are the energy decomposition analysis (EDA) by Kitaura and Morokuma [236] and a similar scheme by Ziegler and Rauk [237].

Morokuma et al. [238] investigated the hydrogen bond with EDA [238] and found that the proton acceptor ability decreases in the order $F > O > N$ due to an increasing term of EDA, which is assigned to exchange repulsion (EX). However, the strength of the hydrogen bond does not solely depend on the exchange repulsion part, but also on the electrostatic part (ES), charge transfer (CT) etc. The electrostatic part and also the other negative components in the EDA are compensating the exchange repulsion part. In the case where the hydrogen bond is dominated by ES and EX, the ratio between ES and EX increases with the electronegativity, which corresponds to the observed ordering of the absolute values of the slope for the different acceptor atoms. We should note that EDA is not free of conceptual difficulties as a small interaction energy is decomposed into comparatively large energy contributions of opposite sign. Detailed insight can be expected by the symmetry-adapted perturbation theory (SAPT) analysis [239, 240]. In SAPT, however, the price of large computation times is paid for the accurate and well-defined analysis of intermolecular interactions, which makes the method difficult to apply to large extended systems and to those whose interactions cannot be properly described in a perturbation theory approach.

As an example for the application of EDA we may refer to the following study: Using quantum chemical calculations for Monte-Carlo-generated conformers of host–guest complexes Johansson and coworkers studied the formation of supermolecules with encapsulated anions [241]. These authors found that the cavity size cannot directly be employed as a descriptor of the encapsulation capacity of the hosts. With the help of the EDA scheme they found that in the investigated anions a decisive role is played by the Coulombic/Pauli balance, with the exception of F^- as host where polarization/charge transfer contributes significantly [241].

12.4.2 A Core-electron Probe for Hydrogen Bond Interactions

Reckien and Peyerimhoff [94] investigated so-called two-fold hydrogen bonds in isophthalic amide complexes as models for amide rotaxane wheel–axle interactions (compare Fig. 12.2 for one example). These complexes possess the same two-fold hydrogen bond donors, while the acceptor molecules are varied. It was found that electron donating substituents like methyl groups strengthen the hydrogen bond, i.e. they increase the binding energy. A decrease in binding energy can be expected if the electron donating NH_2 group is replaced by protons or alkyl groups such as CH_3. These trends can be directly related to the partial charges on the carbonyl oxygen atom that accepts the two hydrogen bonds [94]. Of course, steric effects turned out to also play an important role: Only if the groups can be arranged in an optimal fashion, the complexes feature a maximum bond strength measured in terms of binding energy [94]. Reckien and Peyerimhoff [94] then introduced a promising new descriptor for the assessment of such two-fold hydrogen bridges. This descriptor is the ε_{1s} orbital energy of the acceptor oxygen atom, which is the molecular orbital energy ε_i of Eq. (12.5) of the canonical molecular orbital that closely resembles the 1*s* atomic orbital at the oxygen acceptor. The ε_{1s} energy is parametrized to supermolecularly calculated hydrogen bond energies of a test set of hydrogen bonded complexes. The empirical idea behind this concept is the fact that it is known from core-electron photoelectron spectroscopy such as ESCA that the position (i.e. the energy) of a *K* shell peak of an atom is sensible to the atom's chemical environment, i.e. to its valence state and the number and type of bonds to neighboring atoms. Of course, since the ε_{1s}-descriptor is a scalar quantity, whose value will change if any bonding partner of the oxygen acceptor is changed, parametrization for hydrogen bond energies requires the same acceptor molecule always to be used. We are currently investigating whether this drawback may be circumvented if *differences* of 1*s*-AO-type orbital energies calculated for the isolated acceptor molecule and this acceptor within a particular hydrogen-bonded complex can be used as transferable and easily applicable descriptors [242].

12.4.3 The SEN Approach to Hydrogen Bond Energies

The basic idea of the shared-electron-number (SEN) method [243] is to estimate the strength of a hydrogen bond by means of only one variable. This variable is the two-center shared-electron number σ_{HA}, which is related linearly to the hydrogen bond energy $E_{I,\mathrm{HA}}$,

$$E_{I,\mathrm{HA}}^{\mathrm{SEN}} = \lambda \sigma_{\mathrm{HA}}. \tag{12.41}$$

Qualitatively speaking, σ_{HA} denotes the number of electrons that is shared by the hydrogen atom H and the acceptor atom A. It is calculated in a population analysis [244–246], where electrons are distributed to different atoms according to the contribution of (modified) atomic orbitals (i.e. of a minimum set of basis functions) to the molecular orbitals. The shared-electron number is thus a single-valued descriptor for the electronic density between the atoms H and A.

With the SEN approach it is possible to detect and quantify the interaction energies of individual hydrogen bonds in non-decomposable systems. These energies are quantum mechanically not defined and thus also not accessible by the supramolecular approach that only provides *total* interaction energies. The proportionality constant λ was adjusted to a small test set of hydrogen bonded complexes [243] which was not corrected for the basis set superposition error. These deficiencies were removed in a new and largely extended calibration study for λ [247]. Since the interaction energy as calculated by the supramolecular approach is negative for bound complexes, the slope λ takes negative values.

Furthermore, it could be shown that these large error margins cover deviations that are systematic with regard to the acceptor atom [247]. This gives rise to different ideas on how to improve the accuracy of the method. A promising Ansatz (extension) is the decomposition into sets with the same acceptor atom. With the introduction of this decomposition of the total set, the standard deviation could be considerably lowered [247].

In contrast to the original ε_{1s}-method, the SEN σ_{HA} is able to describe the directional nature of the hydrogen bond and allows one to detect hydrogen bonds which appear hidden or unrecognized when purely geometric criteria (mean bond lengths and angles) are applied for their detection. As mentioned before, the SEN concept is particularly useful for the assignment of an interaction energy for a particular hydrogen bond when there are more than one present in a given hydrogen-bonded complex [231, 247–249]. It has been successfully applied in various cases [14, 175, 231, 250–252].

We shall highlight the usefulness of the wave-function-based SEN method for one example. In a recent communication experimental evidence was found that N1 $\cdots$ N3 hydrogen bonds are stronger in dsRNA A–U than in dsDNA A–T base pairs [253]. The observations were made on the basis of one-bond ^{15}N–^{1}H J-coupling constants $^{1}J_{\mathrm{NH}}$. Table 12.2 contains results from calculations on the isolated base pairs and trimers taken from Ref. [247]. For the base pairs we find no difference in structure or binding energy. Although there is a slight trend for the A–U dimer in the binding energy as well as in the SEN hydrogen bond energy for N1 $\cdots$ N3, this trend is too less pronounced to reveal the difference between the hydrogen bonds, see Table 12.2 first two lines.

Comparing the T·A–T trimer and the A–T dimer no difference in distance and in hydrogen bond energy was detected [247]. The SEN method yields a N1 $\cdots$ N3 hydrogen bond of approximately 33 kJ mol^{-1} for both dimers and for the T·A–T trimer. Although the hydrogen bond distance is still in the same region (changes are of the order of promille) the SEN energy indicates a trend to stronger hydrogen

Tab. 12.2. Total interaction energies E_I for dimers and trimers and SEN hydrogen bond energy $E_{I,\mathrm{HN}}^{\mathrm{SEN}}$ (parameterized for N acceptor atoms) for a single hydrogen bond depicted in Fig. 12.6 taken from Ref. 247 (all energies in kJ mol^{-1}). Intermolecular distance between the two nitrogen atoms and intramolecular N–H distances r_{NN} in pm. The N $\cdots$ HN angles $<$ (N $\cdots$ HN) are given in degrees.

System	E_I kJ mol^{-1}	$E_{I,\mathrm{HN}}^{\mathrm{SEN}}$ kJ mol^{-1}	r_{NN} pm	r_{NH} pm	$<$(N$\cdots$HN)°
A–U dimer	−57.1	−33.4	288.9	104.5	178.6
A–T dimer	−56.1	−32.9	288.9	104.5	178.6
U·A–U trimer	−110.6	−38.4	288.3	104.5	179.5
T·A–T trimer	−109.7	−33.4	287.9	104.5	179.4

bonds in the U·A–U trimer as compared to the A–U dimer and T·A–T trimer (see Fig. 12.6 for an illustration of the structures of the trimers). In Ref. [253] the authors raise the question whether other hydrogen bonds are also stronger in the uracil complex than in the thymine complex. For the trimer an increase in hydrogen bond

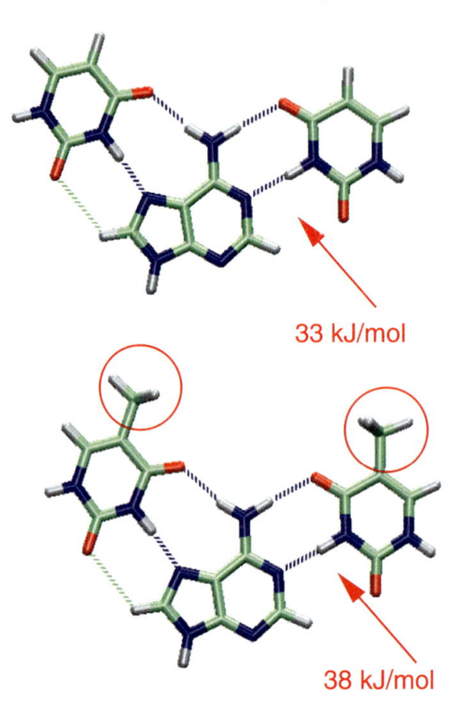

Fig. 12.6. Hydrogen bond energies in an adenine–thymine (top) and adenine–uracil base trimer calculated with the SEN method. The marked hydrogen bond is the N1 $\cdots$ N3 hydrogen bond which is according to the SEN method stronger in the U·A–U trimer (bottom) than in the T·A–T (top) trimer.

strength obtained from the shared-electron number of 13% for the O···N hydrogen bond was observed [247] which is as large as the increase for the N1···N3 bond. The O···C hydrogen bond which is very weak increases by 23% in strength.

12.5 General Mechanism for a Static, Step-wise View on Host–Guest Recognition

A great difficulty for the description of chemical processes and template-assisted reactions in supramolecular aggregates is the huge number of degrees of freedom, which prohibits the drawing of a simplified picture. Nonetheless such a simplified picture is needed in order to devise workable concepts that allow us to describe, quantify and predict host–guest recognition in complicated systems. Molecular dynamics necessarily provides only a statistical, time-averaged point of view, while the quantum chemical approach cannot provide all the relevant conformations that are accessible in a molecular dynamics run. However, the static quantum chemical calculations provide well-defined stationary points obtained in actual calculations on the potential energy hypersurface although the number of these local minima and saddle points is limited in view of the many weak interactions and contacts in large host–guest complexes.

Naturally, the first step in this endeavor would be a deduction based on an Eyring-like transition state theory or absolute rate constant theory, respectively, in combination with specific elements of the chemical process under investigation. According to Eyring we may express the absolute rate of a chemical elementary reaction k solely on the basis of statistical information (i.e. partition functions) calculated with quantum chemical methods at the equilibrium structures of reactants and at the transition structure. In a similar way, thermochemical models have been established, for instance, for unimolecular reactions (compare the RRKM theory [254, 255]) or electron-transfer reactions (where Marcus' theory [256] adds a description of the condensed matter environment to an Eyring-like description of electron transfer). An essential ingredient of such a theory would be the deduction of a most general mechanism of chemical processes within a supramolecular assembly consisting of elementary reaction steps. Reaction energies and entropies for these steps may then be deduced and calculated with quantum chemical methods. As mentioned above, we may call this the TAR model, which is a shorthand notation for a general thermochemical scheme of template-assisted elementary reactions.

Predecessors of some steps of our approach exist in the literature. For instance, Cooks kinetic model [257] relates the equilibrium constant K according to Eq. (12.33) to a protonation energy difference and an entropy difference of relevant channels, i.e. to the free enthalpy change calculated for some model assumptions. It was, for instance, applied by Gozzo et al. [258] in order to understand the hydrogen bonding in supermolecules of imidazolium-based ionic liquids. A condensation model based on estimated free energies of the host–guest complex formation was developed by Dickert et al. in order to predict the sensor signal of supramolecular analyte-receptors [259]. The authors explain in their introduction that computational chemistry allows one to simulate the inclusion processes, which gives access to a quantitative picture of the host–guest complex in terms of the energy associated with its formation. Host–guest chemistry exhibits behavior analogous to both crystal packing and the liquid state, while the state of the analyte (guest) is more that of a liquid than that of a solid.

The classical thermochemical approach has also already been applied to special template-assisted chemical reactions like the template-directed synthesis of oligonucleotides [260], for which a detailed system of kinetic elementary steps has been derived and solved. Other examples are the thermochemical model for the assessment of cooperativity in self-assembly processes proposed in Ref. [261, 262] and for the quantitative description of multicomponent self-assembly processes of polymetallic helicates [263, 264]. More thermochemical approaches applied within particular fields of supramolecular chemistry will be mentioned in the following.

In this account we present a general attempt on how such a classical kinetic description could look for a broad range of host–guest processes and template-assisted reactions. The first step is to define an appropriate reaction mechanism. Then, transition states, reaction barriers, reaction enthalpies and entropies can be assigned and calculated. This second step is straightforward using quantum chemical methods once the elementary reactions are defined. In a final step, it would be useful to decompose these thermodynamic quantities into additive local contributions derived, for example, from a decomposition of the total interaction energy according to schemes presented in the last section. Of course, this may not always be possible but one may hope that certain structural features may dominate others.

In order to assess quantum chemical methods for the construction of a reasonable and sufficiently general thermochemical model of host–guest processes and template-assisted reactions it is necessary to restrict ourselves to reactions which do follow the proposed reaction mechanism as closely as possible. Of course, one may imagine reactions, which deviate in certain details. However, such deviations would not alter the general methodology but would merely define a new class of templates featuring modified elementary steps. To provide a useful classification scheme, which yields solid grounds for the rational design of template-assisted reactions, the unique collaborative research center SFB 624 established at the University of Bonn in 2002 [10, 265], which is devoted to the study of template effects in all areas of chemistry, started out with a pragmatic but very useful classification of templates as *convex* (e.g. when a single metal center acts as a spatially organizing agent), *concave* (e.g. when the template molecule surrounds the reactant in an amoeba like fashion) and *linear/planar* (e.g. when surfaces act as templates[4)]). The

4) Although the presentation so far has been concerned with isolated supramolecules or in homogeneous solution, hetereogeneous effects are, of course, also subject of constant theoretical developments. For instance, the requirements for the formation of a chiral template was addressed only recently [266] on the basis of DFT calculations for propylene oxide on Pd(111) surfaces. Another example: Molecular dynamics simulations on shape-persistent macrocycles revealed that planarization takes place when they are adsorbed on a graphite surface [267]. An extensive study of the effects of guest inclusion on the crystal packing of *p-tert*-butylcalix[4]arenes was carried out by León et al. through a combination of molecular-mechanics-based solid-state calculations and statistical analysis, with a procedure that allowed them to study a variety of classes of organic compounds [268].

SFB 624 in Bonn proposed a definition for template-assisted reactions [265] which states that a template steers a reaction through spatial pre-organization of the reactants via a suitably defined pattern of non-covalent, covalent, or coordinative-reversible bonds. It acts as an entropic sink such that thermochemical criteria can be used to characterize the template-mediated process. Template-mediated reactions may revolve several times but high turn over numbers are not necessarily required. Also, the template does not necessarily need to be recovered once the reaction took place. In this work, we elaborate on this non-formal description, which aims to embrace as many chemical processes involving templates and host–guest interactions as possible. Our aim is to arrive at a more formal basis utilizing well-established thermochemical concepts. The classical thermochemical approach of physical chemistry to chemical reactions may provide an option to define transferable concepts for the plethora of template-assisted processes.

The elementary reaction steps into which a template-assisted chemical process may be decomposed are discussed in the following subsections. Note that with each elementary reaction step, a barrier and a reaction rate is associated, though this is not explicitly mentioned. Nevertheless the transition states can be localized with quantum chemical methods and the corresponding rate can be estimated on the basis of Eyring's absolute rate theory, which is supposed to be valid within the simplifying thermodynamical model proposed here. It is clear from the beginning that the entropic contributions are important but the most difficult to calculate. Molecular dynamics aims at a unified calculation of entropic and enthalpic contributions in form of the free enthalpy, while quantum chemistry provides a very rough estimate of intramolecular entropic contributions on the basis of prominent approximations (particle in a box, harmonic oscillator, rigid rotor) and only for stationary points on the potential energy hypersurface. However, there is evidence from molecular dynamics simulations that one may find dominant structures representing conformationally flexible structures like in the case of polypeptides [279], which is good news for quantum chemical approaches.

12.5.1 Template-free Pre-orientation Processes

In principle, a template-assisted reaction may also occur without the template. In absence of a template, various alternative non-reactive configurations $\{B_i\}$ of the reactants $A_1, A_2, A_3, \ldots$ are possible,

$$A_1 + A_2 + A_3 + \cdots \rightleftharpoons \{B_i\}. \tag{12.42}$$

Here, we used the set braces in order to denote that many reactions may lead to "innumerable" different configurations i, i.e. to many different local minima on the potential energy surface.

An optimum template would guarantee that the reactants are spatially arranged in the relative configuration B_x, which is most appropriate for the sought-for reaction to take place. Note that there may exist in principle more than one such configuration B_x, which may act as a useful starting configuration for the reaction. For the sake of simplicity, we assume that there is only one unique such configuration. As mentioned before, without the template there is a certain probability for the desired reaction,

$$A_1 + A_2 + A_3 + \cdots \rightleftharpoons B_x \rightarrow P \tag{12.43}$$

to take place and to produce the final products P. Note that the final reaction step is written as a unimolecular reaction rather than as a reaction which may involve additional reactants to react with B_x. As the first reaction step in the equation above may be a multi-step process, we may also include any additional reactants from the very beginning into the pre-organization sequence for the generation of the reactive configuration B_x.

12.5.2 Rearrangement Reactions

Depending on the possible number of configurations $\{B_i\}$, several rearrangement reactions will decrease or increase the concentration of the reactive configuration B_x. Consider that the pre-orientation reaction is almost barrierless and that the thermodynamic difference of the complex configurations and the reactants A_i is small (i.e. if the reactants are bound through comparatively weak intermolecular forces, which additionally may be modulated through polar solvents), the rearrangement reaction might be formulated as a two-step process involving a decomposition into the reactants before a new complex configuration is built,

$$B_x \rightleftharpoons A_1 + A_2 + A_3 + \cdots \rightleftharpoons \{B_i\} \tag{12.44}$$

However, the different configurations i may also directly rearrange,

$$\{B_i\} \rightleftharpoons B_x \tag{12.45}$$

so that it is likely to be important to know the barrier heights controlling these rearrangement reactions. In any case, one might simplify the discussion by restricting it either to a "thermodynamic" comparison of local minima $\{B_i\}$ (without taking the barriers into account) or by restricting the rearrangement reaction to the most stable configuration $B_{i,\mathrm{min}}$, which can be assumed to eat up all other configurations $\{B_i\}$, and to the reactive configuration B_x. In this case, it will be feasible to calculate the relevant reaction barriers as well and to get in turn an estimate for the rate constant of this elementary step according to Eyring's theory. However, in a first step it might be sufficient to simply know the relative energies of the various conformers rather than also to compute the transition barriers. Since the optimization of all relevant conformers would be required in any quantum chemical study on the optimization of a host–guest complex.

Extended treatments are, of course, also possible and likely to be found in the MD regime. For instance, Parrinello and coworkers [269] applied the methods that were developed for the study of rare events in simulations to a catenane system (see also Ref. [270] for such methods). With the help of these techniques applied in traditional molecular dynamics simulations they were able to unravel the microscopic behavior of the conformational rearrangements involved in the switching process [269].

12.5.3
The Host-controlled Association Reaction

The action of a template T may now be most conveniently described by an association reaction, in which the template forces a spatial arrangement of the reactants in the reactive configuration,

$$T + A_1 + A_2 + A_3 + \cdots \rightleftharpoons [T - B]_x \qquad (12.46)$$

Note that this association reaction may in principle be a multi-step process in which the action of the template is stepwise and may be controlled through suitable substitution in each individual step. In other words, the template may first form a less stable intermediate $[T - B]_i$, which may rearrange according to the possibilities discussed above before the final reactive configuration $[T - B]_x$ is formed. This would complicate the model and would require an extended treatment. For the sake of simplicity, one may consider only those templates in a first approach, where the association reaction can be understood as an elementary reaction step. Moreover, in such a case it can be expected to be almost barrierless, i.e. barrier heights are of the order of RT. Only a multi-step association reaction may involve significant barriers that may become important for the efficacy of the template and thus for the determination of the rate constant. For a thermochemical model of multi-step binding as developed for a multi-valent guest associated to an immobilized host see Ref. [271] and as developed for typically multi-valent supramolecular assemblies such as rotaxanes see Ref. [272].

Chen [273] proposed a thermochemical model for the molecular recognition step of host–guest interactions which especially emphasizes the role of the enthalpy–entropy relationship in these processes (see also the above-mentioned enthalpy–entropy balancing in Ref. [259] for another recent example). In particular, Chen introduced the term "orientation-based fit" for the recognition via intermolecular orientation. The conceptual basis of this step has been recognized for decades mostly in the context of biochemistry when it comes to the recognition of a substrate by an enzyme. First, a rigid lock-and-key model was proposed by Fischer, which was later softened to the so-called induced-fit model by Koshland in 1958. However, detailed quantum chemical and *first-principles* molecular dynamics studies of these complicated processes are still rare, which is mainly due to the computational effort that is caused by these approaches. Force field studies can be much more easily carried out. The Wipff group were amongst the first to study quantitatively the "best fit" between neutral molecules and the cavity of a rigid neutral receptor using the force field approach [274]. They investigated, by means of molecular dynamics simulations and free-energy perturbation simulations, the macrocyclic ligand cryptophane-E and its complexes with three tetrahedral guests (CH_2Cl_2, $CHCl_3$ and CCl_4) in the gase phase and in solution. Using standard parameters for the force field they obtained trends opposite to the experiment. Based on a systematic study of the relative free energies of binding, new van der Waals parameters could be derived that then did account for the experimental binding behavior [274].

12.5.4 The Transformation Step

Once the template has organized the reactants in the reactive configuration x, the desired reaction can take place,

$$[T - B]_x \rightarrow P_T \tag{12.47}$$

Note that after the reaction the template may still be bound to product P – denoted here as P_T. It may be removed in a subsequent step, which is not relevant for the action of the template. Once the template has been removed from the product, the overall reaction energy for the generation of the product P is, of course, the same as for the template-free process. For the sake of simplicity, one may assume that the activation energy is comparable to the template-free process. If the activation energy would be significantly higher for the template-assisted reaction, the template cannot act as a reaction facilitating and controlling agent. On the other hand, if it were smaller, the template would have an additional catalytic effect.

12.5.5 Inclusion of Environmental Effects

Solvent effects can be easily included in the thermodynamic model. The reactions formulated so far for the isolated species simply need to be formulated with an environment and can be treated in a Born cycle:

$$\begin{array}{ccc} T + A_1(g) + A_2(g) + A_3(g) + \cdots & \rightleftharpoons & [T - B]_x(g) \\ \sum_i \Delta G_i^s \big\downarrow & & \big\downarrow \Delta G^s \\ T + A_1(s) + A_2(s) + A_3(s) + \cdots & \rightleftharpoons & [T - B]_x(s) \end{array} \tag{12.48}$$

The modeling of the environment can be flexible, e.g. it can be modeled through microsolvation by explicitly adding some solvent molecules or through a polarizable electrostatic continuum model.

12.5.6 General Aspects of Template Thermodynamics and Kinetics

From the discussion so far it is clear that the action of a template has a pronounced entropic rather than an enthalpic effect. The reason for this is that the free motion of the substrate complexes B_i, which may lead to various rearrangement reactions, is no longer possible. The entropy loss induced by the pre-organization thus needs to be estimated from the number and structure of the possible configurations B_i. The entropy loss leads to a more positive free energy according to $-T\Delta S$ in the Gibbs–Helmholtz equation. For the association reaction to take place it is necessary to compensate for the entropy loss through a significantly more negative enthalpic contribution so that the driving force for the reaction, i.e. the free energy is still negative. The negative enthalpic contributions are, of course, built up by the attractive intermolecular forces between template and substrate, which may be estimated according to the concepts described in Sections 12.2, 12.3 and 12.4. Note however that the loss of configurational entropy, which stems from the fact that the template reduces the number of possible conformations in configuration space, is not captured within the above-described quantum chemical approach to thermodynamic quantities but may very well be obtained from MD simulations.

Also, the discussion of the entropic contributions so far is valid for the isolated host–guest system. If placed in a solvent, solvent molecules that saturated the attractive contacts of the host and of the guest are liberated upon host–guest complex formation and can account for a significant entropy gain, as in the case of the well known chelate effect in coordination chemistry.

A general model for template-assisted reactions should finally provide answers to various questions some of which are listed in the following:

- Which thermodynamic properties characterize a special, i.e. reactive configuration B_x? This question is a general one for reactions in chemistry and, of course, depends on the free energy surface on which the reaction $B_x \rightarrow P$ takes place. To put it the other way around, once a desired product is known, one may construct B_x from the product valley as the configuration with the smallest barrier.
- When will it be important to know all barriers explicitly for the processes described in Sections 12.5.1 and 12.5.4?
- How does the entropy decrease, induced by the template action, enter the thermodynamic quantities? How can one calculate the entropy loss for a frozen configuration B_x?
- How does one determine whether template T stabilizes exactly B_x as desired and not any other configuration B_i?
- In which cases will it be sufficient to know the relative energetics of the stable conformers B_i?

Once these questions can be answered within the TAR model, a rational design of template effects may become possible. The mechanism presented may already be useful as a helpful classification scheme to incorporate the tremendous variety of empirically found template-assisted chemical processes. The elementary steps sketched so far can be supplemented by reaction rate constants, which can be combined to a final formal-kinetic scheme to yield the total reaction rate. As a consequence, predictive elements of the TAR model will show up.

In the past, this endeavor has often rested on detailed investigations of particular supramolecular systems. However, the general thermochemical view provides a unifying concept and thus deserves a closer inspection in future work. A nice feature of the thermochemical approach is the fact that the thermochemistry of individual reaction steps may very well be measured in experiment (compare also Section 12.2.5).

The design of template molecules by calculations was already explored as early as in the late 1980s and in the early 1990s, respectively [275–277]. One of these approaches was based on two steps. First, optimized structures were collected, second a molecule was cut and "vectors" were added where originally the bonds existed in order to mark their positions and directions. These vectors were then used to find a matching set of vectors in the complement molecule. Thereby, a reconnection to conformationally restricted templates with identical vectors was carried out [275]. Such approaches may very well be combined to better understand elementary reactions of the TAR scheme.

12.6 Conclusions and Perspective

In this account we have attempted to provide a brief overview of the concepts of *first-principles* methods tailored for the calculation of structures, energetics, and properties of supramolecular assemblies. The presentation of the theory focussed on the most essential building blocks in order to provide a general frame to interrelate the various methods available. Thereafter, we discussed the relation of these methods to experiment and to well-known concepts for the description of typical interaction patterns. Also, new methods tailored for tackling problems specific to supramolecular chemistry have been discussed (like the calculation of local dipole moments in CPMD simulations, the Mode-Tracking protocol for the selective calculation of vibrational frequencies and intensities, or the SEN method for the calculation of hydrogen bond energies).

Finally, we argued for the elaboration of a thermochemical model, i.e. the TAR model, to embrace as many of the various empirically found host–guest and template-assisted processes as possible. Apart from the classification-type nature of a decomposition into elementary steps in such a model, it automatically assigns enthalpies and entropies to these steps to be calculated with theoretical methods and to be then compared to experimental measurements as described in Section 12.2.5. Last but not least one may even apply formal kinetics to derive general statements of rate constants on the individual steps as well as on the overall process.

The theoretical study of supramolecular systems has certainly just begun. The technical means for such studies have been provided within the last decades. In the future, these methods will be improved and refined. Also new approaches will be envisaged, which are tailored for the specific purposes of supramolecular chemistry – some of which have already been discussed in this work.

In a report of a workshop called "From Molecules to Materials" in 1998, written by several authors, the problems and goals of the theoretical description of supramolecular chemistry are nicely described [278]: "Learning how to program molecules to self-assemble into finite nanosized objects of high molar mass with well-defined shapes would have great impact on our ability to create self-organizing materials [...] The essential tools to develop the field of supramolecular materials are all the synthetic methodologies of organic and inorganic chemistry, and hopefully our growing abilities to predict structures through computation and theory." The authors suggest that computational chemistry offers the necessary tools to predict cluster energies associated with molecules designed for aggregation on the drawing board and we have seen that this goal has practically been achieved in the meantime. Furthermore, they stress, that tools to predict conformational trends are required, and argued that, therefore, the refinement of molecular force fields via quantum chemistry and the developments of better minimization techniques could be a major contribution. We hope that we have elaborated on these propositions in this account and that we were able to draw a more complete picture of what theoretical methods are capable of and into which directions they should evolve when the focus will be supramolecular chemistry.

Acknowledgments

We thank the collaborative research center SFB 624 "Templates" at the University of Bonn for continuous support. This work has been financially supported by the German research council DFG (SFB 624 and SFB 436) and by the Fonds der Chemischen Industrie (FCI). M.R. thanks the FCI for a Dozentenstipendium. We are grateful to Prof. Christoph Schalley, Dr. Johannes Neugebauer and Dr. Werner Reckien for comments on earlier versions of this manuscript.

References and Notes

1 J. W. Steed, J. L. Atwood, *Supramolecular Chemistry*, Wiley: Chichester, 2000.
2 P. D. Beer, P. A. Gale, D. K. Smith, *Supramolecular Chemistry*; Oxford Chemistry Primers Oxford Science Publications: Oxford, 1999.
3 D. H. Busch, N. A. Stephenson, *Coord. Chem. Rev.* **1990**, *100*, 119.
4 S. Anderson, H. L. Anderson, J. K. M. Sanders, *Acc. Chem. Res.* **1993**, *26*, 469.
5 R. Cacciapaglia, L. Mandolini, *Chem. Soc. Rev.* **1993**, *22*, 221.
6 R. Hoss, F. Vögtle, *Angew. Chem. Int. Ed. Engl.* **1994**, *33*, 375; *Angew. Chem.* **1994**, *106*, 389.
7 T. J. Hubin, D. H. Busch, *Coord. Chem. Rev.* **2000**, *200–202*, 5.
8 N. V. Gerbeleu, V. B. Arion, P. J. Stang, *Template synthesis of macrocyclic compounds*; Wiley-VCH: Weinheim, 1999.
9 F. Diederich, P. J. Stang, (Eds.), *Templated Organic Synthesis*; Wiley-VCH: Weinheim, 2000.
10 C. A. Schalley, F. Vögtle, K. H. Dötz, (Eds.), *Templates in Chemistry I*; Topics in Current Chemistry, Springer-Verlag: Berlin, 2004.
11 C. A. Schalley, F. Vögtle, K. H. Dötz, (Eds.), *Templates in Chemistry II*; Topics in Current Chemistry, Springer-Verlag: Berlin, 2005.

12 B. L. Feringa, *Acc. Chem. Res.* **2001**, *34*, 504–513.
13 C. Ochsenfeld, F. Koziol, S. P. Brown, T. Schaller, U. P. Seelbach, F. Klärner, *Solid State Nucl. Mag.* **2002**, *22*, 128–153.
14 B. Kirchner, M. Reiher, *J. Am. Chem. Soc.* **2005**, *127*, 8748–8756.
15 F.-G. Klärner, J. Panitzky, D. Preda, L. T. Scott, *J. Mol. Model.* **2006**, *6*, 318–327.
16 F.-G. Klärner, B. Kahlert, *Acc. Chem. Res.* **2003**, *36*, 919–932.
17 J. Schatz, *Collect. Czech. Chem. Commun.* **2004**, *69*, 1169–1194.
18 C. A. Hunter, *Angew. Chem. Int. Ed.* **2004**, *43*, 5310–5324; *Angew. Chem.* **2004**, *116*, 5424–5439.
19 T. L. Barr, *Modern ESCA – The Principles & Practice of X-Ray Photoelectron Spectroscopy*; CRC Press, 1994.
20 G. Wipff, *Computational Approaches in Supramolecular Chemistry*; Kluwer: Dordrecht, 1994.
21 N. Attig, K. Binder, H. Grubmüller, K. Kremer, (Eds.), *Computational Soft Matter: From Synthetic Polymers to Proteins*; John von Neumann Institute for Computing: Jülich, 2004.
22 G. Chessari, C. A. Hunter, C. M. R. Low, M. J. Packer, J. G. Vinter, C. Zonta, *Chem. Eur. J.* **2002**, *8*, 2860–2867.
23 G. Ortore, T. Tuccinardi, S. Bertini, A. Martinelli, *J. Med. Chem.* **2006**, *49*, 1397–1407.
24 C. Menke, E. Diemann, A. Müller, *J. Mol. Struct.* **1997**, *436–437*, 35–47.
25 A. Müller, E. Krickemeyer, J. Meyer, H. Bögge, F. Peters, W. Plass, E. Diemann, S. Dillinger, F. Nonnenbruch, M. Randerath, C. Menke, *Angew. Chem. Int. Ed. Engl.* **1995**, *34*, 2122–2124; *Angew. Chem.* **1995**, *107*, 2293–2295.
26 A. Müller, *The Chem. Intelligencer* **1997**, *3*, 58.
27 A. Müller, E. Krickemeyer, H. Bögge, M. Schmidtmann, F. Peters, C. Menke, J. Meyer, *Angew. Chem. Int. Ed. Engl.* **1997**, *36*, 483–486; *Angew. Chem.* **1997**, *109*, 499–502.
28 A. Müller, P. Kögerler, H. Bögge, *Structure and Bonding* **96**, **2000**, *56*, 203–236.
29 M.-M. Rohmer, M. Benard, J.-P. Blaudeau, J.-M. Maestre, J.-M. Poblet, *Coord. Chem. Rev.* **1998**, *581*, 178–180.
30 A. Müller, H. Reuter, S. Dillinger, *Angew. Chem. Int. Ed. Engl.* **1995**, *34*, 2328–2361; *Angew. Chem.* **1995**, *107*, 2505–2539.
31 J. Livage, L. Bouhedja, C. Bonhomme, M. Henry, *Mat. Res. Soc. Symp. Proc.* **1997**, *457*, 13.
32 Y. D. Chang, J. Salta, J. Zubieta, *Angew. Chem. Int. Ed. Engl.* **1994**, *33*, 325; *Angew. Chem.* **1994**, *106*, 347.
33 W. G. Klemperer, T. A. Marquart, O. M. Yagi, *Mater. Chem. Phys.* **1991**, *29*, 97.
34 V. W. Day, W. G. Klemperer, O. M. Yaghi, *J. Am. Chem. Soc.* **1989**, *111*, 5959.
35 J. S. Murray, P. Politzer, *Electrostatic Potentials: Chemical Applications.* In: *Encyc. Comp. Chem.*; P. von Rague Schleyer, (Ed.), Wiley: Chichester, 1998.
36 J. Thar, B. Kirchner, B. L. Guennic, M. Reiher, M. Armbrüster, G. Federwisch, R. M. Gschwind, **2006**, in preparation.
37 T. Schrader, *Chem. Eur. J.* **1997**, *3*, 1537–1541.
38 S. Rensing, M. Arendt, A. Springer, T. Grawe, T. Schrader, *J. Org. Chem.* **2001**, *66*, 5814–5821.
39 T. Schrader, *J. Org. Chem.* **1998**, *63*, 264–272.
40 M. Herm, T. Schrader, *Chem. Eur. J.* **2000**, *6*, 47–53.
41 M. Herm, O. Molt, T. Schrader, *Chem. Eur. J.* **2002**, *8*, 1485–1499.
42 T. Schrader, A. Hamilton, *Functional Synthetic Receptors*; Wiley-VCH: Weinheim, 2005.
43 R. M. Gschwind, M. Ambrüster, I. Z. Zubrzycki, *J. Am. Chem. Soc.* **2004**, *126*, 10228–10229.
44 B. J. Calnan, B. Tidor, S. Biancalana, D. Hudson, A. D. Frankel, *Science* **1991**, *252*, 1167–1171.

45 K. A. Schug, W. Lindner, *Chem. Rev.* **2005**, *105*, 67–113.

46 C. J. Cramer, *Essentials of Computational Chemistry – Theories and Models*; Wiley, 2002.

47 F. Jensen, *Introduction to Computational Chemistry*; Wiley, 1998.

48 J. B. Foresman, Æ. Frisch, *Exploring Chemistry with Electronic Structure Methods*; Gaussian, Inc.: Pittsburgh PA, 1996.

49 H. Hettema, *Quantum Chemistry – Classic Scientific Papers*; volume 8 of *World Scientific Series in 20th Century Chemistry* World Scientific: Singapore, 2000.

50 W. Kohn, *Rev. Mod. Phys.* **1999**, *71*, 1253–1266.

51 J. A. Pople, *Rev. Mod. Phys.* **1999**, *71*, 1267–1274.

52 S. D. Peyerimhoff, The Development of Computational Chemistry in Germany. In *Reviews in Computational Chemistry*, Vol. 18; K. B. Lipkowitz, D. B. Boyd, (Eds.), Wiley-VCH: Weinheim, 2002.

53 A. Szabo, N. S. Ostlund, *Modern Quantum Chemistry*; Dover 1996.

54 F. L. Pilar, *Elementary Quantum Chemistry*, McGraw-Hill, 2nd ed.; 1990.

55 M. Frisch et al., Gaussian03; Gaussian, Inc.: Pittsburgh PA, 2003.

56 R. Ahlrichs, M. Bär, M. Häser, H. Horn, C. Kölmel, *Chem. Phys. Lett.* **1989**, *162*, 165–169.

57 H.-J. Werner, P. J. Knowles, R. Lindh, M. Schütz et al., Molpro, version 2002.5, a package of ab initio programs, 2002 see http://www.molpro.net.

58 http://www.netsci.org/Resources/Software/Modeling/QM/#STU.

59 K. N. Houk, S. Menzer, S. P. Newton, F. M. Raymo, J. F. Stoddart, D. J. Williams, *J. Am. Chem. Soc.* **1999**, *121*, 1479–1487.

60 C. C. J. Roothaan, *Rev. Mod. Phys.* **1951**, *23*, 69–89.

61 R. Buenker, S. Peyerimhoff, *J. Chem. Phys.* **1966**, *45*, 3682.

62 R. J. Buenker, S. D. Peyerimhoff, *Theor. Chim. Acta (Berlin)* **1974**, *35*, 33–58.

63 R. J. Buenker, S. D. Peyerimhoff, *Theor. Chim. Acta (Berlin)* **1975**, *39*, 217–228.

64 B. Roos, *Adv. Chem. Phys.* **1987**, *69*, 399.

65 J. Hinze, C. C. J. Roothaan, *Suppl. Prog. Theor. Phys.* **1967**, *40*, 37.

66 J. Hinze, *J. Chem. Phys.* **1973**, *59*, 6424–6432.

67 H.-J. Werner, *Adv. Chem. Phys.* **1987**, *69*, 1–62.

68 R. Shepard, *Adv. Chem. Phys.* **1987**, *69*, 63–200.

69 T. Helgaker, P. Jørgensen, J. Olsen, *Molecular Electronic-Structure Theory*; Wiley-VCH: Chichester, 2002.

70 P. Hohenberg, W. Kohn, *Phys. Rev.* **1964**, *136*, 864–871.

71 W. Kohn, L. J. Sham, *Phys. Rev.* **1965**, *140*, 1133–1138.

72 M. Filatov, W. Thiel, *Phys. Rev. A* **1998**, *57*, 189–199.

73 J. P. Perdew, S. Kurth, A. Zupan, P. Blaha, *Phys. Rev. Lett.* **1999**, *82*, 2544–2547.

74 C. Adamo, M. Ernzerhof, G. E. Scuseria, *J. Chem. Phys.* **2000**, *112*, 2643–2649.

75 J. Tao, J. P. Perdew, V. N. Staroverov, G. E. Scuseria, *Phys. Rev. Lett.* **2003**, *91*, 146401.

76 B. I. Lundqvist, Y. Andersson, H. Shao, S. Chan, D. C. Langreth, *Int. J. Quantum Chem.* **1995**, *56*, 247–255.

77 P. P. Rushton, D. J. Tozer, S. J. Clark, *Phys. Rev. B* **2002**, *65*, 23503.

78 R. Ahlrichs, R. Penco, G. Scoles, *Chem. Phys.* **1977**, *19*, 119–130.

79 Y. Andersson, D. C. Langreth, B. I. Lundqvist, *Phys. Rev. Lett.* **1996**, *76*, 102–105.

80 M. Kamiya, T. Tsuneda, K. Hirao, *J. Chem. Phys.* **2002**, *117*, 6010–6015.

81 S. Grimme, *J. Comput. Chem.* **2004**, *25*, 1463.

82 Q. Wu, W. Yang, *J. Chem. Phys.* **2002**, *116*, 515.

83 U. Zimmerli, M. Parrinello, P. Koumoutsakos, *J. Chem. Phys.* **2004**, *120*, 2693.

84 G. Seifert, T. Kohler, R. Tenne, *J. Phys. Chem. B* **2002**, *106*, 2497.

85 E. R. Johnson, A. D. Becke, *J. Chem. Phys.* **2005**, *123*, 024101.

86 A. von Lilienfeld-Toal, I. Tavernelli, U. Rothlisberger, D. Sebastiani, *Phys. Rev. Lett.* **2004**, *93*, 153004.

87 O. A. von Lilienfeld, I. Tavernelli, U. Rothlisberger, D. Sebastiani, *Phys. Rev. B* **2005**, *71*, 195119.

88 M. Parac, M. Etinski, M. Peric, S. Grimme, *J. Chem. Theory Comput.* **2005**, *1*, 1110–1118.

89 G. Jansen, A. Heßelmann, *J. Phys. Chem. A* **2001**, *105*, 11156.

90 H. L. Williams, C. F. Chabalowski, *J. Phys. Chem. A* **2001**, *105*, 646.

91 A. Heßelmann, G. Jansen, M. Schütz, *J. Chem. Phys.* **2005**, *122*, 014103.

92 E. B. Wilson, J. C. Decius, P. C. Cross, *Molecular Vibrations*; McGraw-Hill: New York, 1955.

93 J. Neugebauer, M. Reiher, C. Kind, B. A. Hess, *J. Comput. Chem.* **2002**, *23*, 895–910.

94 W. Reckien, S. D. Peyerimhoff, *J. Phys. Chem. A* **2003**, *107*, 9634–9640.

95 D. Frenkel, B. Smit, *Understanding Molecular Simulations*; Academic press: San Diego, 2002.

96 M. P. Allen, D. J. Tildesley, *Computer Simulations of Liquids*; Claredon Press: Oxford, 1987.

97 D. C. Rapaport, *The Art of Molecular Dynamics Simulations*; Cambridge University Press: Cambridge, 1995.

98 W. F. van Gunsteren, et al. *Angew. Chem. Int. Ed.* **2006**, *45*, 4064–4092.

99 E. Ermakova, J. Solca, H. Huber, D. Marx, *Chem. Phys. Lett.* **1995**, *246*, 204.

100 B. Kirchner, E. Ermakova, G. Steinebrunner, A. Dyson, H. Huber, *Mol. Phys.* **1998**, *94*, 257.

101 D. Marx, J. Hutter, *Ab Initio Molecular Dynamics: Theory and Implementation*. In: *Modern Methods and Algorithms of Quantum Chemistry.*, Vol. 3, 2nd ed.; J. Grotendorst, (Ed.), Jülich, 2000.

102 H. Goldstein, *Classical Mechanics*; Addison Wesley, 3rd ed.; 2001.

103 A. Edelman, T. A. Arias, S. T. Smith, *SIAM J. Matrix Anal. Appl.* **1998**, *20*, 303–353.

104 H. C. Andersen, M. P. Allen, A. Bellemans, J. Board, J. H. R. Clarke, M. Ferrario, J. M. Haile, S. Nosé, J. V. Opheusden, J. P. Ryckaert, "New molecular dynamics methods for various ensembles", Technical Report, Rapport d' activité scientifique du CECAM, 1984.

105 W. G. Hoover, *Phys. Rev. A* **1985**, *31*, 1695–1697.

106 S. Nosé, *J. Chem. Phys.* **1984**, *81*, 511–519.

107 G. J. Martyna, M. L. Klein, M. E. Tuckerman, *J. Chem. Phys.* **1992**, *97*, 2635–2643.

108 H. Huber, A. Dyson, B. Kirchner, *Chem. Soc. Rev.* **1999**, *28*, 121.

109 B. Kirchner, E. Ermakova, J. Solca, H. Huber, *Chem. Eur. J.* **1998**, *4*, 379–388.

110 W. S. Oh, M. H. Won, S. Y. Kim, K. T. No, S. K. Chang, *Chem. Lett.* **1996**, *3*, 181–182.

111 R. Car, M. Parrinello, *Phys. Rev. Lett.* **1985**, *55*, 2471.

112 M. E. Tuckerman, *J. Phys.: Condens. Matter* **2002**, *14*, R1297–R1355.

113 O. Edholm, H. J. C. Berendsen, *Mol. Phys.* **1984**, *51*, 1011.

114 J. Schlitter, *Chem. Phys. Lett.* **1993**, *215*, 617–621.

115 H. Schäfer, A. E. Mark, W. F. van Gunsteren, *J. Chem. Phys.* **2000**, *113*, 7809.

116 H. Schäfer, X. Daura, A. E. Mark, W. F. van Gunsteren, *Proteins: Struct., Func., Gen.* **2001**, *43*, 45–56.

117 H. Schäfer, L. J. Smith, A. E. Mark, W. F. van Gunsteren, *Proteins: Struct., Func., Gen.* **2002**, *46*, 215–224.

118 S.-T. D. Hsu, C. Peter, W. F. van Gunsteren, A. M. J. J. Bonvin, *Biophys. J.* **2005**, *88*, 15–24.

119 J. Kästner, W. Thiel, *J. Chem. Phys.* **2005**, *123*, 144104.

120 M. J. S. Dewar, W. Thiel, *J. Am. Chem. Soc.* **1977**, *99*, 4899.

121 M. J. S. Dewar, E. G. Zoebisch, J. P. P. Stewart, *J. Am. Chem. Soc.* **1985**, *107*, 3902.

122 J. P. P. Stewart, *J. Comput. Chem.* **1989**, *209*, 221.

123 R. Puchta, V. Seitz, N. J. R. van Eikema Hommes, R. W. Saalfrank, *J. Mol. Model.* **2000**, *6*, 126–132.

124 P. Ghosh, G. Federwisch, M. Kogej, C. A. Schalley, D. Haase, W. Saak, A. Lützen, R. M. Gschwind, *Org. Biomol. Chem.* **2005**, *3*, 2691–2700.

125 P. Linnartz, S. Bitter, C. A. Schalley, *Eur. J. Org. Chem.* **2003**, 4819–4829.

126 C. A. Schalley, W. Reckien, S. D. Peyerimhoff, B. Baytekin, F. Vögtle, *Chem. Eur. J.* **2004**, *10*, 4777–4789.

127 P. Winget, C. Selcuki, A. H. Horn, B. Martin, T. Clark, *Theor. Chem. Acc.* **2003**, *110*, 254–266.

128 R. Casadesús, M. Moreno, À. Gonzáles-Lafont, J. M. L. M. P. Repasky, *J. Comput. Chem.* **2004**, *25*, 99–105.

129 S. Goedecker, *Rev. Mod. Phys.* **1999**, *71*, 1085–1123.

130 J. M. Millam, G. E. Scuseria, *J. Chem. Phys.* **1997**, *106*, 5569.

131 E. Schwegler, M. Challacombe, M. Head-Gordon, *J. Chem. Phys.* **1998**, *109*, 8764.

132 C. Ochsenfeld, C. A. White, M. Head-Gordon, *J. Chem. Phys.* **1998**, *109*, 1663–1669.

133 K. R. Bates, A. D. Daniels, G. E. Scuseria, *J. Chem. Phys.* **1998**, *109*, 3308.

134 G. E. Scuseria, *J. Phys. Chem. A* **1999**, *103*, 4782–4790.

135 K. N. Kudin, G. E. Scuseria, *Phys. Rev. B* **2000**, *61*, 16440–16453.

136 W. Liang, M. Head-Gordon, *J. Chem. Phys.* **2004**, *120*, 10379.

137 C. A. White, B. G. Johnson, P. M. W. Gill, M. Head-Gordon, *Chem. Phys. Lett.* **1994**, *230*, 8–16.

138 T. Van Voorhis, M. Head-Gordon, *Mol. Phys.* **2002**, *100*, 1713–1721.

139 E. J. Baerends, D. E. Ellis, P. Ros, *Chem. Phys.* **1973**, *2*, 41–51.

140 B. I. Dunlap, J. W. D. Connolly, J. R. Sabin, *J. Chem. Phys.* **1979**, *71*, 3396–3402.

141 C. Hampel, H.-J. Werner, *J. Chem. Phys.* **1996**, *104*, 6286.

142 P. Schuster, W. Mikenda, (Eds.), *Hydrogen Bond Research*; Springer-Verlag: Berlin, Heidelberg, 1999.

143 M. Schütz, H.-J. Werner, *J. Chem. Phys.* **2001**, *114*, 661.

144 M. D. Watson, A. Fechtenkotter, K. Müllen, *Chem. Rev.* **2001**, *101*, 1267–1300.

145 S. Attinger, P. Koumoutsakos, *Multiscale Modeling and Simulation*; volume 39 of *Lecture Notes in Computational Science and Engineering* Springer-Verlag: Berlin, Heidelberg, 2004.

146 S. D. Peroukidis, A. G. Vanakaras, D. J. Photinos, *J. Chem. Phys.* **2005**, *123*, 164904.

147 S. Goedecker, *Rev. Mod. Phys.* **1999**, *71*, 1085–1123.

148 F. L. Gervasio, P. Carloni, M. Parrinello, *Phys. Rev. Lett.* **2002**, *89*, 108102.

149 F. L. Gervasio, A. Laio, M. Iannuzzi, M. Parrinello, *Chem. Eur. J.* **2004**, *10*, 4846–4852.

150 F. L. Gervasio, A. Laio, M. Parrinello, M. Boero, *Phys. Rev. Lett.* **2005**, *94*, 158103.

151 I. Antes, W. Thiel, In: Combined Quantum Mechanical and Molecular Mechanical Methods, J. Gao and M. A. Thompson (Eds.) **1998**, *ACS Symposium Series 712, American Chemical Society, Washington, DC*, 50–65.

152 P. Carloni, U. Rothlisberger, *Simulations of Enzymatic Systems: Perspectives from Car–Parrinello Molecular Dynamics Simulations.* In: *Theoretical Biochemistry – Processes and Properties of Biological Systems*; L. Eriksson, (Ed.), Elsevier 2001.

153 M. Colombo, L. Guidoni, A. Laio, A. Magistrato, P. Maurer, S. Piana, U. Röhrig, K. Spiegel, M. Sulpizi, J. VandeVondele, M. Zumstein, U. Rothlisberger, *Chimia* **2002**, *56*, 13–19.

154 P. Sherwood, et al. *J. Mol. Struct. (Theochem)* **2003**, *632*, 1–28.

155 H. M. Senn, W. Thiel, In: *Computational Tools and Methods for Biology*, M. Reiher, (Ed.), Top. Curr. Chem. Springer-Verlag: Heidelberg, Berlin, 2006.

156 C. Reichardt, *Solvents and Solvent Effects in Organic Chemistry*; Wiley-VCH: Weinheim, 3 ed.; 2003.
157 A. Chaumont, N. Galand, R. Schurhammer, P. Vayssière, G. Wipff, *Russ. Chem. Bull. Int. Ed.* **2004**, *53*, 1459–1465.
158 J. Thar, W. Reckien, B. Kirchner, In *Atomistic Approaches in Modern Biology*; M. Reiher, (Ed.), Top. Curr. Chem. Springer-Verlag: Heidelberg, Berlin, 2006.
159 J. Neugebauer, B. A. Hess, *J. Chem. Phys.* **2004**, *120*, 11564–11577.
160 M. Kaupp, M. Bühl, V. G. Malkin, *Calculation of NMR and EPR Parameters*; Wiley-VCH: Weinheim, 2004.
161 J. Autschbach, T. Ziegler, *Coord. Chem. Rev.* **2003**, *238–239*, 83–126.
162 F. Neese, *Curr. Op. Chem. Biol.* **2003**, *7*, 125–135.
163 C. Ochsenfeld, J. Kussmann, F. Koziol, *Angew. Chem. Int. Ed.* **2004**, *43*, 4485; *Angew. Chem.* **2004**, *116*, 4585.
164 M. Reiher, J. Neugebauer, *J. Chem. Phys.* **2003**, *118*, 1634–1641.
165 C. Herrmann, M. Reiher, First-Principles Approach to Vibrational Spectroscopy of Biomolecules. In: *Computational Tools and Methods for Biology*; M. Reiher, (Ed.), Top. Curr. Chem. Springer-Verlag: Heidelberg, Berlin, 2006.
166 M. Reiher, *New J. Chem.* **2006**, in preparation.
167 M. Reiher, J. Neugebauer, *Phys. Chem. Chem. Phys.* **2004**, *6*, 4621–4629.
168 M. Helmreich, F. Hampel, N. Jux, M. Reiher, **2006**, to be published.
169 T. B. Adler, N. Borho, M. Reiher, M. A. Suhm, *Angew. Chem. Int. Ed.* **2006**, *45*, 3440–3445.
170 J. Neugebauer, M. Reiher, *J. Phys. Chem. A* **2004**, *108*, 2053–2061.
171 J. Neugebauer, M. Reiher, *J. Comput. Chem.* **2004**, *25*, 587–597.
172 A. L. Kaledin, *J. Chem. Phys.* **2005**, *122*, 184106.
173 M. Reiher, J. Neugebauer, *J. Chem. Phys.* **2005**, *123*, 117101.
174 A. L. Kaledin, M. Kaledin, J. M. Bowman, *J. Chem. Theory Comput.* **2005**, *2*, 166–174.
175 B. Kirchner, M. Reiher, *J. Am. Chem. Soc.* **2002**, *124*, 6206–6215.
176 S. Grimme, J. Harren, A. Sobanski, F. Vögtle, *Eur. J. Org. Chem.* **1998**, 1491–1509.
177 C. Siering, S. Grimme, S. R. Waldvogel, *Chem. Eur. J.* **2005**, *11*, 1877–1888.
178 S. J. Rowan, S. J. Cantrill, G. R. L. Cousins, J. K. M. Sanders, J. F. Stoddart, *Angew. Chem. Int. Ed. Engl.* **2002**, *41*, 898–952; *Angew. Chem.* **2002**, *114*, 938–993.
179 I. Dance, *Cryst. Eng. Comm.* **2003**, *5*, 208–221.
180 R. Paulini, K. Müller, F. Diederich, *Angew. Chem. Int. Ed. Engl.* **2005**, *44*, 1788–1805; *Angew. Chem.* **2005**, *117*, 1820–1839.
181 P. Schuster, G. Zundel, C. Sandorfy, (Eds.), *The Hydrogen Bond – Recent developments in theory and experiments*; volume I–III North-Holland Publishing Company: Amsterdam, New York, Oxford, 1976.
182 S. Scheiner, Calculating the Properties of Hydrogen Bonds by *ab Initio* Methods. In: Rev. Comp. Chem., Vol. 2; K. B. Lipkowitz, D. B. Boyd, (Eds.), VCH Publishers: New York, Weinheim, 1991.
183 D. A. Smith, (Ed.), *Modeling the Hydrogen Bond*; Oxford University Press: Oxford, 1999.
184 G. Desiraju, T. Steiner, *The Weak Hydrogen Bond – In Structural Chemistry and Biology*; Oxford University Press: Oxford, 1999.
185 D. Hadzi, (Ed.), *Theoretical Treatments of Hydrogen Bonding*; Wiley Publishers: New York, 1997.
186 S. Scheiner, *Hydrogen Bonding – A Theoretical Perspective*; Oxford University Press: Oxford, 1997.
187 L. Pauling, *The Nature of the Chemical Bond*; Cornell University Press, 3rd ed.; 1960.
188 T. Steiner, G. R. Desiraju, *Chem. Commun.* **1998**, *8*, 891–892.
189 M. J. Calhorda, *Chem. Commun.* **2000**, 801–809.

190 I. Hyla-Kryspin, G. Haufe, S. Grimme, *Chem. Eur. J.* **2004**, *10*, 3411–3422.
191 F. Ugozzoli, C. Massera, *CrystEngComm* **2005**, *7*, 121–128.
192 T. van Mourik, R. J. Gdanitz, *J. Chem. Phys.* **2002**, *116*, 9620.
193 S. Tsuzuki, H.-P. Lüthi, *J. Chem. Phys.* **2001**, *114*, 3949.
194 R. L. Jaffe, G. D. Smith, *J. Chem. Phys.* **1996**, *105*, 2780.
195 P. Hobza, H. L. Selzle, E. W. Schlag, *J. Phys. Chem.* **1996**, *100*, 18790.
196 S. Tsuzuki, T. Uchimaru, K. Matsumura, M. Mikami, K. Tanabe, *Chem. Phys. Lett.* **2000**, *319*, 547.
197 W. B. Schweizer, J. D. Dunitz, *J. Chem. Theory Comput.* **2006**, *2*, 288–291; M. O. Sinnokrot, C. D. Sherrill, *J. Phys. Chem. A.*, **2006**, *110*, 10656–10668.
198 S. Grimme, *Chem. Eur. J.* **2004**, *10*, 3423–3429.
199 B. Kirchner, J. Stubbs, D. Marx, *Phys. Rev. Lett.* **2002**, *89*, 215901.
200 B. Kirchner, J. Hutter, I. W. Kuo, C. J. Mundy, *Int. J. Mod. Phys. B* **2004**, *18*, 1951–1962.
201 W. Blokzijl, J. B. F. N. Engberts, *Angew. Chem. Int. Ed. Engl.* **1993**, *32*, 1545–1579.
202 U. Schröder, *Solid State Commun.* **1966**, *4*, 347.
203 M. Sangster, *J. Phys. Chem. Solids* **1974**, *35*, 195.
204 M. Dixon, *Phil. Mag. B* **1983**, *48*, 13.
205 K. O'Sullivan, M. P., *J. Phys.: Condens. Matter* **1991**, *3*, 8751.
206 P. Lindan, M. Gillan, *J. Phys.: Condens. Matter* **1993**, *5*, 1019.
207 P. Mitchell, D. Fincham, *J. Phys.: Condens. Matter* **1993**, *5*, 1031.
208 A. A. Chialvo, P. T. Cummings, *J. Chem. Phys.* **1996**, *105*, 8274–8281.
209 J. Brodholt, M. Samploi, R. Vallauri, *Mol. Phys.* **1995**, *86*, 149–158.
210 L. X. Dang, T.-M. Chang, *J. Chem. Phys.* **1997**, *106*, 8149–8159.
211 J. W. Caldwell, P. A. Kollman, *J. Phys. Chem.* **1995**, *99*, 6208–6219.
212 D. N. Bernado, Y. Ding, K. Krogh-Jespersen, R. M. Levy, *J. Phys. Chem.* **1994**, *98*, 4180–4187.
213 A. Rowly, P. Jemmer, M. Wilson, P. Madden, *J. Chem. Phys.* **1998**, *108*, 10209.
214 S. R. Gadre, S. S. Pingale, *Curr. Sci.* **1998**, *75*, 1162–1166.
215 A. J. Stone, *The Theory of Intermolecular Forces*; Oxford University Press: Oxford, 2002.
216 J. Israelachvilli, *Intermolecular & Surface Forces*; Academic Press: London, 2nd ed.; 1994.
217 S. F. Boys, *Rev. Mod. Phys.* **1960**, *32*, 296.
218 G. H. Wannier, *Phys. Rev.* **1937**, *52*, 191.
219 N. Marzari, D. Vanderbilt, *Phys. Rev. B* **1997**, *56*, 12847.
220 D. Vanderbilt, R. D. King-Smith, *Phys. Rev. B* **1993**, *47*, 1651.
221 R. Resta, *Rev. Mod. Phys.* **1994**, *66*, 899.
222 D. McQuarrie, *Statistical Mechanics*; Harper and Row: New York, 1976.
223 R. Resta, *Phys. Rev. Lett.* **1998**, *80*, 1800.
224 M.-P. Gaigeot, M. Sprik, *J. Phys. Chem. A* **2003**, *107*, 10344–10358.
225 B. Kirchner, J. Hutter, *J. Chem. Phys.* **2004**, *121*, 5133–5142.
226 A. Laio, J. VandeVondele, U. Rothlisberger, *J. Phys. Chem. B* **2002**, *106*, 7300–7307.
227 B. Kirchner, *Theory of Complicated Liquids – Habilitationsschrift*; University of Bonn, 2005.
228 A. Luzar, D. Chandler, *Phys. Rev. Lett.* **1996**, *76*, 928–931.
229 M. Tarek, D. J. Tobias, *Phys. Rev. Lett.* **2002**, *88*, 138101.
230 D. Chandler, *Nature* **2002**, *417*, 491.
231 M. Reiher, B. Kirchner, *J. Phys. Chem. A* **2003**, *107*, 4141–4146.
232 A. E. Reed, F. Weinhold, L. A. Curtiss, D. J. Pochatko, *J. Chem. Phys.* **1986**, *84*, 5687–5705.
233 A. E. Reed, L. A. Curtiss, F. Weinhold, *Chem. Rev.* **1988**, *88*, 899–926.
234 Y. Mo, S. D. Peyerimhoff, *J. Chem. Phys.* **1998**, *109*, 1687–1697.
235 Y. Mo, J. Gao, S. D. Peyerimhoff, *J. Chem. Phys.* **2000**, *112*, 5530–5538.
236 K. Kitaura, K. Morokuma, *Int. J. Quantum Chem.* **1976**, *10*, 325–340.

237 T. Ziegler, A. Rauk, *Theor. Chim. Acta* **1977**, *46*, 1.
238 K. Morokuma, *Acc. Chem. Res.* **1977**, *10*, 294–300.
239 H. Valdes, J. A. Sordo, *J. Phys. Chem. A* **2003**, *107*, 899–907.
240 B. Jeziorski, K. Szalewicz, *Intermolecular Interactions by Perturbation Theory.* In: *Encyc. Comp. Chem.*; P. von Rague Schleyer, (Ed.), J. Wiley & Sons: Chichester, UK, 1998.
241 P. Johansson, E. Abrahamsson, P. Jacobsson, *J. Mol. Struct.* **2005**, *717*, 215–221.
242 M. Reiher, B. Kirchner, *upublished results*, 2006.
243 M. Reiher, D. Sellmann, B. A. Hess, *Theor. Chem. Acc.* **2001**, *106*, 379–392.
244 E. R. Davidson, *J. Chem. Phys.* **1967**, *46*, 3320–3324.
245 K. R. Roby, *Mol. Phys.* **1974**, *27*, 81–104.
246 R. Heinzmann, R. Ahlrichs, *Theoret. Chim. Acta* **1976**, *42*, 33–45.
247 J. Thar, B. Kirchner, *J. Phys. Chem. A* **2006**, *110*, 4229–4237.
248 M. Reiher, *Found. Chem.* **2003**, *5*, 23–41.
249 M. Reiher, *Found. Chem.* **2003**, *5*, 147–163.
250 M. Reiher, O. Salomon, D. Sellmann, B. A. Hess, *Chem. Eur. J.* **2001**, *7*, 5195–5202.
251 M. Reiher, G. Brehm, S. Schneider, *J. Phys. Chem. A* **2004**, *108*, 734–742.
252 M. Reiher, J. Neugebauer, B. A. Hess, *Z. Physik. Chem.* **2003**, *217*, 91–103.
253 M. N. Manalo, X. Kong, A. LiWang, *J. Am. Chem. Soc.* **2005**, *127*, 17974.
254 P. J. Robinson, K. A. Holbrook, *Unimolecular Reactions*; Wiley: New York, 1972.
255 T. Baer, W. Hase, *Unimolecular Reaction Dynamics: Theory and Experiments*; Oxford University Press: New York, 1996.
256 R. A. Marcus, *Rev. Mod. Phys.* **1993**, *65*, 599–610.
257 R. G. Cooks, J. S. Patrick, T. Kothiaho, S. A. MacLuckey, *Mass Spectrom. Rev.* **1994**, *13*, 287.
258 F. C. Gozzo, L. S. Santos, R. Augusti, C. S. Consorti, J. Dupont, M. N. Eberlin, *Chem. Eur. J.* **2004**, *10*, 6187–6193.
259 F. L. Dickert, M. Reif, R. Sikorski, *J. Mol. Model.* **2000**, *6*, 446–451.
260 S. Assouline, S. Nir, N. Lahav, *J. Theor. Biol.* **2001**, *208*, 117–125.
261 G. Ercolani, *J. Am. Chem. Soc.* **2003**, *125*, 16097–16103.
262 G. Ercolani, *J. Phys. Chem. B* **2003**, *107*, 5052–5057.
263 J. Hamacek, M. Borkovec, C. Piguet, *Chem. Eur. J.* **2005**, *11*, 5217–5226.
264 C. Piguet, M. Borkovec, J. Hamacek, K. Zeckert, *Coord. Chem. Rev.* **2005**, *249*, 705–726.
265 Sonderforschungsbereich 624, *Template – Vom Design chemischer Schablonen zur Reaktionssteuerung*; University of Bonn, since 2002.
266 D. Stacchiola, *J. Phys. Chem. B* **2005**, *109*, 851–856.
267 A. Ziegler, W. Mamdouh, A. Ver Heyen, M. Surin, H. Uji-i, M. M. S. Abdel-Mottaleb, F. C. De Schryver, S. De Feyter, R. Lazzaroni and S. Höger, *Chem. Mater* **2005**, *17*, 5670–5683.
268 S. León, D. A. Leigh, F. Zerbetto, *Chem. Eur. J.* **2002**, *8*, 4854–4866.
269 M. Ceccarelli, F. Mecuri, D. Passerone, M. Parrinello, *J. Phys. Chem. B* **2005**, *109*, 17094–17099.
270 C. Dellago, P. Bolhuis, In: *Atomistic Approaches in Modern Biology*; M. Reiher, (Ed.), Top. Curr. Chem. Springer-Verlag: Heidelberg, Berlin, 2006; T. Huber, A. E. Torda, W. F. van Gunsteren, *J. Comput.-Aided Mol. Design* **1994**, *8*, 695.
271 J. Huskens, A. Mulder, T. Auletta, C. A. Nijhuis, M. J. W. Ludden, D. N. Reinhoudt, *J. Am. Chem. Soc.* **2004**, *126*, 6784–6797.
272 J. D. Badjic, A. Nelson, S. J. Cantrill, W. B. Turnbull, J. F. Stoddart, *Acc. Chem. Res.* **2005**, *38*, 723–732.
273 J. Chen, *J. Phys. Chem. B* **2003**, *107*, 3576–3584.
274 A. Varnel, S. Helissen, G. Wipff, *J. Comput. Chem.* **1998**, *19*, 820–832.
275 S. R. Wilson, W. K. Tam, M. J. Di Grandi, W. Curi, *Tetrahedron* **1993**, *49*, 3655–3663.

276 P. A. Bartlett, G. T. Shea, S. J. Telfer, S. Wattermann. In *Proceedings of an International Symposium on Molecular Recognition*; S. M. Roberts, (Ed.), Royal Society of Chemistry, Special Publication No. 78: University of Exeter, 1989.

277 J. H. van Drie, D. Weininger, Y. C. Martin, *J. Comp.-Aided Mol. Des.* **1989**, *3*, 225.

278 A. P. Alivisatos, P. F. Barbara, A. Welford Castlemann, J. Chang, D. A. Dixon, M. L. Klein, G. L. McLendon, J. S. Miller, M. A. Ratner, P. J. Rossky, S. I. Stupp, M. E. Thompson, *Adv. Mater.* **1998**, *10*, 1297–1336.

279 X. Daura, A. Glättli, P. Gee, C. Peter, W. F. van Gunsteren, *Adv. Prot. Chem.* **2002**, *62*, 341–360; A. Glättli, X. Daura, D. Seebach, W. F. van Gunsteren, *J. Am. Chem. Soc.* **2002**, *124*, 12972–12978; W. F. van Gunsteren, R. Bürgi, C. Peter, X. Daura, *Angew. Chem. Int. Ed.* **2001**, *40*, 351–355.

Index

Analytical Methods in Supramolecular Chemistry. Edited by Christoph Schalley

ISBN: 978-3-527-31505-5

d

q

r

Related Titles

Gauglitz, G., Vo-Dinh, T. (eds.)

Handbook of Spectroscopy

1168 pages in 2 volumes with 407 figures and 59 tables
2003
Hardcover
ISBN-13: 978-3-527-29782-5
ISBN-10: 3-527-29782-0

Gerson, F., Huber, W.

Electron Spin Resonance Spectroscopy of Organic Radicals

479 pages with 79 figures and 108 tables
2003
Softcover
ISBN-13: 978-3-527-30275-8
ISBN-10: 3-527-30275-1

Newkome, G. R., Moorefield, C. N., Vögtle, F.

Dendrimers and Dendrons

Concepts, Syntheses, Applications

635 pages with 429 figures and 4 tables
2001
Hardcover
ISBN-13: 978-3-527-29997-3
ISBN-10: 3-527-29997-1

Lehn, J.-M.

Supramolecular Chemistry

Concepts and Perspectives

281 pages with 51 figures
1996
Softcover
ISBN-13: 978-3-527-29311-7
ISBN-10: 3-527-29311-6